Style Guide™

FOR BUSINESS AND TECHNICAL COMMUNICATION

FIFTH EDITION

Preface

Effective communication is the essence of good business. We serve customers, co-workers, employers, suppliers, and the community well by sharing relevant information clearly and efficiently. We fail to serve them when we communicate in unclear, bland, misleading, or irrelevant ways. That's why this book is essential to all organizations—businesses, government agencies, or educational institutions.

The aim of this fifth edition of the *FranklinCovey Style Guide for Business and Technical Communication* is to help you serve your customers and co-workers in these ways:

- You will be complying with the best current practices in business and technical communication.
- You will solve problems more effectively and make better decisions.

Comply with the best current practices in business and technical communication. Many useful stylebooks serve the needs of professional writers, scholars, editors, and publishers. This book, by contrast, is for people in the business and technical professions. All guidelines, examples, and model documents come from the real "world of work" rather than from the academic world. Everything in this book has been tested and refined in workshops with thousands of professionals literally around the world—from the oilfields of Saudi Arabia and Indonesia to the pharmaceutical industry of Switzerland to the aerospace, engineering, service, and manufacturing centers of North America and Europe.

Solve problems more effectively and make better decisions. Writing in the workplace is far more than pumping out emails, checking grammar, and fixing spelling. It is a problem-solving and decision-making process. Cogent and persuasive business plans allow swift, logical management decisions. Analytic and well-crafted scientific reports lead to robust dialogue and sound policy decisions. Well-designed and clearly written user information builds customer loyalty and prevents costly downtimes.

New in the Fifth Edition

The most current guidelines on email, information management, and online documentation. Learn how to manage the flood of email coming at you and to get results from the email you send. Find out how to add distinctiveness and power to your online presence.

Updated best practices for graphics. Here's the best current thinking on visuals for documents and presentations, charts, color, illustrations, maps, photos, and tables—including all new examples.

Guidance on global English. There is a new section on English as a second language for business professionals, as well as updated guidance on international business English.

Valuable new insights for knowledge workers. Learn new ways to think and process information better in updated sections on thinking strategies and the writing process, as well as practical guidance for managing projects and meetings.

Model documents for today. As email supplants traditional business letters and memos, you need new models to follow. See the MODEL DOCUMENTS section for updated samples of sensitive emails, reports, proposals, procedures, and resumes.

Everything in this fifth edition has been updated to help you meet the communication challenges of the high-tech, high-demand business world of today.

Business Communication Solutions from FranklinCovey

The Challenge

Business professionals devote hours every day to communication tasks in the workplace. Much of this communication is hampered by unproductive thinking, weak attempts at persuasion, poor organization, and a lack of basic writing skill that undercuts credibility. Floods of useless emails swamp and slow the whole organization. Web content lacks distinctiveness and power. Poorly managed, inconclusive meetings eat up time. Weak sales presentations fail to sway customers.

One dramatic way to increase your productivity is to improve your communication processes and skills.

The Solution

FranklinCovey offers tools, training, and services to help people and organizations do the great things they are capable of. Our mission is to enable great performance. We train more than a quarter of a million people every year worldwide in leadership, trust building, execution, and communication. Our unique approach is to challenge the paradigms that hold people back and unshackle them by teaching them new, more effective paradigms.

FranklinCovey training and consulting is available in live and online formats.

Instructor-Led Options. Experienced FranklinCovey consultants or certified facilitators teach our workshops onsite. These workshops can be customized to address the specific needs, challenges, and objectives of your organization.

Online Options. FranklinCovey's LiveClicks™ webinar workshops led by our consultants make our high-quality instruction available online. Engaging and interactive, these two-hour modules offer compelling skills training through award-winning videos, case studies, quizzes, and group discussion.

Training Programs for Effective Communication

- Writing Advantage™: Business Writing Skills for Professionals
- Presentation Advantage™: Professional Presenting Skills
- Meeting Advantage™: How to Lead Great Meetings
- Technical Writing Advantage™: Writing Skills for Technical Professionals
- Leading at the Speed of Trust™
- Working at the Speed of Trust™

For more information on FranklinCovey training and consulting, visit our website at franklincovey.com/tc.

Contents

Alphabetical

Reference Glossary

Contents

Alphabetical

Model Documents

Contents

Topical

Contents

Topical

Punctuation

Sentence Style

Skills/Processes

Word Choice

Foreword

This is a book for the Knowledge Age.

In the 21st century, value is created by knowledge work—the analysis, research, design, and development work done by strategists, scientists, technologists, and service professionals. Knowledge work and writing are roughly the same process: a document or presentation is the means of creating value in a high-tech world. Clearly, the value of the chemicals in a bottle of life-saving pills is negligible, but the value of the research and knowledge documented in the package insert is incalculable. The value of the silicon in a computer chip is slight, but the value of the knowledge embodied in the research reports, patent documents, and procedures is substantial. The documents are your best thinking made visible and sharable.

Although this book gives practical guidance on business grammar and usage, it does far more than that. You will find here guidelines to help you think and communicate more productively: to manage information efficiently, present persuasively, visualize clearly, frame and solve problems, and strategize soundly.

But beyond this practical guidance, this book is imbued with the paradigms and principles of high effectiveness:

- It stresses throughout the key attributes of good character—full honesty, integrity, and high ethics—as the starting point of trustworthy communication.
- "Beginning with the end in mind" is a thread that runs through every section—clearly defining your purpose in every interaction, whether a major presentation or a meeting or the simplest email message.
- First things are always first—priority information takes priority in every business communication.
- Win-win thinking is ever present at the heart of effective proposals, negotiations, presentations, meetings, resumes—in short, in all truly successful business dealings.
- The emphasis is on really listening to the needs of the customer, the co-worker, or the community before making yourself heard. Matching your message to their needs serves your purposes as well as theirs.
- Perhaps the highest form of communication is synergy—when human beings, collaborating with a win-win mindset, truly listening to one another, arrive together at new and better insights. Synergy is central to effective knowledge work.

The guidelines and processes in this book lead to synergistic communication, the kind of knowledge work that unleashes the human capacity to create, to build, and to win in the Knowledge Age.

Stephen R. Covey

Stephen R. Covey

Author Acknowledgments

Larry H. Freeman, teacher, technical writer, and editor, coauthored the first edition of the *Shipley Associates Style Guide* (1985). This edition won an Excellence award from the Society for Technical Communication. Larry became lead author for the second edition (1990) and continued as lead author for the third edition (1997, renamed the *FranklinCovey Style Guide*). Having trained thousands of professionals in technical writing for such clients as Pratt & Whitney, Exxon Mobil, and the U.S. Government, Larry is now a senior environmental consultant for the Shipley Group. A recognized authority in environmental documentation, he holds the Ph.D. in English language and linguistics from the University of Oregon. Larry recently marked 50 years of teaching in both the academic and business worlds.

Breck England, author and consultant, has helped some of the world's leading corporations become more effective in their strategic communication processes. He has directed such projects for Roche, Verizon, Chevron, Aramco, Bristol-Myers Squibb, and many others. Before joining FranklinCovey, he was vice president of consulting for Shipley Associates, an international communication-training firm. A Ph.D. in English from the University of Utah, Breck taught leadership communication for seven years in BYU's graduate school of business. At FranklinCovey, he was a core developer of *The 4 Disciplines of Execution, The 7 Habits for Managers,* and the xQ Survey. A contributor to the third edition of the *FranklinCovey Style Guide*, he is lead author of the fifth edition.

FranklinCovey Co.
2200 West Parkway Blvd.
Salt Lake City, UT 84119
U.S.A.
1-800-599-0238

Printed in U.S.A.

Publishing as Pearson Education, Inc.
Upper Saddle River, New Jersey 07458

Authorized from the original FranklinCovey edition, entitled *FranklinCovey Style Guide*, Fifth Edition, by Stephen R. Covey, published by FranklinCovey Co., ©FranklinCovey Co. 2010.

This edition is published by Pearson Education, Inc., ©2012 by arrangement with Pearson Education Ltd.

12 17

ISBN-10: 0-13-309039-6
ISBN-13: 978-0-13-309039-0

Pearson Education LTD.
Pearson Education Australia PTY, Limited.
Pearson Education Singapore, Pte. Ltd.
Pearson Education Asia, Ltd.
Pearson Education Canada, Ltd.
Pearson Educación de Mexico, S.A. de C.V.
Pearson Education—Japan
Pearson Education Malaysia, Pte. Ltd.

Library of Congress Cataloging-in-Publication Data

Covey, Stephen R.
FranklinCovey style guide for business and technical communication / Stephen R. Covey, Larry H. Freeman, Breck England. — 5th ed.
p. cm.
Includes bibliographical references and indexes.
ISBN 978-0-13-309039-0 (pbk. : alk. paper) 1. English language—Rhetoric—Handbooks, manuals, etc. 2. English language—Business English—Handbooks, manuals, etc. I. England, Breck. II. Title.
PE1115.C674 2012
808.06'65—dc23

2012017283

Reference Glossary

STYLE GUIDE
FIFTH EDITION

Using the Reference Glossary

The Reference Glossary is designed and written to help writers and editors answer routine, yet important, questions about the preparation of business and technical documents. The alphabetical arrangement of the entries allows writers to answer questions easily and rapidly, often without having to search through the Index. The many illustrative phrases, words, and sentences make the various rules and suggestions practical and applicable to real-world situations.

Still, as with any reference book, users need to become familiar with what the Reference Glossary covers and what it doesn't cover. To assist new users, we make the following suggestions about using the Reference Glossary.

- Use the alphabetical arrangement to help you find where a specific topic is addressed. As with any alphabetical list, you may have to try a couple of titles before you find the information you want. If you cannot find a topic, refer to the Index (p. 421).
- After you have found the relevant entry, survey the listed rules or headings previewed in the shaded box at the beginning of the entry. Then turn to the rule or heading that appears to answer your question.
- Read the rule and accompanying text. Be sure to review any illustrative phrases or sentences because they will often help clarify the rule. Remember, also, that many of the rules are suggestions rather than legal requirements.
- Check to see if any notes follow the rule and its examples. Notes begin with the word NOTE and are numbered if there are several notes. Notes often include information about exceptions or options to the stated rule.
- Turn to other entries that are cross-referenced, especially if you still have questions that the entry has not answered. Cross-references have this format: See LETTERS and MEMOS.
- Don't be disappointed if you cannot find the answer to a question. No reference book can answer every question. To help answer difficult or obscure questions, experienced writers and editors usually have several recent references available. For a list of other references, see the entry entitled REFERENCES.

Abbreviations

Abbreviations allow writers to avoid cumbersome repetition of lengthy words and phrases. They are a form of shorthand and are appropriate in technical and business writing, particularly in lists, tables, charts, graphs, and other visual aids where space is limited. See ACRONYMS.

Abbreviations

1. Eliminate periods in and after most abbreviations.
2. Use the same abbreviation for both singular and plural units of measurement.
3. Clarify an unfamiliar abbreviation by enclosing its unabbreviated form within parentheses following its first use in a document.
4. Do not abbreviate a unit of measurement unless it is used in conjunction with a number.
5. Do not abbreviate a title unless it precedes a name.
6. Spell out abbreviations that begin a sentence (except for abbreviated words that, by convention, are never spelled out, like *Mr.* and *Mrs.*).
7. Spell out rather than abbreviate words that are connected to other words by hyphens.
8. Do not abbreviate the names of months and days within normal text. Use the abbreviations in chronologies, notes, tables, and charts.
9. Avoid the symbol form of abbreviations except in charts, graphs, illustrations, and other visual aids.
10. Use a single period when an abbreviation ends a sentence.

1. Eliminate periods in and after most abbreviations.

Formerly, most abbreviations required periods. Today, the trend is to eliminate periods in and after abbreviations, especially in the abbreviated names of governmental agencies, companies, private organizations, and other groups:

AFL-CIO	AMA	CBS	DOE
FTC	IOOF	NFL	NLRB
OPEC	TVA	TWA	YWCA

NOTE 1: The abbreviations covered by this rule do not include informal ones such as *Dept.* and *Mgt.*, which use a final period but no periods between letters.

NOTE 2: By convention, some abbreviations still require periods:

C.E.	a.m.	B.C.E.	Dr.
e.g.	etc.	i.e.	Mr.
Mrs.	Ms.	p.m.	pp.
U.K.	U.S. (*or* U.S.A).		

Retain the period, too, in abbreviations that spell normal words:

in., inches (*not* in)
no., number (*not* no)

A recent dictionary, such as *Webster's New Collegiate Dictionary*, is the best resource for determining if an abbreviation requires periods. See REFERENCES.

NOTE 3: Abbreviations with periods should be typed without spaces between letters and periods:

e.g. (*not* e. g.)
U.K. (*not* U. K.)

2. Use the same abbreviation for both singular and plural units of measurement.

When you abbreviate a unit of measurement, use the same symbol for both the singular and the plural forms:

6 lb and 1 lb
3 m and 1 m
20 ft and 1 ft
23.5 cm and 1.0 cm

If you spell out the abbreviated word, retain the plural when the number is greater than one:

15 kilometers and 1 kilometer
6.8 meters and 1 meter

3. Clarify an unfamiliar abbreviation by enclosing its unabbreviated form within parentheses following its first use in a document:

The applicant had insurance through CHAMPUS (Civilian Health and Medical Program of the Uniformed Services).

The alloy is hardened with 0.2 percent Np (neptunium). Adding Np before cooling alters the crystalline structure of manganese host alloys.

NOTE 1: Some writers and editors prefer to cite the unabbreviated form of the word or words before the abbreviation. We believe that this practice can inhibit, rather than enhance, the reader's comprehension of the abbreviation:

The applicant had insurance through the Civilian Health and Medical Program of the Uniformed Services (CHAMPUS).

The alloy is hardened with 0.2 percent neptunium (Np). Adding Np before cooling alters the crystalline structure of manganese host alloys.

NOTE 2: Do not use an unfamiliar abbreviation unless you plan to use it more than once in the same document.

4. Do not abbreviate a unit of measurement unless it is used in conjunction with a number:

Pipe diameters will be measured in inches.

but

Standard pipe diameter is 3 in.

The dimensions of the property were recorded in both meters and feet.

but

The property is 88 ft by 130 ft.

The southern property line is 45.3 m.

5. Do not abbreviate a title unless it precedes a name:

The cardiac research unit comprises five experienced doctors.

but

Our program director is Dr. Royce Smith.

6. Spell out abbreviations that begin a sentence (except for abbreviated words that, by convention, are never spelled out, like *Mr.* and *Mrs.*):

Oxygen extraction will be accomplished at high temperatures.

not

O_2 extraction will be accomplished at high temperatures.

but

Ms. Jean MacIntyre will be responsible for modifying our subsea sensors.

7. Spell out rather than abbreviate words that are connected to other words by hyphens:

6-foot gap (*not* 6-ft)
12-meter cargo bay (*not* 12-m)
3.25-inch pipe (*not* 3.25-in.)

NOTE: The spelled-out form is preferred. The abbreviated form (as in *6-ft*) is common in some engineering documents, especially those with many numerical values. The hyphen is retained in the abbreviated form. See HYPHENS and FRACTIONS.

8. Do not abbreviate the names of months and days within normal text. Use the abbreviations in chronologies, notes, tables, and charts:

The facilities modernization plan is due January 1985. (*not* Jan 1985 *or* 1/85)

9. Avoid the symbol form of abbreviations except in charts, graphs, illustrations, and other visual aids:

55 percent (*not* 55%)
15 ft (*not* 15′)
32.73 in. (*not* 32.73″)

10. Use a single period when an abbreviation ends a sentence:

To head our laser redesign effort, we have hired the 1994 Nobel prize winner from the U.S.A. (*not* U.S.A..)

NOTE: If the clause or sentence ends with something other than a period, (e.g., comma, semicolon, colon, question mark, exclamation mark), then the other mark of punctuation follows the period at the end of the abbreviation:

Have we hired the 1994 Nobel Prize winner from the U.S.A.?

If you plan to arrive by 6 p.m., you will not need to guarantee your reservation.

List of Abbreviations

Following is a short list of many common abbreviations for words and common measurements. For more complete lists of abbreviations, refer to *The Chicago Manual of Style* and to *Merriam-Webster's Collegiate Dictionary*. See REFERENCES.

In this listing, some abbreviations appear with periods, although the trend is to eliminate the periods (see rule 1). For example, *Ph.D.* appears with periods to assist writers and typists who wish to retain the periods, although many writers today prefer the increasingly more common *PhD* without periods.

In this listing, abbreviations printed without periods are ones that customarily appear without periods—for example, *HF* or *log*.

Abbreviations List

AA, Alcoholics Anonymous
A.B. or B.A., bachelor of arts
abbr., abbreviation
abs., absolute; absent; absence; abstract
acct., account; accountant
A.D. (*anno Domini*), in the year of the Lord
ADP, automated data processing
A.H. (*anno Hegirae*), in the year of the Hijra
a.k.a., also known as
A.M. (*anno mundi*), in the year of the world
A.M. or M.A., master of arts
a.m. (*ante meridiem*), before noon
A/P, accounts payable
app, application
approx., approximately
A/R, accounts receivable
Ave., avenue
a.w.l., absent with leave
a.w.o.l., absent without official leave

BAFO, best and final offer
B.C., before Christ
Bcc: blind courtesy copy
B.C.E., before the common era
bf., boldface
Bldg., building
B.Lit(t). or Lit(t).B., bachelor of literature
Blvd., boulevard
b.o., buyer's option
BPS, basis points
B.S. or B.Sc., bachelor of science

Abbreviations

ca. (*circa*), about
c. and s.c., caps and small caps
cc:, courtesy copy
c.b.d., cash before delivery
C.E., common era
cf. (*confer*), compare or see
Co., company; country
c.o.d., cash on delivery; collect on delivery
COGS, cost of goods sold
COLA, cost-of-living adjustment
con., continued
Conus, continental United States
Corp., corporation
c.p., chemically pure
C.P.A., certified public accountant
CPI, consumer price index
cr., credit; creditor
Ct., court

d.b.a., doing business as
D.D., doctor of divinity
D.D.S., doctor of dental surgery; doctor of dental science
DII, days in inventory
Dist. Ct., District Court
D.Lit(t). or Lit(t).D., doctor of literature
do. (*ditto*), the same
DP, displaced person
D.P.H., doctor of public health
DPO, days payable outstanding
dr., debtor
Dr., doctor; drive
DSO, days sales outstanding
DVD, digital video disc
D.V.M., doctor of veterinary medicine
DVR, digital video recorder

E., east
EBITDA, earnings before interest, taxes, depreciation, and amortization
EDP, electronic data processing
e.g. (*exempli gratia*), for example
EOM, end of message
e.o.m., end of month
EPS, earnings per share
et al. (*et alii*), and others
et seq. (*et sequentia*), and the following one
etc. (*et cetera*), and others
EU, European Union
EVA, economic value added

F., Fahrenheit, farad
f., female, force, forte, frequency
f., ff., and following page (pages)
f.o.b., free on board

GAAP, generally accepted accounting standards
GAAS, generally accepted auditing standards
GDP, gross domestic product
GI, general issue; government issue
G.M.&S., general, medical, and surgical
Gov., governor
Govt., government
gr. wt., gross weight

HD, high definition
HE, high explosive
HF, high frequency

i, interlaced
ibid. (*ibidem*), in the same place
id. (*idem*), the same
ID, identification
i.e. (*id est*), that is
IF, intermediate frequency
Insp. Gen., Inspector General
IOU, I owe you
IP, intellectual property
IQ, intelligence quotient

J.D. (*juris doctor*), doctor of laws
Jr., junior

Lat., latitude
LC, Library of Congress
lc., lowercase
liq., liquid
lf., lightface
LF, low frequency
LL.B., bachelor of laws
LLC, limited liability corporation
LL.D., doctor of laws
loc. cit. (*loco citato*), in the place cited
log, logarithm
long., longitude
Ltd., limited
Lt. Gov., lieutenant governor

M, money supply: M1; M1B; M2
M., monsieur; MM., messieurs
m. (*meridies*), noon
M.D., doctor of medicine
memo, memorandum
MF, medium frequency
MIA, missing in action (*plural*, MIAs)
Mlle., mademoiselle
Mme., madam; Mmes., Mesdames
mo., month
MP, member of Parliament
Mr., mister (*plural*, Messrs.)
Mrs., mistress
Ms., feminine title (*plural*, Mses.)
M.S., master of science
MS., manuscript; MSS., manuscripts
Msgr., monsignor
m.s.l., mean sea level

N., north
NA, not available; not applicable
NE, northeast
n.e.c., not elsewhere classified
n.e.s., not elsewhere specified
net wt., net weight
No., number; Nos., numbers
n.o.i.b.n., not otherwise indexed by name
n.o.p., not otherwise provided (for)
n.o.s., not otherwise specified
n.s.k., not specified by kind
n.s.p.f., not specifically provided for
NW, northwest

OK, OK'd, OK'ing, OK's
op. cit. (*opere citato*), in the work cited

p, progressive
PA, public address system
PAC, political action committee (*plural*, PACs)
Ph.B. or B.Ph., bachelor of philosophy
Ph.D. or D.Ph., doctor of philosophy
Ph.G., graduate in pharmacy
PIN, personal identification number
Pl., place; plural
P&L, profit-and-loss statement
p.m. (*post meridiem*), afternoon
P.O. Box (*with number*), *but* post office box (*in general sense*)
POW, prisoner of war (*plural*, POWs)
Prof., professor
pro tem (*pro tempore*), temporarily
P.S. (*postscriptum*), postscript; public school (*with number*)

QA, quality assurance
QOQ, quarter over quarter

RAM, random-access memory
R&D, research and development
Rd., road
RDT&E, research, development, testing, and evaluation
Rev., reverend
RF, radio frequency
RIF, reduction(s) in force; RIF'd, RIF'ing, RIF's
R.N., registered nurse
ROA, return on assets
ROE, return on equity
ROI, return on investment
ROIC, return on invested capital
RR.,railroad
RSS, rich site summary
Rt. Rev., right reverend
Ry., railway

S., south; Senate bill (*with number*)
S&L(s), savings and loan(s)
sc. (*scilicet*), namely (*see also ss*)
s.c., small caps
s.d. (*sine die*), without date
SE, southeast
2d, second; 3d, third
SG&A, sales, general, and administrative expenses
SHF, superhigh frequency
sic, thus
SMS, short messaging service
SOP, standard operating procedure
SOS, distress signal
sp. gr., specific gravity
Sq., square (*street*)
Sr., senior
SS, steamship
ss (*scilicet*), namely (*in law*) (*see also sc.*)
St., Saint; Ste., Sainte; SS., Saints
St., street
STP, standard temperature and pressure
Supt., superintendent
Surg., surgeon
SW, southwest

T., Tbsp., tablespoon
T., township; Tps., townships

Abbreviations

t., tsp., teaspoon
Ter., terrace
t.m., true mean
TQM, total quality management
TV, television

uc., uppercase
UHF, ultrahigh frequency
U.S.A., United States of America
USA, U.S. Army
U.S. 40; U.S. No. 40; U.S. Highway No. 40

v. or vs. (*versus*), against
VAR, value-added reseller
VAT, value-added tax
VHF, very high frequency
VIP, very important person
viz (*videlicet*), namely
VLF, very low frequency

W., west
w.a.e., when actually employed
wf, wrong font
w.o.p., without pay

YOY, year over year

ZIP Code, Zone Improvement Plan Code (Postal Service)
ZIP+4, 9-digit ZIP Code

Abbreviations of Units of Measurement

A, ampere
Å, angstrom
a, are
a, atto (*prefix*, one-quintillionth)
aA, attoampere
abs, absolute (*temperature and gravity*)
ac, alternating current
AF, audiofrequency
Ah, ampere-hour
A/m, ampere per meter
AM, amplitude modulation
asb, apostilb
At, ampere-turn
at, atmosphere
atm, atmosphere
at wt, atomic weight
au, astronomical unit
avdp, avoirdupois

b, born
B, bel
b, bit
bbl, barrel
bbl/d, barrel per day
Bd, baud
bd. ft., board foot
Bé, Baumé
Bev (*obsolete*); see GeV
Bhn, Brinell hardness number
bhp, brake horsepower
bm, beam
bp, boiling point
Btu, British thermal unit
bu, bushel

c, ¢, ct; cent(s)
c, centi (*prefix*, one-hundredth)
C, coulomb
c, cycle (*radio*)
°C, degree Celsius
ca, centiare (1 square meter)
cal, calorie (*also*: cal_{IT}, International Table; cal_{th}, thermochemical)
cc. (*obsolete*), use cm^3
cd, candela (*obsolete*: candle)
cd/in^2, candela per square inch
cd/m^2, candela per square meter
c.f.m. (obsolete), use ft^3/min
c.f.s. (obsolete), use ft^3/s
cg, centigram
Ci, curie
cL, centiliter
cm, centimeter
c/m, cycles per minute
cm^2, square centimeter
cm^3, cubic centimeter
cmil, circular mil
cp, candlepower
cP, centipoise
cSt, centistokes
cu ft (*obsolete*), use ft^3
cu in (*obsolete*), use in^3
cwt, hundredweight

D, darcy
d, day
d, deci (*prefix*, one-tenth)
d, pence
da, deka (*prefix*, 10)
dag, dekagram
daL, dekaliter
dam, dekameter
dam^2, square dekameter
dam^3, cubic dekameter
dB, decibel
dBu, decibel unit
dc, direct current
dg, decigram
dL, deciliter
dm, decimeter
dm^2, square decimeter
dm^3, cubic decimeter
dol, dollar
doz, dozen
dr, dram
dwt, deadweight tons
dwt, pennyweight
dyn, dyne

EHF, extremely high frequency
emf, electromotive force
emu, electromagnetic unit
erg, erg
esu, electrostatic unit
eV, electronvolt

°F, degree Fahrenheit
f, farad
f, femto (*prefix*, one-quadrillionth)
F, fermi (*obsolete*); *use* fm, femtometer
fc, footcandle
fL, footlambert
fm, femtometer
FM, frequency modulation

ft, foot
ft^2, square foot
ft^3, cubic foot
ftH^2O, conventional foot of water
ft-lb, foot-pound
ft-lbf, foot pound-force
ft/min, foot per minute
ft^2/min, square foot per minute
ft^3/min, cubic foot per minute
ft-pdl, foot poundal
ft/s, foot per second
ft^2/s, square foot per second
ft^3/s, cubic foot per second
ft/s^2, foot per second squared
ft/s^3, foot per second cubed

G, gauss
G, giga (*prefix*, one billion)
g, gram; acceleration of gravity
Gal, gal cm/s^2
gal, gallon
gal/min, gallons per minute
gal/s, gallons per second
GB, gigabyte
Gb, gilbert
g/cm^3, gram per cubic centimeter
GeV, giga-electron-volt
GHz, gigahertz (gigacycle per second)
gr, grain; gross

h, hecto (*prefix*, 100)
H, henry
h, hour
ha, hectare
HF, high frequency
hg, hectogram
hl, hectoliter
hm, hectometer
hm^2, square hectometer
hm^3, cubic hectometer
hp, horsepower
hph, horsepower-hour
Hz, hertz (cycles per second)

id, inside diameter
ihp, indicated horsepower
in., inch
in^2, square inch
in^3, cubic inch
in/h, inch per hour
inH^2O, conventional inch of water
inHg, conventional inch of mercury
in-lb, inch-pound
in/s, inch per second

J, joule
J/K, joule per kelvin

K, kayser
K, Kelvin (*degree symbol improper*)
k, kilo (prefix, 1,000)
k, thousand (7k = 7,000)
kc, kilocycle; *see also* kHz (kilohertz), kilocycles per second
kcal, kilocalorie
keV, kilo-electron-volt
kG, kilogauss
kg, kilogram
kgf, kilogram-force

Abbreviations

kHz, kilohertz (kilocycles per second)
kl, kiloliter
klbf, kilopound-force
km, kilometer
km^2, square kilometer
km^3, cubic kilometer
km/h, kilometer per hour
kn, knot (*speed*)
ký, kilohm
kt, kiloton; carat
kv, kilovolt
kVa, kilovoltampere
kvar, kilovar
kw, kilowatt
kwh, kilowatt-hour

L, lambert
L, liter (*also* l)
lb, pound
lb ap, apothecary pound
lb avdp, avoirdupois pound
lbf, pound-force
lbf/ft, pound-force foot
lbf/ft^2, pound-force per square foot
lbf/ft^3, pound-force per cubic foot
lbf/in^2, pound-force per square inch
lb/ft, pound per foot
lb/ft^2, pound per square foot
lb/ft^3, pound per cubic foot
lct, long calcined ton
ldt, long dry ton
LF, low frequency
lin ft, linear foot
l/m, lines per minute
lm, lumen
lm/ft^2, lumen per square foot
lm/m^2, lumen per square meter
lm•s, lumen second
lm/W, lumen per watt
l/s, lines per second
l/s, liter per second
lx, lux

M, mega (*prefix*, 1 million)
M, million (3M = 3 million)
m, meter
m, milli (*prefix*, one-thousandth)
M^1, monetary aggregate
m^2, square meter
m^3, cubic meter
μ, micro (*prefix*, one-millionth)
μ, micron (*obsolete*); *use* μm, micrometer
mA, milliampere
μA, microampere
mbar, millibar
μbar, microbar
Mc, megacycle; *see also* MHz (megahertz), megacycles per second
mc, millicycle; *see also* mHz (millihertz), millicycles per second
mcg, microgram (*obsolete*); use μg
mD, millidarcy
meq, milliquivalent
MeV, mega electron volts
mF, millifarad
μF, microfarad
mG, milligauss
mg, milligram
μg, microgram
Mgal/d, million gallons per day
mH, millihenry
μH, microhenry
mho, mho (*obsolete*); *use* S, siemens
MHz, megahertz
mHz, millihertz
mi, mile (statute)
mi^2, square mile
mi/gal, mile(s) per gallon
mi/h, mile per hour
mil, mil
min, minute (*time*)
μin, microinch
ml, milliliter
mm, millimeter
mm^2, square millimeter
mm^3, cubic millimeter
mμ, (*obsolete*); *see* nm, nanometer
μm, micrometer
$μm^2$, square micrometer
$μm^3$, cubic micrometer
μμ, micromicron (*use of compound prefixes is obsolete*); *use* pm, picometer
μμf, micromicrofarad (*use of compound prefixes is obsolete*); *use* pF
mmHg, conventional millimeter of mercury
μmho, micromho (*obsolete*); use μS, microsiemens
MW, megohm
mo, month
mol, mole (*unit of substance*)
mol wt, molecular weight
mp, melting point
ms, millisecond
μs, microsecond
Mt, megaton
mV, millivolt
μV, microvolt
MW, megawatt
mW, milliwatt
μW, microwatt
MWd/t, megawatt-days per ton
Mx, maxwell

n, nano (*prefix*, one-billionth)
N, newton
nA, nanoampere
nF, nanofarad
nm, nanometer (millimicron, obsolete)
N-m, newton meter
N/m^2, newton per square meter
nmi, nautical mile
ns, nanosecond
$N\text{-}s/m^2$, newton second per square meter

od, outside diameter
Oe, oersted (*use of* A/m, amperes per meter, *preferred*)
oz, ounce (*avoirdupois*)

p, pico (*prefix*, one-trillionth)
P, poise
Pa, pascal
pA, picoampere
PB, petabyte (1 million gigabytes)
pct, percent
pdl, poundal
pF, picofarad (micromicrofarad, *obsolete*)
pF, water-holding energy
pH, hydrogen-ion concentration
ph, phot; phase
pk, peck
p/m, parts per million
ps, picosecond
pt, pint
pW, picowatt

qt, quart
quad, quadrillion (10^{15})

°R, degree rankine
R, roentgen
rad, radian
rd, rad
rem, roentgen equivalent man
r/min, revolutions per minute
rms, root mean square
r/s, revolutions per second

s, second (*time*)
s, shilling
S, siemens
sb, stilb
scp, spherical candlepower
s•ft, second-foot
shp, shaft horsepower
slug, slug
sr, steradian
$stdft^3$, standard cubic foot (feet)
Sus, saybolt universal second(s)

T, tera (*prefix*, 1 trillion)
Tft^3, trillion cubic feet
T, tesla
t, tonne (metric ton)
TB, terabyte (1,000 gigabytes)
tbsp, tablespoonful
thm, therm
ton, ton
tsp, teaspoonful
Twad, twaddell

u, (unified) atomic mass unit
UHF, ultrahigh frequency

V, volt
VA, voltampere
var, var
VHF, very high frequency
V/m, volt per meter

W, watt
Wb, weber
Wh, watt-hour
W/(m•K), watt per meter kelvin
W/sr, watt per steradian
$W/(sr \bullet m^2)$, watt per steradian square meter

x, unknown quantity

yd, yard
yd^2, square yard
yd^3, cubic yard
yr, year

Acronyms

Acronyms are abbreviations that are pronounced as words:

ALGOL (ALGOrithmic Language)

ARAMCO (ARabian AMerican oil COmpany)

ASCII (American Standard Code for Information Interchange)

BIT or bit (BInary digiT)

BAC (blood alcohol content)

CAD (Computer-Aided Design)

CAM (Computer-Aided Manufacturing)

COAD (chronic obstructive airways disease)

FASB (Financial Accounting Standards Board)

FIFO (first in, first out)

GUI (Graphical User Interface)

IMAP (Internet message access protocol)

LAN (Local Area Network)

loran (LOng-RAnge Navigation)

MIPS (million instructions per second)

Nasdaq (National Association of Securities Dealers Automated Quotations)

NATO (North Atlantic Treaty Organization)

NAV (net asset value)

NSAID (non-steroid anti-inflammatory drug)

OEM (original equipment manufacturer)

PERT (Program Evaluation and Review Technique)

radar (RAdio Detecting And Ranging)

RAM (Random Access Memory)

secant (SEparation Control of Aircraft by Nonsynchronous Techniques)

SEO (search engine optimization)

sonar (SOund NAvigation Ranging)

TIFF (tagged image file format)

UNICEF (United Nations International Children's Emergency Fund)

ZIP (Zone Improvement Plan)

Acronyms

1. When you introduce new or unfamiliar acronyms, use the acronym and then, in parentheses, spell out the name or expression.
2. Avoid overusing acronyms, especially if your readers are unlikely to be very familiar with them.

Acronyms may be written in all capitals if they form proper names. However, some acronyms are conventionally uppercase and lowercase:

Amtrak　　Nasdaq

The most common acronyms, those representing generic technical concepts rather than organizations or programs, are typically all lowercase:

laser　　radar　　sonar

Some acronyms appear either in capitals or in lowercase:

BIT *or* bit

See ABBREVIATIONS.

1. When you introduce new or unfamiliar acronyms, use the acronym and then, in parentheses, spell out the name or expression:

Our program fully complies with the provisions of STEP (the Supplemental Training and Employment Program). To implement STEP, however, we had to modify subcontracting agreements with four components suppliers.

NOTE: Some writers and editors prefer to introduce unfamiliar acronyms by first spelling out the component words and then placing the acronym in parentheses. We believe that readers should see the acronym first because that is how they will see it on later pages. See ABBREVIATIONS.

2. Avoid overusing acronyms, especially if your readers are unlikely to be very familiar with them.

Until readers learn to recognize and instantly comprehend an acronym (like *laser*), the acronym hinders reading. It creates a delay while the reader's mind recalls and absorbs the acronym's meaning. Therefore, you should be cautious about using acronyms, especially unfamiliar ones. Overloading a text with acronyms makes the text unreadable, even if you have previously introduced and explained the acronyms.

Acronyms are good shorthand devices, but use them judiciously. See SCIENTIFIC/TECHNICAL STYLE.

Active/Passive

Active- and passive-voice sentences each convey actions. They differ in how they convey these actions by their different grammatical structures. Both types of sentences are good sentences, but you should use active-voice sentences when you can. Passive-voice sentences can seem weak-willed, indecisive, or evasive.

Active- and passive-voice sentences usually have three basic elements:

- The actor—the person or thing performing the action
- The action—the verb
- The receiver—the person or thing receiving the action

When the structure of the sentence has the actor in front of the action, the sentence is in the **active voice**:

> Australian companies manufacture millions of precision machine tools.

Companies is the actor; *manufacture* is the action; and *tools* receives the action. Because the actor comes before the action, the sentence is active. The subject of the sentence performs the action.

When the structure of the sentence has the receiver in front of the action, the sentence is in the **passive voice**:

> Millions of precision machine tools are manufactured by Australian companies.

In this sentence, the subject (*tools*) is not doing the manufacturing. The tools are being manufactured. They are being acted upon; they are receiving the action. Therefore, the subject—and the sentence—is passive.

Active/Passive

1. Prefer active sentences.
2. Use a passive sentence when you don't know or don't want to mention the actor.
3. Use a passive sentence when the receiver is more important than the actor.
4. Use a passive sentence when you need to form a smooth transition from one sentence to the next.
5. Do not use passive sentences to avoid using first person pronouns.
6. Make sentences active by turning the clause or sentence around.
7. Make sentences active by changing the verb.
8. Make sentences active by rethinking the sentence.

1. Prefer active sentences.

Active sentences are usually shorter and more dynamic than passive sentences. They generally have more impact and seem more "natural" because readers expect (and are accustomed to) the actor-action-receiver pattern. Active writing is more forceful and more self-confident.

Passive writing, on the other hand, can seem weak-willed, indecisive, or evasive. In passive sentences, the reader encounters the action before learning who performed it. In some passive sentences, the reader never discovers who performed the action. So passive sentences seem static.

Passive sentences are useful—even preferable—in some circumstances, but you should prefer active sentences.

When to Use Passives

2. Use a passive sentence when you don't know or don't want to mention the actor:

> The failure occurred because metal shavings had been dropped into the worm-gear housing.
>
> Clearly, the site had been inspected, but we found no inspection report and could not identify the inspectors.

In the first example above, a passive sentence is acceptable because we don't know who dropped the metal shavings into the housing. In the second example, we might know who inspected the site but don't want to mention names because the situation could be sensitive or politically charged.

3. Use a passive sentence when the receiver is more important than the actor.

The strongest part of most sentences is the opening. Therefore, the sentence element appearing first will receive greater emphasis than those elements appearing later in the sentence. For this reason, a passive sentence is useful when you wish to emphasize the receiver of the action:

> Cross-sectional analysis techniques—the most important of our innovations—are currently being tested in our Latin American Laboratory.

> Minimum material size or thickness requirements will then be established to facilitate recuperator weight, size, and cost estimates.

In both examples, we wish to emphasize the receiver of the action. Note how emphasis changes if we restructure the first example:

> The most important of our innovations (cross-sectional analysis techniques) is currently being tested in our Latin American Laboratory.

> Our Latin American Laboratory is currently testing the most important of our innovations—cross-sectional analysis techniques.

> Our Latin American Laboratory is currently testing cross-sectional analysis techniques—the most important of our innovations.

The emphasis in each sentence differs, depending on sentence structure. The first revision emphasizes *innovations*, and it is still a passive sentence. The last two revisions are active, and both stress *our Latin American Laboratory*.

The ending of a sentence is also emphatic (although not as emphatic as the beginning), so the sentence ending with *techniques* does place secondary emphasis on *techniques*. However, the best way to emphasize *cross-sectional analysis techniques* is by opening the sentence with that phrase.

4. Use a passive sentence when you need to form a smooth transition from one sentence to the next.

Occasionally, writers must arrange sentence elements so that key words appearing in both sentences are near enough to each other for readers to immediately grasp the connection between the sentences. In the example below, for instance, the writer needs to form a smooth transition between sentences by repeating the key words *work packages*:

> We will develop a simplified matrix of tasks that will include all budgetary and operational work packages. These work packages will be scheduled and monitored by individual program managers.

The second sentence is passive. It would be shorter and stronger as an active sentence:

> Individual program managers will schedule and monitor these work packages.

However, the active version does not connect as well with the previous sentence:

> We will develop a simplified matrix of tasks that will include all budgetary and operational work packages. Individual program managers will schedule and monitor these work packages.

For a brief moment, the second sentence seems to have changed the subject. Not until readers reach the end of the second sentence will they realize that both sentences deal with work packages. Therefore, making the second sentence passive creates a smoother transition and actually improves the passage. See TRANSITIONS.

Passives and First Person

5. Do not use passive sentences to avoid using first person pronouns.

Some writers use passives to avoid using first person pronouns (*I*, *me*, *we*, or *us*). These writers mistakenly believe that first person pronouns are inappropriate in business or technical writing. In fact, the first person is preferable to awkward or ambiguous passive sentences like the example below:

> It is recommended that a state-of-the-art survey be added to the initial redesign studies.

Who is recommending it? You? The customer? Someone else? And who is supposed to add the survey?

In the following sentences, things seem to be happening, but no one seems to be doing them:

> Cost data will be collected and maintained to provide a detailed history of the employee hours expended during the program. This tracking effort will be accomplished by the use of an established employee-hour accumulating system.

Writers who overuse the passive to avoid first person pronouns convey the impression that they don't want to accept the responsibility for their actions. This implication is why passive sentences can seem evasive even when the writer doesn't intend them to be.

Passive sentences allow you to eliminate the actor. In some cases, eliminating the actor is appropriate and desirable. In other cases (as in the previous examples), eliminating the actor creates confusion and doubt. Active versions of these examples, using first person pronouns, are much better:

We recommend that the initial redesign studies include a state-of-the-art survey.

Using our employee-hour accumulating system, we will collect and maintain cost data to provide a detailed history of the employee hours expended during the program.

How to Convert Passives

Technical and scientific writers generally use too many passives. They use them unnecessarily, often more from habit than choice. Converting unneeded passives to actives will strengthen the style of the document, making it appear crisper and more confident. The following guidelines present three techniques for converting passives to actives.

6. Make sentences active by turning the clause or sentence around:

These methods are described in more detail in section 6.

Section 6 describes these methods in more detail.

A functional outline of the program is included in the Work Breakdown Structure (figure 1.1–2).

The Work Breakdown Structure (figure 1.1–2) includes a functional outline of the program.

Brakes on both drums are activated as required by the control system to regulate speed and accurately position the launcher.

The control system activates brakes on both drums as required to regulate speed and accurately position the launcher.

After these requirements are identified, we will develop a comprehensive list of applicable technologies.

After identifying these requirements, we will develop a comprehensive list of applicable technologies.

7. Make sentences active by changing the verb:

The solutions were achieved only after extensive development of fabrication techniques.

The solutions occurred only after extensive development of fabrication techniques.

The Gaussian elimination process can be thought of as a means of "decomposing" a matrix into three factors.

The Gaussian elimination process "decomposes" a matrix into three factors.

The Navy recuperator requirements are expected to bring added emphasis to structural integrity.

The Navy recuperator requirements will probably emphasize structural integrity.

Coalescence was always observed to start at the base of the column.

Coalescence always started at the base of the column.

8. Make sentences active by rethinking the sentence:

Special consideration must be given to structural mounting, heat exchanger shape, ducting losses, and ducting loads.

Structural mounting, heat exchanger shape, ducting losses, and ducting loads are especially important.

To ensure that a good alternate design approach is not overlooked, a comparison between plate-fin and tubular designs will be made during the proposed study program.

Comparing plate-fin and tubular designs during the proposed study program will ensure that we thoughtfully consider alternate design approaches.

This study will show what can be done to alleviate technology failure by selectively relaxing requirements.

This study will show how selectively relaxing requirements can alleviate technology failure.

It must be said, however, that while maximum results are gained by a design-synthesis approach such as we propose, the area to be covered is so large that it will still be necessary to concentrate on the most important technologies and their regions of interest.

Our proposed design-synthesis approach will yield maximum results. Nevertheless, the area of interest is very large. Concentrating on the most important technologies and their regions of interest will still be necessary.

Adjectives

Adjectives describe or modify nouns or pronouns. They typically precede nouns or follow either verbs of sense (*feel*, *look*, *sound*, *taste*, *smell*) or linking verbs (*be*, *seem*, *appear*, *become*):

> The slow process . . . (*or* The process is slow.)
>
> Warm weather . . . (*or* The weather seems warm.)
>
> The cautious superintendent . . . (*or* The superintendent became cautious.)
>
> The news seemed bad. *(not* badly, *which is an adverb)*

Adjectives also tell which one, what kind, or how many people or things are being discussed.

NOTE: As in the preceding examples, most adjectives potentially occur between an article and a noun (*a bad message*) or following a linking verb (*the message is bad*). In both of these positions, adjectives are describing a noun. Adjectives can also describe a pronoun:

> He is slow.
> They are ignorant.

Or less likely, but still possible:

> An arrogant somebody decided to speak up before the meeting ended.

Finally, an adjective may seem to describe a following adjective rather than the main noun in a phrase:

> low moral character
> pale yellow flowers

We consider *low* and *pale* to be adjectives that are describing or modifying *moral* and *yellow*. Optionally, *low* and *pale* might functionally be labeled adverbs. Native users of English still intuitively know that *low* and *moral* work together to describe *character*. For such users, the grammatical terminology is unimportant.

Adjectives

1. Use adjectives, not adverbs, following verbs of sense *(feel, look, sound, taste, smell)* and linking verbs *(be, seem, appear, become)*.
2. Use the comparative *(–er/more)* forms when comparing two people or things and the superlative *(–est/most)* forms when comparing more than two.
3. Avoid noun strings unless you are sure your readers know what each string means.
4. Arrange nouns used as adjectives in technical expressions so that the more general nouns are closest to the word they are modifying.
5. For the names of an organization or a company modifying a noun, choose to use either a possessive form (with an apostrophe) or an unchanged descriptive form. Once you choose, stay with your choice throughout a document.

Adjectives and Adverbs

Adjectives and adverbs are similar. They both describe or modify other words, and they both can compare two or more things. Sometimes they appear in similar positions in sentences:

> Harry felt **cautious**. *(adjective)*
>
> Harry felt **cautiously** along the bottom of the muddy stream. *(adverb)*

> The guard remained **calm**. *(adjective)*
>
> The guard remained **calmly** at his post. *(adverb)*

> The car was **close** to the building. *(adjective)*
>
> The car came **close** to me. *(adverb)*
>
> The corporal watched the prisoner **closely**. *(adverb)*

NOTE: Not all adverbs end in *–ly* (for example, the adverbs *deep*, *fair*, *fast*, *long*, *wide*). Some forms can be both an adjective or an adverb (for example, *early* or *monthly*). Other adverbs have two forms: an *–ly* form and another form that is identical to the adjective (*deep/deeply*, *fair/fairly*, *hard/hardly*, *wide/widely*). You can determine whether most words are adjectives by trying to put them in front of a noun. In the previous examples, *cautious Harry* and *the calm guard* both make sense, so *cautious* and *calm* are adjectives. In the third example, *close* is an adverb in the second context, but in the phrase *a close friend*, the word *close* is an adjective. See ADVERBS.

1. Use adjectives, not adverbs, following verbs of sense (*feel, look, sound, taste, smell*) and linking verbs (*be, seem, appear, become*):

> The engine sounded **rough**. (*not the adverb* roughly)
>
> The surface of the mirror felt **smooth**. (*not the adverb* smoothly)
>
> The programmer was **cautious** about saving each new electronic file.
>
> *but*
>
> The programmer **cautiously** saved each new electronic file. (*adverb preceding the verb* saved)

The auditor appeared **eager** to assist our division.

but

The auditor volunteered **eagerly** to assist our division. (*adverb following the verb* volunteered)

See ADVERBS.

NOTE: *Harold felt badly because of the flu.* This use of *badly* is currently acceptable, especially in spoken English. The older parallel form with *bad* is still correct and widely used. *Harold felt bad because of the flu.* See *bad/badly* in WORD PROBLEMS.

Comparatives and Superlatives

Adjectives have different forms for comparing two objects (the comparative form) and comparing more than two objects (the superlative form):

Our networking system is slower than the new WebWare system. (Slower *is the comparative form.*)

The Gemini software package was the slowest one we surveyed. (Slowest *is the superlative form.*)

Stocks are a likelier investment than bonds if long-term growth is the goal. (*or* more likely)

Nissan's likeliest competitor in the suburban wagon market is General Motors. (*or* most likely)

The 2011 budget is more adequate than the 2010 budget.

The cooling provisions are the most adequate feature of the specifications.

NOTE: One-syllable words use *–er/–est* to form comparatives or superlatives. Two-syllable words use either *–er/–est* or *more/most.* Three-syllable words use *more/most.* A few adjectives have irregular comparative forms: *good* (*well*), *better, best*; *bad, worse, worst*; *many, more, most.*

2. Use the comparative (*–er/more*) forms when comparing two people or things and the superlative (*–est/most*) forms when comparing more than two:

Of the two designs, Boeing's seems more efficient.

The Pinnacle Finance proposal is the most attractive. (*More than two options are implied, so the superlative is proper.*)

Weekly deductions are the best method for financing the new hospital insurance plan.

Weekly deductions are better than any other method for financing the new hospital insurance plan. (*The comparative* better *is used because the various options are being compared one by one, not as a group.*)

Nouns Used as Adjectives

Nouns often behave like adjectives, especially in complex technical phrases. Turning nouns into adjectives can reduce verbiage:

percentage of error
error percentage (*the noun* error *becomes an adjective*)

reduction in weight
weight reduction

function of the liver
liver function

Such nouns are useful because English often does not have an adjective form with the same meaning as the noun.

3. Avoid noun strings unless you are sure your readers know what each string means.

You should beware of noun strings, which are groups of nouns strung together as adjectives. Here is an example from an aircraft manual: *C-5A airframe weight calculation error percentage.* The first five words in this phrase are a noun string. Such strings often cloud meaning. Breaking up noun strings clarifies the meaning: *percentage of error in calculating C-5A airframe weight.*

Although useful and often necessary, nouns used as adjectives in a noun string may be clear only to technically knowledgeable people:

aluminum honeycomb edge panels

What is aluminum—the honeycomb, the edges, or the panels? Only a knowledgeable reader can tell for sure. Sometimes, the order of the words suggests an interpretation:

aluminum edge honeycomb panels

From this phrase, we may expect the edges, and not the honeycomb, to be aluminum, but we still can't know for sure if *aluminum edge* and *honeycomb* equally modify *panels,* or if *aluminum edge* and *honeycomb* combine to become a single modifier of *panels,* or if *aluminum* modifies something called *edge honeycomb*:

(aluminum + edge) + honeycomb panels

or

(aluminum + edge + honeycomb) panels

or

aluminum + (edge + honeycomb) panels

In alphabetical lists of parts, the main noun being modified must be listed first. Therefore, the modifying words appear afterwards, usually separated by commas. The modifying words are typically listed in reverse order, with the most general modifiers closest to the main noun:

panels, honeycomb, aluminum edge

or

panels, edge, aluminum honeycomb

or

panels, aluminum edge honeycomb

Adjectives

A helpful technique for discovering or clarifying the structure of noun strings is to ask the question, *What kind*? Begin with the main noun being modified and proceed from there to build the string of modifying nouns:

panels

What kind of panels?

honeycomb

What kind of honeycomb?

aluminum edge

In this case, we have assumed that *aluminum edge* describes a particular type of honeycomb. Because *aluminum* and *edge* jointly modify *honeycomb*, they act as one word. We usually show that two or more words are acting together as joint or compound modifiers by hyphenating them:

aluminum-edge honeycomb panels

See HYPHENS.

4. Arrange nouns used as adjectives in technical expressions so that the more general nouns are closest to the word they are modifying:

semiautomatic slat worm gear

automatic slat worm gear

semiautomatic strut backoff gear

automatic strut backoff gear

NOTE 1: The structure of such phrases (as well as the logic behind this rule) appears in catalogued lists. You can display the structure by reversing the order of the noun string and using indentation to show levels of modification:

gear
- backoff
  - automatic strut
  - semiautomatic strut
- worm
  - automatic slat
  - semiautomatic slat

NOTE 2: Some technical writers and editors rarely use internal punctuation (either hyphens or commas) to separate nouns in noun strings. In many scientific and technical fields, hyphens that would normally connect parts of a unit modifier are eliminated:

methyl bromide solution (*not* methyl-bromide solution)

black peach aphid (*not* black-peach aphid *or* black peach-aphid)

grey willow leaf beetle

swamp black currant seedlings

Hyphens in many technical words are, however, very hard to predict: *horse-nettle* vs. *horseradish* or *devilsclaw* vs. *devils-paintbrush*. In instances where the first word is capitalized, the compound is often hyphenated: *China-laurel*, *Queen Anne's-lace, Australian-pea*, etc. See HYPHENS.

NOTE 3: Commas are not used to separate nouns in noun strings. However, we use commas to separate true adjectives when the adjectives **equally** modify the same noun:

grey, burnished, elliptical sphere

sloppy, poorly written, inadequate proposal

See COMMAS.

5. For the names of an organization or a company modifying a noun, choose to use either a possessive form (with an apostrophe) or an unchanged descriptive form. Once you choose, stay with your choice throughout a document.

For most organizations or companies, you can choose between two types of phrases:

Possessive form

Shell's corporate benefit package

General Motors' financial officer

The Fitness Committee's recommendations

Descriptive (noun used as an adjective)

The Shell corporate benefit package

A General Motors financial officer

The Fitness Committee recommendations

Both the possessive and the descriptive versions are acceptable. Some companies, however, have firm policies about which version to use in their documents. When a policy exists, a company frequently chooses to avoid the possessive form on the grounds that the company does not possess or own something.

NOTE 1: As in the descriptive phrases above, a common sign that the possessive is not appropriate is the use of *a*, *an*, or *the* before the organizational or company name. This practice is not 100 percent reliable as a sign because, as in the Fitness Committee example, an organization may have an attached *the* or *a*/*an* and still use the possessive.

NOTE 2: Deciding which form to use is especially difficult when the organizational or company name looks like a collection of individuals.

Green, Hancock, Blaine, and Jestor

Goodmark Consultants

In cases like the preceding, choose one pattern for your correspondence and stay with your choice:

Green, Hancock, Blaine, and Jestor's acquittal rate is . . .

Goodmark Consultants' fee structure is . . .

or

The Green, Hancock, Blaine, and Jestor acquittal rate is . . .

The Goodmark Consultants fee structure is . . .

See AGREEMENT and APOSTROPHES.

Adverbs

Adverbs are modifiers that give the how, where, when, and extent of the action within a sentence. Most adverbs end in *–ly*, but some common adverbs do not: *so, now, later, then, well,* etc. Adverbs often modify the main verbs in sentences:

The engineer **slowly** prepared the design plan. *(How?)*

The supply ship moved **close** to the drilling platform. *(Where?)*

They **later** surveyed all participants in the research project. *(When?)*

The abdominal pain was **clearly** evident in all treatment groups. *(Extent?)*

Adverbs can also modify adjectives or other adverbs:

Their proposal was **highly** entertaining.

Costs were **much** lower than expected.

The well was **so** deep that its costs became prohibitive.

The board of directors cut costs **more** severely and **more** rapidly than we anticipated.

Adverbs

1. Place the adverbs *only, almost, nearly, merely,* and *also* as close as possible to the word they modify.
2. Choose adverbs, not adjectives, to modify main verbs.

1. Place the adverbs *only, almost, nearly, merely,* and *also* as close as possible to the word they modify:

The bank examiners looked at only five accounts. (*not* The bank examiners only looked at five accounts.)

The engineer had almost finished the specifications. (*not* The engineer almost had finished the specifications.)

Adverbs and Adjectives

Adverbs and adjectives are quite similar. They each modify or describe other words, and they often appear in similar positions in sentences, but they have quite different meanings:

The lab technician **carefully** smelled the sample. *(adverb)*

The cheese smelled **bad**. *(adjective)*

The Internet connection worked **badly** the first day. *(adverb)*

Not knowing the language, they stayed **close** to the interpreter. *(adverb)*

We **closely** studied the blueprints. *(adverb)*

The election was so **close** that no one was a clear winner. *(adjective)*

2. Choose adverbs, not adjectives, to modify main verbs:

Our accountants predicted **accurately** that cash flow would be a problem.

The manager asked **quickly** for the up-to-date estimates.

The test engineers calculated **roughly** the expected power.

NOTE 1: Some adverbs have two forms, one without the regular *–ly* and one with it: *close/closely, deep/deeply, late/lately, loud/loudly, quick/quickly, slow/slowly, wide/widely.* Sometimes the two adverbial forms have different meanings:

We submitted the invoice **late**.

We were involved **lately** in some takeover discussions.

The loose flywheel moved very **close** to its housing.

The flywheel is monitored **closely** during the trial run.

In other instances, the two forms mean almost the same thing, so the choice depends on personal preference or individual idiom (based on the surrounding words):

Go **slow**. *vs.* Go **slowly**. *(Either form is correct.)*

The evaluation team wanted to play **fair**. *(The phrase* play fairly *means the same but sounds a little stiff and overly formal.)*

The evaluation team wanted to respond **fairly**. (Fair *would sound awkward with the verb* respond.)

See ADJECTIVES.

NOTE 2: Adjectives, not adverbs, follow verbs of sense (*feel, look, sound, taste, smell*) and linking verbs (*be, seem, appear, become*):

The adhesive felt **cool** and **rubbery** when dry.

The surface of the wing appeared **uneven**.

See *bad/badly* in WORD PROBLEMS.

Comparative and Superlative Forms

Adverbs, like adjectives, have different forms to show comparison of two things (the comparative form) and comparison of more than two things (the superlative form). The comparative uses an *–er* form or *more*, but not both; the superlative uses an *–est* or *most*, but not both.

The counselor left sooner than expected. *(comparative)*

The fluid returned more slowly to its original level. *(comparative)*

They debated most successfully the wisdom of expanding into the West Coast market. *(superlative)*

The most rapidly moving car turned out to be the new Ford high-performance model. *(superlative)*

NOTE 1: Some adverbs have irregular comparative and superlative forms: *well, better, best; badly, worse, worst; little, less, least; much, more, most.*

NOTE 2: See ADJECTIVES for a discussion of using *–er* or *more* for comparatives and *–est* or *most* for superlatives. The rules for adverbs are similar to those for adjectives.

Agreement

Agreement is a basic grammatical rule of English. According to this rule, subjects of sentences must agree in number with their verbs:

The proposal was finished. (*not the plural* were finished)

She is the engineer who designed the valve. (*not the plural* are)

The boilers have become corroded. (*not the singular* has become)

They are our competitors on most major procurements. (*not the singular* is)

This rule also includes gender agreement (between pronouns and the persons or objects to which they refer):

Jane Swenson submitted her report. (*The pronoun* her *agrees with its antecedent* Jane Swenson.)

See NOUNS, PRONOUNS, and VERBS.

Agreement

1. The subject of a sentence (nouns or pronouns) should agree in number with the sentence verb.
2. Subjects connected by *and* require a plural verb.
3. Singular subjects connected by *either . . . or, neither . . . nor,* and *not only . . . but also* require a singular verb.
4. When used as a subject or as the modifier of the subject, *each, every, either, neither, one, another, much, anybody, anyone, everybody, everyone, somebody, someone, nobody,* and *no one* require singular verbs.
5. When used as a subject or as the modifier of a subject, *both, few, several, many,* and *other(s)* require plural verbs.
6. *All, any, more, most, none, some, one-half of, two-thirds of, a part of,* and *a percentage of* require either a singular or a plural verb, depending upon the noun they refer to.
7. Collective nouns and expressions with time, money, and quantities take a singular or a plural verb, depending upon their intended meaning.
8. Choose either a singular or plural verb for subjects that are organizational names, and then be consistent in all other contexts with the name.

1. The subject of a sentence (nouns or pronouns) should agree in number with the sentence verb:

The investigator is analyzing the analgesic efficacy. (*singular noun and singular verb*)

The employees are discussing the benefit package. (*plural noun and plural verb*)

I am going to attend the international conference in June.

We are designing a light-sensitive monitoring system.

The Elements of Geometry is the basic textbook.

Our textbooks are usually translated into Russian, French, and German.

Midwest states normally include Kentucky and Missouri.

A list of Midwest states normally includes Kentucky and Missouri.

NOTE: A noun ending with an *–s* or *–es* is usually plural. A verb ending with an *–s* or *–es* is usually singular. *Employees* is plural. The verbs *is* and *includes* are both singular.

Some verbs do not change their form to reflect singular and plural: *will include, included, had included, will have included,* etc. See NOUNS and VERBS.

Agreement problems sometimes occur because the subject of the sentence is not clearly singular or plural:

None of the crew is going to take leave.

or

None of the crew are going to take leave.

Both versions are correct. Some writers become confused, too, when the subject is separated from the verb by words or phrases that do not agree in number with the subject:

Only one of the issues we discussed is on the agenda for tomorrow's meeting.

Few aspects of the problem we are now facing are as clear as they should be.

The availability of rice, as well as of medical supplies, determines the life expectancy of a typical adult in Hong Kong.

Normal wear and tear, along with planned obsolescence, is the reason most automobiles provide only an average of 6.5 years of service.

The number of the subject must still agree with the verb even when following the verb:

What are your arguments for creating online access to the database?

There are five new pumps in the warehouse.

Discussed are the basic design flaws in the preliminary specifications and the lack of adequate detail in the drawings.

Finally, some noun subjects look plural because they end in *–s* or *–ics*, but they are still singular:

Politics has changed drastically with the advent of television.

The news from Algeria continues to be discouraging.

Measles rarely occurs in adults.

2. Subjects connected by *and* require a plural verb:

The ceiling panels and the fasteners have been fabricated.

The software designer and the graphic artist agree that we should market the new instructional manual immediately.

A personal computer and a photo copier are essential business tools today.

NOTE: Sometimes words connected by *and* become so closely linked that they become singular in meaning, thus requiring a singular verb:

Bacon and eggs is my favorite breakfast.

My name and address is on the inside cover.

Simon & Schuster is an excellent publishing firm.

3. Singular subjects connected by *either . . . or, neither . . . nor,* and *not only . . . but* also require a singular verb:

Either the post-operative therapy or the inflammation is causing the acute pain.

Neither the district engineer nor the superintendent has approved the plans.

Not only the cost but also the design is a problem.

NOTE: When one of a pair of subjects is plural, the verb agrees with the subject closest to it:

Either the tail assembly or the wing struts are causing excessive fuel consumption.

Either the wing struts or the tail assembly is causing excessive fuel consumption.

4. When used as a subject or as the modifier of the subject, *each, every, either, neither, one, another, much, anybody, anyone, everybody, everyone, somebody, someone, nobody,* and *no one* require singular verbs:

Every proposal has been evaluated.

Each engineer is responsible for the final proofing of engineering proposals.

Everyone has received the pension information.

Somebody was responsible for the drop in production.

No one but the design engineer knows the load factors used in the calculations.

NOTE: Although words ending with *–one* and *–body* require a singular verb, sentences with such words often become awkward when a pronoun refers to those words:

Everyone turns in his report on Monday.

Using the singular pronoun *his* maintains the agreement with the subject, but if the *everyone* mentioned includes women, the expression may be considered sexist. Some writers and editors argue that male pronouns (*he, his, him, himself*) are generic, that they refer to both males and females. Others maintain that this convention discriminates against women. Writers and editors who share this view prefer to include both men and women in their sentences:

Everyone turns in his or her report on Monday.

Finally, some liberal editors argue that *everyone* implies a plurality, so the plural *their* becomes the acceptable pronoun. For example:

Everyone turns in their reports on Monday.

The sexism problem is avoidable in most sentences simply by making the subject plural and eliminating such troublesome words as *everyone*:

All engineers turn in their reports on Monday.

See PRONOUNS and BIAS-FREE LANGUAGE.

5. When used as a subject or as the modifier of a subject, *both, few, several, many,* and *others* require plural verbs:

Both proposals were unsatisfactory.

Several were available earlier this month.

Few pipes were still in service.

6. *All, any, more, most, none, some, one-half of, two-thirds of, a part of,* and *a percentage of* require either a singular or a plural verb, depending upon the noun they refer to:

All of the work has been assigned. *(singular)*

All of the trees have been removed. *(plural)*

Most sugar is now made from sugar beets.

Most errors were caused by carelessness.

Some of the report was written in an ornate style.

Some design features were mandatory.

One-half of the project has been completed.

One-half of the pages have been proofed.

A percentage of the room is for storage.

A percentage of the employees belong to the company credit union.

7. Collective nouns and expressions with time, money, and quantities take a singular or a plural verb, depending upon their intended meaning:

The committee votes on pension policy when disputes occur. (Committee, *a collective noun, is considered singular. In British English* committee *is often used as a plural.*)

The committee do not agree on the interpretation of the mandatory retirement clause. (Committee, *a collective noun, is considered plural.*)

The audience was noisy, especially during the final act.

The audience were in their seats by 7:30 p.m.

Two years is the usual waiting period. (Two years *is an expression of time considered as a single unit.*)

The 2 years were each divided into quarters for accounting purposes. (Two years *is an expression of time considered as a plural of* year.)

Six dollars is the fee.

Six dollars were spread out on the counter.

Five liters is all the tank can hold.

Five liters of wine were sold before noon.

Note: Sometimes sentences with collective nouns become awkward because they seem both singular and plural. In such cases, rephrasing often helps:

Audience members were in their seats by 7:30 p.m.

8. Choose either a singular or plural verb for subjects that are organizational names, and then be consistent in all other contexts with the name.

Problems arise because organizational names often look plural even though they are the names of single organizations:

Kraus, Jones, and Blackstone

FranklinCovey

The Money Group

Thomas & Sons, Inc.

These names take a plural verb if you intend to stress the individual members or partners:

Kraus, Jones, and Blackstone have their law offices in the Tower Center Building.

FranklinCovey present training courses throughout the world.

The Money Group are uniquely qualified to advise you on your investments.

Thomas & Sons have been in business since 1950, and their reputation is unexcelled.

Otherwise, use a singular verb, which is the preferred pattern in business writing, probably because readers usually view an organization as a single entity:

Kraus, Jones, and Blackstone has signed a lease for a suite in the Tower Center Building.

FranklinCovey is a leading training firm, and its materials have won national awards.

The Money Group is licensed in Michigan, and its corporate offices are in Detroit.

Thomas & Sons has the city contract for all plumbing work.

Note 1: As in two of the examples with plural verbs, sentences with an organizational name often include a pronoun that refers back to the organization. If you have chosen a plural verb, this pronoun will be *they* or *them*. See Nouns.

As in two of the examples with singular verbs, sentences often contain the singular pronoun *its* to refer back to the organization. See Pronouns.

Note 2: Often you must decide whether to use an apostrophe with an organizational name, as in contexts such as these:

Kraus, Jones, and Blackstone's personnel policy requires . . .

FranklinCovey's proposal includes . . .

The Money Group's line of credit exceeds . . .

Thomas & Sons' vans have . . .

An apostrophe is more common if you are considering the names plural. If you choose to make the names singular, then an apostrophe is usually unnecessary, especially if you precede the name with *a*, *an*, or *the*:

The Kraus, Jones, and Blackstone personnel policy requires . . .

A FranklinCovey proposal includes . . .

The Money Group line of credit exceeds . . .

The Thomas and Sons vans have . . .

Again, be consistent within a single document. Either use the possessive forms with their apostrophes or use descriptive forms without apostrophes. See Adjectives, Apostrophes, and Possessives.

Apostrophes

Apostrophes signal omitted letters, possession, and the plural of letters and symbols. In possessive forms, an apostrophe (') can appear with or without a following *–s*.

> **Apostrophes**
>
> 1. Use apostrophes to signal omitted letters in a contraction.
> 2. Use apostrophes to show possession.
> 3. Use apostrophes to show the passage of time in certain stock phrases.
> 4. Use only the *–s* to form the plural of letters, signs, symbols, figures, acronyms, and abbreviations, unless the absence of the apostrophe would be confusing.
> 5. Distinguish between true possessives and merely descriptive uses of nouns (especially with company names).

1. Use apostrophes to signal omitted letters in a contraction:

It's not going to be easy. (It is not going to be easy.)

It won't be easy. (It will not be easy.)

We will coordinate with the manufacturer who's chosen to supply the semiconductors. (who is chosen)

NOTE: Use contractions in letters and memos to help establish an informal tone. Avoid contractions in more formal, edited documents. See CONTRACTIONS.

2. Use apostrophes to show possession:

Microsoft's design capabilities are world-renowned.

The unit's most unique capability is its amplification of weak echoes.

- When the possessive word is singular, the apostrophe comes before the *–s*:

Rockwell International's process for budgeting is one of the most progressive in the industry.

The circuit's most unusual capability is its error detection and correction function.

- When the possessive word is singular and already ends with an *–s*, the apostrophe follows the *–s* and may itself be followed by another *–s* (although most writers prefer the apostrophe alone):

General Dynamics' (*or* Dynamics's) management proposal is very project-specific.

Our project manager will be Dr. Martin Jones. Dr. Jones' (*or* Jones's) experience with laser refractors has made him a leader in the field.

- When the possessive word is plural and ends in *–s*, the apostrophe follows the *–s*:

The suppliers' requests are not unreasonable considering the amount of time required for fabrication.

We consider the states' environmental quality offices to be our partners in reclamation.

NOTE 1: Irregular plurals that do not end in *–s* require an *'s*:

The report on women's status in the executive community is due next Friday.

Materials for children's toys must conform to Federal safety standards.

NOTE 2: The possessive form of the pronoun *it* is *its*, not *it's* (*it's* is the contraction of *it is* or *it has*):

Possessive: Its products have over 10,000 hours of testing behind them.

Contraction: It's (It is) in the interests of economy and efficiency that we pursue atmospheric testing as well.

Similarly, the possessive form of *who* is *whose*, not *who's*. *Who's* is a contraction for *who is* or *who has*. See POSSESSIVES and *who's/whose* in WORD PROBLEMS.

3. Use apostrophes to show the passage of time in certain stock phrases:

a month's pay

an hour's time

4 days' work

3 years' study

5 years' experience

NOTE: Distinguish between the preceeding examples with apostrophes and the following unit modifiers with hyphens:

a 4-day work week

his 3-year study

See HYPHENS.

4. Use only the *–s* to form the plural of letters, signs, symbols, figures, acronyms, and abbreviations, unless the absence of the apostrophe would be confusing.

As in the following examples, the forms without apostrophes are preferred, but forms with apostrophes are acceptable:

The Xs indicate insertable material. (*or* x's *or* X's *but not* xs)

All of our senior staff have PhDs. (*or* Ph.D.'s *or* Ph.D.s)

Simplex Pharmaceuticals coded all experimental drug runs with A's and I's. (*not* As *or* Is *nor* as *or* is)

Symmetek began making microchips in the 1990s. (*or* 1990's)

The tracer tests will be run on all APOs in Europe. (*or* APO's)

Apostrophes

The Bureau of Land Management has prepared three EAs (Environmental Assessments) for those grazing allotments. (*or* EA's)

In the following instances, the absence of the apostrophe produces confusing forms:

The manufacturer indicates fragile material by placing #'s in any of the last three positions in the transportation code.

Our risk management process is designed to eliminate the if's and but's.

5. Distinguish between true possessives and merely descriptive uses of nouns (especially with company names):

Exxon's response (possessive)
an Exxon response (descriptive)

General Motors' news release
(*or* a General Motors news release)
The General Motors sales staff

FranklinCovey's proposal
FranklinCovey *Style Guide*

Verdi's first opera
an early Verdi opera

The teachers' testimony
A teacher guide
Teachers Guide
(*or* Teacher's Guide)
(*or* Teachers' Guide)

As in these examples, competing forms are common. The traditional use of the possessive (with an apostrophe) is less common today, especially with corporate names. See ADJECTIVES and AGREEMENT.

As in the examples with *teachers'/ teacher/teachers*, a number of options are possible. In *the teachers' testimony*, the possessive signals testimony from several different teachers.

The name for a guide for teachers is open to all sorts of possibilities. In *the teacher guide*, the noun *teacher* functions as an adjective, not a possessive. When both nouns are capitalized, the most common form is *Teachers Guide.* This form, without the apostrophe, would appear in titles and in news headlines (which often omit apostrophes). But notice that other options are possible (and correct).

The best advice is to decide for a single document whether you want to use a descriptive or possessive. Then be consistent throughout that document.

Appendices

Appendices (often informally referred to as attachments) are more and more common in documents, especially those intended for busy peers, supervisors, and managers who do not have time to wade through pages of data and analysis. Appendices and attachments are acceptable (often desirable) in letters and memos as well as in reports.

The following types of information can and often do appear in appendices or attachments:

—Background data

—Case studies

—Computations

—Derivations

—Detailed component descriptions

—Detailed test results

—Excerpts from related research

—Histories

—Lengthy analyses

—Parts lists

—Photographs

—Raw data

—Sources of additional information

—Supporting letters and memos

—Tables of data

The word *appendix* has two acceptable plurals: *appendices* and *appendixes*. *Appendices* is still widely used by educated speakers and writers, but *appendixes* is growing in popularity because it follows the regular method for making English words plural. The style used by the U.S. Government is *appendixes*.

Appendices

1. Use appendices to streamline reports and memos that would otherwise be too lengthy.
2. Avoid making appendices a dumping ground for unnecessary information.
3. Number or letter appendices and attachments sequentially.
4. Refer to all appendices and attachments in the body of the document.

1. Use appendices to streamline reports and memos that would otherwise be too lengthy.

In business and technical reports and memos, assess your readers' need to know the background and analysis behind the relevant conclusions and recommendations.

Relevant conclusions and recommendations should appear very early in most business and technical reports, often as part of an executive summary. Busy readers can therefore receive a streamlined report of 8 to 10 pages (instead of the traditional formal report of 30 to 50 pages) with appendices containing appropriate background information, detailed results, and lengthy analyses. See SUMMARIES.

Would a knowledgeable reader need the information in the appendices to interpret the conclusions and recommendations? If so, then the appendices are justified. In writing your document, determine who the readers will be and ask yourself what additional information these readers will need to better understand your approach, analysis, results, conclusions, and recommendations.

One rule of thumb is that appendices should contain only information prepared for the project in question. Background information from files and tangential reports (general background information) should not appear in appendices. Often readers know such background information anyway.

To summarize, if a reader needs certain information to understand a report, this information belongs in the body of a report. All other information belongs either in appendices or in backup files.

2. Avoid making appendices a dumping ground for unnecessary information.

Because the appendices are not part of the body of the report, some writers believe they have the license to include in the appendices every scrap of information they know about the subject. This practice leads to massive, often confusing appendices that discourage readers.

Appendices

3. Number or letter appendices and attachments sequentially.

Sequential numbering or lettering is essential: Appendix A, Appendix B, etc.; or Attachment 1, Attachment 2, etc. Numbers and letters are both correct, so either is acceptable. In longer documents, your choice may depend upon whether you have numbered or lettered the sections or chapters. If your sections or chapters are numbered, then use letters to label appendices. Conversely, if your sections or chapters are lettered, use numbers for the appendices. The system you use to label appendices should indicate a clear distinction between the appendices and the body of the document.

Typically, appendices are numbered in the order in which the references to them appear in the body of the report. So the first appendix mentioned in the report becomes appendix A (or appendix 1), the second one mentioned is appendix B (or appendix 2), and so on.

NOTE 1: As in the preceding sentence, you need not capitalize the initial A in appendix in ordinary text. Some authorities, however, prefer a capital: Appendix A. Choose one pattern and be consistent. See CAPITALS.

NOTE 2: Give each appendix or attachment a title. Referring to appendices or attachments only by number is not informative and can be confusing. In the text, refer to the appendix or attachment by both number and title.

4. Refer to all appendices and attachments in the body of the document.

Refer to all appendices or attachments in the body of the document so that readers know that the information within them is available.

Your references should be informative rather than cryptic. A cryptic reference (such as *See appendix C*) does not tell readers enough about the appended information. The following references are informative:

> Particulate counts from all collection points in the study area appear in appendix C, Particulate Data.

> Attachment 5, A Report on Reserve Faulting in the Boling Dome, provides further evidence of the complex faulting that may control production.

> See appendix A (Prescription Trends Since the 1970s) for further analysis of Valium use and abuse since its introduction.

Articles are the simple structure words *a*, *an*, and *the*. These words always precede a following noun, but not every noun can accept both articles, and some nouns need neither of them. See Nouns.

Articles

1. Choose indefinite *a* or *an* to precede a singular countable noun.
2. Use definite *the* to precede (point out) either countable or mass nouns, both singular and plural.
3. Choose *a/an* and *the* (different pronunciations) to match the way an acronym is pronounced, not how it is spelled.
4. Alphabetize acronyms without using the customary article that would appear in the written-out titles.
5. For English as a Second Language (ESL) questions regarding articles, use a dictionary designed for ESL users.

Common nouns that name countable things can use both *a/an* and *the* and have both a singular and a plural form:

a book *(singular indefinite)*
the book *(singular definite)*
the books *(plural)*

not

a books

Common nouns that name uncountable things are mass nouns. They have no plural forms, and they don't commonly accept *a* or *an*:

the rice/rice
the gold/gold

not

the rices
a rice
the golds
a gold

Some common nouns are both countable and uncountable (mass):

cake *(mass)*
a cake *(countable)*
the cake *(either mass or countable)*
cakes *(countable)*

---

milk *(mass)*
a milk *(countable)*
the milk *(mass)*
milks *(rare use—as in milk from cows, goats, horses, humans, etc.)*

Proper nouns usually do not require articles:

Susan B. Anthony
Senator Robert Dole
Kansas City
Stone Mountain

but

The Mississippi River
The Great Lakes

***A/An* is the indefinite article** because it points toward a single indefinite (unspecified) object:

A candle (*meaning* "any single candle")

An elephant (*meaning* "any single elephant")

***The* is the definite article** because it points toward a definite (specified) object, either singular or plural:

The candle (*meaning* "one specific candle")

The candles (*meaning* "more than one specific candle")

The elephant (*meaning* "one specific elephant")

The elephants (*meaning* "more than one specific elephant")

1. Choose indefinite *a* or *an* to precede a singular countable noun:

a tube
a leaf
a ring
an ear
an apple
an owl

Note 1: As in the above examples, the choice between *a* and *an* is based on the initial sound—not spelling—of the noun (or word) that follows the article. Words pronounced with a vowel sound require *an* (pronounced as in the initial sound of *ant*). Words pronounced with a consonant sound require *a* (pronounced as in the hesitation sound *uh*).

an amber leaf
an emerald ring
a good apple
a tiny owl

Note 2: Nouns beginning with *h* can potentially have either *a* or *an*. If the *h* is pronounced, as in *heap* or *hair*, *a* is correct:

a heap
a hair

If the *h* is silent (unpronounced), as in *honor* or *hour*, *an* is correct:

an honor
an hour

The word *history* has competing options. Most editors would routinely choose *a history* (not *an history*) as the written form. But speakers often use *an*, as in *an historical* event, perhaps because this *h* is nearly silent.

Note 3: *A/an* has an implied meaning of "one," so *a/an* could not precede the plurals of countable nouns. The plurals do not need any article, or *the* can precede them:

tubes/the tubes
ears/the ears
leaves/the leaves
apples/apples

2. Use definite *the* to precede (point out) either countable or mass nouns, both singular and plural:

the engineer *(singular countable)*
the bee *(singular countable)*

the engineers *(plural countable)*
the bees *(plural countable)*

---

the butter (mass)
the air (mass)

Note 1: Although *the* in the above examples does not change its spelling, its pronunciation often changes just as *a/an* changes. Note, however, that native speakers of English do not always choose pronunciations that follow these rules. Sometimes they even reverse the rules, for emphasis.

When a vowel begins the following word, *the* is usually pronounced as the word *thee*:

the elevator
the almond tree

When a consonant begins the following word, *the* is pronounced as in the beginning of the word *thus*:

the director
the symposium

For emphasis, either *the director* or *the symposium* could begin with a *the* pronounced like *thee*.

As with *a/an*, in some words beginning with *h*, as with *historical*, either pronunciation of *the* is correct:

the historical profile *(with either pronunciation)*

Note 2: In all the above examples, using *the* points to a definite object or thing. This pointing function is, however, not as strong as in some other structure words—for example, in the demonstrative pronouns *this, that, these,* and *those*. See Pronouns. Using *the* points back to a prior sentence or phrase where the object or objects have been identified.

3. Choose *a/an* and *the* (different pronunciations) to match the way an acronym is pronounced, not how it is spelled:

a NEPA requirement

the NEPA requirement *(with* the *pronounced as in the beginning of* thus*)—for the National Environmental Policy Act (NEPA), which is pronounced as in a word, not by separate letters*
an NFL player

the NFL player *(pronounced as in* thee*)—for National Football League (NFL), which is pronounced letter by letter, not as a word*

an/the AFL-CIO publication

a/the DoD proposal

an/the MIS supervisor

a/the KUED television program

NOTE: Pronunciation also influences whether the article is needed or not. For example, with the *National Environmental Policy Act (NEPA)*, the article *the* often drops when NEPA (pronounced as a word) appears in sentences:

NEPA requires that each agency . . .

A provision of NEPA is . . .

In either of these cases, the written-out version would include the article: *the National Environmental Policy Act.*

4. Alphabetize acronyms without using the customary article that would appear in the written-out titles:

CIA
NAFTA
NASA
SOP

NOTE 1: Each of these abbreviations would ordinarily use an article if written out:

the Central Intelligence Agency

the General Agreement on Tariffs and Trade

the North American Free Trade Agreement

NOTE 2: *SOP* illustrates the principle discussed in rule 3. The acronym—read letter by letter—begins with a vowel sound, so it would require *an* if an article is used with it. But the written-out version opens with a consonant sound, so it uses *a*:

a/the standard operating procedure

As in this example, the original words for an acronym may not require capitalization. See CAPITALS.

5. For English as a Second Language (ESL) questions regarding articles, use a dictionary designed for ESL users.

Articles are some of the most unpredictable words in English, despite their frequency. Most native speakers of English choose articles without thinking. A nonnative English speaker has to work with rules. Unfortunately, the rules presented above have innumerable exceptions.

We recommend, therefore, the most current version of A.S. Hornby's *Oxford Advanced Learner's Dictionary of Current English,* (Oxford, England: Oxford University Press). See REFERENCES.

This excellent dictionary is designed especially for non-English-speaking users. As such, it has a wealth of examples and, in the case of articles, it marks nouns as countable (marked with a [C]) or uncountable (marked with a [U]).

For example, the simple noun *cover*, as discussed in this dictionary, has a variety of uses, some as a countable noun; others as an uncountable (mass) noun.

a cover on a typewriter [C]

design a front cover for the book [C]

provide cover from a storm [U]

a spy's cover (assumed identity) [U]

This dictionary also identifies special uses when a noun is only plural:

The covers on the bed [Plural only]

Finally, this dictionary lists special idioms, many of which either use or avoid articles, often without apparent logic:

under cover of darkness *(No article is used before* cover.*)*

Bias-free language is an increasingly important issue for business and technical writers and speakers. The list of forbidden words and phrases grows longer each week, and the legal penalties for violations are increasingly severe.

United States Federal laws and many state regulations mandate that an employee may not be discriminated against based on race, creed, sex, age, or national origin. These laws have various provisions, but penalties for violations apply both to individual employees and to their companies.

Bias-Free Language

Cultural Awareness

1. Do not use words that unnecessarily identify a person's race, religion, physical status, social status, age, national origins, or gender.
2. Don't rely on the stereotypes often implied by the categorizations included in rule 1.
3. When appropriate, choose terms and designations that are neutral and acceptable to the group you are discussing.
4. Be sure to choose graphics—especially photographs—that fairly represent all groups and types of people within society.

Gender-Neutral Language

5. Use words that do not unnecessarily distinguish between male and female.
6. Avoid unnecessary uses of *he, him,* or *his* to refer back to such indefinite pronouns as *everyone, everybody, someone,* and *somebody.*
7. Avoid unnecessary uses of *he, him, his* or *she, her, hers* when the word refers to both males and females.
8. Avoid the traditional salutation *Gentlemen* if the organization receiving the letter includes males and females.
9. Do not substitute *s/he, he/she, hisorher,* or other such hybrid forms for standard personal pronouns.
10. Avoid demeaning or condescending gender terms for either females or males.

Cultural Awareness

Cultural awareness (rules 1, 2, 3, and 4), in the following discussion, covers language choices dealing with race, religion, physical status, social status, age, and national origins. Issues relating to gender are covered in a separate section (rules 5, 6, 7, 8, 9, and 10).

Many of the cultural awareness issues are changing social conventions. Writers and speakers should be constantly alert to changes in what is acceptable because words and phrases fall in and out of favor so rapidly. Also, as appropriate, check the status of terms in up-to-date dictionaries and other references. See REFERENCES.

1. Do not use words that unnecessarily identify a person's race, religion, physical status, social status, age, national origins, or gender.

In most instances, write documents giving people's names and, if appropriate, their job titles. Do not include, for example, references indicating that a person is a Native American, a Methodist, a woman, the user of a wheelchair, a vegetarian, nearly 65, or born in Puerto Rico.

These categorizations are irrelevant to any serious business discussion. They are also often insulting to the person referred to, especially if the category is being used as a shorthand way of implying something about the person.

For instance, mentioning someone's age can be a way of suggesting that the person is too close to retirement to be considered for a promotion or a special task team. This reference to the person's age is irrelevant, unfair, and likely illegal.

The golden rule is a good rule to follow when answering questions about cultural awareness. Would you appreciate someone identifying one or more traits about you if the trait had no relevance to the topic at hand? So follow this rule: Treat others as you would want to be treated.

NOTE: In some contexts, such personal categorizations are appropriate for discussion and documentation. Census surveys routinely ask for such information. Or the Human Resources Department for a company may develop survey information about employees to comply with Federal guidelines relating to Equal

Employment Opportunity or Affirmative Action programs.

Usually, personal information about any of these categories should be kept confidential.

2. Don't rely on the stereotypes often implied by the categorizations included in rule 1.

Stereotypes are a classic fault in logic because the stereotype for any group of people always fails when matched up to the traits for a single individual. Misuse of stereotypes is the basic reason why using group or category terms about an individual is wrong (see rule 1).

An argument based on stereotypes would be, for instance, that a specific auto mechanic was cheating you because all auto mechanics are crooks and out to gouge customers. Based on this stereotype, for example, people also make supposedly innocent jokes about the typical mechanic's shrug—a shrug implying that the mechanic doesn't have a clue as to what might be wrong with the car.

Neither the use of the stereotype nor the joke is innocent if you happen to be the mechanic in question. No legal issues are likely involved, but using a stereotype in this manner is probably offensive to the mechanic.

On the other hand, a serious assessment of a single auto mechanic would need to use facts and data about the individual. Such topics as the mechanic's rate of pay, quality of work, and solicitude for the customer are proper items for a company to document and discuss. They have nothing to do with the stereotype of the typical mechanic.

Stereotypes are common about any group of people who share a category or several categories:

Religion: Buddhist, Methodist, Latter-day Saint, Moslem, etc.

Race: Black, Native American, Latino, etc.

Age: retired, middle-aged, yuppie, etc.

National Origin: French, Nigerian, Peruvian, Canadian, etc.

Profession: lawyer, doctor, banker, etc.

Sectional Origin: Southerner, Down Easter, Midwesterner, etc.

Physical Features: sightless, user of a wheelchair, diabetic, anorexic, etc.

Sexual Preference: heterosexual, homosexual, bisexual

Economic Status: homeless, fixed (or limited) income, well-to-do, etc.

Gender: female, male

A single individual can share more than a single category. For example, a middle-aged Nigerian female lawyer might be diabetic and on a fixed income. Given this complexity, which stereotypes apply? Probably none of them. And as noted above, most of them are irrelevant and likely illegal to mention in any business discussion or business document.

3. When appropriate, choose terms and designations that are neutral and acceptable to the group you are discussing.

News articles, research reports, and other serious documents properly and legally analyze and discuss data for various groups. Such discussions, which would not be based on stereotypes, include terms for a group or category of persons. These terms are, however, often the problem because they carry unfortunate echoes, and these echoes change rapidly.

For example, stylebooks for both the Associated Press and the United Press identify *blacks* as the term to choose when you need a category term for black Americans. *African-American* is also common. Some people now object to its use of the hyphen, which implies less than full status as an American. Even earlier, the terms *Negro* and *colored person* had seasons of use. Each of them collected negative echoes and fell out of favor.

When you choose any term for a group of people, be careful to choose the one that is most current and acceptable. If in doubt, don't assume that your choice makes no difference. Your choice may be insulting and even a legal issue.

For example, a recent court case on sexism included among other points, the assertion that the defendant called the women on his staff "girls." Of course, to wind up in court, he did much more than call them names. The moral is that the wrong word or phrase can be a costly mistake, especially to your reputation, if not to your pocketbook.

A second example comes from the terms dealing with people who have disabilities. Choose terms and phrases that do not emphasize the negative features of the disabilities, nor should the features be seen as more significant than they may be.

not these

crippled

blind

mentally defective

dumb

afflicted with MS

unfortunately has a speech problem

but these

person who uses a wheelchair
person without sight/partially sighted
person with mental disability
person unable to speak
person with multiple sclerosis
person with speech impediment

4. Be sure to choose graphics—especially photographs—that fairly represent all groups and types of people within society.

Both major graphics and incidental figures need to be balanced as to their representations of typical people. The graphics should not rely on either an overt or covert use of stereotyping. (See rule 2.)

A recent court case dealt with a realty firm that consistently pictured in advertisements middle-aged Caucasians who seemed to have money. In court, the firm was challenged with sending a signal that first-time black buyers would not be welcome. The realty firm lost.

Much the same sort of problem exists in training materials or procedures that present women in secretarial roles and men in managerial roles. Again, this is a covert reliance on stereotypes.

Finally, be sure that your representations of a racial or physical appearance do not accentuate facial features or dress to the point where the picture is more parody than reality.

Gender-Neutral Language

Gender signals are an integral part of our language. English from its earliest history has often marked words as either male or female (and even sometimes neuter). Pronouns are the commonest surviving examples: *he, him, his* vs. *she, her, hers* vs. *it, its*. A number of nouns also have had different male and female forms: *waiter/waitress, stewardess/steward, heir/heiress, countess/count, host/hostess, actress/actor, usher/usherette*. And some words used for both genders seem to include only males: *mankind, layman, manpower*, and so on.

Many such distinctions, called gender distinctions, have become objectionable, especially in recent years with the debate about equal rights for women. So, many publishing firms and most writers routinely remove unnecessary and objectionable gender distinctions from published writing. This trend is the basis for the following rules, most of which require little effort from writers.

5. Use words that do not unnecessarily distinguish between male and female:

these	*not these*
flight attendant	stewardess
people, humans	mankind
workforce	manpower
layperson	layman
employee	workman
heir	heiress
chair, chairperson	chairman
serving person or server	waitress

NOTE 1: The use of female forms such as *waitress* and *heiress* has declined. *Heir* now includes both male and female; *waiter* still has male echoes, but these may fade soon. The best advice is to be sensitive to this issue and avoid female designations.

NOTE 2: Historically the word *man* (especially used in compound words like *chairman* or *layman*) could include both males and females; its closest modern equivalent would be, for instance, the indefinite pronoun *one* or *person*. This historical meaning has, however, been forgotten, so much so that many women now argue that they are silently being left out when compounds with *man* are used.

6. Avoid unnecessary uses of *he, him,* or *his* to refer back to such indefinite pronouns as *everyone, everybody, someone,* and *somebody.*

The problem sentences are often ones where the indefinite pronouns introduce a single person and then a later pronoun refers to that person:

Everyone should take (his? her?) coat.

Someone left (his? her?) report.

Unless we clearly know who *everyone* and *someone* refer to, we cannot pick the proper singular pronoun. We thus have to choose among several options:

—Make the sentences plural, if possible:

All employees should take their coats.

—Remove the pronoun entirely:

Someone left a (*or* this) report.

—Use both the male and female pronouns:

Each employee should take his or her report.

Someone left her or his report.

—Use the plural pronoun *their* (or maybe *they* or *theirs*):

Each employee should take their coat.

Someone left their report.

NOTE: This last option is fine for informal or colloquial speech, but many editors and writers would object to the use of the plural pronouns to refer back to the singular *everyone* and *someone*. See PRONOUNS and AGREEMENT.

7. Avoid unnecessary uses of *he, him, his* or *she, her, hers* when the word refers to both males and females:

not these

An assistant should set her (his?) priorities each day.

The engineer opened her (his?) presentation with a slide presentation.

A writer should begin his (her?) outline with the main point.

As with rule 6, writers have several options:

—Change the sentences to plurals:

Assistants should set their priorities each day.

Writers should begin their outlines with the main point.

—Remove the pronouns:

The engineer began the presentation with a slide presentation.

An assistant should set firm priorities each day.

NOTE: A third option is to use the phrase *his* or *her*, but this becomes clumsy in a text of any length, so it is better to use one of the two options given above.

8. Avoid the traditional salutation *Gentlemen* if the organization receiving the letter includes males and females.

Omit the salutation if your letter is to an organization, not to an individual. Your letter would then have an inside address, a subject line, followed by your text. This format is called a simplified letter. See LETTERS.

Whenever writing to people whose gender you don't know, use the title or the name without a title:

Dear Personnel Manager:

Dear G. L. Branson:

NOTE 1: We do not recommend *Ladies and Gentlemen* or *Gentlemen and Ladies*. The term *Ladies* (and maybe *Gentlemen*) seems old-fashioned. Similarly, *Dear Sir or Madam* is old-fashioned and overly formal.

NOTE 2: In recent years a number of unusual salutations have appeared. Avoid them:

Dear Gentlepersons:

Dear Gentlepeople:

Dear People:

Dear Folks:

See LETTERS.

9. Do not substitute *s/he, he/she, hisorher,* or other such hybrid forms for standard personal pronouns.

These hybrid forms are unpronounceable and are not universally accepted by English users, so avoid them. Instead, either remove pronouns or change the sentences to plurals, as suggested under rule 7.

Where you must use singular personal pronouns, use *he and she, his or her,* or *him and her*. Or as an option in longer documents, alternate between male and female pronouns.

10. Avoid demeaning or condescending gender terms for either females or males.

Gender terms such as *girls/gals* or *boys/guys* carry echoes of immaturity or irresponsibility. As such, they are condescending or, at the very least, humorous. So do not use them in a business context, either in speech or writing.

For slightly different reasons, the more formal terms *ladies* and *gentlemen* have also become questionable.

The term *ladies* seems to belong to another era, when ladies wore white gloves, attended garden parties, and talked about the social scene. A lady of that era did not work or worry about business.

The term *gentlemen* is not quite so demeaning, but it still has echoes of past formality.

Use *ladies* or *gentlemen* only in formal speech and probably only when people addressed are well over 30. Similarly, we recommend avoiding the old-fashioned letter salutation: *Ladies and Gentlemen*. See the note in rule 8.

The best advice is to be sensitive to the echoes or implications of your language. As a final example, the common phrase *man and wife* identifies only the sex of the man, but for the wife, both the sex and the marital role are marked. A better choice: *man and woman, husband and wife, woman and man,* or *wife and husband*.

Bibliographies

Bibliographies appear at the end of chapters, articles, and books. Whatever the exact format, complete bibliographic entries include the name of the author, the title, and the full publication history (including the edition, the publisher or press, the city of publication, the date of publication, and the online source, if any).

The forms of bibliographic entries vary greatly, depending on the professional background of the author, the profession's needs and traditions, the type of publication, and the publisher. The bibliographic form that we recommend represents a standard format useful for a variety of professions and publishers.

However, we advise you to find out the specific format requirements (including bibliographic format) of the publisher or community to whom you are submitting a document.

NOTE: In the following rules, the titles of publications in bibliographic entries are italicized. Underlining replaces italics when documents are typed or when italics is not available. See UNDERLINING and ITALICS.

Bibliographies

1. For a book, give the name of the author or authors, the date of publication, the full title, the volume number, the edition, the city of publication, the publisher, and the online format or source (if any).
2. For a journal or a magazine article, give the name of the author or authors, the year of publication, the full title of the article (in quotation marks), the name of the journal or magazine, the volume, the month or quarter of publication, and the pages (if available). If citing an online source, add the URL and, in brackets, the date you accessed the source.
3. For publications available only on electronic media, give the author, date, title, name of publication, URL and, in brackets, the date you accessed the site.
4. For unpublished material, give the author or authors, the title (in quotation marks), and as much of its history as available.
5. For public documents, give the country, state, county, or other government division; the full title; and complete publication information.

1. For a book, give the name of the author or authors, the date of publication, the full title, the volume number, the edition, the city of publication, the publisher, and the online format or source (if any):

Book by one author

Apter, Andrew. 2005. *The Pan-African Nation: Oil and the Spectacle of Culture in Nigeria*. Chicago: University of Chicago Press. Kindle e-book.

Book by two authors

Gallo, George, and L. J. Lane. 2008. *Paper and Paper-Making*. Third Edition. Baltimore: The Freedom Press & Co. Google Book search [accessed July 31, 2009].

Book by three authors

Covey, Stephen R., Robert A. Whitman, Breck England. 2009. *Predictable Results in Unpredictable Times*. Salt Lake City: FranklinCovey Press. Kindle e-book.

Book by more than three authors

Nestoras, Ezequiel, et al. 2008. *La Evolución del Internet*. Los Angeles: The Hispanic Press.

Book by one editor

Nfusi, Claire, ed. 2008. *Sourcebook of Fonts*. New York: Simon & Schuster.

Book by two editors

Ibanez, Charlotte, and Fred Stein, eds. 2010. *Streaming Online Media*. Boston: JMap E-Publishing. E-audio book.

Two volumes by an organization

Modern Language Association of America. 2009. *MLA Handbook for Writers of Research Papers*, Seventh Edition.

Chapter of a book

Williams, Clive. 1979. "The Opacity of Ink." In *The Art of Printing*, edited by Jason Farnsworth. New York: Holt, Rinehart & Winston.

NOTE 1: In these entries, the date directly follows the name of the author or authors. This convention complements the author/date style of citations in the text. See CITATIONS. In this style, the text of a document contains parenthetical references:

> A 1981 study revealed that fleas transmit the virus (Babcock 1981). This study relied on two earlier studies (Duerdun 1976 and Abbott 1973).
>
> *or*
>
> A 1981 study revealed that fleas transmit the virus (Babcock). This study relied on two earlier studies (Duerdun 1976 and Abbott 1973).

Because the date of Babcock's study is already in the sentence, including the date in the citation is unnecessary.

Bibliographies

NOTE 2: Publications in the humanities usually cite the publication date following the name of the publisher:

Ke-Wen, Chung. *The History of Modern China*. New York: Simon & Schuster, 2006.

This bibliographic form complements the footnoting pattern of citations routinely used by most scholars in the humanities. For more information on this style, see the most recent edition of *The Chicago Manual of Style*. See also FOOTNOTES.

NOTE 3: Bibliographic entries in the physical and biological sciences often capitalize only the first word of the title:

Ke-Wen, Chung. *The history of modern china*. New York: Simon & Schuster, 2006.

2. For a journal or a magazine article, give the name of the author or authors, the year of publication, the full title of the article (in quotation marks), the name of the journal or magazine, the volume, the month or quarter of publication, and the pages (if available). If citing an online source, add the URL and, in brackets, the date you accessed the source:

Article by one author

Broward, Charles Evans. 1981. "Traveling the Southern California Desert." *UCLA Chronicle* 15 (Spring): 45–54. http://www.uclachronicle. edu/pdfs/broward_traveling.pdf [accessed March 5, 2010].

Article by two authors

Calleston, Dwight R., and James Buchanan. 2008. "The Desert Tortoise: Its Vanishing Habitat." *The Californian* 7 (April): 23–28.

For articles with multiple authors, follow the example shown of the book format that deals with two or more authors:

Article appearing in more than one issue

Mathis, Stéphanie, and Jason Okolong. 2007. "Dyslexia and Hearing." *Learning Disabilities Online Newsletter* 8 (Fall) and 9 (Winter): 34–35, 28–31. http://www.ldablonline.org/article/34352822.html [accessed June 11, 2010].

Article from a popular magazine

Walt, Vivienne. 2010. "School Lunches in France: Nursery School Gourmets." *Time*, February 23. http://www.time.com/time/world/article/0,8599,1967060,00.html [accessed March 5, 2010].

Review of a published book

Hamash, Sorya. 2010. Review of *The Coming Generational Storm* by Laurence J. Kotlikoff and Scott Burns. *Population Studies* (July): 24–31.

See QUOTATION MARKS.

NOTE 1: As with books, these entries cite the year of publication immediately after the name of the author or authors. In publications for the humanities, the date appears (with the month) after the volume of the journal or magazine:

Stillman, Wendy. "Photographing Desert Sunsets." *Photo Chronicle* 15 (Spring 2006): 4–8.

NOTE 2: Some editors, especially in the biological and physical sciences, prefer to omit the quotation marks around the title of the article and to capitalize only the first word of the title:

Stillman, Wendy. 2006. Photographing desert sunsets. *Photo Chronicle* 15 (Spring): 4–8.

3. For publications available only on electronic media, give the author, date, title, name of publication, URL and, in brackets, the date you accessed the site.

If the month, day, and year of distribution are available, include them. If the source is in a format other than text, provide the format, such as video file, podcast, audio recording, or electronic database. Give as much publication information as you can; because online content is transient, your goal should be to give users the best possible chance of retrieving the source.

Online article

Aziz, A., & Brooks, G. 1995. Pharmacokinetics of synthesized serotonin in fasted and non-fasted subjects. *Clinical Pharmaceutics* [Online serial]. synth_sero_aziz_brooks/publicfolder/ftp.clincalpharma.net. Also available http://www.clinicalpharma.org/cgi/pdf_extract/299/39299/76?ct= [accessed May 7, 2010].

Govindarajan, Vijay. "The Case for 'Reverse Innovation' Now." *Business Week*, October 26, 2009. http://www.businessweek.com/innovate/content/oct2009/id20091026_724658.htm [accessed November 30, 2009].

Online video

Gregerson, Hal. "Innovation in the Islamic World." *INSEAD Knowledge*, February 10, 2010. Windows Media Player video file. http://knowledge.insead.edu/video/index.cfm?vid=385 [accessed March 8, 2010].

The Economist. "A Brilliant Solution: Better Lighting with Quantum Dots." Video recording. http://www.economist.com [accessed March 8, 2010].

Blog entry

Stephen R. Covey, "My New Book—How to Win, Even in Unpredictable Times." Stephen R. Covey Community Blog, August 28, 2009. http://www.stephencovey.com/blog/?p=43 [accessed September 2, 2009].

Online database

Survey of General-Obligation Bond Ratings [Electronic database]. 2008. New York, NY: D&F Bond Clearinghouse. Available ProfNet: *bndpfnt.unibas.ny* Directory: */archive_bnd/2008.volume.1*

Computer software

EasyChart [computer software]. 2010. Baltimore, MD: PlumStone Software.

NOTE 1: Describe the pathway to online information as precisely as possible. Indicate source or address, directory, and file or volume number if available.

NOTE 2: Do not end a pathway description with a period, which will hamper retrieval if inadvertently used. Use lowercase for pathway addresses.

4. For unpublished material, give the author or authors, the title (in quotation marks), and as much of its history as available:

Dissertation or thesis

Johnson, Dugdale. 2006. "The Habitat of the Desert Tortoise: Its Inter-Relationship with Man." D.Sc. diss., University of Southern California.

Professional paper

Miskas, Lana, and Gullaug Nordstrand. 2009. "Applications of Quantum Dot Technology." Paper presented at the annual meeting of the International Nanochemistry Society, Oslo, 24–26 May.

Personal communication

Turgott, Edward. 2005. Letter to the author, 31 May.

NOTE: The formats for other unpublished documents (television shows, radio shows, interviews, duplicated material, diaries, etc.) should supply as much bibliographic information as possible. The bibliographic form should allow readers to locate the document easily.

5. For public documents, give the country, state, county, or other government division; the full title; and complete publication information:

Ministry of Finance, Government of India. *Creating an Enabling Environment for State Projects*. 2008. Report to the Public-Private Partnerships Committee.

U.S. Congress. House. Committee on Ways and Means. 1945. *Hearings on Import Duties on Shellfish*. 79th Cong., 1st sess.

U.S. Bureau of the Census. 2004. *Gross and Net Fishing Revenues*, 2000. Prepared by the Commerce Division in cooperation with the Commodity Division. Washington, D.C.: United States Government Printing Office.

Boldface

Boldface type uses thicker and darker letters—for example, **boldface type** vs. normal type.

> **Boldface**
>
> 1. Use boldface type to emphasize key words and phrases and to complement the design and appearance of the page.
> 2. Do not overuse boldface and do not use boldface randomly or in isolation from other emphasis techniques.

1. Use boldface type to emphasize key words and phrases and to complement the design and appearance of the page.

Integrate boldface into your overall design of the page or pages in your document. In particular, use boldface for the following key words and phrases:

—Titles of a book or chapter

—Headings and subheadings

—Listed items

—Steps in a procedure

—Rules or notes

—Subject lines (in letters and memos)

—Names or titles of graphics

—Titles or sections in an index

In each of these instances, the boldface type complements other emphasis techniques: placement, spacing, different type sizes, and the organization of content. See EMPHASIS and PAGE LAYOUT.

So even without boldface type, the document would already be designed to emphasize major points and key words. Boldface just makes the items even more emphatic.

2. Do not overuse boldface and do not use boldface randomly or in isolation from other emphasis techniques.

Overusing boldface destroys its value and impact. Also, several lines of boldface or even a paragraph entirely in boldface can be more difficult to read than if it were in normal type.

Random and unplanned use of boldface occurs when, for example, a writer has written a paragraph and later decides to boldface a phrase or several words in the middle of the paragraph:

not this

> During our recent audit of the Production Department, we identified a major problem arising from a **shortage of crucial fasteners**. The vendor who supplies fasteners often takes as long as 2 days to resupply our production line. When the production line fails to project ahead, **downtimes** can occur, often **as long as 2 days** while fasteners are in transit.

The writer of the preceding paragraph should have decided what she wanted to emphasize before writing the paragraph. The following is one possible rewrite, using an initial subheading for the major ideas:

> **Shortage of Fasteners (2-Day Delay)**
>
> During our audit, we discovered that the vendor for certain fasteners often required 2 days to resupply our line. Such delays would occur when workers on the production line failed to project accurately when they would need more fasteners.

Or as an option, the following sentence could replace the heading in the preceding example. In this case, no boldface would be necessary, although the phrase *up to 2 days* might be boldfaced.

> Without crucial fasteners, downtimes on the production line can last for up to 2 days.

Brackets

Brackets are a pair of marks [ ] used to set off comments, corrections, or explanatory material. Although similar to parentheses, brackets do have different uses, as the following rules indicate. See PARENTHESES.

Brackets

1. Use brackets to insert comments or corrections in quoted material.
2. Use brackets to enclose parenthetical or explanatory material that occurs within material that is already enclosed within parentheses.
3. For mathematical expressions, place parentheses inside brackets inside braces inside parentheses.
4. No other marks of punctuation need to come before or after brackets unless the bracketed material has its own mark of punctuation or the overall sentence needs punctuation.

1. Use brackets to insert comments or corrections in quoted material:

> "Your quoted price [$3,750] is far more than our budget allows."

> "Our engineers surveyed the cite [site] for its suitability as a hardware assembly cite [site]."

See QUOTATIONS.

NOTE 1: In these examples, the brackets indicate that the material quoted did not have the information included within the brackets or was an error.

NOTE 2: A common use of brackets, especially in published articles, is to insert *sic* in brackets following an error:

> "We studied the affect [sic] of the new design on production outputs."

Sic, borrowed from Latin, means "thus" or "so." It tells readers that the text quoted appears exactly as it did in the original, including the error. In the example above, the word preceding [*sic*] should have been *effect*.

2. Use brackets to enclose parenthetical or explanatory material that occurs within material that is already enclosed within parentheses:

> We decided to reject the bid from Gulf Industries International. (Actually the bid [$58,000] was tempting because it was far below our estimate and because Gulf Industries usually does good work.)

See PARENTHESES.

NOTE: You can sometimes use dashes instead of the outer parentheses and then replace the brackets with parentheses:

this

> The Board of Directors—or more accurately, a committee of the actual owners (Hyatt, Burke, and Drake)—are answerable to no one but themselves.

not this

> The Board of Directors (or more accurately, a committee of the actual owners [Hyatt, Burke, and Drake]) are answerable to no one but themselves.

Some writers and editors consider the version without brackets preferable because having both parentheses and brackets in the same sentence can look clumsy and can be confusing. See DASHES.

3. For mathematical expressions, place parentheses inside brackets inside braces inside parentheses:

({[()]})

See MATHEMATICAL NOTATION.

4. No other marks of punctuation need to come before or after brackets unless the bracketed material has its own mark of punctuation or the overall sentence needs punctuation:

this

> The procedure was likely to be costly. (Actually, the cost [$38 per unit] included some of the research and development expenses.)

not this

> The procedure was likely to be costly. (Actually, the cost, [$38 per unit], included some of the research and development expenses.)

British English

British English is standard throughout the British Commonwealth of Nations (for example, Canada, Australia, Singapore) and among most English-speaking Africans, Asians, and Europeans. In former British colonies such as India, Pakistan, and Nigeria, British English is a common standard where many languages are spoken. American English is standard in the United States and among most English-speaking Pacific Rim Asians.

Business English does not vary greatly from one country to another. However, American and British writing styles differ somewhat because of 400 years of geographical and cultural separation. Because American and British English are the main forms of English, this discussion focuses on the differences between them.

Writers addressing both cultures should avoid terms specific to each. For example, baseball terminology means little to British readers just as the language of cricket baffles Americans.

For complete guidance on British and American usage, see the most recent versions of *Oxford Advanced Learner's Dictionary,* (Oxford: Oxford University Press), and *The Economist Style Guide* (London: The Economist Books Ltd.). For a glossary of words that differ in British and American usage, see *American English/English American,* (London: Abson Books).

Choose between British or American usage depending on the needs and culture of the audience you are addressing. Be consistent about your choice. The following rules present the main traits of British English.

British English

1. Eliminate periods in and after most abbreviations.
2. Take care to make subjects agree with verbs.
3. Do not generally use commas before *and.*
4. Use a hyphen to separate a prefix ending with a vowel from a word beginning with the same vowel.
5. Use single quotation marks to mark off short quotations or sayings.
6. Take particular care to avoid gobbledygook in a British context.
7. Be aware of differences between British and American word choice, numbers, and spelling.

1. Eliminate periods in and after most abbreviations.

The trend to eliminate periods in abbreviations is even more marked in British than in American usage.

American	British
A.D.	AD
e.g.	eg
etc.	etc
i.e.	ie
Mrs.	Mrs

See ABBREVIATIONS.

2. Take care to make subjects agree with verbs.

Some nouns that are singular in American English are often treated as either plurals or singulars in Britain:

American

the government denies the rumor
the department is changing the policy

British

the government denies/deny the rumour
the department is/are changing the policy

See AGREEMENT.

3. Do not generally use commas before *and*:

Please revise the deadlines, budgets and specifications of the project.

Profits exceeded expectations and employment figures are up for the first quarter.

As in American usage, commas in British English are sometimes used to separate long, complete thoughts linked by a conjunction *(and, but, or, for, nor, so, yet):*

We had anticipated finishing product trials this month, but several important quality issues arose early and had to be dealt with.

See COMMAS.

4. Use a hyphen to separate a prefix ending with a vowel from a word beginning with the same vowel:

re-educate, pre-eminent, co-ordinate

This practice is now becoming less prevalent, but many Britons still hold to it.

See HYPHENS.

5. Use single quotation marks to mark off short quotations or sayings:

'We can deliver,' he said.

Double quotation marks are used to enclose quotations within quotations:

'It was Shakespeare who said, "To thine own self be true"'.

Only the British observe this rule; in the U.S.A. and the Commonwealth, double quotation marks enclose quotations, while single quotation marks enclose quotations within quotations.

Although enclosed by quotation marks in American English, commas and periods (full stops) usually appear outside the quotation marks in British English:

'We can deliver', he said.

Long quotations in British usage are usually indented as a block. See QUOTATIONS and QUOTATION MARKS.

6. Take particular care to avoid gobbledygook in a British context.

The British tend to be more economical and concise in their language than Americans are. Americans who are given to gobbledygook should take a lesson in British restraint.

American
This problem is an ongoing situation.

British
The problem continues.

American
The project team is violating accepted norms.

British
The project team is breaking the rules.

Sometimes Americans lengthen words, creating a pretentious effect to a British ear: *oblige* becomes *obligate*; *orient* becomes *orientate*; *need* becomes *necessitate*. See GOBBLEDYGOOK.

7. Be aware of differences between British and American word choice, numbers, and spelling.

The British and American vocabularies differ slightly. Some examples:

American	British
aluminum	aluminium
trunk (of a car)	boot
windshield	windscreen
thumbtack	drawing pin
freight train	goods train
elevator	lift
ZIP code	post code
fender	wing
highway	motorway
overpass	flyover
to mail	to post
a raise (in price)	a rise
gasoline	petrol
canned	tinned

Most technical terms are identical in British and American English.

Some words have different meanings: *homely* means "simple" or "comfortable" in British English, but "unattractive" in American English.

Some verb forms differ. Here are the main British forms for some common irregular verbs:

get, got, got (*instead of* gotten)
prove, proved, proved (*instead of* proven)
dive, dived (*instead of* dove), dived

British English uses –*t* instead of –*ed* with some verbs:

burn, burnt, burnt
leap, leapt, leapt
learn, learnt, learnt
spell, spelt, spelt

See VERBS.

Style

British writing and speaking style differs from American in terms of tone and pacing. Although generalizations are risky, usage experts generally agree that British style tends to be more impersonal and restrained, particularly in word choice. The American tendency to use hyperbolic language (*fantastic, excellent, terrific*) discomfits the British, as well as most Europeans.

Numbers

The terms for very large numbers differ in British and American usage:

American	British
one billion	one thousand million(s)
one trillion	one billion
one quadrillion	one thousand billion(s)
one quintillion	one trillion
one sextillion	one thousand trillion
one septillion	one quadrillion
one octillion	one thousand quadrillion
one nonillion	one quintillion
one decillion	one thousand quintillion

Spelling

For historical reasons, many British and American spellings differ. In the early 1800s the American lexicographer Noah Webster campaigned to simplify English spelling, and his influential dictionary led to American spelling patterns:

American	British
color	colour
candor	candour
labor	labour
center	centre
meter	metre
gray or grey	grey
defense	defence
mold	mould
check	cheque
traveling	travelling
counselor	counsellor
program	programme
tire	tyre
organize	organise/organize
analyze	analyse
paralyze	paralyse

Note that some words ending in –*ise* are the same for both British and Americans, for example:

advertise
compromise
exercise
premise
televise

Capitals

Capitalization follows two basic rules—the first two rules cited below. Unfortunately, these two rules cannot begin to account for the number of exceptions and options facing writers who have to decide whether a word should be capitalized.

Because of the number of exceptions and options, this section includes many minor rules that supplement the two basic rules. Together, the basic and supplementary rules provide guidance, but you should also check an up-to-date dictionary for additional guidance if the proper choice is still not clear. See References.

Once you have decided whether to capitalize a word, record your decision in a list of editing reminders. Such a list will help you maintain consistent capitalization, especially if your document is long and complex. See Editing and Proofreading.

Capitals

1. Capitalize the first letter of proper names—that is, those specific, one-of-a-kind names for a person, place, university or school, organization, religion, race, month or holiday, historic event, trade name, or titles of a person or of a document.
2. Do not capitalize the first letter of common nouns—that is, those nouns that are general or generic.
3. Capitalize the first letter of the first word of sentences, quotations, and listed items (either phrases or sentences).
4. Capitalize the first letter of the names of directions when they indicate specific geographical areas. Do not capitalize the first letter of the names when they merely indicate a direction or a general or unspecified portion of a larger geographical area.
5. Capitalize the first letter of names for the Deity, names for the Bible and other sacred writings, names of religious bodies and their adherents, and names denoting the Devil.
6. Capitalize the first letter of the first word and all main words of headings and subheadings and of titles of books, articles, and other documents. Do not capitalize the first letter of the articles (*a, an*, and *the*), the coordinate conjunctions (*and, but, or, nor, for, so, yet*), or the short prepositions (*to, of*, etc.) unless they appear as the first word.
7. Capitalize in titles and headings the first letters of the initial word and of all later words in a hyphenated compound except for articles, short prepositions, and short conjunctions.
8. Capitalize the first letter of the geological names of eras, periods, systems, series, epochs, and ages.
9. In text, do not capitalize the first letter of a common noun used with a date, number, or letter merely to denote time or sequence.
10. Capitalize the first letter of proper nouns combined with common nouns, as in the names of plants, animals, diseases, and scientific laws or principles.

1. Capitalize the first letter of proper names—that is, those specific, one-of-a-kind names for a person, place, university or school, organization, religion, race, month or holiday, historic event, trade name, or titles of a person or of a document:

John F. Kennedy
Angela Merkel
Bill Gates
Michaëlle Jean
Henry Ford
Cynthia Carroll
Sally

the Far East
China
the Eastern Shore (Maryland)
Massachusetts
Grove County
Baltimore City (or Baltimore)
United States of America
Lake Baikal
the Rhine River

the University of Glasgow
Western High School
the Golden Daycare Center

Shell Oil Company
the Prudential Life Insurance Co.
the Legion of Honour
the Elks
the United Mine Workers
the Conservative Party

Baptists
Judaism
Japanese
Hindus

May
September
Fourth of July
New Year's Day

the Reformation
World War I
Battle of Waterloo
the Crucifixion

Cyclone® (fence)
Xerox® copier
Band-Aid bandage®
Kodak®
Coca-Cola®

Mrs. Louise Lopez
Mr. Wing Phillips
Dr. Georgia Burke
Professor Robert Borson
Lieutenant Jeb Stuart

Handbook of Chemical Terms
The New York Times
The American Heritage Dictionary
"Time-Sharing" in *McLean's* magazine

See TITLES.

NOTE 1: In everyday English usage, the terms *capitalization* and *to capitalize* mean that only the initial letter of a word appears with a capital letter. So rule 1 above would usually be clear to English speakers if written as follows: "Capitalize proper names . . . ." The expanded versions of rule 1 and the other rules in this discussion of capitals are intended to remove any ambiguities as to which letters in a word should be capitalized.

NOTE 2: As the many instances of lowercase *the* above indicate, *the* is usually not capitalized unless it has become part of the full official name:

The Hague
The Johns Hopkins University

NOTE 3: In lengthy proper names, conjunctions, short prepositions, and articles *(a, an,* and *the)* are not capitalized:

the Federal Republic of Germany
Johnson and Sons, Inc.
"Recovery of Oil in Plugged and Abandoned Wells"

See rule 6 in the following discussion.

NOTE 4: Capitalize the first letter of an individual's title only when it precedes the individual's proper name:

Professor George Stevens (*but* George Stevens, who is a professor . . .)

Captain Ellen Dobbs (*but* Ellen Dobbs, who is the captain of our company . . .)

President Henry Johnson (*but* Henry Johnson, president of Johnson Motors . . .)

See TITLES.

NOTE 5: Adjectives derived from proper names are capitalized only when the original sense is maintained:

a French word (*but* french fries)
Venetian art (*but* venetian blinds)

Even in these cases, dictionaries often differ; for instance, the current *American Heritage Dictionary* recommends *French fries* rather than the form preferred above. So you often have to use your judgment. However, be consistent throughout a document.

2. Do not capitalize the first letter of common nouns—that is, those nouns that are general or generic:

a geologist
my accountant
the engineers
your secretary

a country
a planet
a river
north
the city

a trade school
college
high school
a holiday
the swing shift

a copier
the facial tissue

a foreman
my supervisor
the doctor

spring
fall

twentieth century
the thirties (*however*, the Gay Nineties)

See ARTICLES.

NOTE 1: One useful test to determine whether a noun is common is to ask if *a* or *an* does or can precede the noun in your context. If *a* or *an* makes sense before the noun, then the noun is common:

a pope (*but* the Pope)
an attorney
a U.S. senator

but

a President *(referring to any President of the United States)*

Because of special deference, the word *President* is always capitalized when it refers to any or all of the Presidents of the United States. This supersedes rule 2 above.

NOTE 2: Titles that follow a noun rather than precede it are not capitalized:

Theo Jones, who is our comptroller . . .
Betty Stevens, my assistant . . .
Rene Leon, who is a staff geophysicist . . .

NOTE 3: Common nouns separated from their proper nouns (or names) can occasionally be capitalized:

—Titles of high company officials, when the titles take the place of the officials' names:

We spoke to the President about the new labor policy.

The State Director has to sign before the plan goes into effect.

—Names of departments when they replace the whole name of the department:

We sent the letter to Accounting.

According to Maintenance, the pump had been replaced just last month.

—Names of countries, national divisions, governmental groups when the common noun replaces the full name (often in internal government correspondence):

From the beginning of the Republic, a balance of powers was necessary.

The State submitted a brief as a friend of the court.

The Department has a policy against overtime for employees at professional levels.

The House sent a bill to the Conference Committee.

—Names of close family members used in place of their proper names, especially in direct address:

Please understand, Mother, that I intend to pay my fair share.

Before leaving I spoke to Mother, Father, and Uncle George.

NOTE 4: Capitalize plural common nouns following two or more proper nouns unless the common nouns represent topographical features (such as lakes, rivers, mountains, oceans, and so on):

West and South High Schools
the Korean and Vietnam Wars
the State and Defense Departments

but

the Fraser and Moose rivers
the Wasatch and Uinta mountains
the Pacific and Indian oceans

3. Capitalize the first letter of the first word of sentences, quotations, and listed items (either phrases or sentences):

Researchers propose to complete eight projects this year.

The technical specifications stated: "All wing strut pins should have a 150 percent load factor."

The accountant discussed the following issues:

—Budget design
—Cost overruns
—Entry postings

NOTE: The first letter of a word following a colon or a dash within a sentence is often capitalized, although some editors prefer not to. A good rule of thumb is that full sentences and long quotations (usually sentences) begin with a capital letter after a colon or dash:

The Bible states: "The race is not to the swift."

We followed one principle: Short-term investments must be consistent with long-term goals.

or

We followed one principle—Short-term investments must be consistent with long-term goals.

4. Capitalize the first letter of the names of directions when they indicate specific geographical areas. Do not capitalize the first letter of the names when they merely indicate a direction or a general or unspecified portion of a larger geographical area:

the Deep South
the Midwest
the Near East
the North
the Northwest

blowing from the southeast
eastern Romania
southern Italy
the northern Midwest
toward the south
traveling north

5. Capitalize the first letter of names for the Deity, names for the Bible and other sacred writings, names of religious bodies and their adherents, and names denoting the Devil:

Christ
God
He, Him
Messiah
Son of Man
the Almighty
Thee
Allah
Zeus

God's Word
the Good Book
the Old Testament
the Word
the Koran

Baha'is
a Lutheran
an Episcopalian
Episcopal Church
Lutheran Church
a Muslim
Buddhists

Flying Spaghetti Monster

His Satanic Majesty
Satan
the Great Malevolence
Loki
Pluto

6. Capitalize the first letter of the first word and all main words of headings and subheadings and of titles of books, articles, and other documents. Do not capitalize the first letter of the articles (a, an, and the), the coordinate conjunctions (and, but, or, nor, for, so, yet), or the short prepositions (to, of, etc.) unless they appear as the first word:

"An Examination of Church-State Relations"
Declaration of Independence
MacUser
The Geology of East Texas
"The Greening of Panama" in *Scientific American*

See HEADINGS.

7. Capitalize in titles and headings the first letters of the initial word and of all later words in a hyphenated compound except for articles, short prepositions, and short conjunctions.

See rule 6 above.

A Report on Tin-Lined Acid Converters
Up-to-Date Power-Driven Extraction Methodologies
State-of-the-Art Technology
Seventy-Five World Leaders as Voting Representatives

8. Capitalize the first letter of the geological names of eras, periods, systems, series, epochs, and ages:

Jurassic Period
Late Cretaceous
Little Willow
Paleozoic Era
Upper Triassic

NOTE 1: Unless they are part of a proper name, do **not** capitalize structural terms such as *arch, basin, formation, zone, field, pay, pool, dome, uplift, anticline, reservoir,* or *trend* when they combine with geological names:

Cincinnati arch
Delaware basin
East Texas field
Ozark uplift

Note 2: Experts disagree about the capitalization of *upper, middle, lower* and *late, middle, early.* The following list from the *United States Government Printing Office Style Manual* (1984) provides the best summary of the difficult capitalization conventions in this technical area. Note that both *upper Oligocene* and *Upper Devonian* are correct, although the capitalization of *upper* is inconsistent:

Alexandrian
Animikie
Atoka
Cambrian:
  Upper, Late
  Middle, Middle
  Lower, Early
Carboniferous Systems
Cayuga
Cenozoic
Cincinnatian
Chester
Coahuila
Comanche
Cretaceous:
  Upper, Late
  Lower, Early
Des Moines
Devonian:
  Upper, Late
  Middle, Middle
  Lower, Early
Eocene:
  upper, late
  middle, middle
  lower, early
glacial:
  interglacial
  postglacial
  preglacial
Glenarm
Grand Canyon
Grenville
Guadalupe
Gulf
Gunnison River
Holocene
Jurassic:
  Upper, Late
  Middle, Middle
  Lower, Early
Keweenawan
Kinderhook
Leonard
Little Willow
Llano
Meramec
Mesozoic:
  pre-Mesozoic
  post-Mesozoic
Miocene:
  upper, late
  middle, middle
  lower, early
Mississippian:
  Upper, Late
  Lower, Early
Missouri
Mohawkian
Morrow
Niagara
Ochoa
Ocoee
Oligocene:
  upper, late
  middle, middle
  lower, early
Osage
Ordovician:
  Upper, Late
  Middle, Middle
  Lower, Early
Pahrump
Paleocene:
  upper, late
  middle, middle
  lower, early
Paleozoic
Pennsylvanian:
  Upper, Late
  Middle, Middle
  Lower, Early
Permian:
  Upper, Late
  Lower, Early
Pleistocene
Pliocene:
  upper, late
  middle, middle
  lower, early
Precambrian:
  upper
  middle
  lower
Quaternary
red beds
Shasta
Silurian:
  Upper, Late
  Middle, Middle
  Lower, Early
St. Croixan
Tertiary
Triassic:
  Upper, Late
  Middle, Middle
  Lower, Early
Virgil
Wolfcamp
Yavapai

Note 3: Topographical terms are usually capitalized, but the general terms *province* and *section* are not:

Hudson Valley
Interior Highlands
Middle Rocky Mountains
Ozark Plateaus
Uinta Basin

but

Navajo section
Pacific Border province

9. In text, do not capitalize the first letter of a common noun used with a date, number, or letter merely to denote time or sequence:

appendix A
collection 3
drawing 8
figure 5
page 45
paragraph 2
plate VI
section c
volume III

Note 1: The first letter of these common nouns should be capitalized if they appear in headings, titles, or captions: *Appendix A* or *Figure 5.*

Note 2: In these cases, *no.*, *#*, or *No.* (for *Number*) is unnecessary:

Appendix A (not Appendix No. A)
page 45 (not page no. 45)
site 5 (not site #5)

Note 3: Some technical and scientific fields do capitalize a common noun used with a date, number, or letter. For example, The Society of Petroleum Engineers recommends in its *Style Guide* that writers observe the following style or capitalization in text:

Method 3
Sample 2
Table 4
Wells A22 and B7

10. Capitalize the first letter of proper nouns combined with common nouns, as in the names of plants, animals, diseases, and scientific laws or principles:

Boyle's law
Brittany spaniel
Cooper's hawk
Down's syndrome
Fremont silktassel
Gunn effect
Hodgkin's disease
Virginia clematis

Captions

Captions for graphics include the title and any explanatory material immediately under (or sometimes over) a graphic. Readers initially identify graphics using conventional references: *Figure 14, Map 3, Chart 16,* etc. However, readers' eyes need to be guided beyond mere identification.

Good captions are what guide readers not only to see but also to understand. Good captions both label graphics with titles and explain to readers what they are seeing and how to interpret the information captured in the graphic.

See Graphics for Documents and Graphics for Presentations. See also Charts, Graphs, Illustrations, Maps, Photographs, and Tables.

Good captions are interpretive and informative. In some types of writing, such captions are even called "action captions" because they urge readers to make a decision or take an action.

A graphic and its caption should be clear and understandable without requiring readers to search for clarifying information in the text.

As an example, for some years now, *Scientific American* has used "stand alone" captions for their graphics. *Scientific American* has also discontinued even referencing graphics by title and number in their text. So a graphic in *Scientific American* does not have a label and title such as *Figure 4—A Peruvian Burial Site.* Instead, the caption begins by explaining and interpreting: *In a typical Peruvian burial site, the funereal jewelry signified . . . .*

Captions

1. Use interpretive captions whenever possible.
2. Avoid using short, often ambiguous, titles to replace interpretive captions.
3. Number figures and tables sequentially throughout the document, and place the number before the caption.
4. Use periods following interpretive captions, but no punctuation following short captions that are not sentences.
5. Captions may appear below or above the visual, but be consistent throughout a document.

1. Use interpretive captions whenever possible.

Interpretive captions provide both a title and explanatory information (usually expressed in a complete sentence) to help readers understand the central point(s) that you want the graphic to convey:

Figure 4. Cabin-Temperature Control System. *Constant cabin-temperature control is maintained by the system's modulated cabin sensor.*

Figure 23. Check Valve. *The risk of bad air entering the chamber is near zero because the check valve permits air flow in one direction only.*

Table 17. Air Intersect Data. *Only stations 23 and 45 experienced significant increases in CO levels during the study period.*

The figure 4 interpretive caption gives the title of the figure and then emphasizes that the cabin has a constant temperature—a benefit provided by the feature (modulated cabin sensor) described in the figure. The caption states clearly what the writer wants the reader to learn from the graphic.

The figure 23 interpretive caption gives the title and then tells the reader the principal message—that the check valve provides near-zero risk. Further, it states **how** the check valve provides near-zero risk.

The table 17 interpretive caption names the data in the table and then highlights significant data. By pointing out what is most important in the table, the caption helps the reader interpret the table, which might be very complex, and reinforces the major points that the writer will make in the text.

The following are additional examples of interpretive captions for figures and tables:

Figure 1. Axial-Flow Design. *The axial-flow design has the greatest performance potential.*

Figure 2. Life-Cycle Cost Projections. *General Framitz's design-to-cost strategy will guarantee low life-cycle costs.*

Figure 3. Project Management System. *Our project management software ensures maximum responsiveness to user requirements at all levels.*

Table 1. Population Impacts. *If population redistribution trends continue, the Southwest will exceed baseline figures by 2015.*

Table 2. Cost of Mitigation Measures. *The socioeconomic mitigation measures proposed are effective but very costly.*

Figure 4. Seasonal Streamflow Patterns. *Keta River and White Creek peak in mid-October during Snow Geese migration and provide the only suitable feeding and resting habitat within a radius of 150 miles.*

Table 3. Projections of New Jobs During the Next 10 Years. *Contrary to media opinion, the Thunder Basin Project will create—not destroy—jobs in the Sequaw Valley: at least 500 within the next 10 years.*

Figure 5. Test Results of Thermal Model 2. *Flame-envelope thermal Model 2 results in lower ambient temperatures but produces diffuse radiation.*

Figure 6. Limits of Multiplexer Design. *The analog input multiplexer design limits components that are not included in self-testing to a few passive components.*

NOTE: You can create interpretive captions without using complete sentences or a separate title—if the captions tell readers how to read and interpret the graphic:

Figure 14. Declining Production Through the 1980s

or

Figure 14. *Declining production through the 1980s*

In these examples (both acceptable versions of a caption), the title for the figure is combined with a phrase telling readers about the production trend during the 1980s. As the two examples illustrate, you can use either boldface type or italics to highlight the caption.

A longer version of the figure 14 caption would include a separate title and then add a complete sentence interpreting the figure:

Figure 14. Production (1980–1989). *Production declined steadily through the 1980s.*

Whether or not an interpretive caption is a complete sentence, it should provide a point of view on the graphic. It tells readers not only what the graphic is about but also what the graphic means.

2. Avoid using short, often ambiguous, titles to replace interpretive captions.

In the past, many technical and scientific documents used only short, simple titles (captions) for graphics.

Figure 3. A Horse

Figure 4. Schematic 23–A

In figure 3, surely a reader could see that the image was of a horse, so this title is worthless. The title should give some interpretive information: *Figure 3. A typical Arabian stallion in full gallop.*

Similarly, the title for figure 4 may be somewhat useful, especially if the figures include a number of different schematics. However, the writer probably should include information about how schematic 23–A relates to 23–B or how 23–A is different from 22.

Titles that provide minimum information are often so short and cryptic that they sound telegraphic:

Table 2. Problem Options

Table 14. Water and Soil Impacts

Telegraphic captions are only appropriate when the graphics are self-explanatory, thus requiring no interpretation. But even then, the telegraphic style can be confusing. In the above examples, for instance, are the table 2 options really problems, or are the options solutions to a single problem? Similarly, does table 14 include water impacts as well as soil impacts, or is the subject some sort of data about water and a separate presentation about impacts on soil types? As these examples illustrate, avoid telegraphic captions.

3. Number figures and tables sequentially throughout the document, and place the number before the caption.

Figures and tables should be numbered sequentially as they appear in the document. If you present an important figure or table twice, treat it as two separate visuals and number each according to its position in the sequence. See GRAPHICS FOR DOCUMENTS.

Figure and table numbers should be whole numbers: *Figure 1, Figure 2, Figure 3,* etc. If you are numbering graphics by chapter, then use an en dash to separate chapter number from graphics number: *Figure 14–2* (chapter 14, figure 2), T*able 2–8* (chapter 2, table 8). You can also use the decimal numbering system with an en dash to designate graphics numbers within a section of a report if the report's sections have been numbered decimally: *Figure 23.2–1* (section 23.2, figure 1), *Table 7.4–13* (section 7.4, table 13). If the graphic has several parts and you need to identify all parts, use parentheses and lowercase alphabetical characters to designate subparts: *Figure 34(a), Figure 34(b), Figure 34(c).* See DASHES.

NOTE 1: The style of punctuation for captions varies. Some editors prefer a colon or a period and a dash following the number. Some editors prefer no punctuation and leave three or four spaces between the number and the caption. Each version is acceptable.

Figure 14–2: Federal Shipbuilding and Repair Budget

Figure 14–2.—Federal Shipbuilding and Repair Budget

Figure 14–2   Federal Shipbuilding and Repair Budget

Captions

NOTE 2: In captions, the words *figure* and *table* should be capitalized. However, when you are referring to graphics in the text, even if you refer to a specific graphic, do not capitalize *figure* or *table* unless it begins a sentence:

> As shown in figure 33, . . .
>
> According to table 14.2–4, . . .
>
> *however*
>
> Figure 33 shows . . .
>
> Table 14.2–4 presents . . .

See CAPITALS.

4. Use periods following interpretive captions, but no punctuation following short captions that are not sentences.

Interpretive captions are usually complete sentences and should therefore end with a period. Short captions, on the other hand, are like titles or headings and are normally not complete sentences, so they require no punctuation.

If you mix interpretive and telegraphic captions, end all of them with periods. See LISTS and PERIODS.

5. Captions may appear below or above the visual, but be consistent throughout a document.

Some graphics specialists argue that captions should always appear above graphics because a caption (which comes from the word *head*) is like a title. Others argue that captions should always appear below graphics because the graphics are more important than the captions, and placing captions below their graphics is more aesthetic. Still others argue that captions for tables should appear above tables, but captions for figures should appear below figures.

When captions are used with slides and other projected visual aids, put the caption above the visual for better visibility. Captions placed at the bottom may be blocked by the heads of those seated in front. See GRAPHICS FOR PRESENTATIONS.

Be consistent. Treat tables no differently from figures. If you're going to place captions above or below graphics, then do so throughout your document.

Charts are some of the most valuable and frequently used types of graphics. Unfortunately, the term *chart* has different meanings for different people.

Webster's New Collegiate Dictionary defines *chart* as a map, table, graph, or diagram. That definition is why people are confused over the names of different types of graphics. Originally, *chart* meant a document, although most charts were maps. When maps were combined with tabular data (e.g., the mileage charts on modern road maps), *chart* came to mean a table or matrix display.

In this *Style Guide*, charts are graphics that do not rely on numerical interpretations. These include **organizational charts, flow charts**, and **schedule charts.**

In contrast to charts, graphs are graphics that do rely on numerical interpretations, especially showing how one variable (such as time) relates to another variable (e.g., production or dollars). The graphics discussed in the GRAPHS entry include line (or coordinate) graphs, bar graphs, and circle (pie) graphs. In everyday language, however, bar graphs and pie graphs are often called bar charts and pie charts. Graphics software usually includes all of these graphics in a "charts" menu.

Don't let confusion between the terms *chart* and *graph* bother you. The best graphics are those that rapidly and effectively communicate their central ideas, regardless of what they are called. See GRAPHS, GRAPHICS FOR DOCUMENTS, and GRAPHICS FOR PRESENTATIONS.

Charts

1. Use an organizational chart, a flow chart, or a schedule chart to help readers visualize easily the major points in a document.
2. Orient horizontally all letters, numbers, words, and phrases in headings, legends, and labels.
3. Number each chart sequentially, provide a title, and then add an informative caption to identify the purpose of the chart.
4. Place footnotes and source information below charts.

Organizational Charts

5. Use squares or rectangles to indicate divisions and subdivisions within the organization.
6. Structure an organizational chart from the top down.
7. Use solid lines between boxes to indicate direct relationships and dotted or dashed lines to indicate indirect relationships.

Flow Charts

8. Choose a standard system of symbols to indicate the types of activities and the control or transfer points.
9. Place the starting activity in the upper left corner of the chart and proceed to the right and down. Place the ending activity in the lower right corner of the chart.
10. Break large or complicated flow charts into smaller, simpler flow charts.
11. Use arrows to show the sequence or direction of flow within flow charts.

Schedule Charts

12. List activities in chronological order beginning at the top of the page and moving down.
13. Clearly label or identify the bars.
14. If appropriate, indicate milestones on the chart.
15. Use bar colors to identify groups of related activities. If you do so, explain in the legend or caption what the colors represent.

1. Use an organizational chart, a flow chart, or a schedule chart to help readers visualize easily the major points in a document.

These charts should not be mere decoration. If they capture for readers the major points or themes in a document, they will serve their intended purpose.

Remember that a chart is only as good as the story it tells. Charts should communicate quickly and simply. They should be integrated with the text and should convey information more forcefully or dramatically than is possible in text.

Charts

In many documents, using one of these three charts will allow you to replace unnecessary text. A flow chart, for example, can help readers visualize the steps in a manufacturing process or in a political negotiation. A chart can replace text or provide a visual road map that can help readers with dense and complicated text. A chart also helps readers recall or review what the document said.

Let your purpose, readers, medium, data, and ideas dictate the form of chart (or other graphics) that would be most effective.

2. Orient horizontally all letters, numbers, words, and phrases in headings, legends, and labels.

All of the letters, numbers, and words on a chart should be readable from one reading perspective. Readers should not have to rotate the page to read different parts of a chart.

See the figures later in this discussion for examples of how to present letters, numbers, and words.

3. Number each chart sequentially, provide a title, and then add an informative caption to identify the purpose of the chart.

Charts and all other graphics are sequentially numbered and labeled as figures, so the numbering would be *Figure 1*, *Figure 2*, etc. Or, if the document has chapters, *Figure 1–1*, *Figure 1–2*, *Figure 1–3*, where the first number is the chapter number.

Following *Figure* and the number comes the title of the chart:

> **Figure 3–1. Organizational Structure of the Finance Department**

Then, provide an informative or interpretive caption to help readers understand the message of the chart:

> **Figure 3–1. Organizational Structure of the Finance Department.** *Following the 2010 reorganization, all accounting personnel report to a single assistant departmental manager.*

See Captions.

4. Place footnotes and source information below charts.

Footnotes typically explain or clarify the information appearing in the entire chart or in one small part of it. Often footnotes tell what the data apply to, where they came from, or how accurate they are:

> [1] All project times are rounded up to full days.
>
> [2] For the two regional offices only; the head office uses a different support organization.
>
> [3] According to the Food and Agricultural Organization. The solid lines reflect the current policy for who reports to whom. The dotted lines reflect the earlier lines of authority, which still are important even though outdated.

The footnotes for each chart are numbered independently from footnotes in the text and from footnotes in other graphics. Begin with footnote 1 and proceed sequentially for that single graphic; for the next graphic, start the numbering again. Within the body of the chart, use superscripted footnote numbers. Place the footnote explanations (in numerical order) immediately below the chart and flush with the left margin. Repeat the superscripted footnote number and then provide the appropriate explanation, followed by a period.

If the chart covers more than one page, place the appropriate footnotes with each page. If both a caption and footnote appear below a chart, place the footnotes below the caption.

If footnote numbers would be confusing in the body of the chart, use letters (a, b, c, d, etc.), asterisks (*, **, ***), or other symbols.

Source information may appear in footnotes if the referenced source provides only the data indicated by the footnote and not the data for the rest of the chart:

> [4] From *The Wall Street Journal*, May 14, 2010.
>
> [5] Source: International Union of Geological Sciences.

Source information may also appear within parentheses in the caption (regardless of where the caption appears) or within brackets under the caption if the caption appears ahead of the chart:

> **Figure 1. U.S. Aerospace Mergers and Acquisitions, 1960–2010** (U.S. Department of Commerce)

> **Figure 1. U.S. Aerospace Mergers and Acquisitions, 1960–2010**
>
> [U.S. Department of Commerce]

See Citations, Footnotes, and Intellectual Property.

Organizational Charts

Organizational charts depict the structure of an organization. These charts typically show the divisions and subdivisions of the organization; the hierarchy and relationship of the groups to one another; lines of control (responsibility and authority); and lines of communication and coordination. Organizational charts help readers visualize the structure of an organization and the relationships within it. Figure 1 is an example of an organizational chart showing how an engineering department is organized.

Software templates make it easy to create organizational charts (often called "hierarchy charts"). Make sure the template you use serves your purpose and aligns with these guidelines.

5. Use squares or rectangles to indicate divisions and subdivisions within the organization.

Indicate divisions, subdivisions, groups, project teams, functional areas, etc., by enclosing the name of the organizational unit within a square or rectangle. Optionally, the squares can also contain the names of people filling the positions, but such a person-by-person chart for large organizations can be too long and complex, and it is quickly out of date because of transfers and reassignments.

You might distinguish between higher and lower units by changing the size of the rectangle or by changing its border (from boldface or thick lines representing upper-level units to thinner, normal lines representing lower-level units). In figure 1, for example, the box around the V.P. rectangle is a heavier ruled line.

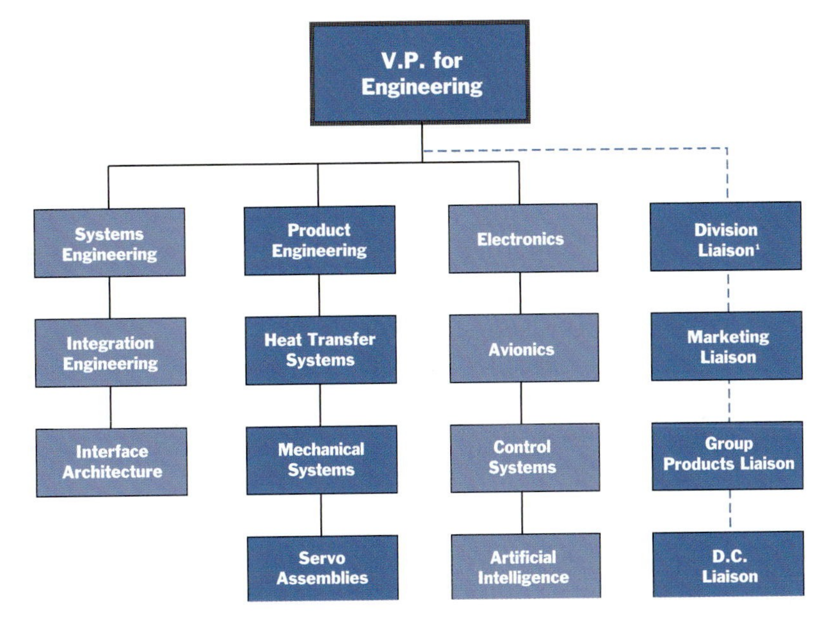

Figure 1. Engineering Department Organization. *Besides those who report directly to her (shown in the boxes connected with solid lines), the Vice President oversees the Division Liaison office (shown in the boxes connected with dashed lines) to ensure that all lobbying in Washington, D.C., complies with Federal Acquisition Regulations.*

6. Structure an organizational chart from the top down.

This rule reflects the way we read and our perception and understanding of typical organizational structure.

Readers of English read from left to right and from top to bottom. Therefore, an organizational chart should be structured from left to right and top to bottom. The left-to-right progression might be useful in depicting a flat organizational structure. However, the top-to-bottom progression is typical, simply because most organizations are based on a hierarchy.

In this case, you should display the structure of the organization in descending order of authority—with the highest authority or level at the top of the chart and the lowest authorities or levels at the bottom of the chart. This structure reflects the metaphor of top-down management and thus reinforces the readers' expectations about organizational structure.

If the organization you are describing does not operate on a top-down basis, be inventive and create an organizational display that does reflect the organization's operational style and structure. To depict an organizational structure based on teams, you might want to use a radial chart with lines showing how these teams interact (see rule 7).

7. Use solid lines between boxes to indicate direct relationships and dotted or dashed lines to indicate indirect relationships.

Solid lines usually show direct lines of control. Dashed or dotted lines usually indicate lines of communication or coordination. Keep this in mind when using templates.

In figure 1, the dashed lines through the rectangles along the right side of the chart indicate that the Vice President for Engineering coordinates with the Division Liaison office and has lines of communication and coordination down through the Division Liaison organization, but the Vice President's direct authority extends only through Systems Engineering, Product Engineering, and Electronics.

Flow Charts

Flow charts depict a process and are often known as process charts. They show readers the parts of a process and how those parts are related. Processes include manufacturing processes (as shown in figure 2) or procedures (as in figure 3), decision-making processes, audit processes (with checks and balances), biological or physical processes (as shown in figure 4), or any other process that moves from a start to a finish.

Again, software templates make it easy to create flow charts that serve many purposes. For example, there are different templates for continuous, parallel, or staggered processes, among others. Make sure the template you use serves your purpose and aligns with these guidelines.

8. Choose a standard system of symbols to indicate the types of activities and the control or transfer points.

Squares and rectangles typically indicate activities or steps in the process. In figure 2, for instance, the upper left rectangle represents "ore crushing," the first activity in ore processing. The arrow linking this rectangle to the rectangle directly to the right indicates that, after being crushed, the ore undergoes a chemical bath.

The arrows indicate the sequence of activities in the w and show a chronological (and sometimes cause-and-effect) relationship between linked activities or control points.

Circles typically indicate control or transfer points. Control points indicate where the activities are monitored, started, stopped, or in some other way controlled. Transfer points indicate where the sequence of activity leaves one flow chart and continues on to another. In figure 2, the two right-most circles indicate that the ore has been processed and is ready for packaging. To continue following the process, the reader must go to the packaging flow chart, which is shown in figure 3.

The "pack" circles in figure 2 represent all of figure 3. Consequently, figure 2 is the more general flow chart. If "ore crushing" involved a series of steps, the writer could have turned the "ore crushing" symbol in figure 2 into a circle and then constructed another subordinate.

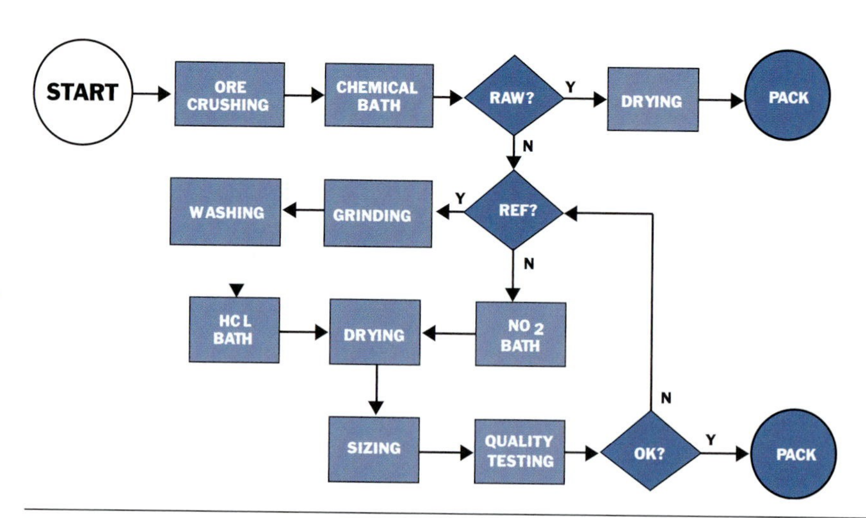

Figure 2. Ore Processing. *The quality and quantity of the ore is checked at three decision points, indicated by blue diamonds and Y (yes) and N (no) labels. Customers can opt to receive raw (semifinished) ore directly without additional refining or extra packaging.*

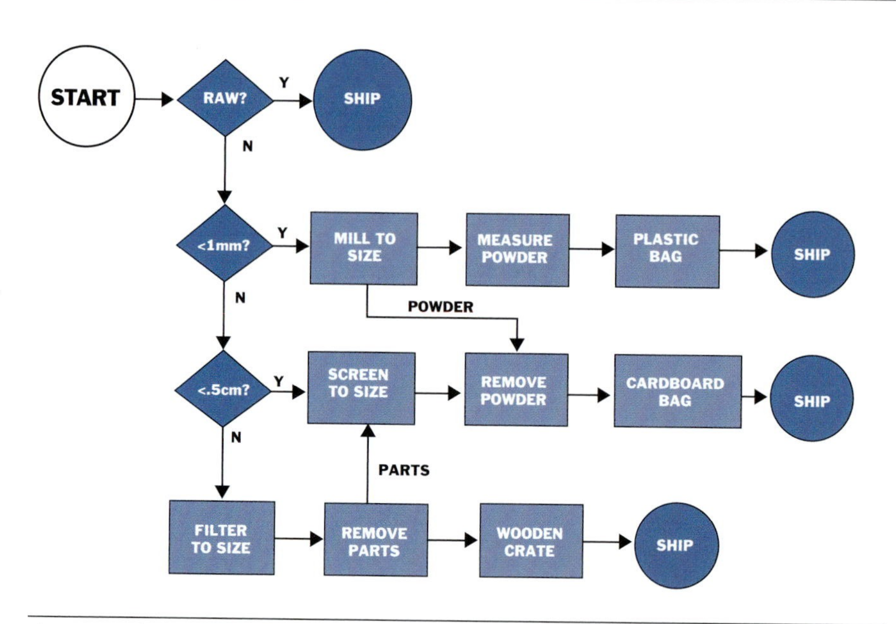

Figure 3. Packaging Options for Ore. *Customers can choose four packaging options. Raw (semifinished) ore is shipped in bulk by railway cars. Refined ore (in varying sizes) is shipped in three ways: (1) in sealed plastic bags, (2) in cardboard boxes, and (3) in wooden crates.*

Note that by putting all of the packaging activities into one subordinate flow chart, the writer has avoided significant repetition in figure 2.

Diamonds typically represent decision points. Often, as in figures 2 and 3, decisions are represented by simple yes/no questions. Normally, three lines link diamonds to other symbols: one incoming line indicating what precedes the decision, one outgoing line indicating "yes," and the other outgoing line indicating "no."

The flow chart in figure 4 differs from figures 2 and 3. In figure 4, the boxes represent a conceptual sequence of possible effects, originating with the initial cause (the harvesting of a stand of Douglas-fir timber). Flow charts can be used to depict any cause-and-effect sequence. For instance, a flow chart might present the sequence of

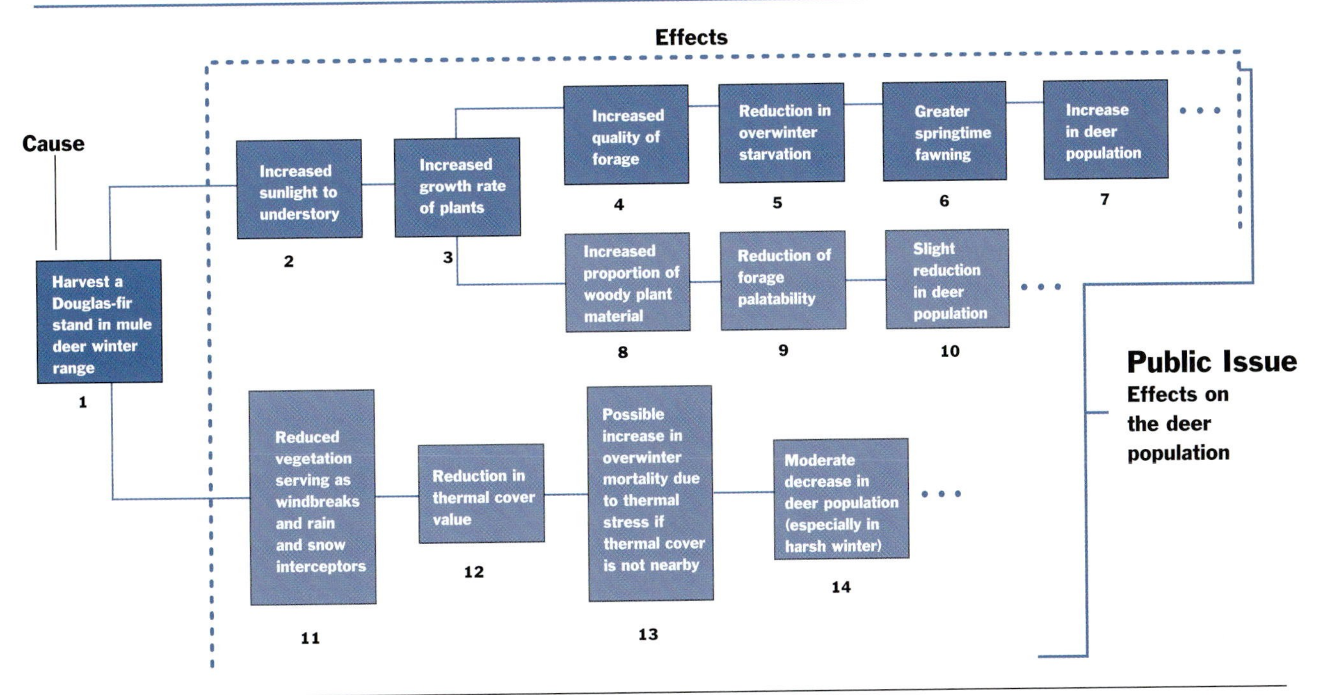

Figure 4. Potential Effects of a Timber Harvest on Deer Populations. *The most positive scenario (dark blue boxes) would increase the deer population. The other two scenarios (light blue boxes) would each decrease the deer population. Further analysis in the text would lead to a resolution between opposing scenarios. Note that the numbers on the boxes are for reference only, not to reflect the sequence of causes and effects.*

events if a government decided to subsidize automobiles with engines that did not use gasoline. Effects would flow through consumers to manufacturers, with possible feedback loops to government.

Other symbol systems for flow charts are possible, particularly in software development and other specialized fields, such as architecture and electrical engineering. These symbols often have very specific meanings and have become conventional in particular scientific and technical applications. If you need to create flow charts for these specialized areas, consult appropriate trade publications and websites.

9. Place the starting activity in the upper left corner of the chart and proceed to the right and down. Place the ending activity in the lower right corner of the chart.

Readers will expect the flow chart to begin in the upper left corner. Don't disappoint them. If possible, try to end the flow chart in the lower right corner.

10. Break large or complicated flow charts into smaller, simpler flow charts.

Flow charts that become too large or too complicated are unreadable as well as intimidating. Provide users with only as much of the process as is relevant to them. To avoid repetition and to keep flow charts from becoming too long, break them into general (overview) charts and subordinate (component) charts.

After breaking up a flow chart, you must also decide just which parts of it to use in your document. Do your readers need the subordinate charts to understand your information, or would the overall chart be sufficiently detailed?

In many instances, you can retain the overall chart in your document while putting the subordinate charts in an appendix or attachment.
See Appendices.

11. Use arrows to show the sequence or direction of flow within flow charts.

Flow charts with activities or control points linked only by lines are often confusing. Place an arrowhead on the end of the line to indicate the sequence or direction of flow (see figures 2 and 3). Most flowchart templates use arrows for this purpose.

Arrows are sometimes unnecessary. Notice in figure 4 that arrows are unnecessary because the flow is obviously from left to right; in addition, each square is numbered for easy reference and discussion.

Schedule Charts

Schedule charts are horizontal charts used to schedule tasks, projects, and programs.

Schedule charts are also called *time charts* or *timelines*. Other terms are also possible, as in various scheduling software programs. For example, critical-path schedules are merely schedule charts that show whether items are concurrent or sequential.

Schedule charts help readers visualize a sequence of activity occurring over time. They help readers see how sequential and concurrent activities relate to one another in time and how activities depend on one another for completion on schedule.

The horizontal (x) axis in schedule charts is always time, in decades, years, quarters, months, weeks, days, or hours. To make schedule charts easier to read, use vertical dotted or dashed lines to mark major time periods (see figures 5 and 6).

Project management software programs typically contain templates for schedule charts. If such a program isn't available, you can create a schedule chart using a template for a bar chart. Make sure the template you use follows these guidelines.

12. List activities in chronological order beginning at the top of the page and moving down.

Schedule charts require strict chronology. Do not violate the readers' expectation that the events listed from top to bottom along the vertical (y) axis will appear in chronological order.

13. Clearly label or identify the bars.

Always identify what each bar represents, either with a bar label along the left margin or with an explanation block (as in figure 5). Traditionally, the bar labels appear on the left side of the chart.

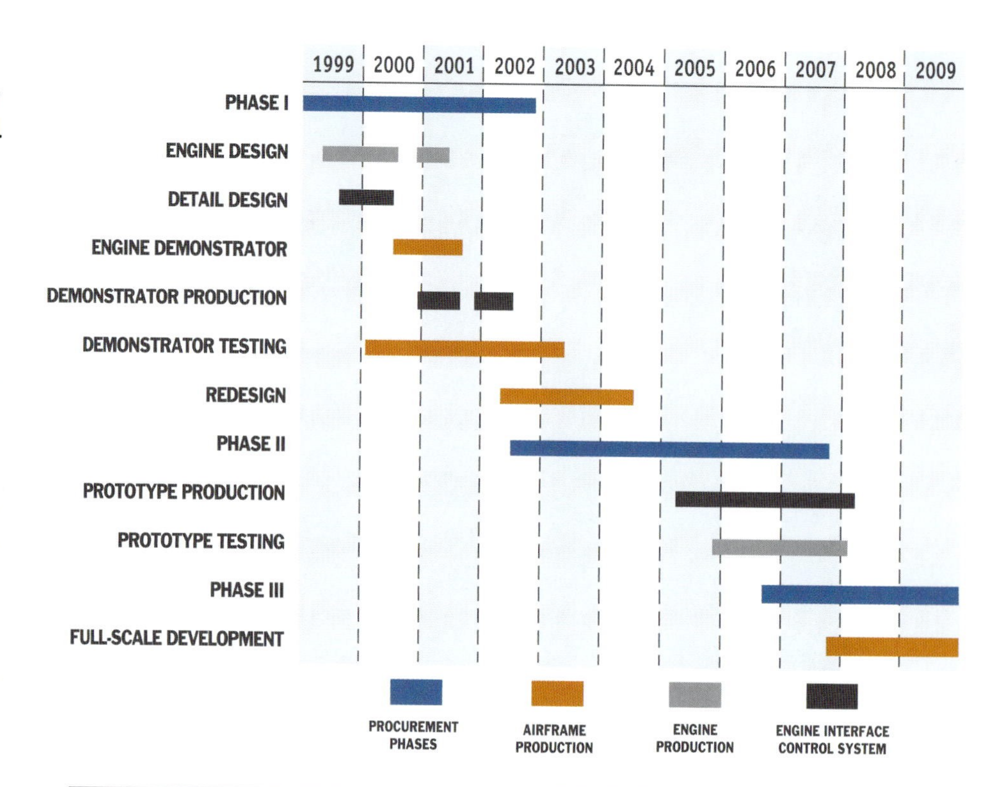

Figure 5. AAW2 Production Schedule. *To shorten production time, we have scheduled each phase (shaded orange) to begin before the prior phase is finished. These savings in time are possible because of our self-funded R&D work with the AAW2 program.*

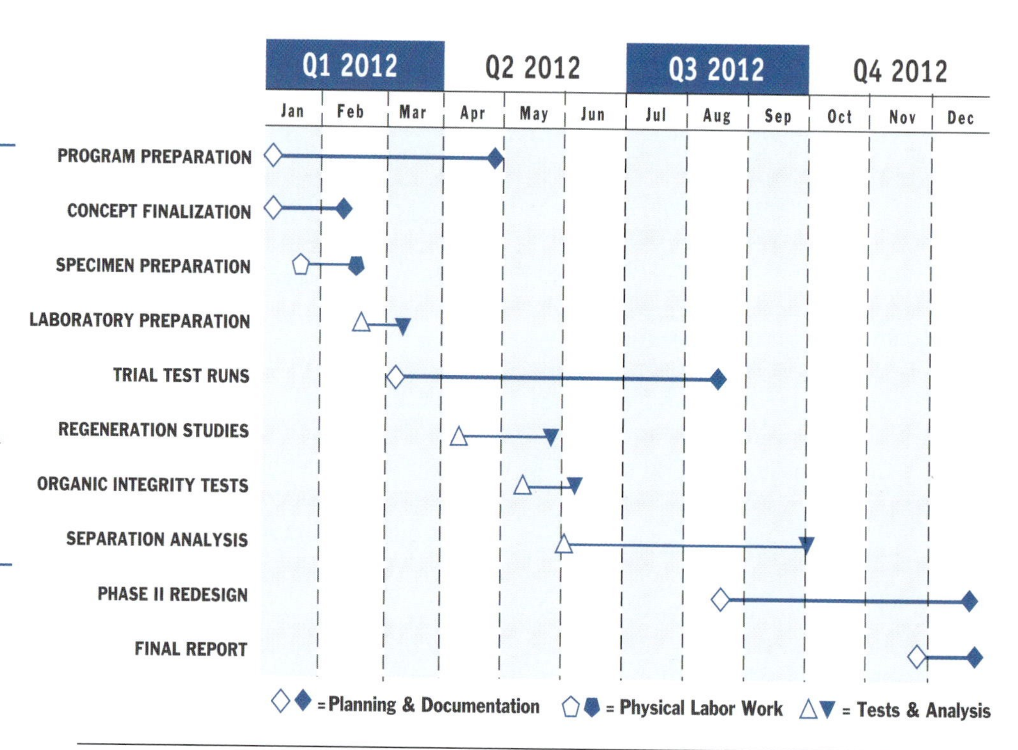

Figure 6. Schedule of Recombinant Tests and Documentation. *As the bars with the diamonds indicate, planning and documentation efforts are concurrent with the lab work (pentagons) and tests (triangles). As this schedule suggests, documentation should never be left until all the research work is finished.*

14. If appropriate, indicate milestones on the chart.

Milestones may be indicated with small circles, dots, or triangles. If you have reporting, control, or performance deadlines that constitute milestones for tracking and monitoring progress, then indicate them on the chart.

15. Use bar colors to identify groups of related activities. If you do so, explain in the legend or caption what the colors represent.

Bar colors (see the discussion of bar graphs in GRAPHS) can be used to indicate similar or identical activities that occur at different times. In figure 5, for instance, all airframe-production activities are shown with orange bars, while engine-production activities are in grey. This way readers can see how related activities fit within the whole sequence.

Using colors as in figure 5 is increasingly desirable. Remember, however, that colors will not reproduce adequately in a black-and-white copy, so you might use patterns or textures to help readers distinguish among bars. See COLOR.

Citations

Citations (bibliographic entries) enable writers to identify in the text itself the sources of their information. The methods of citation vary, depending on the technical field and its traditions, the type of publication, and the publisher.

The method of citation that we recommend represents the standard convention in the physical sciences and engineering disciplines. However, if you are writing for a particular professional society or technical journal, you should follow its method of citation.

Citations

1. Enclose the author's name and the date of the publication in parentheses following the material quoted or the ideas referred to. Attach at the end of the text an alphabetical list of the cited works.
2. Use a consistent format for citing the name of the author and the date of publication.
3. Include a full alphabetized list of cited sources following the article or chapter.

1. Enclose the author's name and the date of the publication in parentheses following the material quoted or the ideas referred to. Attach at the end of the text an alphabetical list of the cited works:

> One critic called the whole dispute a "galaxy of confusion" (Jameson 2007). In reply, the spokesman for the conservative wing rebutted Jameson point by point (S. Clarke 2008).

See BIBLIOGRAPHIES.

NOTE 1: This system, or some variation of it, is favored by physical and biological scientists, as well as many social scientists. Scholars in the humanities sometimes still use footnotes. See FOOTNOTES.

NOTE 2: An alternative method uses only numbers in the text, not the author's last name and date:

> One critic called the whole dispute a "galaxy of confusion" (1). In reply, the spokesman for the conservative wing rebutted Jameson point by point (2).

or

> One critic called the whole dispute a "galaxy of confusion" [1]. In reply, the spokesman for the conservative wing rebutted Jameson point by point [2].

The numbers appearing within parentheses or brackets are keyed to an unalphabetical list of sources at the end of the text.

NOTE 3: With only a few citations and if you want to omit the final alphabetical list, follow the footnote format within the text:

> Sinclair Jameson in *Age of Change* (New York: Freedom Press, 2007) called the whole dispute a "galaxy of confusion."

See FOOTNOTES.

2. Use a consistent format for citing the name of the author and the date of publication:

> (Collins 2001)
> (Mihelcic and Zimmerman 2009)
> (United Nations Committee on Science and Technology 2007)
> (Covey, Colosimo 2010)
> (U.S. Department of Agriculture 1994)

NOTE: Some authors and journals prefer to place a comma between the author and the date:

> (Collins 2001)

Another common variation is to include the page number or volume and page number following the date:

> (Collins 2001, 43–48)

The abbreviations *p.* and *pp.* (for *page* and *pages*) are unnecessary:

> (Mihelcic and Zimmerman 2009, 156)
> (UN Committee on Science and Technology 2007, 2:34–36)

In the last example, *2* is the volume number.

Whatever the format used, you should be consistent in your method of citation within the same document.

3. Include a full alphabetized list of cited sources following the article or chapter:

Book by one single author

Bricke, Larry N. 2008. *Canadian Political Parties*. Toronto: New Country Press.

Book by two authors

Azuma, Shoji and Ryo Sanbongi. 2001. *Business Japanese*. Washington, D.C.: Georgetown University Press.

Journal or magazine article

Kumar, Nirmalya. 2003. "Kill a Brand, Keep a Customer." *Harvard Business Review* 81:12:86-95.

NOTE 1: In these bibliographic entries, the date of publication comes immediately after the name of the author. In a common optional format, the date appears after the name of the publisher:

Bricke, Larry N. *Canadian Political Parties*. Toronto: New Country Press, 2010.

See BIBLIOGRAPHIES.

NOTE 2: The titles may be underlined rather than italicized if italics is not available.

Cliches are worn-out phrases that were originally effective, even vivid:

innocent bystander
irony of fate
too funny for words
cool as a cucumber
moot point
far and wide

Such phrases are so common that writers and speakers use them habitually, without thinking. Their familiarity makes cliches convenient. So when writers are struggling to express an idea, using a cliche becomes a tempting alternative to serious thought. See WORDY PHRASES.

Cliches

1. Use cliches sparingly, if at all.

The Origin of Cliches

The word *cliche* comes from French and is often still written with the French accent: *cliché*. Writers of English typically omit the accent mark because the word has been anglicized. See SPELLING.

Many cliches were originally metaphors and were therefore vivid. Their wittiness and sparkle made them memorable, so they were repeated often. However, the vividness of the original metaphor is dulled by repetition and the expression becomes a cliche.

The first person in medieval Europe to associate the concept of avoidance with the Black Death must have created a vivid image in listeners' minds. But today's users of the expression *to avoid it like the plague* experience little if any of the original effect. Today the cliche means little. We've heard it too often.

Other cliches developed and survived because they sounded good:

bag and baggage
rack and ruin
not wisely but too well
snug as a bug in a rug
willy-nilly

The alliteration (repetition of consonants) and repetition of *bag* in the expression *bag and baggage* likely ensured the phrase's survival. Similarly, *willy-nilly* has survived so long that its original meaning has been lost: "whether you are willing or unwilling." Now *willy-nilly* implies haphazard or weak actions. The logical choice in its original meaning has faded. Now the cliche has more sound than substance.

1. Use cliches sparingly, if at all.

Writers would find eliminating **all** cliches a hard row to hoe. Reasonable cliches are fine in certain contexts, but the one in the preceding sentence is clearly inappropriate. The context of this discussion makes a cliche based on farm work ridiculous.

However, some cliches—in the right context—are valuable:

> Following the testimony, the judge had to sift through a thousand pages of unreadable and often contradictory testimony.

Here, the cliche of *sifting through* many pages is not objectionable. It is, in fact, a fine metaphor, given the circumstance to which it applies. In this case, the cliche does not clash with the context.

So one test of a cliche's acceptability is the degree to which it is relevant to the context in which it occurs and the extent to which it goes unnoticed. If the cliche does not call attention to itself, it is probably acceptable. The moment a writer (or reader) knows the expression is a cliche, it is unacceptable in most contexts.

NOTE: Sometimes a cliche can be used to advantage if its meaning or phrasing allows the writer to play against the cliche itself, as in the following quote from Oscar Wilde:

> "Truth is never pure, and rarely simple."

By rephrasing the cliche, Wilde asks readers to reexamine the cliche about *pure and simple truth.*

Some Common Cliches

English contains hundreds of cliches. The following list includes some of the more common ones currently in use:

A bad scene
a can of worms
acid test
active consideration
add insult to injury
agree to disagree
all things considered
all too soon
along these lines
among those present
ample opportunity
an end run
armed to the teeth
as a matter of fact
as the crow flies
at a loss for words
at the end of the day
attached hereto
at this point in time
auspicious moment
avoid it like the plague
awaiting further orders

Back at the ranch
back to the drawing board
bag and baggage
bated breath
beat a hasty retreat
be at loggerheads
beginning of the end
benefit of the doubt
best-laid plans
better late than never

Cliches

better left unsaid
beyond the shadow of a doubt
bite the bullet
bitter end
blissful ignorance
block out
bloody but unbowed
bolt from the blue
bone of contention
bottom line
brain dump
bright and shining faces
broad daylight
brook no delay
brute force
budding genius
built-in safeguards
burning question
burning the midnight oil
busy as a bee
by leaps and bounds
by the same token

Calm before the storm
capacity crowd
cast a pall
casual encounter
chain reaction
charged with emotion
checkered career/past
cherished belief
chief cook and bottle washer
circumstances beyond my control
city fathers
clean bill of health
clear as crystal/day
colorful display
come full circle
common/garden variety
confirming our conversation
conservative estimate
considered opinion
consigned to oblivion
conspicuous by its absence
contents noted
controlling factor
cool as a cucumber
crying need
curiously enough
cut a long story short
cut down in his prime

Dark horse
date with destiny
days are numbered
dazed condition
dead as a doornail
deadly earnest
deafening crash
deficits mount
deliberate falsehood
depths of despair
diamond in the rough
dig in your heels
discreet silence
do not hesitate to
doom is sealed
doomed to disappointment
dramatic new move
drastic action
drink the Kool-Aid
due consideration
dynamic personality

Each and every
easier said than done
eight-hundred-pound gorilla
eloquent silence
eminently successful
enclosed herewith
engage in conversation
enjoyable occasion
entertaining high hopes of
epic struggle
equal to the occasion
errand of mercy
even tenor
exception that proves the rule
existing conditions
express one's appreciation
eyeball to eyeball

Failed to dampen spirits
fair sex
fall on bad times
fall on deaf ears
far and wide
far be it from me
far cry
fateful day
fate worse than death
feedback loop
feel free to
feel vulnerable
festive occasion
few and far between
few well-chosen words
fickle finger of fate
final analysis
fine-tune one's plans
finishing touches
fire on all cylinders
fit as a fiddle
food for thought
fools rush in
foregone conclusion
foul play
from the sublime to the ridiculous

Gala occasion
generation gap
generous to a fault
gild the lily
give the green light to
glowing cheeks
go down the drain
goes without saying
goodly number
good team player
grateful acknowledgement
grave concern
green with envy
grim reaper
grind to a halt

Hale and hearty
hands across the sea
happy pair
hastily summoned
have the privilege
heartfelt thanks/appreciation
heart of the matter
heart's desire
heated argument
heave a sigh of relief
height of absurdity
herculean efforts
hook, line, and sinker
hook or crook
hope springs eternal
hot pursuit
house divided
how does that grab you?
hunker down
hurriedly retraced his steps

Ignominious retreat
ignorance is bliss
ill-fated
immaculately attired
immeasurably superior
impenetrable mystery
in close proximity
inextricably linked
infinite capacity
inflationary spiral
innocent bystander
in no uncertain terms
in reference/regard to
in short supply
internecine strife
in the limelight
in the nick of time
in the same boat with
in the twinkling of an eye
in this day and age
into full swing
iron out the difficulty
irony of fate
irreducible minimum
irreparable/irreplaceable loss
it dawned on me

Just desserts
just for openers

Keep options open
knock your socks off

Labor of love
lashed out at
last analysis
last but not least
last-ditch effort
leaps and bounds
leave no stone unturned
leaves much to be desired
leave up in the air
lend a helping hand
let well enough alone
like a bolt from the blue
limped into port
line of least resistance

little woman
lit up like a Christmas tree
lock, stock, and barrel
logic of events
long arm of the law
low-hanging fruit

Make good one's escape
man the barricades
marked contrast
masterpiece of understatement
matter of life and death
mecca for travelers
method to/in his madness
milk of human kindness
miraculous escape
moment of truth
momentous decision/occasion
monumental traffic jam
moot point
more in sorrow than in anger
more sinned against than sinning
more than meets the eye
more the merrier
motley crew

Narrow escape
nearest and dearest
needs no introduction
never a dull moment
never before in the history of
nipped in the bud
none the worse for wear
no sooner said than done
not wisely but too well

One and the same
ongoing dialogue
on more than one occasion
on unimpeachable authority
open kimono
order out of chaos
other things being equal
outer directed
overwhelming odds
own worst enemy

Pales into insignificance
paralyzed with fright
paramount importance
part and parcel
patience of Job
pay the piper
peer group
pet peeve
pick and choose
pie in the sky
pinpoint the cause
pipe dream
place in the sun
play hardball
play it by ear
point with pride
poor but honest
powder keg
powers that be
pretty kettle of fish
pros and cons
proud heritage
pull one's weight
push the envelope

Rack and ruin
ravishing beauty
red-letter day
regrettable incident
reigns supreme
reliable source
remedy the situation
right on
riot-torn area
ripe old age
round of applause
rude habitation

Sadder but wiser
saw the light of day
scathing sarcasm
sea of faces
seat of learning
second to none
seething mass of humanity
select few
selling like hotcakes
shattering effect
shift into high gear
shot in the arm
sigh of relief
silence broken only by
silhouetted against the sky
simple life
skeleton in the closet
snug as a bug in a rug
social amenities
something hitting the fan
spectacular event
spirited debate
steaming jungle
stick out like a sore thumb
stick to one's guns
straight and narrow path
structure one's day
such is life
sum and substance
superhuman effort
supreme sacrifice
sweat of his brow
sweeping changes
swim with the sharks

Take the bull by the horns
teaching moment
telling effect
tender mercies
terror stricken
thanking you in advance
there's the rub
think outside the box
this day and age
those present
throw a monkey wrench
throw a party
throw caution to the winds
thrust of your report
thunderous applause
tie that binds
time immemorial
time of one's life
tongue in cheek
too funny for words
too numerous to mention
tough it out/through
tower of strength
trials and tribulations
trust implicitly
tumultuous applause

Uncharted seas
unprecedented situation
untimely end
untiring efforts
up tight

Vale of tears
vanish into thin air
viable alternative

Watery grave
wax eloquent/poetic
weaker sex
wear and tear
wend one's way
whirlwind tour
wide open spaces
words fail to express
word to the wise
work one's wiles
worse for wear
wrought havoc

X-ray vision/view

Yea verily yea
yeasty blend/mix
yellow journalism
yen for . . .
Young Turk

Zero hour
zest for life

Colons

Colons signal readers to keep reading because related thoughts or a list will follow. In this role, colons differ from periods, semicolons, and even commas, all of which signal a pause or even a full stop.

> **Colons**
>
> 1. Colons link related thoughts, one of which must be capable of standing alone as a sentence.
> 2. Colons introduce lists or examples.
> 3. Colons separate hours from minutes, volumes from pages, and the first part of a ratio from the second.
> 4. Colons follow the salutation in a formal letter.
> 5. Colons separate titles from subtitles.

1. Colons link related thoughts, one of which must be capable of standing alone as a sentence.

Colons emphasize the second thought (unlike semicolons, which emphasize both thoughts equally, and dashes, which emphasize the break in the sentence and can emphasize the first thought).

Colons shift emphasis forward: They tend to make the second thought the most important part of the sentence. When such is the case, the colon indicates that explanation or elaboration follows:

> The Franklin Shipyard needed one thing to remain solvent: to win the Navy's supercarrier contract.
>
> The Franklin shipyard needed one thing to remain solvent: It had to win the Navy's supercarrier contract.

See Capitals.

Note: The two complete thoughts in the second example could also appear as two sentences:

> The Franklin Shipyard needed one thing to remain solvent. It had to win the Navy's supercarrier contract.

However, linking these thoughts with a colon emphasizes their close connection. Writing them as two sentences is less emphatic if the writer wishes to stress that the **one thing** Franklin needs is to win the contract.

2. Colons introduce lists or examples:

> Our management-development study revealed the need for greater monitoring during these crucial phases:
>
> 1. Initial organization
> 2. Design and development
> 3. Fabrication and quality control

> The Mars Division's audit of field service-personnel centers found the following general deficiencies:
>
> 1. Service personnel do not fully understand the new rebate policy.
> 2. Parts inventories are inadequate.
> 3. The centralized customer records are not operational, although the computer terminals have all been installed.

Note 1: A colon need not follow a heading or subheading that introduces a list. The heading itself is sufficient; a colon is redundant.

Note 2: The items listed do not require periods unless they are complete sentences. See Lists.

3. Colons separate hours from minutes, volumes from pages, and the first part of a ratio from the second:

> The deadline is 3:30 p.m. on Friday.
>
> See *Government Architecture* 15:233.
>
> The ratio of direct to indirect costs is 1:1.45.

4. Colons follow the salutation in a formal letter:

> Dear Ms. Labordean:
>
> Dear President Crouch:
>
> Dear Clarence Johns:

See Letters.

5. Colons separate titles from subtitles:

> *Government Architecture: Managing Interface Specifications*

Color helps readers and viewers acquire and interpret information. Appropriate use of contrasting or complementary colors clarifies the structure and emphasis of a visual message. For example, if main headings are black and subheadings are blue, readers can easily grasp the organization of a document.

Color originates when an object emits or reflects different wavelengths of light. Light (as in a beam of sunlight) is made up of the colors of the spectrum. See figure 1. Light shining through a prism bends to a different degree depending on its wavelength, thus revealing all the different colors in the spectrum.

Color

1. Establish a color scheme and then add color standards to the project styles.
2. Use contrasting, bright colors to show opposing concepts or major changes; use shades or tints of one color to show minor variations.
3. Match your color choices to your goal or purpose in designing a document or making a presentation.
4. Remember that color perception varies greatly among individuals.
5. For more legibility, use a light background with dark text, and use colors sparingly.
6. Combine colors and textures to improve legibility and understanding.

Figure 1. Colors of the Spectrum Displayed With a Prism. *As in this figure, the rainbow colors in a spectrum always appear in the same order.*

Colors are categorized as primary or additive. Additive color starts with the light of three primary colors: red, green, and blue. Mixing light of these colors in equal amounts makes white light (imagine three overlapping colored spotlights in a darkened room). See figure 2. Changing the mixture produces any color. For example, equal parts of red and green light make yellow light; red and blue make magenta; and green and blue make cyan.

Figure 2. Colors of the Spectrum Displayed by Mixing Primary Light Colors. *Visualize multiple filters overlapping each other to produce different hues.*

That is why color in a computer-generated graphic is designated by its RGB ratio (i.e., the ratio of red to green to blue). A graphic with an RGB ratio of 0-255-0 is pure green, where 0-255-255 is cyan. You can manipulate the RGB ratio with the color wheel or the custom color field in your software program.

1. Establish a color scheme and then add color standards to the project styles.

The styles for your document or presentation should include a color scheme, with text color(s), when or where they will be used, and background colors listed by topic, section, subject, or other logical grouping. For example, you might decide to use blue lettering for important rules and a light gray-shaded background for quoted passages. To predefine colors, you can choose the standard colors of your software program or select custom colors. Often you must use colors branded by your organization. The formula for these colors should be available as an RGB ratio. See PAGE LAYOUT for examples of styles. As you select colors, try to establish a color scheme that makes sense to potential readers. For example, if your readers

Color

Figure 3. The Color Wheel. *This schematic presentation shows which colors are complementary (opposite) of each other and which are only minor shades and tints.*

are from the United States, green suggests prosperity (as in money), yellow suggests caution (as in a traffic signal), and red shows failure (as in red ink) or danger.

Color associations are not universal, so avoid assuming that a certain color always has a particular meaning. For example, in several Eastern countries, the color white is associated with death and mourning; in the West, white is traditionally a sign of purity and innocence.

2. Use contrasting, bright colors to show opposing concepts or major changes; use shades or tints of one color to show minor variations.

The color wheel in your software program is a basic diagram showing the relationships of colors, hues, shades, and tints, as well as complementary and harmonious relationships. A shade is a darkened hue—the decrease of light or the addition of black. A tint (the opposite of a shade) is a lightened hue—the increase of light or the addition of white.

The color wheel in figure 3 shows the relationship between hues, tints, and shades.

Complementary colors are opposite one another on the color wheel. As the top row of figure 4 shows, yellow and violet are complementary colors. Other complementary pairs are blue and orange, or red and green. As complementary colors, yellow and violet contrast sharply, as shown in the middle circle on the top row. Choosing a tint of yellow or a shade of violet produces less contrast, as in the third box on that row.

Graphic artists often choose complementary colors to create

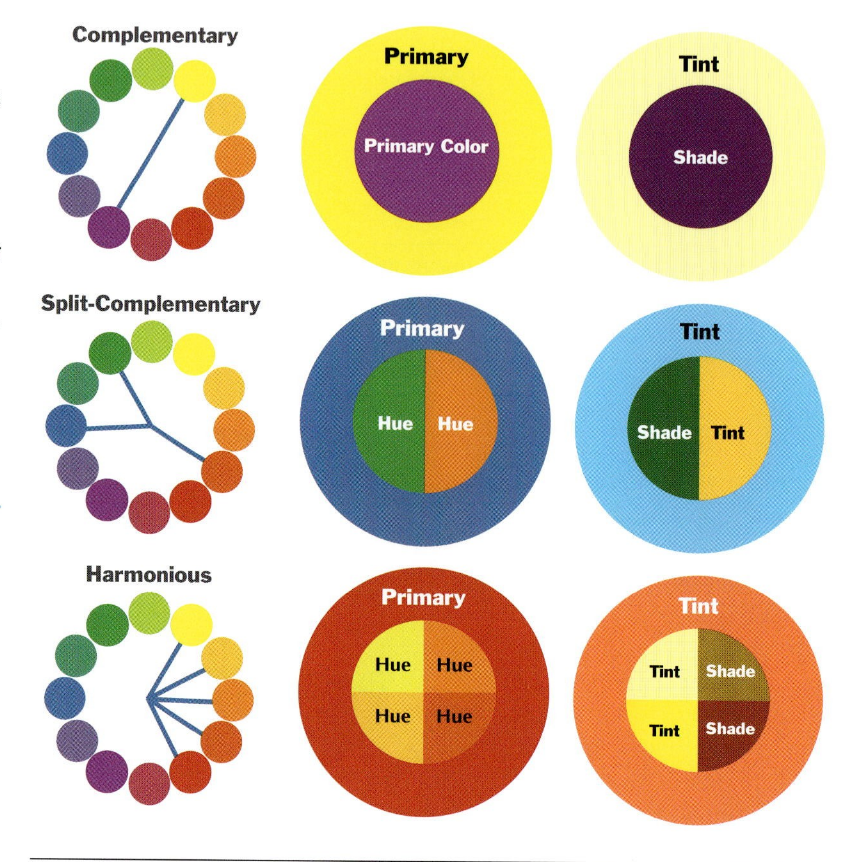

Figure 4. Complementary, Split-Complementary, and Harmonious Combinations. *The left-hand column shows how these three combinations are defined. The middle and right-hand columns demonstrate different combinations as they would appear in graphics.*

high contrasts. But this choice must depend on the purpose of each graphic. Some graphics—such as the advertising on a website—need to catch a buyer's eye immediately, so high contrast is valuable. In other contexts—for instance, a high-definition computer projection—complementary colors are often too bright, maybe even annoying.

Split-complementary colors are two colors positioned adjacent to a single color on the wheel. Their complement is opposite their adjacent color on the wheel. As the second row in figure 4 shows, red-orange is complementary to both blue and green. Split-complementary colors provide less contrast than complementary colors, but combinations of them are still bright, as the middle box on row 2 indicates. Using tints and shades decrease the contrast, as shown in the third box on row 2.

Harmonious colors lie between two primary colors on the color wheel. As figure 4 shows, the harmonious colors between red and yellow clearly relate to each other; thus they provide less contrast than complementary combinations. Tints and shades will further decrease the low contrast between harmonious colors. Graphic artists use harmonious colors when they want to convey related ideas within a graphic or a document.

3. Match your color choices to your goal or purpose in designing a document or making a presentation.

Which color combinations should you choose? No set answers exist. Assess the purpose of your graphic and your text. If you need to communicate highly contrasting ideas or create a strong impact, choose complementary or split-complementary colors, as explained in rule 2 above. Conversely, if your message needs to be more subtle, then choose tints, shades, or even harmonious combinations.

Figure 5 shows one use of high-contrast colors. On road signs, the goal is to use colors so vividly that no one can miss seeing the sign.

A more subtle use of color is desirable, however, in most business and technical documents and in graphics for business or technical presentations. Figure 6 shows how three business graphs would look with different color combinations. As in this figure, try printing sample graphics with different color combinations to judge what will be

Figure 5. Road Signs. *Road signs use bright, contrasting color combinations for high visibility.*

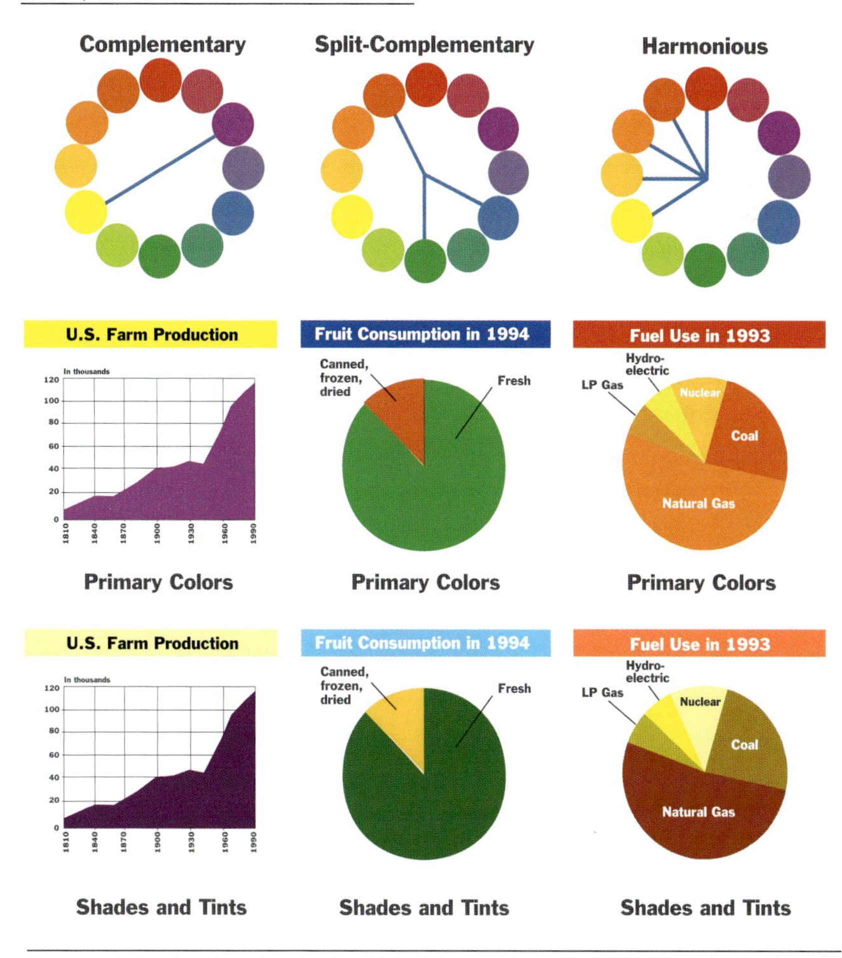

Figure 6. Color Combinations in Graphics. *Color combinations in the second row are bright and might be distracting. The bottom row shows more subdued combinations using tints and shades.*

effective. Also, remember that the colors you see on a computer screen may not print accurately. You must often adjust colors to the parameters of the printer.

If possible, test your color combinations by asking colleagues to review actual versions of your materials. Do they find the combinations to be appropriate, given the intended use of the materials? What changes would they suggest?

After some experimentation and after comments from colleagues, you will be ready to settle on a color scheme that works for a particular project. This is the color scheme you should include in your project style sheet (see rule 1 above).

4. Remember that color perception varies greatly among individuals.

Some combinations, such as orange-blue and red-green, appear to vibrate and disturb many readers. Although rare in women, red-green color blindness affects one in every 10 men. This combination is very difficult for color-blind men to interpret and should not be used. A red-blue combination does not provide enough contrast for a clear message. See GRAPHICS FOR PRESENTATIONS.

5. For more legibility, use a light background with dark text, and use colors sparingly.

Text in documents is usually printed in black ink on white paper because black ink generally costs less, so don't make understanding your message dependent on text color.

But color is useful to separate sections or emphasize important points. Using colored paper is an inexpensive way to add color. For example, the USDA Forest Service manual, which contains rules governing Forest Service management activities, has traditionally used different colored papers for national, regional, and local sections of the manual.

Different combinations of text color and background vary widely in readability. Figure 7 on the next page shows some examples of colored text and backgrounds.

Remember that any of these examples may be less legible if you are using a computer projection or a video screen. Test the circumstances. Stand halfway back in the audience. A person in the back receives about one-fourth the image quality you see halfway back. See GRAPHICS FOR PRESENTATIONS.

In computer presentations, the high contrast of a dark background and light letters creates impact, but also a darker, more sober emotional effect.

6. Combine colors and textures to improve legibility and understanding.

Colors and textures can make almost any graphic easier to understand. But you must choose carefully to ensure that the viewer interprets textural elements as you intend.

Use similar colors to bring together elements in groups. Black-and-white patterns are almost as effective as color in grouping elements.

Be cautious in using textures, gradients, and embossing. The purpose of such elements is to help readers interpret information, but overuse can detract from your message and confuse readers. Avoid faddish use of these elements, such as swirls, vectors, splatters, or smoky effects.

Textures such as crosshatching, dots, or other shapes or lines should be used only when color is not available to distinguish between elements of a visual. Neighboring areas should not be too similar in texture.

Also, when selecting colors and tones or textures for graphs, consider how the finished graphic will photocopy. You may want to use both color and texture to create a pleasing color graphic while ensuring improved legibility of a photocopy. See also CHARTS, GRAPHICS FOR DOCUMENTS, GRAPHICS FOR PRESENTATIONS, GRAPHS, ILLUSTRATIONS, MAPS, and PAGE LAYOUT.

Color	Not Recommended	Recommended
Dark Red	Blue, green, black	White, yellow, lt. yellow, lt. blue, lt. red, lt. green
Red	Brown, blue, green, orange	White, yellow, lt. yellow, lt. red, lt. orange, black
Medium Red	Brown, orange, yellow, red	White, black, dk. blue, brown, blue violet, dk. green
Light Red	White, yellow, lt. yellow, lt. blue	Black, dk. red, dk. blue, brown, blue violet, dk. green
Dark Green	Black, brown, med. brown, red, blue	White, yellow, lt. yellow, lt. red, lt. blue
Green	Blue, red, violet, blue violet, dk. green, med. brown	White, yellow, lt. yellow, lt. red, black
Light Green	lt. red, lt. blue, lt. yellow, med. brown	Black, white, brown, dk. blue, dk. green
Dark Blue	Red, green, orange	White, yellow, lt. yellow, lt blue
Blue	Red, green, orange, violet	White, yellow, lt. yellow, lt. red
Light Blue	Lt. red, lt. blue, lt yellow	Black, yellow, dk. blue, violet
Dark Yellow	Red, green, blue	Black, yellow, white, lt yellow
Light Yellow	White, lt. blue, lt. green, orange	Black, red, dk. red, blue, dk. blue, dk. violet, dk. green, brown
Dark Brown	Black, green, blue, dk. blue, red	White, yellow, lt. yellow, lt. red, lt. green, lt. blue
Medium Brown	Blue, green, red	White, black, yellow, lt. yellow,
Black	Red, blue, green, violet	White, yellow, lt. yellow, lt. blue, lt. red

Figure 7. Colored Text on Different Colored Backgrounds. *The recommended combinations (right column) still need to be verified in your particular context. Your printer or your paper may be just different enough to produce readability problems.*

Commas

Commas keep English sentences readable, especially long, involved sentences. Without commas, readers wouldn't know when to pause. But as the following rules show, correct placement of commas reflects the grammar and syntax of the language, not merely places to pause. See PUNCTUATION for information on mandatory and optional uses of commas.

Commas

1. Commas separate complete thoughts joined by these simple conjunctions: *and, but, or, for, nor, so, yet.*
2. Commas separate items in a series consisting of three or more words, phrases, or even whole clauses.
3. Commas separate long introductory phrases and clauses from the main body of a sentence.
4. Commas enclose parenthetical expressions.
5. Commas separate nonessential modifying and descriptive phrases and clauses from a sentence, especially those clauses beginning with *who, which,* or *that.*
6. Commas separate two or more adjectives that equally modify the same noun.
7. Commas separate items in dates and addresses.
8. Commas separate titles and degrees from names.
9. Commas follow the salutation in informal letters and the complimentary closing in all letters.
10. Commas enclose in text the names of people addressed.
11. Commas (or a comma and a semicolon) set off (enclose) the following transitional words and expressions when they introduce sentences or when they link two complete thoughts: *accordingly, consequently, for example, for instance, further, furthermore, however, indeed, nevertheless, nonetheless, on the contrary, on the other hand, then, thus.*
12. Commas, like periods, always go inside closing quotation marks. Commas go outside parentheses or brackets.

1. Commas separate complete thoughts joined by these simple conjunctions: *and, but, or, for, nor, so, yet:*

> He was a Russian linguist in communications intelligence, and he has logged over 5,000 hours as a C-130 navigator in the Air Force.
>
> Visophane has been marketed abroad since 2005, but it was not approved for the local market until May of last year because of insufficient clinical trials.

NOTE 1: You may omit this comma if both complete thoughts are short:

> The chairman resigned and the company failed.

The simple conjunctions cited above are called coordinating conjunctions. When they link two complete thoughts, the resulting sentence is called a compound sentence. See CONJUNCTIONS, SENTENCES, and BRITISH ENGLISH.

NOTE 2: If you use any other transitional or connecting word (*however, furthermore, consequently,* and so on) to join two complete thoughts, use a semicolon. See SEMICOLONS and TRANSITIONS.

2. Commas separate items in a series consisting of three or more words, phrases, or even whole clauses:

> Control Data's Integrated Support Software System provides compatibility between tools and workers, consistent tool interfaces, ease of learning, user friendliness, and expandability.
>
> The user may also return to the control program to perform such other functions as database editing, special report generation, and statistical analyses.
>
> The Carthage-Hines agreement contained provisions for testing the database, cataloguing the findings, creating a more user-friendly software package, and marketing any new software developed jointly.

NOTE 1: A comma separates the last two items in a series, even though these items are linked by a conjunction (*and* in the above examples, but the rule applies for any conjunction). This comma was once considered optional, but the trend is to make it mandatory, especially in technical and business English. Leaving it out can cause confusion and misinterpretation. See PUNCTUATION and BRITISH ENGLISH.

Note 2: If all of the items in the series are linked by a simple conjunction, do not use commas:

> The user may also return to the control program to perform such other functions as database editing and special report generation and statistical analyses.

Note 3: In sentences containing a series of phrases or clauses that already have commas, use semicolons to separate each phrase or clause:

> Our legal staff prepared analyses of the Drury-Engels agreement, which we hoped to discontinue; the Hopkinson contract; and the joint leasing proposal from Shell, Mobil, and Amoco.

See Conjunctions and Semicolons.

3. Commas separate long introductory phrases and clauses from the main body of a sentence:

> Although we are new to particle scan technology, our work with split-beam lasers gives us a solid experiential base from which to undertake this study.

> For the purposes of this investigation, the weapon will be synthesized by a computer program called RATS (Rapid Approach to Transfer Systems).

> Oil production was down during the first quarter, but when we analyzed the figures, we discovered that the production decline was due to only two of our eight wells.

Note 1: In the last example, the *when we analyzed* clause does not open the sentence, but it must still be separated from the main clause following it. It introduces the main thought of the last half of the sentence.

Note 2: If the introductory thought is short and no confusion will result, you can omit this comma:

> In either case the Carmichael procedure will be used to estimate the current requirements of the preliminary designs.

4. Commas enclose parenthetical expressions.

Parenthetical expressions are words or groups of words that are inserted into a sentence and are not part of the main thought of the sentence. These expressions describe, explain, or comment on something in the sentence, typically the word or phrase preceding the parenthetical expression:

> The transport will, according to our calculations, require only 10,000 feet of runway.

> The survey results, though not what we had predicted, confirm that the rate of manufacturer acceptance will exceed 60 percent.

Parentheses and dashes may also enclose parenthetical expressions. Use commas most of the time, but when you want to make the expression stand out, enclose it with parentheses (which are more emphatic than commas) or dashes—which are more emphatic than parentheses. See Parentheses and Dashes.

5. Commas separate nonessential modifying and descriptive phrases and clauses from a sentence, especially those clauses beginning with *who, which,* or *that:*

> These biocybernetic approaches, which merit further investigation, will improve performance of the man/machine interface.

In this sentence, *which merit further investigation* is not essential because the reader will already know which biocybernetic approaches the sentence refers to. The clause beginning with *which* is nonessential and could be left out:

> These biocybernetic approaches will improve performance of the man/machine interface.

If several biocybernetic approaches were listed, however, and if the writer needed to identify only those meriting further investigation, the clause would be essential, could not be left out, and would **not** take commas:

> Improving the performance of the man/machine interface meant identifying those biocybernetic approaches that merit further investigation.

The *that* in the preceding example commonly introduces essential clauses, although *which* sometimes appears. See *that/which* in Word Problems.

Modifying or descriptive clauses should always follow the words they modify. If they cannot be removed from the sentence without changing the meaning, they are essential and must not be separated by commas from the word they modify. If they can be removed, they are nonessential and must be separated by commas from the main thought in the sentence:

> *Essential:* She is the Dr. Gruber who developed analytical engine compressor stability models for NASA.

She is the Dr. Gruber does not make sense as an independent statement. The descriptive clause beginning with *who* is essential and therefore cannot be separated by a comma from *Gruber*.

> *Nonessential:* Our Design Team Leader will be Dr. Janet Gruber, who developed analytical engine compressor stability models for NASA.

Our Design Team Leader will be Dr. Janet Gruber does stand alone as a complete and independent thought. In this case, the descriptive clause beginning with *who* is nonessential. Separating it from *Gruber* with a comma shows that it is additional and nonessential information. Note that a comma would follow *NASA* if the sentence continued. See Pronouns for a discussion of relative pronouns.

Commas

6. Commas separate two or more adjectives that equally modify the same noun:

This design features an advanced, multidose oral therapy.

NOTE: If two or more adjectives precede a noun, however, and one adjective modifies another adjective—and **together** they modify the noun—you must use a hyphen:

They had designed a no-flow heat exchange.

A good test for determining whether two or more adjectives equally modify a noun is to insert *and* between them. If the resulting phrase makes sense, then the adjectives are equal, and you should use commas to replace the *ands*:

old and rusty pipe (*therefore,* old, rusty pipe)

however

old and rusty and steam pipe *(The* and *between* rusty *and* steam *makes no sense. Therefore, the phrase should be* old, rusty steam pipe.)

See HYPHENS and ADJECTIVES.

7. Commas separate items in dates and addresses:

The proposal was signed on March 15, 2007.

Contact Benson Pharmaceuticals, Lindsay, Indiana, for further information.

NOTE: A comma follows the day and the year when the month **and day** precede the year. However, when the date consists only of month and year, a comma is not necessary:

The final report will be due January 14, 2011, just a month before the board meeting.

but

The final report will be due in January 2011.

When the date appears in the day-month-year sequence, no commas are necessary:

The report is due 14 January 2011.

See PUNCTUATION.

8. Commas separate titles and degrees from names:

The chief liaison will be Roger Hillyard, Project Review Board Chairman.

Mary Sarkalion, PhD, will coordinate clinical studies.

Clinical studies will be the responsibility of Mary Sarkalion, PhD.

NOTE: When the degree or title appears in the middle of a sentence, commas must appear before and after it.

9. Commas follow the salutation in informal letters and the complimentary closing in all letters:

Dear Joan,

Sincerely,

See COLONS and LETTERS.

10. Commas enclose in text the names of people addressed:

So, Bob, if you'll check your records, we'll be able to adjust the purchase order to your satisfaction.

11. Commas (or a comma and a semicolon) set off (enclose) the following transitional words and expressions when they introduce sentences or when they link two complete thoughts: *accordingly, consequently, for example, for instance, further, furthermore, however, indeed, nevertheless, nonetheless, on the contrary, on the other hand, then, thus:*

Consequently, the primary difference between CDSP and other synthesis programs is development philosophy.

Synthesis programs are now common in industry; however, CDSP has several features that make it especially suitable for this type of study.

or

Synthesis programs are now common in industry; CDSP has, however, several features that make it especially suitable for this type of study.

See SEMICOLONS.

NOTE: A few of these transitional words *(however, thus, then, indeed)* are occasionally part of the main thought of the sentence and do not form an actual transition. When such is the case, omit the punctuation before and after the words:

However unreliable cross-section analysis may be, it is still the most efficient means of scaling mathematical models.

Thus translated, the decoded message can be used to diagram nonlinear relationships.

12. Commas, like periods, always go inside closing quotation marks. Commas go outside parentheses or brackets:

The specifications contained many instances of the phrase "or equal," which is an attempt to avoid actually specifying significant features of a required product.

Thanks to this new NSAID (non-steroidal anti-inflammatory drug), posttraumatic or postoperative conditions were significantly reduced.

NOTE: British usage places commas and periods inside or outside the quotation marks, depending on whether they are or are not part of the quotation. See SPACING, BRITISH ENGLISH, and QUOTATION MARKS.

Compound Words

1. Write compounds as two words when the compounds appear with the words in their customary order and when the meaning is clear.
2. Write compounds as single words (no spaces between joined words) when the first word of the compound receives the major stress in pronunciation.
3. Hyphenate compounds that modify or describe other words.
4. Treat compounds used as verbs as separate words.

Compound words are formed when two or more words act together. The compound may be written as a single word (with no space between the joined words), with a hyphen between the joined words, or with spaces between the joined words:

footnote
ourselves
right-of-way
3-minute break
delayed-reaction switch
land bank loan
parcel post delivery

The form of the compound varies with custom and usage as well as with the length of time the compound has existed.

Compound words usually begin as two or more separate, often unrelated words. When writers and speakers begin using the words together as nouns, verbs, adjectives, or adverbs, the compound generally has a hyphen or a space between words, depending on custom and usage. As the new compound becomes more common, the hyphen and space might drop, and the compound might be written as one word:

on-site *has become* onsite
co-operate *has become* cooperate
rail road *has become* railroad
auto body *has become* autobody

However, because of custom or usage, some compounds retain the hyphen or space between words:

all-inclusive
deep-rooted
living room
middle-sized
re-cover *(to cover again)*
re-create *(to create again)*
rough-coat *(used as a verb)*
sand-cast *(used as a verb)*
satin-lined
steam-driven
sugar water
summer school
terra firma
throw line
under secretary

Because new compound words are continually appearing in the language and because even familiar compounds might appear in different forms, depending on how they are used in a sentence, writers might have difficulty deciding which form of a compound to use. Recent dictionaries can often help by indicating how a word or compound has appeared previously.

However, for new compounds and for compounds not covered in dictionaries, use the principles of clarity and consistency, as well as the following guidelines, to select the form of the compound.

1. Write compounds as two words when the compounds appear with the words in their customary order and when the meaning is clear:

test case	report card
sick leave	barn door
flood control	social security
real estate	civil rights

Note 1: Many such combinations are so common that we rarely think of them as compounds (especially because they do not have hyphens and are written with spaces between words). In many cases, writing them as a single word would be ridiculous: *floodcontrol, realestate.*

Note 2: We continue to pronounce such compounds with fairly equal stress on the joined words, especially when one or more of the words has two or more syllables (as in *social security*).

2. Write compounds as single words (no spaces between joined words) when the first word of the compound receives the major stress in pronunciation:

airplane
cupboard
doorstop
dragonfly
footnote
nightclerk
seaward
warehouse

Note 1: The stress often shifts to the first word when that word has only one syllable, as in the preceding examples.

Note 2: Words beginning with the following prefixes are not true compounds. Such words are usually written without a space or a hyphen:

*after*birth
*Anglo*mania
*ante*date
*bi*weekly
*by*law
*circum*navigation
*co*operate
*contra*position
*counter*case
*de*energize
*demi*tasse
*ex*communicate
*extra*curricular
*fore*tell
*hyper*sensitive
*hypo*acid
*in*bound
*infra*red
*inter*view
*intra*spinal
*intro*vert
*iso*metric

Compound Words

*macro*analysis
*meso*thorax
*meta*genesis
*micro*phone
*mis*spelling
*mono*gram
*multi*color
*neo*phyte
*non*neutral
*off*set
*out*back
*over*active
*over*flow
*pan*cosmic
*para*centric
*parti*coated
*peri*patetic
*plano*convex
*poly*nodal
*post*script
*pre*exist
*pro*consul
*pseudo*scientific
*re*enact
*retro*spect
*semi*official
*step*father
*sub*secretary
*super*market
*thermo*couple
*trans*onic
*trans*ship
*tri*color
*ultra*violet
*un*necessary
*under*flow

NOTE 3: Words ending with the following suffixes are not true compounds. Such words are usually written without a space or hyphen:

port*able*
cover*age*
oper*ate*
plebis*cite*
twenty*fold*
spoon*ful*
kilo*gram*
geo*graphy*
man*hood*
self*ish*
meat*less*
out*let*
wave*like*
procure*ment*
partner*ship*
lone*some*
home*stead*
north*ward*
clock*wise*

3. Hyphenate compounds that modify or describe other words:

rear-engine bracket
tool-and-die shop
two-phase engine-replacement program
down-to-cost model
two- or three-cycle process
4-year plan
20-day turn around
2- or 3-week vacation

See HYPHENS and ADJECTIVES.

NOTE 1: Such compounds are hyphenated only when they come before the word they modify. If the words forming the compound appear after the word they are describing, leave out the hyphens:

bracket for the rear engine (*but* rear-engine bracket)

a shop making tools and dies (*but* tool-and-die shop)

a program with two phases (*but* two-phase program)

NOTE 2: When the meaning is clear, such compound modifiers may not need hyphens:

sick leave policy
land management plan
life insurance company
per capita cost
production credit clause
speech improvement class

NOTE 3: Do not hyphenate if the first word of the compound modifier is an adverb ending with –*ly*:

barely known problem
eminently qualified researcher
highly developed tests
gently sloping range

however

well-developed tests
well-known problem
well-qualified researcher

4. Treat compounds used as verbs as separate words:

to break down
to check out
to follow up
to get together
to go ahead
to know how
to run through
to shut down
to shut off
to stand by
to start up
to take off
to trade in

The parallel compound nouns are usually either written as one word or hyphenated:

breakdown
checkout
follow-up
get-together
go-ahead
know-how
run-through
shutdown
shutoff
standby
start-up
takeoff
trade-in

However, some verb phrases are identical to the compound noun form:

cross-reference *(both a noun and a verb)*

When in doubt, check your dictionary.

Conjunctions connect words, phrases, or clauses and at the same time indicate the relationship between them. Conjunctions include the simple coordinating conjunctions *(and, but, or, for, nor, so, yet)*, the subordinate conjunctions *(because, since, although, when, if, so that,* etc.), the correlative conjunctions *(either ... or, neither ... nor, both ... and)*, and the conjunctive adverbs (*however, thus, furthermore*, etc.).

Conjunctions

1. Ensure that in choosing *and* and *or* you select the conjunction that conveys exactly what you mean.
2. Occasionally, sentences can begin with a coordinating conjunction.
3. Do not use *and* or *but* before *which* (or *that, who, whose, whom, where*) unless you use a preceding parallel *which* (or *that, who, whose, whom, where*).
4. Subordinate conjunctions can begin sentences.
5. Distinguish between some subordinate conjunctions that have overlapping or multiple meanings (especially *because/since/as* and *while/although/as*).
6. Make the constructions following each coordinating conjunction parallel.
7. Use a semicolon before and a comma after conjunctive adverbs used to join two complete thoughts.
8. Use a comma following conjunctive adverbs at the beginning of a sentence.

Coordinating Conjunctions

The simple coordinating conjunctions are *and, but, or, for, nor, so,* and *yet*. They often connect two independent clauses (complete thoughts):

> The program designer established the default settings, and the programmer built them into the system.
>
> Our proposal was a day late, but we were not eliminated from competition.
>
> The pump will have to be replaced, or we will continue to suffer daily breakdowns.
>
> We rejected his budget, yet he continued to argue that all contested items were justified.

See SENTENCES.

These simple connectors establish the relationship between the thoughts being coordinated:

—*And* shows addition
—*Or* shows alternative
—*Nor* shows negative alternative
—*But* and *yet* show contrast
—*For* and *so* show causality

NOTE 1: When you use a coordinating conjunction to connect two independent clauses or complete thoughts, place a comma before the conjunction, as in the sentences above. However, you may omit the comma when the two clauses are short and closely related. Also, a semicolon can replace both the comma and the conjunction. See SEMICOLONS and COMMAS.

NOTE 2: The conjunctions *and* and *or* (preceded by a comma) also connect the last two items in a series:

> The engineer designed an emergency exit door, a narrow outside stairway, and a concrete support pad.
>
> She requested full written disclosure, an apology, or financial compensation.

See COMMAS.

1. Ensure that in choosing *and* and *or* you select the conjunction that conveys exactly what you mean.

At first glance, *and* and *or* merely join two or more items, but they can and often do imply much more.

And

In the following sentences *and* does more than merely connect the ideas. What *and* implies is stated in parentheses following each example:

> He saw the accident, and he called the police. *(therefore)*
>
> My boss is competent, and David is not. *(contrast)*
>
> He changed the tire, and he replaced the hub cap. *(then)*
>
> Explain the cost savings, and I'll approve your proposal. *(condition)*

Or

The conjunction *or* usually means one of two possibilities:

> *I want either a Ford or an Acura.*

However, *or* sometimes has other, occasionally confusing, implications:

> The faulty part or the worm gear seemed to be causing our problem. *(Are the faulty part* and *the worm gear* the same? *Only knowledgeable readers would know for sure.)*
>
> Add to the bid, or I'll reject your offer. *(negative condition)*
>
> He began doing the schematics, or at least he appeared to be doing them. *(correction)*

See AND/OR in WORD PROBLEMS.

Conjunctions

2. Occasionally, sentences can begin with a coordinating conjunction.

This advice contradicts the rule that many of us learned in school: "Never begin a sentence with *and*." Some writers and editors still offer this advice, but most have now recognized that this so-called rule has no basis. Even Shakespeare began some of his sentences with coordinating conjunctions.

A coordinating conjunction at the beginning of a sentence links the sentence to the preceding sentence or paragraph. Sometimes, the linking is unnecessary:

> We objected to the proposal because of its length. And others felt that it had errors in its facts.

The *and* at the beginning of the second sentence is simply unnecessary. It adds nothing to the thought and may easily be omitted:

> We objected to the proposal because of its length. Others felt that it had errors in facts.

Using a conjunction to begin a sentence is not grammatically incorrect. Sometimes, it is good stylistic variation. But it tends to look and sound informal, so avoid this practice in formal documents.

3. Do not use *and* or *but* before *which* (or *that, who, whose, whom, where*) unless you use a preceding parallel *which* (or *that, who, whose, whom, where*):

> We explored the DeMarcus itinerary, which you explained in your letter but which you failed to mention in Saturday's meeting.
>
> The meetings should take place where we met last year or where we can arrange for equally good facilities.

The following sentence violates this principle. Consequently, it is awkward and nonparallel:

> The plans called for a number of innovative features, especially regarding extra insulation, and which should save us much in fuel costs. *(Deleting the* and *would solve the lack of parallelism in this sentence.)*

See PARALLELISM.

Subordinate Conjunctions

In contrast to the limited set of coordinating conjunctions, subordinate conjunctions are a varied and diverse group:

> after, although, as, because, before, if, once, since, that, though, until, when, where, while
>
> in that, so that, such that, except that, in order that, now (that), provided (that), supposing (that), considering (that), as far as, as long as, so long as, sooner than, rather than, as if, as though, in case
>
> if . . . (then)
> although . . . yet/nevertheless
> as . . . so
> more/–er/less . . . than
> as . . . as
> so . . . (that)
> such . . . as
> such . . . (that)
> no sooner . . . than
> whether . . . or (not)
> the . . . the

Subordinate conjunctions introduce subordinate clauses and phrases (dependent clauses and phrases that do not convey complete thoughts and are therefore not independent):

> After the engineer gave her talk
>
> Because of the voltage loss
>
> When the test results come in
>
> While still producing fluids
>
> In that you had already made the request
>
> Except that the procedure was costly
>
> Provided that you calculate the results
>
> As though it hadn't rained enough
>
> If we fail
>
> As aware as he is
>
> So expensive that it was prohibitive
>
> Whether or not you submit the report

These subordinate clauses and phrases must be attached to independent clauses (complete thoughts) to form sentences:

> After the engineer gave her talk, several colleagues had questions.
>
> In that you had already made the request, we decided to omit the formal interview.
>
> If we fail, the project stops. (*or* If we fail, then the project stops.)
>
> As aware as he is, he must be sensitive to the personnel problems.

See SENTENCES.

NOTE 1: A subordinate clause or phrase that opens a sentence should be followed by a comma. The preceding sentences illustrate this rule. See COMMAS.

NOTE 2: When the subordinate clause or phrase follows the independent clause or main thought of the sentence, no commas are necessary:

> The experiment failed because of the voltage loss.
>
> We would have denied the request except that the procedure was so costly.
>
> We wondered whether you would turn in your report.

NOTE 3: Occasionally, the subordinate clause or phrase interrupts the main clause and must have commas on both sides of it to indicate where the clause or phrase appears:

> The President and the Joint Chiefs of Staff, after receiving the latest aerial reconnaissance photos of the area, decided on a naval blockade of all ports.
>
> Our budgetary problems, regardless of the Madiera Project expense, would have taken care of themselves if the prime rate hadn't gone up three points.

4. Subordinate conjunctions can begin sentences:

When the test results come in, we'll have to analyze them carefully.

Because the project manager was unfamiliar with the budget codes, we failed to expense the costs of fabrication.

NOTE: The old-school rule "Never begin a sentence with *because*" was and remains a bad rule. You may begin a sentence with *because* as long as the dependent clause it introduces is followed by an independent clause or complete thought.

5. Distinguish between some subordinate conjunctions that have overlapping or multiple meanings (especially *because/since/as* and *while/although/as*).

Avoid using *since* and *as* to mean "because":

Because the Leiper Project failed, several engineers were reassigned to electro-optics. (*not* Since the project failed . . .)

Because we had ample supplies, no new batteries were ordered. (*not* As we had ample supplies . . .)

Avoid using *while* and *as* to mean "although":

Although many employees begin work at 8 a.m., others begin at 7 a.m. (*not* While many employees begin work at 8 a.m. . . .)

Although the value of the test results declined, we still felt we could meet the deadline. (*not* As the value of the test results declined . . .)

Correlative Conjunctions

Correlative conjunctions are pairs of coordinating conjunctions:

both . . . and
either . . . or
neither . . . nor
not only . . . but also

6. Make the constructions following each coordinating conjunction parallel:

The committee was interested in both real estate holdings and stock investments. (*not* . . . both in real estate holdings and the stock investments.)

The investigation revealed that either the budget was inaccurate or our records had gaps. (*not* The investigation revealed either that the budget was inaccurate or our records had gaps.)

NOTE: Faulty parallelism problems occur when the same phrase structure or word patterns do not occur after each coordinating conjunction:

He was aware that not only was the pipe too small but also that the pipe supports were made of aluminum instead of stainless steel.

This sentence is confusing because the two *thats* are not parallel. The first *that* comes before *not only,* and the second *that* comes after *but also.* A parallel version of the sentence is much smoother:

He was aware not only that the pipe was too small but also that the pipe supports were made of aluminum instead of stainless steel.

See PARALLELISM.

Conjunctive Adverbs

Conjunctive adverbs are adverbs that function as conjunctions, typically by connecting independent clauses or complete thoughts. Usually, a semicolon appears along with the conjunctive adverb. The most common conjunctive adverbs are *accordingly, also, besides, consequently, further, furthermore, hence, however, moreover, nevertheless, otherwise, then, therefore, thus,* and *too.* See TRANSITIONS.

NOTE: Conjunctive adverbs and the accompanying semicolons lengthen sentences and convey a heavy, formal tone. If possible, replace conjunctive adverbs with *and, but, or, for, nor, so,* and *yet.*

7. Use a semicolon before and a comma after conjunctive adverbs used to join two complete thoughts:

Motherboard assembly is a lengthy production process; however, the individual assembly steps must still be tightly controlled.

Increasing pressure in the T-valves is potentially dangerous; nevertheless, we will not be able to monitor effluent discharge without increasing the pressure.

See SEMICOLONS and COMMAS.

NOTE: You can omit the comma following the conjunctive adverb if the sentence is short:

I think; therefore I am.

8. Use a comma following conjunctive adverbs at the beginning of a sentence:

Therefore, I am recommending that Pharmaco reconsider the baseline scores for the principal efficacy parameters.

However, sulfur compounds might not be the answer either.

NOTE 1: You may omit this comma if the sentence is short:

Thus the plan failed.

NOTE 2: If the adverb appears at the beginning of the sentence but does not behave as a conjunction, it is part of the sentence and cannot be followed by a comma:

Then the seam split at the forward discharge valve, and the boiler lost pressure rapidly.

Regardless of how we examined the problem, we could not resolve the fundamental dispute between the software designers and the copyright holders.

See COMMAS.

Contractions

Contractions are words formed by joining two words and dropping letters. An apostrophe marks the dropped letters:

cannot
can't

we are
we're

there is
there's

> **Contractions**
>
> 1. Use contractions to establish a personal, informal tone.
> 2. Don't confuse contractions with possessive pronouns.

1. Use contractions to establish a personal, informal tone.

Contractions are not appropriate for very formal or ceremonial documents such as contracts or legal notices. However, contractions lead to a conversational, friendly tone in most other business correspondence. Contractions are common in electronic mail (email) messages because they cut the amount of screen space required.

We're excited that you'll be joining our sales force!

It's been a long time since you've come to see us.

See Letters, Tone, and contractions in Word Problems.

2. Don't confuse contractions with possessive pronouns.

Writers often confuse contractions, which require apostrophes, with possessive pronouns, which need no apostrophes:

It's (it is) a regulatory issue.

The company lost *its* lease.

There's (there is) no time to waste.

The contract is *theirs* if they want it.

Watch out for these pronouns commonly mistaken for contractions:

hers, *not* her's
yours, *not* your's
ours, *not* our's
theirs, *not* their's

Distinguish between:

it's (it is)
its (belonging to it)

you're (you are)
your (belonging to you)

they're (they are)
their (belonging to them)

See Apostrophes and Pronouns.

Dashes are excellent devices for emphasizing key material and for setting off explanatory information in a sentence. They can also be used to indicate where each item in a list begins and to separate paragraph headings from succeeding text. See HEADINGS, LISTS, and PUNCTUATION.

Dashes primarily appear as an em dash—meaning that the dash is about as wide as the letter "m." Dashes also appear as an en dash, which is as wide as a letter "n." The en dash has only a few uses:

1959–1960
Appendix D–2
pages 120–122

Most word-processing software programs have a special code for dashes so that dashes appear as a solid line, not two separate hyphens. Using this code makes your text appear to be typeset, not typed on an old-fashioned typewriter. When you use a dash between two words, leave no space on either side of the dash. See SPACING.

NOTE: Traditionally, hyphens are even shorter than en dashes, but many software programs have the same code for hyphens and en dashes. See HYPHENS.

Dashes

1. Dashes link introductory or concluding thoughts to the rest of the sentence.
2. Dashes interrupt a sentence for insertion of thoughts related to, but not part of, the main idea of the sentence.
3. Dashes emphasize explanatory information enclosed in a sentence.
4. Dashes link particulars to a following summary statement.

1. Dashes link introductory or concluding thoughts to the rest of the sentence.

Dashes linking thoughts emphasize the break in the sentence. Dashes often make the first thought the most important part of the sentence:

Winning the Navy's supercarrier contract—that's what the Franklin Shipyard needed to remain solvent.

Dashes can act like colons, however, and throw emphasis to the last part of the sentence:

We subjected the design to rigorous testing—but to no avail because stress, we discovered, was not the problem.

Often, the information following the dash clarifies, explains, or reinforces what came before the dash:

We consider our plan bold and unusual—bold because no one has tried to approach the problem from this angle, unusual because it's not how one might expect to use laser technology.

Dashes can also link otherwise complete sentences:

The technical problem was **not** the design of the filter—the problem was poor quality assurance.

2. Dashes interrupt a sentence for insertion of thoughts related to, but not part of, the main idea of the sentence.

Octoronase had been undergoing clinical tests—all these were done abroad—for 3 years before the patients were withdrawn from the trial.

In this example, parentheses could replace the dashes; with parentheses, the sentence becomes slightly less emphatic. See PARENTHESES.

3. Dashes emphasize explanatory information enclosed in a sentence:

Two of Barnett's primary field divisions—Industrial Manufacturing and Product Field Testing—will supervise the construction and implementation of the prototype.

In this example, commas or parentheses could replace the dashes. The commas would not be as emphatic as dashes; the parentheses would be more emphatic than commas, but less emphatic than dashes. See PARENTHESES and COMMAS.

4. Dashes link particulars to a following summary statement:

Reliability and trust—this is what Bendix has to offer.

Developing products that become the industry standard, minimizing the risk of failure, and controlling costs through aggressive management—these have become the hallmarks of our reputation.

Decimals

Decimal numbers are a linear way to represent fractions based on multiples of 10. The decimal *0.45* represents the following fraction:

45/100

See FRACTIONS.

The decimal point (period) is the mark dividing the whole number on the left from the decimal fraction on the right:

504.678

In some countries, writers use a comma for the decimal point:

504,678

Decimals

1. Use figures for all decimals and do not write the equivalent fractions.
2. If the decimal does not have a whole number, insert a zero before the decimal point.
3. Retain the zero after the decimal point or at the end of the decimal number only if the zero represents exact measurement (or a significant digit).
4. Use spaces but not commas to separate groups of three digits in the decimal fraction.
5. In columns, line up the decimal points.
6. Do not begin a sentence with a decimal number.

1. Use figures for all decimals and do not write the equivalent fractions:

4.5 (*not* 4 5/10)

0.356 (*not* 356/1000)

0.5 (*not* 5/10)

0.4690 (*not* 4690/10,000)

2. If the decimal does not have a whole number, insert a zero before the decimal point:

0.578 (*not* .578)
0.2 (*not* .2)

NOTE: This rule has a few exceptions, including:

Colt .45
A batting average of .345
A probability of p =.07

3. Retain the zero after the decimal point or at the end of the decimal number only if the zero represents exact measurement (or a significant digit):

0.45 *or* 0.450
28.303 *or* 28.3030

NOTE: Also retain the final zero in a decimal if the zero results from the rounding of the decimal:

23.180 *for* 23.1789 *(if the decimal number is supposed to be rounded to three digits in the decimal fraction)*

4. Use spaces but not commas to separate groups of three digits in the decimal fraction.

In the metric system, the decimals may be broken into groups of three digits by inserting spaces:

56.321 677 90
707.004 766 321

but 567.4572 (*not* 567.457 2)

You can use commas to separate groups of three digits that appear in the whole number part of the decimal:

56,894.65
500,067.453 467

However, do not use commas to separate groups of three digits in the decimal fraction:

4.672 34 (*not* 4.672,34)

2344.000 567 (*not* 2344.000,567)

See METRIC SYSTEM.

5. In columns, line up the decimal points:

```
 56
  0.004
115.9
 56.24445
  0.6
```

NOTE: Whole numbers without decimals (e.g., *56* above) do not require a decimal point.

6. Do not begin a sentence with a decimal number:

this

The timer interrupts the processor 14.73 times a second.

not this

14.73 times a second, the timer interrupts the processor.

See NUMBERS.

Anyone who works with documents must have a system for indicating changes to text. Much editing and proofreading is now done on a computer screen, and your software program has those capabilities. But because screen resolution is typically one third or so less than print, it's wise to proofread printed text to catch errors difficult to see on a screen. The following rules apply mostly to proofreading hard copies of documents. You will also find suggestions for proofreading onscreen.

Standard editing and proofreading symbols (listed in most dictionaries) are more numerous and complex than most of us need unless we are copy editors, typesetters, or printers.

A simplified set of editing and proofreading symbols listed in rule 1 addresses the needs of most business and technical writers who must communicate suggestions and editorial corrections to others. If you need the complete set of proofreading symbols, see the most recent editions of *The Chicago Manual of Style* or the *United States Government Printing Office Style Manual.*

The example under rule 1 illustrates the simplified method of editing and proofreading printed text. This example also follows the rules cited below.

Editing and Proofreading

1. Use consistent proofreading symbols to indicate changes or corrections to text.
2. Use marginal marks to indicate corrections made within lines.
3. Use different colors of ink for different proofreadings (either by the same person or several people).
4. Keep a list of editorial or proofreading decisions so you can be consistent and so you can summarize for the writer the changes you routinely make.
5. Follow effective proofreading strategies.

1. Use consistent proofreading symbols to indicate changes or corrections to text:

Delete or take out.

Insert a phrase, word, or punctuation mark.

Transpose letters, words, or phrases.

Move to the right.

Move to the left.

Use capital letter(s).

Use lower case letter(s).

Close up a space.

Add a space.

Make a new paragraph.

NOTE 1: Professional proofreaders sometimes use a different symbol in the margin than they use in the text. For instance, the # sign in the margin indicates that a space should be added. In text, a slash mark indicates where the space should be added:

The incorrect/proposal

NOTE 2: Some reviewers also use the symbol *sp* in the margin to indicate a spelling error.

Original

Writers and Secrtaries of word proccessing specailists have to agree on what to use when editingand proofreading drat materials. without, such an agreement, and a consistent convention, erros kreep in and quality writing is impossible.

Corrected

Writers and secretaries or word processing specialists have to agree on what symbols to use when editing and proofreading draft materials.

Without such an agreement, errors creep in and quality writing is impossible.

Editing and Proofreading

2. Use marginal marks to indicate corrections made within lines.

Changes to a text are sometimes difficult to see, particularly those changes made in pencil or black ink, which readers may have trouble distinguishing from surrounding print. To highlight changes or corrections, you should use a red or green pencil or pen for changes. Even the change in color is sometimes difficult to see, however, particularly for color-blind reviewers.

So indicate changes by marking the change within the text but also inserting a check mark to show that a change appears in the text beside the mark.

Be consistent in using these standard proofreading symbols.

3. Use different colors of ink for different proofreadings (either by the same person or several people).

Printed text going through multiple revisions can become difficult to decipher if readers can't distinguish between versions. A very good system is to change the color of the reviewer's or proofreader's pencil or pen (as in the example below).

The first reviewer might indicate changes in blue ink, the second in red, the third in green, and so on. The color of the suggestion thus indicates when and by whom the suggestion was made. This system is particularly effective during peer or group review.

4. Keep a list of editorial or proofreading decisions so you can be consistent and so you can summarize for the writer the changes you routinely make.

The list of editorial or proofreading decisions is sometimes called an editorial style sheet. Writers themselves sometimes develop it, or they wait for an editor or a proofreader to develop one. The earlier it can be developed, the better. Your software program can be set to follow your rules so you don't have to track these decisions.

Items on this list would include all decisions about punctuation, capitalization, spelling, or word usage. To illustrate, a proofreader working with the preceding example could make these sorts of decisions:

Grayson plant (*not* Grayson Plant)

MOGO (*not* Mogo)

An Example of Multiple Proofreadings

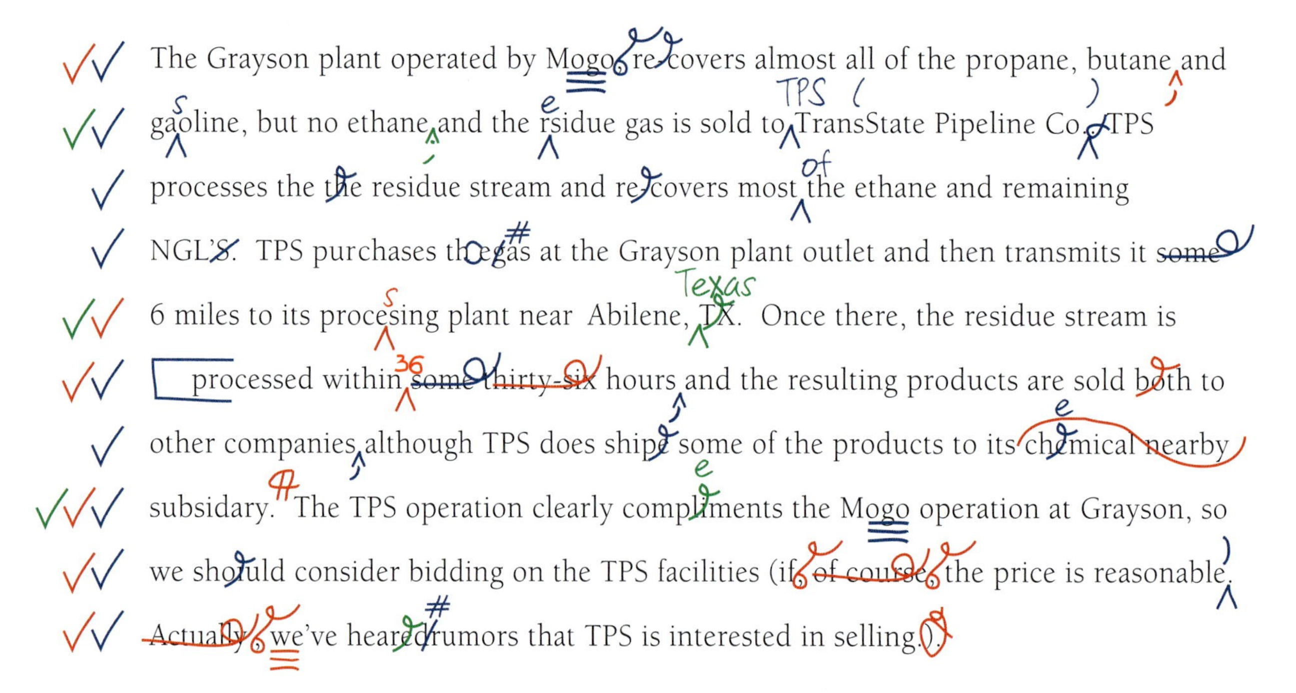
The Grayson plant operated by Mogo re-covers almost all of the propane, butane and
gaoline, but no ethane and the rsidue gas is sold to TransState Pipeline Co. TPS
processes the the residue stream and re-covers most the ethane and remaining
NGL'S. TPS purchases the gas at the Grayson plant outlet and then transmits it some
6 miles to its procesing plant near Abilene, TX. Once there, the residue stream is
processed within some thirty-six hours and the resulting products are sold both to
other companies although TPS does ship some of the products to its chemical nearby
subsidary. The TPS operation clearly compliments the Mogo operation at Grayson, so
we should consider bidding on the TPS facilities (if, of course, the price is reasonable.
Actually, we've heard rumors that TPS is interested in selling.)

Comma in a series precedes and:
propane, butane, and gasoline

TransState Pipeline Co. (*not* Company)

TPS (TransState Pipeline Co.) *rather than*
TransState Pipeline Co. (TPS)

When proofreading on a computer screen, use the feature that records and tracks changes so others can see your suggestions. You can find this feature in the review or revision menu of your software program. Once activated, the computer marks changes you make in the document. You can also indicate which changes you want marked. For example, you can order the computer to underline insertions, strike through deletions, or bold changes in format. You can also identify which reviser is making these changes.

NOTE: As the above examples suggest, some language decisions about a document are not clearly right or wrong. Instead, an editor or proofreader has to pick the preferred form and then stay with that choice throughout the document. See STYLE and PUNCTUATION.

5. Follow effective proofreading strategies.

To ensure the effectiveness of your document and to protect your image and your organization's, you should proofread every important document—even emails if they have significant implications. Documents differ in importance, however, and require different proofreading strategies. See WRITING AND REVISING.

When hurried or faced with a low-priority document, read through the document once, paying special attention to important points such as headings, topic sentences of paragraphs, visuals, and captions.

For more important documents, consider some of these proofreading strategies:

- **Check format.** Does the document look good—with uniform spacing, heading styles, and lists? Are emphasis techniques (boldface, italics, etc.) consistently applied?
- **Check content.** Is the information correct? For example, if the invitation says the meeting is on Monday, June 15, will the meeting actually take place then? And does June 15 really fall on a Monday? Are figures such as monetary amounts or percentages correct? Do you find facts contradicted from one page to another?
- **Check for errors.** Double-check spelling of names. Question every capitalization, punctuation, and word division. Note that typographical errors often occur in groups. Question every number and add up figures to make sure sums are accurate.

When proofreading onscreen:

- **Magnify the text** so you can spot problems that would otherwise go unnoticed. Enlarge the text to 150 percent or whatever suits you.
- **Correct errors** flagged by spell check and grammar check. Do not rely on these features to find every error. Although a spell check will flag spellings it does not recognize, you must often decide if words are spelled correctly for the context (for example, *their* vs. *there*). Even advanced grammar checking software can be wrong. If it flags a sentence as a fragment, for example, refer to this *Style Guide* for help in deciding if the sentence truly is a fragment or not.
- **Turn on hidden formatting symbols** such as paragraph and space marks so you can see if the spacing between lines and words is proper.

For particularly important documents, consider these strategies:

- **Read backwards** so the content does not distract you from watching for errors.
- **Read aloud** so you slow your reading speed and are more alert to flaws in grammar and sense.
- **Read in groups.** For long, complex documents, some readers can mark changes on the text or look up words or facts while another reads aloud. In this way, you get more than one viewpoint and speed up the work.

Electronic Mail

Electronic mail (email or e-mail) is the exchange of digital messages through a network server.

A boon to business, email has made communication inexpensive, virtually instantaneous, and—most important—far less time consuming than regular mail or even the telephone. It permits immediate communication but also allows people to respond to messages at a convenient time instead of having to be present. Email reduces paperwork and enables more efficient, more rapid decision making.

Still, email can be a hindrance to business as well. Documentation becomes weak and incomplete because email is a shorthand form of communication. You might have trouble explaining or defending a decision if the record of it is a long chain of fragmentary emails. Because writers give much less thought to emails than to, say, traditional letters, messages can mislead recipients. The tone of an email can give the wrong impression.

But the main problem with email is the sheer quantity of it. Hundreds of billions of email messages are sent each day. Adding to the problem are instant messaging, texting, and streaming social-networking services such as Twitter. These services enable people to hold billions of conversations every day via text and images in real time anywhere they may be. Managing this tidal wave of information is a major productivity challenge for many people. For guidance, see Managing Information.

Despite these problems, email is essential to the high-tech business world. The following rules will help you write effective e-messages for these various media and avoid the pitfalls.

Electronic Mail

Using Email Effectively

1. Choose email when you want to communicate information rapidly and when the information is better conveyed digitally than by phone or hard (printed) copy.
2. Write an informative subject line
3. Preview key content up front and limit your document to one screen (page) if possible.
4. Use business-appropriate tone of voice in an email.
5. Review and revise (as necessary) your email before sending it to readers.
6. Signal clearly the end of your message.
7. Control the distribution of your email.

Using Social Media Effectively

8. Contribute value to the ongoing conversation of social media.
9. Follow high ethical standards in online conversations.

Using Voice Mail Effectively

10. Be sure to identify yourself and give your listener the date, time, and your phone number.
11. Think before you speak.
12. Speak clearly and repeat important information.

Using Email Effectively

1. Choose email when you want to communicate information rapidly and when the information is better conveyed digitally than by phone or hard (printed) copy.

Email is especially efficient when the persons you want to contact are unavailable. Email allows you to send the message so that it will be available when the recipients log in.

Electronic mail is also valuable when the data or information would be inconvenient to deliver in other ways. For instance, a long list of names, addresses, and phone numbers are time-consuming to dictate over the phone. Hard copy is, of course, an option, but hard copy might take several days to arrive if it has to go by outside mail or even through an internal mail system. A fax is another option, but it often requires the sender and receiver to go to fax stations somewhere else in their buildings.

Use the phone when you want to get immediate feedback or response to your message. For instance, if your message requires extra tact and the personal touch, use the phone. Email can seem cold and dismissive, for example, when the writer has to send unpleasant or negative messages.

Print and send hard copies when you want the recipient to have a record of your message. Email is not always delivered, and even an archived email can disappear for many reasons; so you will want to forward and retain hard copies of certain documents. For example, you might want to summarize a meeting where important departmental decisions were made. A second example would be personnel decisions, which potentially become part of an employee's personnel file.

2. Write an informative subject line.

Enter your entire message in the subject line, if possible, so readers do not have to open your email. They will appreciate the convenience, and you will be more likely to get the result you want. One good practice is to type in EOM for "end of message," signaling that there's no need to open the email.

For longer emails, make sure your subject line will stand out from a long list of subject lines that appear on the reader's screen. Hundreds of entries can confront a reader who calls up a list of emails. If your subject line doesn't catch a reader's attention, your file might never be opened!

Write subject lines that get your message across in a few words. See "Subject Line" in LETTERS.

this

—Agenda for scoping meeting 10 p.m. Nov. 9
—Review cost overruns of 20% on A-345 Prototype
—Please sign divisional budget by July 5 EOM

not this

—Scoping meeting
—Cost overruns
—Divisional budget

See HEADINGS.

3. Preview key content up front and limit your document to one screen (page) if possible.

If your message is long, list your conclusion and main points first so readers will know what is coming. Email readers do not like being forced to scroll through several screens to get to the point.

If possible, limit your document to one screen (page).

Whenever possible, design this one screen using emphasis techniques such as lists, headings, and single-sentence paragraphs. See EMPHASIS.

this

We propose increasing the division's supplemental budget for July by $15,000 to account for cost overruns on the XYZ project. Here's why:

1. Labor rates are going up from January through July.
2. Several additional fact-finding trips will be needed during July.
3. Managers are now very interested in XYZ.

not this

As you know, during the recent managerial coordination meeting (January 15), the subject of XYZ came up. Concerns expressed included the timing of the project, especially work during July. Also, the engineering representatives indicated that several extra trips might be necessary during July ...

See ORGANIZATION.

For longer documents, consider writing a separate executive summary for the first page (screen) and then include other data as necessary. In many cases, the executive summary might be sufficient by itself, with the background or supporting data merely referenced or transmitted in hard copy to follow up the electronic version. See SUMMARIES.

4. Use business-appropriate tone of voice in an email.

Email invites informal language—unguarded, casual, and personal in tone. At the same time, you need to make sure that a too familiar or offhand tone of voice doesn't offend readers. You should adopt a conversational, businesslike tone. See TONE.

Depending on your familiarity with the reader, you can vary your tone. Don't be flippant, terse, or abrupt with someone you don't know well and whose business you need.

this

Thank you for the opportunity of submitting our ideas for your new artwork.

not this

Here's the artwork you wanted.

Avoid using breezy abbreviations like "plz 4ward yr specs 4 new artwork." Emailing a client is not the same as texting a close friend.

Avoid fancy fonts, patterned backgrounds, or gimmicky animations unless your branding requires them.

5. Review and revise (as necessary) your email before sending it to readers.

The immediacy of email is both its strength and its weakness. An important message will profit from review, both for errors and undesirable or misleading content. See WORD PROCESSING.

Depending on your potential readers, take time to clean up your email. A few minor errors will detract from the message; a glaring error or many errors will damage your credibility and the impact of your message.

Fix flagged misspellings, but remember that a spell check will often not identify wrong words (for example, *there* instead of *their*).

With all business documents—especially those written under time pressure or in anger—a cooling period has always been desirable. Consider allowing a cooling period before you send certain emails to recipients. Give yourself a few minutes (or longer) to reconsider a sensitive message. Often you will change the message, and sometimes you may even decide not to send it.

6. Signal clearly the end of your message.

Give your documents a quick, complimentary close—*Sincerely*, *Thanks*, *See you Thursday*, etc.

End a long email with a brief summary or review of the content. You might restate a request or a deadline, or you might even list again the reasons for your request.

If the end of the message is not obvious, signal it with *EOM* (end of message).

Automate your signature line, and include all the contact information a recipient needs to get in touch with you.

7. Control the distribution of your email.

Keep in mind that your email might be forwarded to others, so your audience is potentially larger than you think. Even messages marked private are easy to transfer to others and can spread around the world in seconds. If the email contains information you would not want others besides your addressee to read, don't send it. Use a more private medium.

Don't copy recipients unless they need to know the content of your email. People will learn to ignore your emails if they repeatedly get marginally relevant messages from you.

Don't ask for "return receipt" (RR) unless you specifically need to know if the recipient received the message. Replying can be inconvenient and even intrusive to recipients.

When sending an email to a large group of recipients, say, your entire contact list, do not include them all in the "To:" field. Rather, address the email to yourself and then insert the large list in the "Bcc:" field (blind courtesy copy). There are several reasons for this:

- Some recipients will reply to everyone on that lengthy list with their opinions, quips, or anecdotes, thus wasting people's time.
- Some recipients must abide by company policies about personal use of email. Receiving inappropriate email can be contrary to those policies.
- Recipients concerned about the volume of spam messages and the danger of viruses do not want their email addresses exposed to the world.

When replying to an email, take care to send your reply to the proper audience. Is it only for the sender or for the entire group of addressees?

See LETTERS and MEMOS.

Using Social Media Effectively

Electronic media such as blogs, podcasts, and networking sites are now a primary means of communication and marketing. Unlike email, social media are usually open to any subscribers who want to participate. Many businesses and government agencies now sponsor online communities for the use of their clients and the interested public.

Unlike the old industrial media that communicated only one way, social media are ongoing conversations among organizations and their clients and communities. Websites, blogs, and community sites can be wonderful tools for promoting an organization.

Thus, business and technical professionals have a serious stake in the use of social media. Like other media, they can be used effectively or ineffectively—and can even become destructive.

Social media can severely hinder an organization's success. People typically spend from half an hour to three hours during the workday accessing social media, wasting a tremendous amount of time and bandwidth. Also, frivolous, defamatory, or obscene blog entries or e-messages can cause you and your organization real trouble. One well-known restaurant chain suffered a good deal of bad publicity when an employee posted on the Internet a video of himself shoving French fries up his nose.

The following rules will help you use social media effectively:

8. Contribute value to the ongoing conversation of social media.

Social media encourage informal, spontaneous writing with little thought or planning behind it. As a result, much online content is banal or valueless. If you write online a blog entry or comment that represents your organization—as more and more people do—use the same thoughtful process you would follow for a more formal document (i.e., planning, revising, and so forth). See WRITING AND REVISING.

Consider carefully your purpose for writing. What is the job that needs to be done? Who will read this? What do you want readers to know, do, and feel? What kind of response do you want from them?

9. Follow high ethical standards in online conversations.

Social media invite anonymity. As a result, many people misuse the media to defame others or to spread falsehoods. Often inadvertently, people post misleading or confidential information.

Be honest, open, and respectful in online conversations. Correct inaccuracies as soon as possible. Know and strictly follow your organization's policies governing disclosure of confidential information about people, financials, trade secrets, strategic initiatives, and intellectual property.

If you have any doubts about the appropriateness of your writing, ask a trusted colleague to review it with you.

Avoid being negative in a public online space. Private conversations can be taken offline.

Using Voice Mail Effectively

The main benefits of voice mail are to avoid "telephone tag" and to get your message across quickly and efficiently.

10. Be sure to identify yourself and give your listener the date, time, and your phone number.

Don't assume that your listener will recognize your voice; also, not every system will automatically record the date, time, and your phone number. Give your phone number even if it's listed; this way the respondent won't have to look it up.

11. Think before you speak.

Take a minute before dialing to review mentally your main points and your intent in making the call—perhaps even jot a list of points to cover. Unless you do one or both of these things, you are likely to ramble and to confuse your listener. Rambling is a problem if you are limited in the time you have to record your message.

12. Speak clearly and repeat important information.

Misunderstandings are inevitable, so work to reduce them in your recorded messages. Speak clearly and slightly slower than you would normally. As necessary, spell out difficult words—for example, people's names or the names of places, because names often have unusual spellings. Technical terms and associated numbers are also easy for a listener to confuse.

Repetition of meeting times, deadlines, and other important details is a courtesy. You might, for example, conclude by repeating your key request or recommendation, including any associated date or meeting time. See REPETITION.

Ellipses

Ellipses consist of three periods (. . .) separated by single spaces and preceded and followed by a single space. Ellipses indicate omissions, primarily in quoted material. See QUOTATIONS.

Ellipses are the opposite of brackets, which indicate insertions in quoted material. See BRACKETS.

Ellipses

1. Use an ellipsis within quoted material to indicate omissions of words, sentences, or paragraphs.
2. Do ***not*** use an ellipsis to omit words if such omissions change the meaning or intent of the original quotation.
3. Do ***not*** use an ellipsis to open or close a quotation if the quotation is clearly only part of an original sentence.
4. Use a line of periods to indicate that one or more entire lines of text are omitted.
5. Use an ellipsis to indicate omitted material in mathematical expressions.
6. Use an ellipsis to indicate faltering speech.

1. Use an ellipsis within quoted material to indicate omissions of words, sentences, or paragraphs:

"Labor costs . . . caused an operating loss for January of nearly $10,000."

Original: Labor costs, which our executive committee has been studying, caused an operating loss for January of nearly $10,000.

"No tax increases for 2012 . . . will occur."

Original: No tax increases for 2012 in personal withholding will occur.

NOTE 1: If omitted material comes at the beginning of a sentence, the quoted material opens with an ellipsis, especially if the material appears to be a complete sentence:

" . . . the printed budget will remain unchanged."

Original: Despite a few inconsistencies, the printed budget will remain unchanged.

NOTE 2: If the omitted material comes at the end of a sentence, the quoted material ends with an ellipsis plus the ending punctuation of the sentence:

"The Department of Energy denied our request for an energy subsidy . . . . "

Original: The Department of Energy denied our request for an energy subsidy, even though we felt our request would be cost-effective.

NOTE 3: If one or more sentences are omitted following a complete sentence, retain the ending punctuation for the sentence followed by an ellipsis:

"The submittee adjourned at noon . . . . Shortly after they reconvened, the Majority Leader requested that they adjourn until the following Monday."

2. Do *not* use an ellipsis to omit words if such omissions change the meaning or intent of the original quotation:

"Chairman James Aubrey indicated that financing the debt load would . . . seriously undermine efforts to recover delinquent loans."

Original: Chairman James Aubrey indicated that financing the debt load would not detract from or seriously undermine efforts to recover delinquent loans.

3. Do *not* use an ellipsis to open or close a quotation if the quotation is clearly only part of an original sentence:

this

We discussed the "three legal loopholes" mentioned in the last Supreme Court decision on school busing.

not this

We discussed the " . . . three legal loopholes . . . " mentioned in the last Supreme Court decision on school busing.

4. Use a line of periods to indicate that one or more entire lines of text are omitted:

Friends, Romans, countrymen, lend me your ears;

I come to bury Caesar, not to praise him.

. . . .

He was my friend, faithful and just to me:

NOTE: The line of periods does not tell a reader how much was omitted. The writer is responsible for retaining the intent and meaning of the original material.

5. Use an ellipsis to indicate omitted material in mathematical expressions:

$a_1, a_2, \ldots, a_n$

$1 + 2 + \ldots + n$

See MATHEMATICAL NOTATION.

6. Use an ellipsis to indicate faltering speech:

I protest . . . or maybe I should only suggest that you have made a mistake.

I wonder . . . perhaps . . . if . . . that is a wise choice.

Emphasis includes any techniques writers use to make their messages both readable and unambiguous. Readers must immediately see your major point or points, and rereading should be unnecessary if you have designed your document using proper emphasis techniques.

Every misleading sentence or disputed fact represents a waste of time and money in today's technical society. Organizations cannot afford ambiguous procedures, unclear memos, or inconclusive reports.

Textual options and a rich, varied vocabulary become liabilities when writers need to remove ambiguities. The traditional academic essay or scholarly paper is often inefficient and unproductive precisely because it deliberately includes multiple meanings and different interpretations.

A well-designed, emphasis-driven document forces every reader to come away with the same message. A simple example would be the instructions for assembling an exercise bike. Every customer assembling the bike should be 100 percent successful. No calls to the bike company's customer service center should be necessary, and the final bike should have no unfastened parts or other unsafe features.

Successful world-of-work documents have a 100 percent usability rating! See Project Management.

Use the following format and emphasis principles to help you generate your initial content. Your intended format should guide your content. You will be writing to fit the format, not imposing a format on content that is already created. See Writing and Revising.

Emphasis

1. Control your readers' eyes by controlling the position and appearance of your ideas.
2. Open with important ideas.
3. Subordinate minor ideas.
4. Repeat important ideas.
5. (Optional) Close with repetition of important points.
6. Use space and page design to highlight important ideas.
7. Use headings and lists to highlight important ideas.
8. Use graphics to emphasize important ideas.
9. Use single-sentence paragraphs to emphasize important ideas.
10. Use typographical features and color to emphasize words and ideas.

1. Control your readers' eyes by controlling the position and appearance of your ideas.

Effective writing is document design, not the spinning out of words and phrases. Effective writing is a visual activity, not an oral/aural one.

Yes, writing begins with and relies on the oral/aural resources of language, but these resources must be presented in an emphatic manner. Proper emphasis techniques will help guarantee that readers take one and only one meaning from a passage.

Writers achieve emphasis in documents by working with position and appearance.

Position refers to the placement of words within a sentence, paragraph, or section. Language, both written and oral, moves from word to word, so the order and sequence of language is important, even if it is not always speedy or efficient.

Appearance refers to the visual presentation of ideas and to the physical character of the words or ideas—for example, page layout, spacing, indentation, boldface type, italics, type size, type style, color, and production quality. See Page Layout and Graphics for Documents.

2. Open with important ideas.

The beginnings of documents, sections, paragraphs, and sentences are the most visible and memorable parts of any document. Make the first page, the first line, the major subheading count by using them to record important ideas. Lead from important ideas, not up to them. See Organization.

Test each document, section, paragraph, or sentence. The opening page, the opening paragraph, the opening sentence, and the opening words should capture the most important ideas. If they don't, you will have misled your readers. Or at the very least, you will be forcing your readers to read and reread your information. The more rereading you require, the more ambiguous your document will be.

Emphasize your most important ideas by placing them in opening positions.

3. Subordinate minor ideas.

Arrange your document so that minor ideas receive less emphasis than major ones. Minor ideas should receive both less time and space than major ones. Minor ideas should not appear in headings, in graphics, or in boldface.

Place minor ideas in the middle of documents, paragraphs, and sentences. This will happen automatically if you have followed rule 2 above, which tells you to open with major ideas. See ORGANIZATION.

And in some cases, minor ideas need only be referenced because they should not appear in the document. These referenced ideas become part of the legal record or organizational documentation. See APPENDICES.

Minor ideas include backup data and their interpretation; lengthy analyses leading up to conclusions; routine explanations of your methodology; summaries of prior research; and any other peripheral information.

4. Repeat important ideas.

Repetition helps readers remember earlier messages as new words and ideas appear. Repetition signals important ideas and guarantees that readers remember the ideas. See REPETITION.

However, repetition can be either ineffective or effective.

Ineffective repetition states the same thing over and over again. Students often use ineffective repetition because they are trying to fill pages and to generate a certain number of words. They may write a dozen sentences, each saying they had an exciting summer vacation!

Effective repetition occurs when the design of a document repeats important ideas. The title, subject line, and major headings introduce major ideas. Graphics and accompanying captions further illustrate these same ideas. And if you include an executive summary, it will present the highlights of these ideas. See SUMMARIES.

Each mention of a major idea should use the same wording and the same technical data, such as acreage or the annual output. This type of deliberate repetition means that the most important ideas should be 100 percent unambiguous and consistent each time they appear.

Effective repetition is one indication that a document has been carefully designed and well edited.

5. (Optional) Close with repetition of important points.

This rule is optional because the visual nature of documents has decreased the role of the closing section, the closing paragraph, or the closing sentence. See ORGANIZATION and REPETITION.

Some documents still may require closings that repeat important points. Letters and memos quite often should end with a repetition of the requested meeting, the follow-up action, or other activity. Textbooks usually wrap up chapters with a final summary, which is often identical to the chapter overview that begins the chapter.

Of course, oral language still relies on the final statement, the final word. In oral presentations, speakers must lead up to their clincher idea, their major point, their most memorable phrase. This emphasis technique has become unnecessary in most well-designed documents.

6. Use space and page design to highlight important ideas.

Make your page design, especially its open spaces, support and reinforce your important ideas. See PAGE LAYOUT and GRAPHICS FOR DOCUMENTS.

Initially, readers visually perceive the overall appearance, not the details. They derive a general impression of your intent and your content before they ever begin reading sentence by sentence.

If readers scan your first page, what should they see? If you've designed the page well, the title and major headings have the important ideas, and these are always set off by surrounding space. Perhaps the page has a graphic on it, with an interpretive caption; this graphic should also present important ideas.

Make sure, however, that your pages are not so cluttered and busy that the important ideas are lost.

7. Use headings and lists to highlight important ideas.

Writing today is increasingly less linear and sequential. Passages of language do move word by word, but the use of headings and lists means that the sequence of words is not always the primary tool for conveying meaning.

Headings and lists are schematic ways of presenting bits of information so readers can select what they read. Readers using headings and lists can easily skip and scan your document for crucial content, and they often will not even read the entire document unless they need to. See HEADINGS and LISTS.

8. Use graphics to emphasize important ideas.

Graphics are increasingly important. Today's readers expect, even demand, that well-designed documents use graphics. See GRAPHICS FOR DOCUMENTS and GRAPHICS FOR PRESENTATIONS.

The availability of graphics software has revolutionized routine reports and other everyday documents. Readers are now accustomed to seeing bar graphs, illustrations, and other common graphics. What was innovative some years ago in *USA Today* is now commonplace in many world-of-work documents and business presentations.

9. Use single-sentence paragraphs to emphasize important ideas.

Paragraphs are increasingly less important as ways of organizing ideas. In particular, a document with a sequence of long paragraphs is not very readable, even if the paragraphs are well written and the language interesting. Most of the time such a document is a costly waste of time. See PARAGRAPHS.

The single-sentence paragraph is, however, still a valuable emphasis tool. Readers can quickly read a single-sentence paragraph and scan back easily to find such a paragraph. Of course, too many short paragraphs make a document choppy and hard to use.

So reserve single-sentence paragraphs for important ideas and don't overuse this technique.

10. Use typographical features and color to emphasize words and ideas.

CAPITALS, underlining, **boldface**, *italics*, color, and different type styles stand out when used within ordinary text. Don't overuse such techniques, but design your pages to draw on these tools when appropriate. See PAGE LAYOUT.

Color is also a valuable tool, especially if used in graphics. It is probably the most visually effective of the document-design tools available to writers, but it is still a costly option and time-consuming, given the need for multiple printing runs. As with typeface styles, don't overuse color or mix too many colors into a single page or graphic. See COLOR.

English as a Second Language

English is a global language of business. An engineer in Singapore talking to a customer in Qatar is likely to use English. The spoken language is richly diverse, with dialects and accents ranging from Alaska to India, from London to Johannesburg.

Nevertheless, the written language is much more standardized, with little variation worldwide. The picture is somewhat complicated by differences between British and American English, which make international English a mix of both. But the differences are minor, mostly in spelling and a few common expressions (see BRITISH ENGLISH). A mastery of standard written English is an advantage to any professional person of any nationality.

This section helps those who write ESL (English as a Second Language) in a business context. Rules address problems common to ESL speakers that usually make little difficulty for native speakers.

English as a Second Language

1. Avoid using idioms (clichés).
2. Introduce countable nouns with articles or other qualifying or quantifying words.
3. Never use the "–ing" verb form with a state-of-being verb.
4. Ensure that verbs agree with subjects.
5. Be aware that verbs change meaning unpredictably when paired with prepositions.
6. Do usability testing with native speakers before publishing technical information.

1. Avoid using idioms (clichés).

ESL speakers should avoid trying to express themselves in informal or colloquial ways when writing in a business context. Statements like "Your salesman rubbed me the wrong way" or "Keep your nose to the grindstone!" sound pretentious or ridiculous in business. No effective writer uses clichés in any case. Avoid the temptation to sound glib to the ear of a native English speaker. Use plain, simple language and precise terminology, especially in a technical context. See CLICHES.

2. Introduce countable nouns with articles or other qualifying or quantifying words.

ESL speakers often have difficulty deciding when to use the articles *a* and *the* before nouns. Traditionally, articles or some other qualifying or quantifying word must introduce countable nouns such as *boss*, *computer*, *office*, but must not introduce uncountable (mass) nouns such as *commerce*, *marketing*, *work*. Although this distinction can be helpful, it is not always reliable. For example, the word *paper* can be both countable and uncountable depending on the context:

> Paper is getting more costly.
>
> We faxed the papers to the attorney this morning.

A more helpful rule (and easier to remember) is to distinguish between general and specific nouns. (See figure 1.) When referring to a specific noun, use the articles *a* or *the*:

> Kim gained an education in project management.
>
> The education of project managers is an essential part of our business.

When referring to a noun in a general sense, do not use an article:

> Education is crucial to the success of a nation.
>
> The people of that country value education highly.

In the absence of an article, a countable noun must be preceded by some qualifying or quantifying word:

each employee
this boss
those computers
five offices
some books
few people
many customers

Because they imply numbers, the following quantifying words can precede only countable nouns:

a, an
both
each
every
few, fewer, fewest
many
one (or any number)
several
these, those

By contrast, the following quantifying words can precede only uncountable nouns:

less, lesser, least

little (*to describe quantity rather than size* [*e.g.*, little *money*])

much

General	Specific
business	the business we bought
economics	the economics of the deal
email	the email you sent me
financing	the financing they arranged for us
food	the food of India
money	the money you deposited
software	the software we ordered
technology	the technology we need

Figure 1. *The nouns in the left column above are usually considered uncountable nouns. Still, as in the right column, they are often countable nouns when specifying a certain object or objects.*

3. Never use the "–ing" verb form with a state-of-being verb.

English has two kinds of verbs: those that express action and those that express a state of being.

Action: They are accepting the agreement.

State of being: The agreement is acceptable to them.

(See figure 2.)

Correct	Incorrect
He is in charge of the company.	He is being in charge of the company.
The cargo weighs 40 metric tons.	The cargo is weighing 40 metric tons.
These are last month's invoices.	These are being last month's invoices.

Figure 2. *The "–ing" verb form works only in describing an action, not a state of being.*

4. Ensure that verbs agree with subjects.

As in many languages, English verbs must agree in form with their subjects. Fortunately, that agreement in English is simple because it affects only the third-person singular form of the verb, which always ends with an *s*. Other forms never end with an *s* and remain stable, as seen in figure 3.

Person	Singular	Plural
1st	I work	we work
2nd	you work	you work
3rd	he/she works	they work

Figure 3. *As in many languages, English verbs must agree in form with their subjects; but in English, this rule affects only the third-person singular form of the verb, which always ends with an* s.

The only exception to this rule is the verb *be*, which is so commonly used that the ancient forms for the various persons still persist.

Fortunately, *are* is the plural form for all persons and for the second-person singular. (See figure 4.)

Person	Singular	Plural
1st	I am	we are
2nd	you are	you are
3rd	he/she is	they are

Figure 4. *The only exception to this rule is the verb* be, *which is so commonly used that the ancient forms for the various persons still persist.*

Subject-verb agreement can be tricky when the subject and the verb are distant from each other:

> The teams that provide our environmental analysis *come/comes* from the local universities.

In this sentence, the subject of the verb come is *teams*, not *analysis*. Therefore, the correct verb form is *come*—the third-person plural.

5. Be aware that verbs change meaning unpredictably when paired with prepositions.

A single English verb can express many different ideas when paired with different prepositions. For example, *to hand out* means to distribute something to a group and *to hand in* means to give something to someone, while *to hand over* means to give it unwillingly.

Many of these so-called "phrasal verbs" make obvious sense: *to report up* means to respond to someone with more authority, while *to report down* means to respond to someone with less authority. Still others, however, are unpredictable in meaning, as in figure 5, a table of pairings for the verb *to put*:

to put by	to save money or supplies
to put down	to insult someone
to put on	to dress; to mislead
to put over on	to trick or deceive someone
to put through	to seek approval of something
to put under	to hypnotize; to anesthetize a person with a drug
to put up	to house someone; to contribute something
to put up with	to tolerate someone or something offensive

Figure 5. *Many English verbs like* put *change meaning unpredictably when paired with prepositions.*

This table only begins to account for all possible pairings of the verb *to put*. Other verbs that are commonly paired with prepositions include *come, go, get, call, stay, fill, pay, pull, push, look, wait, lay/lie, speak, talk, sit, play, buy, sell,* and *give.*

Phrasal verbs present many pitfalls to ESL speakers. While native English speakers use verbal pairings with confidence, ESL speakers should avoid using them unless they are sure of the meaning. Use more precise verbs instead, as in the right-hand column in figure 5.

For more information, see PREPOSITIONS and VERBS.

6. Do usability testing with native speakers before publishing technical information.

Because of a weak grasp of idiomatic English, ESL writers often produce unclear, misleading, or even nonsensical user manuals, procedures, and product guides. For example, the following set of instructions came with a headphone manufactured in a non-English-speaking country:

> For you easily to use our products please read it through before using:
>
> 1. Please tune up power volume to apropositу, before put into the jeck.
> 2. Take for you clothing the plastic.
> 3. Don't dranght in the jeck.
>
> If our products have quality problem indeed, protection time for 6 months.

Although this example is extreme, similar poor English perplexes millions who buy and use products manufactured in our global economy.

Automated English spelling and grammar checkers can help solve these problems. In the example above, spell check will flag the odd words *aproposity, dranght,* and *jeck* and suggest alternatives. Still, there is no automated solution for this kind of bizarre phrasing.

Therefore, you should always test technical instructions and information with native English speakers before publishing them. Natives can usually spot problems with sense and grammar that are invisible to nonnatives. The ideal test subject is a native English speaker who is also a natural user of the product or procedure. Follow the guidelines for usability testing in PROJECT MANAGEMENT.

Additionally, a useful reference for ESL writers is a current edition of the *Oxford Student's Dictionary of American English*. The entries are brief and clear, and special attention is given to idioms (phrases that have special meanings). See REFERENCES.

Ethics

Ethical dilemmas will confront every professional during his or her career.

As in the following cases, professionals must make hard choices about what to say and how best to say it. The line between right and wrong may be fairly clear, but what if jobs—especially your job—depend on what you choose to do?

- Your supervisor tells you that your business product plan is too negative in its financial projections. You feel that the new product will not break even for at least 18 months. Your supervisor argues that the break-even point could be as early as 6 months after product introduction.
- Your current project parallels previously published information. You wonder how closely you can follow the ideas and phrasing of this source. Do you need to write for permission?
- A technical colleague suggests that you soften your write-up of negative conclusions about some tests.
- Your boss tells you to write a proposal using the résumé of an employee who retired long ago, and you know he'll never work on the project.

To help you confront such questions and make the hard choices, we present the following rules (or suggestions). Only you can make the necessary decisions, based on the realities of your own job and your own professional code of ethics.

Ethics

1. Use full, accurate, and honest information in everything you communicate.
2. Analyze who (including yourself) will be affected and how and when they will be affected by ethical or professional choices you make.
3. Assess your options, given your analysis of the situation (in rule 2), then pick a course of action.
4. Prepare documents to support your decisions on ethical issues.

1. Use full, accurate, and honest information in everything you communicate.

This rule says that all readers and listeners have the right to trust what you tell them. Anything less would violate your own expectations about communication. So this rule becomes the golden rule: "Communicate with others as you expect them to communicate with you." See INTELLECTUAL PROPERTY.

This rule (or some version of it) appears in every code of sound professional practice or ethics. Under the Hippocratic oath, a doctor promises to deal ethically with patients—thus neither by word nor deed harming them. The Code of Ethics for the Society of Professional Journalists states: "Truth is our ultimate goal."

A person's word must be true and firm. That is, you must be able to depend on what someone tells you; otherwise, sound relationships, agreements, and decisions are impossible. Someone whose word is untrue and unreliable risks disclosure and sanctions, either personal or legal.

What are the legal sanctions from violations of this rule? Consider the case where you violate your professional code of practice (ethics) by not revealing a design flaw you've discovered.

From a legal perspective, you personally, as well as your company, could be liable for tens of thousands of dollars. Liability exists even if you fail to test for obvious flaws. The law assumes that sound product engineering—that is, good professional ethics—would provide for tests to discover reasonably foreseeable problems. Failure to record tests is, therefore, no protection from legal sanctions.

2. Analyze who (including yourself) will be affected and how and when they will be affected by ethical or professional choices you make.

In ethical questions about past events, decisions of right or wrong are easy. George Washington chose not to tell a lie about cutting down the cherry tree. He was ethically right (according to rule 1).

Ethical questions become more ambiguous, however, when future events are involved. Will a proposed computer system pay for itself? How safe and effective is a new antibiotic? Is the new antilock braking system foolproof in ordinary driving?

Consider the following problem situation, where ethical or professional questions arise in relation to a business plan.

Problem: Your supervisor tells you that your business product plan is too negative in its financial projections. You feel that the new product will not break even for at least 18 months. Your supervisor argues that the break-even point could be as early as 6 months after product introduction.

Here is a brief analysis of what could happen if you go ahead with your financial projections of 18 months and if management decides to cancel or delay the introduction of the new product:

- **Who (including yourself) will be affected?**

  You, your supervisor, other employees, upper management, shareholders—maybe even customers.

- **How and when will they be affected?**

  You, as the source of negative news, risk your chances for advancement, future assignments and, perhaps, even your job, if the new product group is disbanded.

  Your supervisor and other employees may also receive new assignments, or perhaps they might be laid off. Top management risks missing a business opportunity, and this decision may affect the company's dividends for several quarters or even several years. Shareholders will complain about the lost dividends, driving company stock prices down.

  Customers may be only indirectly affected. They may not have the option of buying the product. Or, if management rushes the product to market to meet financial goals, quality problems may occur.

3. Assess your options, given your analysis of the situation (in rule 2), then pick a course of action.

In the simplest terms, you can either choose to revise your projections (the 18-month break-even point) or you can refuse to back down. Neither of these answers is clearly right or wrong because you are dealing with a forecast.

If you revise your projections, how much can you change them before they are dishonest? Can you in good conscience write a report stating that 6 months is a possible break-even point?

If your projections are open to negotiation, how hard do you want to fight for them? This question is usually the crucial one when an employee considers options dealing with ethical or professional questions.

Every organization answers such questions in different ways. Our suggestion is that you address these sorts of questions as honestly and as carefully as you can. And as the next rule states, prepare documents to support your decisions on ethical or professional issues.

4. Prepare documents to support your decisions on ethical issues.

Written documentation is essential when you have to confront ethical dilemmas that might have legal implications for you and for your organization.

What should you record? No quick or easy answer exists. With the high frequency, even likelihood, of litigation today, some might be tempted to record very little or even nothing that might reveal internal debates. This view is wrong. As noted above under rule 1, professional standards mandate good documentation.

For example, in the area of U.S. environmental law, the mandate of the National Environmental Policy Act (NEPA) is to disclose to the public and any other parties (including the courts) the pros and cons of a proposed government action. Honest and professional differences of opinion are encouraged, not prohibited.

We recommend that you carefully document your professional views. Remember also that from a legal perspective, documentation extends beyond formal printed documents to include, for example, your computer files, research logs, and email archives. Careful documentation means that you have analyzed and cataloged all the pertinent records.

Recall the old saying "Honesty is the best policy." The saying reflects the reality that, if your honesty is called into question, nothing you say or do will be credible.

Exclamation Marks

Exclamation marks (!) follow words, phrases, or sentences that are surprising or unusual enough to merit the use of an exclamation mark.

Some style guides or grammars refer to *exclamation points* rather than to *exclamation marks*. The two terms mean the same thing.

Exclamation Marks

1. Use an exclamation mark to end a sentence that expresses surprise or excitement.
2. Use an exclamation mark in direct quotations that are exclamations, with the mark itself usually coming inside the quotation mark.
3. Place exclamation marks outside closing quotation marks if the quoted portion of an exclamatory sentence is itself not an exclamation.

1. Use an exclamation mark to end a sentence that expresses surprise or excitement:

We completed the proposal 2 weeks ahead of the deadline!

That error will cost them their jobs!

Excellent work! No one in the department could have done as well. (*Excellent work* is a sentence, despite not having a verb.)

See SENTENCES.

NOTE: Use exclamation marks sparingly; they lose their power if overused, and they can make your language sound too informal, almost breathless.

2. Use an exclamation mark in direct quotations that are exclamations, with the mark itself usually coming inside the quotation mark:

"I sent you that check over 2 weeks ago!"

"No! You do not have permission to take the car tonight," responded Mother.

Bill exclaimed: "Ouch! That hurts!"

optionally

"No, you do not have permission to take the car tonight!" exclaimed Mother.

Bill exclaimed, "Ouch, that hurts!"

See QUOTATION MARKS.

3. Place exclamation marks outside closing quotation marks if the quoted portion of an exclamatory sentence is itself not an exclamation:

I'm speechless that all you can say is, "We forgot"!

We were amazed to discover that "all earthquake damage is excluded"!

False Subjects

False subjects are words like *it* and *there* that have no concrete antecedents; that is, the words do not refer to anything real. They are abstractions.

False subjects often occur at the beginning of a sentence and displace the true subject:

It is this phase that is important.

In this sentence, *it* seems to stand for *phase*, but replacing the pronoun with its apparent antecedent creates nonsense:

This phase is this phase that is important.

The true subject of the sentence is *phase*. Beginning with the true subject creates a shorter, much crisper sentence:

This phase is important.

False subjects can also appear within sentences:

We decided that there were some costs that needed to be explained.

The false subject *there* weakens the middle of the sentence and adds unnecessary additional words. The sentence is far stronger without the false subject:

We decided that some costs needed to be explained.

or

We decided to explain some costs.

False Subjects

1. Eliminate false subjects.

1. Eliminate false subjects.

Whenever possible, eliminate constructions with *it* and *there* when those words have no clear antecedents. Getting rid of false subjects makes your writing more concise and often clearer.

Sometimes, however, the false subject is necessary, as in the expression *it is raining*. The word *it* is an abstraction, but what could you say in its place? Ask yourself, what is raining? No other word or words will be as functional or as expedient as the false subject. In such limited circumstances as *it is raining* or *it is noon*, false subjects are acceptable. Otherwise, try to eliminate them.

Below are "before and after" examples of false subjects. Note how much simpler, more direct, and shorter the sentences are when we eliminate the false subjects along with certain accompanying words:

In designing a thermal protection system, it is possible to meet the 1,100-degree fire requirement yet not be reliable. (*20 words*)

Thermal protection system designs can meet the 1,100-degree fire requirement without being reliable. (*14 words*)

Within the family of hydrates, there are several solid metal oxides that both chemically combine with water (hydration) and mechanically retain water (capillary condensation). (*24 words*)

Within the family of hydrates, several solid metal oxides both chemically combine with water (hydration) and mechanically retain water (capillary condensation). (*21 words*)

It will also be possible, as the study proceeds, to identify and extract important performance degradations resulting from failure to improve a given technology. (*24 words*)

As the study proceeds, identifying and extracting important performance degradations resulting from failure to improve a given technology will also be possible. (*22 words*)

or

As the study proceeds, we can also identify and extract important performance degradations resulting from failure to improve a given technology. (*21 words*)

From table 5.3, it appears that the use of relatively compact heat-transfer surfaces in the 10- to 20-fins/in. range will provide the compactness necessary to achieve the Navy goal of reduced size and weight. (*34 words*)

As table 5.3 suggests, using relatively compact heat-transfer surfaces in the 10- to 20-fins/in. range will apparently provide the compactness necessary to achieve the Navy goal of reduced size and weight. (*31 words*)

There are five significant factors that influenced the assessment of disease severity. (*12 words*)

Five factors influenced the assessment of the disease severity. (*9 words*)

It will be the responsibility of the team manager to ensure that the total required time-phased quantity and skill mix can be supplied from the onsite pool. (*27 words*)

The team manager will ensure that the required time-phased quantity and skill mix can be supplied from the onsite pool. (*20 words*)

Faxes

Faxes (derived from the word *facsimiles*) are documents transmitted via telephone lines. The use of faxes declined as email became the primary communication tool between businesses and organizations. Still, faxes remain useful in some cases; for example, legal documents that require signatures continue to be faxed where electronic signatures are unacceptable. For security reasons, people still prefer to fax certain documents, such as financial or personnel information. In some professions, such as law and real estate, fax machines are in constant use.

In one sense, a good fax is no different from a well-written letter except that it is sent by telephone line instead of in an envelope. See LETTERS.

Faxes and email, when sent in haste, share many of the same problems. Faxes can increase misunderstandings if they are not prepared carefully. The preparer of the fax may spend too little time gathering the information or making sure pages to be sent are both appropriate and readable. The result is that the customer gets pages that frustrate or annoy rather than communicate. See ELECTRONIC MAIL and WRITING AND REVISING.

NOTE: Although we are using *fax* or *faxes* in this discussion, some writers choose to use all caps, similar to an acronym: *FAX* or *FAXs*. This use of capitals is especially common on cover sheets. Both versions of the word are acceptable.

Faxes

1. Check pages for appropriateness and usefulness before faxing them.
2. Send pages that are clean and likely to be readable after transmission.
3. Be sure to include a cover sheet both with your customer's name and phone numbers and with your own organization clearly identified.
4. List on the cover sheet the documents transmitted and, if appropriate, add a personal message.

1. Check pages for appropriateness and usefulness before faxing them.

Ask yourself: Would these pages be appropriate for and useful to the recipient or customer?

Often documents to be faxed are prepared for other purposes and for other customers. At best, they may only indirectly address the current customer's issues. At the worst, they may mislead or annoy the current customer.

For example, a faxed contract originally intended for one recipient might be totally inapplicable to another recipient. Make sure the document to be faxed is completely tailored to the person at the other end.

2. Send pages that are clean and likely to be readable after transmission.

The quality of a fax transmission copy is still a possible problem. Faxes are typically more fuzzy and indistinct than printouts from a good photocopy machine.

Often pages with a small typeface, handwritten comments, or shaded boxes will not be very readable when they are received. Handwritten comments often extend to the edge of the paper, thus risking being cut off during transmission, or the handwriting is so unclear as to be unreadable.

You and your organization owe it to yourselves to guarantee that documents sent are readable and professional looking. The documents you send, by whatever means, reflect your professional image and your organization's commitment to quality.

3. Be sure to include a cover sheet both with your customer's name and phone numbers and with your own organization clearly identified.

In haste, we are often tempted to send documents without a cover sheet. Often, these are sensitive documents that should not sit out in the open visible to anyone who passes by the fax machine. In many cases, you merely need to take a couple of minutes to fill in your organization's standard cover sheet. Your personal computer probably features a cover-sheet template. See the sample cover sheet below.

A "headless" transmission means that pages arrive at a fax machine without anyone knowing where they should be directed. Pages can stay in the fax machine overnight or even for days before someone troubles to find out who should receive the fax.

This delivery problem would be partially solved if you and the recipient agreed that you would receive a call either confirming the receipt of the pages or notifying you that the pages had not arrived as intended.

4. List on the cover sheet the documents transmitted and, if appropriate, add a personal message.

A list of documents transmitted is helpful if you have sent more than two or three separate documents. This list of titles is more helpful than the number of pages transmitted, although a page count is also a good idea.

This list of titles becomes an ad hoc table of contents for the fax and a helpful tracking tool beyond. As you list items, be sure to give their full title or an identifying phrase, such as their authors or the date prepared. See TABLES OF CONTENTS.

Finally, use the cover sheet to personalize your message. Make this message direct and specific:

> Janice—
>
> You'll find the contract and signature pages attached. Please sign and fax them back to me by Monday. We appreciate your business!

This message can be typed, or a handwritten note is appropriate if you have a last-minute request or comment to make.

FAX COVER SHEET

FranklinCovey
Professional Services Group
2200 W. Parkway Boulevard, Third Floor
Salt Lake City, UT 84119

DATE: May 6, 2010 **TIME:** 2:45 p.m.

TO: Terri G. Langdon, Account Manager
COMPANY: FranklinCovey, Gaithersburg
PHONE: 303/757-1200, ext. 1257
FAX #: 303/757-4444

FROM: Alex J. Bowen, Finance
COMPANY: FranklinCovey, SLC
PHONE: 801/299-4700, ext. 4801
FAX #: 801/299-4701

SUBJECT: Information From Management Meeting, May 5

NUMBER OF PAGES INCLUDING COVER SHEET: 12

MESSAGE:

Terri,

Here is the information distributed during management meeting, including:

1. Financial statements for the Eastern Office (April Report) 6 p.
2. Travel expense reimbursement record, May 4 2 p.
3. Company Credit Card statements, May 1 1 p.
4. Personnel changes 1 p.
5. Stock purchase application (new) 1 p.

Call if you have any questions about these items. We look forward to seeing you at the Sales Training Retreat in June.

Alex

Footnotes

Footnotes are a common method of citing sources, especially on the internet. As their name implies, footnotes usually appear at the bottom of a page or a screen. In printed materials, sources are more often cited in endnotes that appear at the end of a chapter or a book.

Your word-processing program allows you to insert footnotes on each page. Consider using this feature, assuming you wish to use footnotes, because readers find footnotes at the bottom of the page easier to read while they are reading the text.

Instead of using footnotes, writers in the physical and biological sciences usually cite the author and the date of publication by enclosing them within parentheses in the text. These citations are developed fully in bibliographies that appear at the end of the text. See CITATIONS and BIBLIOGRAPHIES.

Choose parenthetical citations for most business and technical documents. Do not use footnotes unless you are writing for the internet or for a publisher or a journal that requires that you use footnotes.

Footnotes

1. Use superscript (raised) Arabic numerals immediately following a quotation or paraphrase to indicate that the quotation or paraphrase has a footnote.
2. In the first footnote to a source, include the author or authors, the full title, complete publishing information, and the pages being referred to. If citing an electronic book, indicate the format. If citing an online source, add the URL and, in brackets, the date you accessed the source.
3. In second and subsequent references, make footnotes brief. Generally, include only the author's last name and the page number of the material referred to.

1. Use superscript (raised) Arabic numerals immediately following a quotation or paraphrase to indicate that the quotation or paraphrase has a footnote:

Within your text, you may have a quotation from a published book or article: "Writers should always use quotation marks for exact quotations."[1] Sometimes you may be paraphrasing someone's ideas.[2] In these cases, your footnote number should come as close to the idea as possible,[3] even if the particular sentence goes on to discuss a second source.[4] Naturally, in a normal document, you should avoid having footnotes after every sentence or phrase.

NOTE 1: The footnote number comes after all punctuation, except for a dash. Footnotes are numbered sequentially within a chapter of a book and within an article. See QUOTATIONS and QUOTATION MARKS.

NOTE 2: Footnotes in graphics sometimes use an asterisk (*) or a dagger (†) when an Arabic numeral might be confusing.

2. In the first footnote to a source, include the author or authors, the full title, complete publishing information, and the pages being referred to. If citing an electronic book, indicate the format. If citing an online source, add the URL and, in brackets, the date you accessed the source:

Book by one author

1. Gunjan Bagla, *Doing Business in 21st Century India* (Paris: Hachette Livre, 2009), Gemstar e-book, 90.

Book by two authors

2. Stephen R. Covey and Jennifer Colosimo *Great Work, Great Career* (Salt Lake City: FranklinCovey Press, 2010), Kindle e-book, 34–35.

Book by more than three authors; information from several pages

3. Gavin Schmidt, et al., *Climate Change: Picturing the Science* (New York: W.W. Norton & Co., 2009), 126–127, 175, 189.

Journal article

4. Janice Kumar, "Metal Matrix Alignment in Fiber Production," *Delhi Institute of Technology Journal*, 16 (October 2007): 45–46. http://nopr.niscair.res.in/kumar/123456789/7257.html [accessed May 7, 2010].

Public document

5. U.S. Congress, Senate, Foreign Affairs Committee, Report on Two Chinas in the Coming Decade, 110th Cong., 2d sess., 2008 (Washington, D.C.: United States Government Printing Office, 2010), 187–188.

Dissertation or thesis

6. O. X. Jones, "The Influence of Congressional Resolutions on Trade with China: A Study of Inconsistencies" (PhD dissertation, University of Maryland, 2009), 87–88. http://library.dialog.com/jones/trade/china/bl0035.html [accessed June 1, 2010].

Personal letter

7. Senator Jeanne Shaheen of New Hampshire to O. X. Jones, 23 November 2009. Personal files of O. X. Jones, Cambridge, UK.

Interview

8. Wang Sung, interview during the annual meeting of the American Committee on U.S.-China Relations, Seattle, Washington, October 2008.

NOTE 1: Footnotes are similar in form to paragraphs. The first line is indented, and all items are punctuated as if the information were the first sentence in the paragraph. The items are usually separated by commas.

NOTE 2: Titles of books, journals, and newspapers are italicized, even if cited from an electronic source. Titles of blogs are not italicized. See TITLES.

NOTE 3: Writers can add a comment or additional facts to a typical footnote:

9. Wang Sung, interview during the annual meeting of the American Committee on U.S.-China Relations, Seattle, Washington, October 2008. Mr. Sung,a cultural attaché for the People's Republic of China, granted this interview with the understanding that all of his comments would be off the record until after the 2008 U.S. Presidential election.

NOTE 4: Superscripts are a common option for numbering footnotes presented either at the bottom of the page or at the end of the chapter or article:

[10.] Jason K. Bacon, *The Two-China Policy* (New York: Columbia University Press, 2005), 85.

NOTE 5: The examples in rule 2 illustrate only the most common types of footnotes. For other publications, check a source such as Kate L. Turabian, *A Manual for Writers of Term Papers, Theses, and Dissertations*, (Chicago: University of Chicago Press), which is updated frequently (a quick guide is available online). Turabian devotes over 40 pages to detailed examples of footnotes.

3. In second and subsequent references, make footnotes brief. Generally, include only the author's last name and the page number of the material referred to:

Second or subsequent footnote for one author

11. Bacon, 56–57.

Second or subsequent footnote for two authors

12. Sloan and Seymour, 18–21.

Second or subsequent footnote in which two or more works by one author are cited

13. Bacon, *Two-China Policy*, 76.

NOTE: Using the author's name and a short title for second and subsequent footnotes eliminates the need for such traditional Latin abbreviations as *ibid.*, *op. cit.*, and *loc. cit.* These abbreviations make footnotes difficult to read.

Fractions

Fractions are mathematical expressions for the quotient (division) of two quantities: $^1/_2$. In this fraction, the slash mark means *1* divided by *2*. Strictly speaking, decimals are also fractions. See DECIMALS.

Fractions

1. Spell out and hyphenate fractions appearing by themselves in ordinary text, especially if they are followed by *of a* or *of an*.
2. Use figures for fractions when they are combined with abbreviations or symbols.

1. Spell out and hyphenate fractions appearing by themselves in ordinary text, especially if they are followed by *of a* or *of an*:

two-thirds of an inch (*not* $^2/_3$ of an inch)

decreased by one-third

one-half foot

one-fourth inch

one-tenth

one-hundreth of a mile

two one-thousandths

eighty-four one-thousandths (*better* 0.084)

NOTE 1: The longer a fractional expression becomes, especially if whole numbers are involved, the more desirable it is to express the fraction in figures (or a decimal):

$^{56}/_{64}$

$^{98}/_{100}$ (*or* 0.98)

$2^1/_2$ times

$6^3/_4$ (*or* 6.75)

$29^1/_3$

NOTE 2: Measurements, especially in scientific and technical documents, require figures:

$^1/_3$-foot step

$^1/_2$-inch pipe

$^2/_3$-inch-diameter pipe

$7^1/_2$ meters

$8^1/_2$- by 11-inch paper

See NUMBERS and HYPHENS.

2. Use figures for fractions when they are combined with abbreviations or symbols:

$34^1/_3$ km

$8^1/_2$ hr

$5^1/_2$″ x 6 $^2/_3$″

Such fractions and abbreviations are most common in figures or in field reports. See NUMBERS.

Gobbledygook is not a recent problem. Well over 200 years ago, opponents of Benjamin Franklin argued that to vote, a man had to own property. Franklin's supporters disagreed and stated their case as follows:

> It cannot be adhered to with any reasonable degree of intellectual or moral certainty that the inalienable right man possesses to exercise his political preferences by employing his vote in referendums is rooted in anything other than man's own nature, and is, therefore, properly called a natural right. To hold, for instance, that this natural right can be limited externally by making its exercise dependent on a prior condition of ownership of property, is to wrongly suppose that man's natural right to vote is somehow more inherent in and more dependent on the property of man than it is on the nature of man. It is obvious that such belief is unreasonable, for it reverses the order of rights intended by nature.

Franklin agreed with this argument but knew that people wouldn't be moved by such pompous oratory. So he explained his position as follows:

> To require property of voters leads us to this dilemma: I own a jackass; I can vote. The jackass dies; I cannot vote. Therefore the vote represents not me but the jackass.

Gobbledygook is language that is so pompous, long-winded, and abstract that it is unintelligible. Some dictionaries trace the term to the verb *gobble*, describing the sounds made by turkeys, and it is tempting to believe that writers of gobbledygook resemble this vocal bird. Actually, such writers are usually well-intentioned. They might even take pride in writing what they consider to be sophisticated and complex language.

Perhaps the best way to appreciate gobbledygook is to read a couple of more recent samples:

> This office's activities during the year were primarily continuing their primary functions of education of the people to acquaint them with their needs, problems, and alternate problem solutions, in order that they can make wise decisions in planning and implementing a total program that will best meet the needs of the people, now and in the future.

> Because the heavy mistletoe infestation in the Cattle Creek drainage area has rendered the residual timber resources useless for timber production, the ultimate goal is to establish a healthy, viable new stand of Douglas fir.

The average reader has to read these passages several times before beginning to decipher such nonsense. Why are the passages so difficult?

—**Words and phrases are abstract.** What does *alternate problem solutions* mean? Similarly, are *residual timber resources* the same thing as *trees*? If so, the writer should say *trees*.

See Wordy Phrases and Redundant Words.

—**Words and phrases are pompous-sounding.** Are the *office's activities* primarily their primary functions? Is a *healthy timber stand* different from a *viable timber stand*? If not, then the writer should stick with the simpler word: *healthy*.

—**Sentences are long and clumsy.** By themselves, the 57 words in the first paragraph would make reading difficult, but the clumsy phrasing makes the reading impossible. The 35 words in the second paragraph are closer to a reasonable number, but the writer delays the major thought in the sentence with a massive introductory clause (beginning with *because*). As written, the sentence demands that readers remember a long opening condition while they try to absorb the main thought. The sentence would be clearer if the main and introductory clauses were reversed.

See Sentences, Strong Verbs, Active/Passive, and Parallelism.

Gobbledygook

1. Use concrete and specific words and phrases whenever possible.
2. Avoid pompous words and phrases.
3. Make sentences short, direct, and clear.

How to Avoid Gobbledygook

1. Use concrete and specific words and phrases whenever possible:

this

> To complete the recreational plan, we will need pictures of all tables, fireplaces, and other existing camping facilities in the state park.

not this

> In order to bring the proposed recreational plan to completion; to evaluate existing recreation-site appurtenances and facilities; and to include applicable facilities such as tables, fireplaces, etc., in the proposed new recreational plan, it will be necessary to receive photographs of all current appurtenances and facilities located within the state park area.

See Wordy Phrases and Redundant Words.

2. Avoid pompous words and phrases.

The word *appurtenance* in the preceding example is an excellent example of a pompous word. Most readers will not understand *appurtenance*, and forcing them to look up the word in a dictionary might not clarify the passage. Two recent desk dictionaries define *appurtenance* quite differently: "something added to another, more important thing; accessory" (The *American Heritage Dictionary*) and "an incidental right (as a right-of-way) attached to a principal property right and passing in possession with it" (*Webster's New Collegiate Dictionary*). Which meaning should readers choose? More to the point, why make them choose?

If a word is not in common usage, avoid it or use it in such a way that the context provides a definition. In the example above, *appurtenance* surely fails the test. Here are some other pompous words and phrases with possible substitutes in parentheses:

Accordingly (so)
acquaint (inform or tell)
activate (start)
additional (more)
adhere (stick)
ameliorate (improve)
apprise (tell or inform)

Cognizant (aware)
commence (begin)
compensation (pay)
component (part)
concur (agree)
configuration (shape, design)
conflagration (fire)
curtail (slow, shorten)

Demonstrate (show)
descend (fall, climb down)
donate (give)

Encounter (meet)
evacuate (leave, empty, clear)
exhibit (show)

Fabricate (make)
factor (cause)
feasible (likely, possible)
fracture (break)
function (work, act)

Implicate (involve)
impotent (weak)
incinerate (burn)
increment (amount, gain)
indubitably (doubtless, undoubtedly)
inform (tell)
in isolation (alone, by itself)
initiate (begin)

Locality (place)
locate (find)

Major (chief, main)
manifest (show)
manipulate (operate)
manufacture (make)
modification (change)
moreover (besides)

Necessitate (compel)
necessity (need)

Paramount (main, chief)
perspective (view)
phenomenal (unusual)
philosophy (belief, idea)
potent (strong)
practically (nearly, most, all but)
proceed (go)
purchase (buy)

Ramification (result)
render (make)
request (ask)
reside (live)
residence (home)

Sophisticated (complex)
spotlight (stress)
state (say)
stimulate (excite)
succor (help)
sufficient (enough)

Thoroughfare (aisle, street)
terminate (end, fire)
transmit (send)

Utilization (use)
utilize (use)

Vacillate (waver)
veracious (true)
visualize (imagine, picture)

3. Make sentences short, direct, and clear.

Sentence length is only one sign of complexity. A 10-word sentence can be unclear because it is poorly structured and difficult to read:

these

Cost overruns continued throughout the second quarter. They continued even though we had audited all production and remanufacturing processes.

We need to approve their financial proposal before the end of our fiscal year. If we don't, problems will arise.

not these

Despite a careful audit of all pertinent processes, cost overruns, especially those related to production and remanufacturing, continued throughout the second quarter.

There are certain problems likely to develop if approval of their financial proposal is deferred until after the close of our fiscal year.

See SENTENCES, STRONG VERBS, ACTIVE/PASSIVE, and PARALLELISM.

Gobbledygook and Jargon

Gobbledygook and jargon can both make reading difficult, but they are not the same.

Jargon includes terms used by a specific technical or professional group. Carpenters, for instance, have a number of jargon terms: *stud, joist, sill plate, header, cap plate, trip-L-grip, etc.* See JARGON.

Gobbledygook can include technical jargon, but gobbledygook also includes nontechnical words that are unfamiliar, unnecessary, or too long. Good writing can include some jargon, particularly if the words are defined or understandable within the context, but good writing never includes gobbledygook.

Graphics for Documents

Graphics are essential for conveying key information in documents. Because graphics are visual rather than verbal (as writing is), they are much more emphatic than the text around them. Readers will recall one impactful graphic long after they have forgotten a thousand words of text.

Today's readers expect to see graphics in documents, even in routine newsletters and business reports. Readers also expect these graphics to be professional-looking. After all, high-definition imagery now dominates all media, from big-screen presentations to small handheld devices.

Fortunately, graphics software has made creating high-quality, professional-looking graphics possible for everyone. Creating an effective graphic is not, however, always as easy or simple as selecting a preformatted bar chart or pie chart. Sometimes a program can create a pretty graphic, but it isn't exactly what you want, or the graphic may be the wrong size or shape. Figure 1 contains some tips that apply to creating almost any type of graphic.

For the following discussion, graphics refers only to visual information used within documents. Many of the principles discussed, however, apply also to slides and other presentation graphics. See Charts, Color, Graphics for Presentations, Graphs, Illustrations, Maps, Photographs, and Tables.

Graphics for Documents

1. Choose graphics to emphasize your primary purpose or major concepts.
2. Create your graphics before you write the text.
3. Select graphics that are appropriate for your topic and your audience.
4. Design or modify graphics so readers can easily interpret them.
5. Introduce graphics in the text before they appear.
6. Design graphics that are simple, uncluttered, focused, and easy to read.
7. Choose emphatic devices to stress important ideas graphically.
8. Number graphics in the order of appearance.
9. Use clear and interpretive captions.

1. Choose graphics to emphasize your primary purpose or major concepts.

Graphics should capture your key recommendation, a surprising trend, the unexpected financial problem, or the most convincing data. Choose graphics for the highlights of your message. Save the supporting details for your text (or better yet, for an attachment or appendix).

In some cases, just one primary graphic may convey the main point of your entire document. If so, the written text serves merely as a "deeper dive" into the meaning or implications of the visual message. Effectively designing the primary graphic becomes your most important job. For example, if your purpose is to advocate one business strategy over another, one simple bar graph that contrasts the revenues associated with each strategy may be enough to persuade your readers.

In other cases, your document may require several key graphics to serve your purpose. For example, if your purpose is to recommend a new mobile phone provider, you might present a series of bar graphs, each showing how four competing providers compare on these criteria: (1) monthly charges for a service contract; (2) global coverage; and (3) user satisfaction ratings for the last three years. Figure 2 shows the bar chart for the third criterion (user satisfaction ratings).

Because of their impact, graphics often *are* the message of a document. Never use graphics for mere decoration. Reserve their use for visual information central to your message. If you waste graphics on unimportant or unnecessary information, you waste a valuable opportunity, and you might distract your readers from a key point.

A graphic and its caption should be clear and understandable without requiring readers to search for clarifying information in the text. The caption should tell readers not only what they are looking at but also why it is important. See Captions.

2. Create your graphics before you write the text.

Graphics should never be an afterthought. Visualization drives the text. Create visuals first; write text last. Graphics are more emphatic than text and should therefore be designed early in the writing process.

This rule is fundamental to an effective writing process. Writing a document begins when you visually

Tips for Creating Computer Graphics

Graphics software is more accessible and sophisticated all the time. Still, using graphics software to generate just what you want is not always easy. Use the following tips for working with graphics software. These tips are generic, not referring to any single software program.

- **Practice developing graphics of the sort you expect to need.** Are file formats compatible, and does your software program allow easy transfer to another program or format such as an email message? Actually create or copy a graphic into a dummy document so you'll learn what the problems are.
- **Verify the size and format of the graphic.** Design and size your graphic for your intended use and audience, not for your own computer screen. What you see on the screen may not be what you get. Use the "print view" function or print a hard copy to see the results.
- **Check to see how the insertion or the enlargement of a graphic affects the surrounding text.** Does the text wrap (move) automatically into the position you want?
- **Practice inserting a legend (key to symbols in a graphic), a caption, or explanatory notes to the graphic.** It might take you time and practice to understand how to use the menus and various features of your program.
- **Use existing graphics if possible, rather than starting from scratch.** Sources include downloadable clip art (some of which allows free use) or collections of photos and other graphics on CD.
- **Use the graphic templates in your software rather than designing a graphic yourself.** A preformatted line graph or process flowchart can save you hours.
- **Choose the simplest template that will enable you to get your message across.** You don't need fancy gradients or three-dimensional bars if they will distract from your message.

Figure 1. Tips for Creating Computer Graphics. *Remember to allow extra time for graphics preparation, especially if you are unfamiliar with the software.*

design the whole document as a prototype (or mock-up), which will include all projected graphics. SEE WRITING AND REVISING and PAGE LAYOUT.

Even if you have the most sophisticated graphics software available, you should start your design with thumbnail sketches on paper of the key graphics you might use. If you prepare your graphics early, you may discover that the graphics make some of your text unnecessary. If so, you want to discover this redundancy even before you have actually written the text in question.

3. Select graphics that are appropriate for your topic and your audience.

Graphics come in many forms: charts, graphs, illustrations, maps, photographs, and tables. These forms—and all of their variations and combinations—give you many choices for organizing ideas and data into graphic visual representations. Use the matrix in figure 3 to help you determine which graphics to use in a given document.

- Charts show relationships. Charts can display organizational relationships (organizational charts) and illustrate the flow and relationship of steps in a process (flow charts). SEE CHARTS.
- Graphs show the correlation of two or more variables and how they change. The most common graph is a simple line graph. Graphs also include bar graphs and pie graphs, both of which show comparisons and trends. SEE GRAPHS.
- Illustrations and diagrams help show abstract concepts as objects or processes and provide a perspective on these concepts that verbal descriptions cannot capture. Illustrations can also show exploded views of smaller parts of a larger object or assembly. SEE ILLUSTRATIONS.
- Maps show topographical relationships and indicate scale and distance. SEE MAPS.

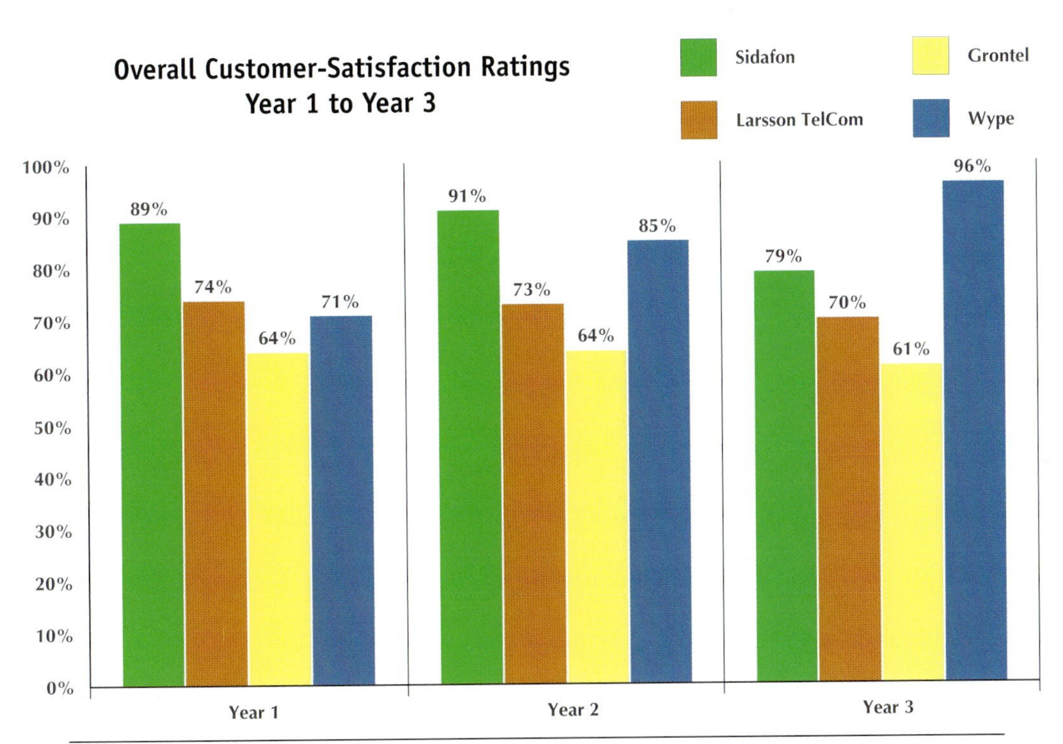

Figure 2. Overall Customer-Satisfaction Rating for the Last Three Years. *Wype is the most promising vendor, especially in light of its most recent rating and its continuous improvement trend.*

Type of Graphic	Document Purposes										
	Costs	Causes/ Effects	Trends	Organizational Relationships	Policies Procedures	Decision Alternatives	Work Flow	Chronology	Design Parts Apparatus	Comparison Contrast	Advantages Disadvantages
Tables	X	X	X	X	X	X		X		X	X
Line Graphs	X		X					X		X	
Bar Graphs	X		X			X		X		X	X
Pie Graphs	X		X			X				X	X
Schematic Diagrams									X		
Flow Charts		X		X			X	X			
Maps/Site Plans			X			X					
Photographs		X							X	X	X
Tree Diagrams				X	X	X	X				
Illustrations									X		
Blueprints									X		
Combination		X	X	X	X	X	X	X	X	X	X

Figure 3. Graphics for Different Documents. *An X appears in a square to show when a specific type of graphic would likely implement a particular document's purpose.*

- Photographs convey realism and authenticity. They purport to show readers exactly what something looks like. Retouched photos are illustrations. SEE PHOTOGRAPHS.
- Tables display data in rows and columns. They allow for quick comparisons of data with great precision. SEE TABLES.

The graphic you select depends on your readers, but also on what you are trying to achieve with it. The table in figure 3 shows how various information graphics can be used to accomplish your purposes.

4. Design or modify graphics so readers can easily interpret them.

The graphics you choose should be consistent with the orientation, skill, and educational or technical level of your readers. For nontechnical readers, you would not use logarithmic graphs (readers may not understand them); on the other hand, for highly technical readers you should not use oversimplified bar graphs or pie graphs to present complex scientific data.

Remember, however, that your graphics (and your text) should be accessible to the least technical of your readers. Keep these readers in mind as you generate your graphics and your text.

If you must include a graphic that is little too difficult for such nontechnical readers, explain the key point of the graphic in the caption.

The best technical writers and editors intentionally vary the readability of their text and graphics because they know that different readers will focus on different sections of a document. Technical readers may survey only the key table or the most technical graph. Less technical readers may read only the introduction and summary (or abstract) and then skip sections or scan the rest of the document for key points.

As a writer, you should carefully analyze your prospective readers—all of your readers. Once you know who your readers are and their technical levels, you can adapt your document and your graphics to these readers.

5. Introduce graphics in the text before they appear.

Always precede graphics with a clear introduction in the text. A graphic appearing suddenly, without an introduction or explanation, generally confuses the reader.

The introduction in the text should be informative and specific:

Informative and specific

As figure 3 shows, produced water from the 2nd Langley is much more acidic than produced water from the 1st Langley.

or

This project is estimated to cost $356,200. The cost breakdown in table 15 shows that hardware costs account for nearly 65 percent of the total costs, while labor costs constitute only 12 percent of the total.

Not informative

See Figure 3.

or

The total cost of this project is estimated to be $356,200. Table 15 provides a cost breakdown.

A good introduction indicates not only what the graphic is about but also explains how the reader should interpret it. The introduction in the text will repeat some of the ideas and even the language in the caption for the graphic. See CAPTIONS.

NOTE: Sometimes a graphic must appear in the text before you can introduce it. This may occur, for instance, when a small table or chart appears at the top of a new page and is followed by a column of text. Rather than break up the text, you might keep the graphic at the top of the page with the introduction following the graphic.

6. Design graphics that are simple, uncluttered, focused, and easy to read.

When you are designing a graphic, ask yourself key questions:

—What is the point?

—What is the most important idea that I am trying to convey visually?

—What is my central concept?

Keep things simple. Don't ask a graphic to do too many things at once. When a graphic fails to communicate, it is often because the author tried to make it do too much (making it cluttered or too complex). Sometimes an author has not designed a graphic with a clear concept in mind (making the point of the graphic unfocused).

Focus on your key purpose or concept and then build graphics around it. Like a good paragraph, a good graphic focuses on one idea and conveys it sharply and purposefully.

Make graphics uncluttered by eliminating all extraneous

information. A graph full of coordinates or a full table of data readings contains a lot of valuable information, but either of these is likely to be more effective if you pare it down to the essentials.

If your readers cannot tell you what the point is in 5 or 10 seconds, the graphic is not effective.

Eliminate all unnecessary borders, lines, and shadings for an open and simple look (*and* to save ink).

Choose a simple typeface for lettering in graphics, and use capitals and lowercase letters for all headings and labels. (Lettering all in capitals is hard to read, especially if it extends beyond two or three words.)

Be sure that lettering is large enough so that if the graphic is reduced, the lettering will still be readable.

Make lines and lettering that indicate axes, scales, notes, and explanations lighter than the lines and lettering indicating data points, curves, areas, or bars.

7. Choose emphatic devices to stress important ideas graphically.

Create styles for graphics so that they will be consistent throughout each project. Choose the same emphatic device to stress the same kind of word or data. See Page Layout.

Emphatic devices include:

- Underlining
- *Italics*
- **Boldface type**
- Larger font size
- Outline type
- Shadow type
- Small caps
- ~~Strikethrough~~
- Line patterns
- Symbols √ ☐ •
- Dingbats ✳❖✖✢☆
- Color

See Emphasis and Page Layout.

Color is particularly useful in helping readers understand graphics, locate key information, and distinguish between different parts, phases, or configurations. Use color to highlight a row, line, slice, area, column, circle, or data point to focus the reader's attention on that item.

Highlight special conditions in color: cautions are usually yellow and black; warnings are often red and white. See Color.

As you select colors, try to establish a color scheme that makes sense: green seems prosperous, yellow is transitional, red is for failure. Or red and yellow are warm, while blue is cool.

Use contrasting colors to show contrasting concepts or major changes; use shades of one color to show minor variations. Use the brightest colors to emphasize the most important ideas, the most muted shades to subdue the less important ideas.

Beware of too much color, making the graphic look kaleidoscopic. Remember that downloaded graphics might be printed only in black and white, so don't depend only on color to make your point; use dotted lines or other appropriate textures. Finally, consider that, on average, 10 percent of your readers may be color blind. See Color and Graphics for Presentations.

8. Number graphics in the order of appearance.

Figures include all graphics that are not tables, including graphs, charts, diagrams, illustrations, photographs, and maps.

Traditionally, tables and figures were numbered separately. So table 5 could appear in a document after figure 20. But a current trend is to call all graphics figures, regardless of their type, and to number them consecutively.

In a short document or a document with few figures, number them sequentially through the entire text. In a document with chapters or numbered sections, give figures numbers with an en dash, such as figure 3–2. The first number is the chapter or section, and the second number denotes the number of the graphic (figure) within a chapter or section.

In longer documents, especially formal reports and publications (pamphlets, books, research studies, etc.), include a list of figures following the table of contents.

If the document has a large number of specialized figures, then create a separate list of each special type. A lengthy report with a number of maps might include a list of maps. Another report, in which photography plays an important role, might include a separate list of photographs. If the document does not warrant such special listings, however, don't create them.

Graphics for Documents

Lists of tables and figures appear immediately after the table of contents (usually on separate pages) and should themselves be listed in the table of contents. If you create both a list of tables and a list of figures, either may appear first. See TABLES OF CONTENTS.

9. Use clear and interpretive captions.

Interpretive captions usually require one or more sentences. But they should be informative without becoming too lengthy. See CAPTIONS.

A list of figures should provide, in sequence, each figure number, a title, and page number for each graphic.

Within text, captions for graphics should be informative and specific.

Captions that are only titles for a graphic are not very helpful, as these examples of weak captions show:

Figure 17. Amphibian Population Trends.

Table 2. Particulates in Sonoma Valley, California

Interpretive captions give specific information. They not only name the graphic, but they also tell readers what the point or importance of the graphic is. Notice how the following examples improve on the captions above:

Figure 17. Amphibian Population Trends in Central America. *African Clawed Frog population has declined 43 percent in the last decade.*

Table 2. Particulates in Sonoma Valley, California. *From May to October, particulates in Sonoma Valley, California, remained well below EPA emission standards.*

See CAPTIONS for further examples.

Graphics for Presentations

Graphics for presentations refer to graphics used in oral presentations (either recorded or live) and created with any media or technology. The following suggestions apply to any of the possible media: slides, videos, or Web presentations. For graphics used within documents, see GRAPHICS FOR DOCUMENTS.

Graphics for Presentations

1. Choose and design graphics to emphasize important ideas and a major theme and to match your audience's needs.
2. Limit your graphics to no more than you can easily show in the time allotted for your presentation.
3. Design graphics so they are clear and easy for all members of the audience to read.
4. Use graphical special effects only for emphasis.
5. Use color to focus attention, but don't overuse it.
6. Plan enough time to complete and revise the graphics.
7. Use graphics to communicate the structure of your presentation.
8. Use graphs, charts, tables, and illustrations instead of text if they are easier to understand.
9. Make handouts of your primary presentation graphic.

Nearly all business and technical presentations rely on graphics, usually computer-driven slides. A presenter may choose to focus on one or two primary graphics, tell a story through a series of slides, or rapidly talk through a large amount of visual data to a highly technical audience, depending on the presenter's purpose and the audience's level of understanding.

Presentations fall into two categories: to inform or to incite action. Therefore, the visual message must be carefully designed to do one or the other. A visual that merely informs might not motivate listeners to act. For example, some analysts of the events surrounding the Columbia space-shuttle disaster believe that technical presentations made to the authorities, while informative, did not move them to the urgent action that might have saved the shuttle.

At bottom, virtually all presenters want to incite action. Therefore, they should choose graphics that will make the audience feel like doing what needs to be done. A simple line graph that shows the competition rapidly gaining ground may well be more likely to motivate action than 10 slides full of contrasting data about product categories.

Any message is stronger if it is concise. Graphics show data in a concise and effective manner (in tables); they quickly reveal important trends (in graphs); and they show key configurations that would be difficult, if not impossible, to describe with words (in charts and illustrations). Therefore, preparing an effective presentation is largely about preparing an impactful visual message. See CHARTS, GRAPHICS FOR DOCUMENTS, GRAPHS, ILLUSTRATIONS, MAPS, PHOTOGRAPHS, and TABLES.

1. Choose and design graphics to emphasize important ideas and a major theme and to match your audience's needs.

Graphics for presentations should capture the key recommendation, the surprising trend, the unexpected financial problem, or the most convincing data. Use graphics that will have an impact on and be memorable to your audience.

Always profile your potential audience and their needs. In a single presentation, you may face senior executives, corporate accountants, sales representatives, industry bloggers, and engineers, each with only partial knowledge of the issues you are presenting.

Remember also that some of your audience may not be conversant with your industry or jargon, or share your cultural assumptions. In such cases, analyze the audience to assess whether your message is missing part of your audience. Design graphics that include rather than exclude your listeners.

Begin planning your presentation by designing a primary graphic or several graphics showing the key points or the major theme you want to get across to your audience.

Figure 1 is an example of a graphic designed for a specific presentation. The graphic is an illustration designed to show the three sources of products returned to a retail store chain. This illustration would be more memorable and effective than an ordinary bulleted list of the three sources.

Figure 1. Three Sources of Products Returned. *About the same volume of product is returned through each of the three channels. This graphic is an illustration designed to show the three sources, not just list them. This graphic would be more memorable than an ordinary bulleted (or numbered) list of sources.*

The presenter designed this graphic to show that about the same volume of products was returned through each of the three channels. Her purpose was to change a general impression that returns came primarily through stores. A bulleted list would not have made the case as clearly.

Use a primary graphic, such as figure 1, at the beginning of your presentation, particularly if top managers are in the audience. (They tend to be time-conscious and may not stay for the whole presentation.) You might also use the graphic at the end as a summary.

See Graphics for Documents for a discussion of how to select charts, diagrams, graphs, illustrations, maps, photographs, and tables. Any one of these different graphics will work equally well in a presentation, assuming that the format and font size allow an audience to read or view the content easily.

As you are planning your presentation, list possible graphics before or concurrently with your script. Make a thumbnail sketch of your presentation slides, or your graphics will be an afterthought. If graphics are not an integral part of your presentation, you will weaken your case.

2. Limit your graphics to no more than you can easily show in the time allotted for your presentation.

A good rule of thumb for presentations is that you should have no more than two or three graphics per minute. And this number is excessive if you plan on taking time to explain to your audience what they are seeing and what the implications of the graphics are. Even in a series of slides that are self-explanatory, you will find that you have to give viewers 10 or 15 seconds to analyze and absorb the message of a single slide.

These time estimates assume a presentation framework where your goal is to share information. If, as in a fast-cut video, you want to create only a mood or feeling, then the images can and should move quickly. Similarly, if you are preparing an informational video, you may find that some segments may require 15 or 20 seconds of screen time, while other segments can follow a fast-cut sequence of images.

Far too often, presenters make the mistake of using too many slides. The trend is toward short presentations that go straight to the point. The more slides, the less the impact of each. Avoid inflicting "death by PowerPoint" on your audience, subjecting them to an endless sequence of slides. A good strategy is to focus on a primary graphic and a handful of supporting slides that are essential to your purpose. If other graphics remain important as background, consider including copies of such graphics in a download or handout the audience can review later (see rule 9).

3. Design graphics so they are clear and easy for all members of the audience to read.

The ideal graphic for a presentation is designed so that the message can be perceived in less than 5 seconds. **Don't try to squeeze in too much on each graphic.** Figure 2, for example, would be fine as a page in a document, but as a slide, it would not be visible much beyond the second row in a small room. Often presenters settle for a screen projection of a page such as in figure 2. They fail to plan far enough ahead to prepare graphics especially for their presentation.

Figure 3 is an adaptation of one part of the page shown in figure 2. An adaptation is necessary so that the size of type used will be readable.

Type size and type design are major concerns when you are designing graphics. Figure 4 is a page of sample type sizes. You may want to make a test slide of figure 4; then try projecting it onscreen to see how legible text is at different sizes. Remember to stand as far back from the screen as the most distant viewer in your audience. Which line can you read clearly from there?

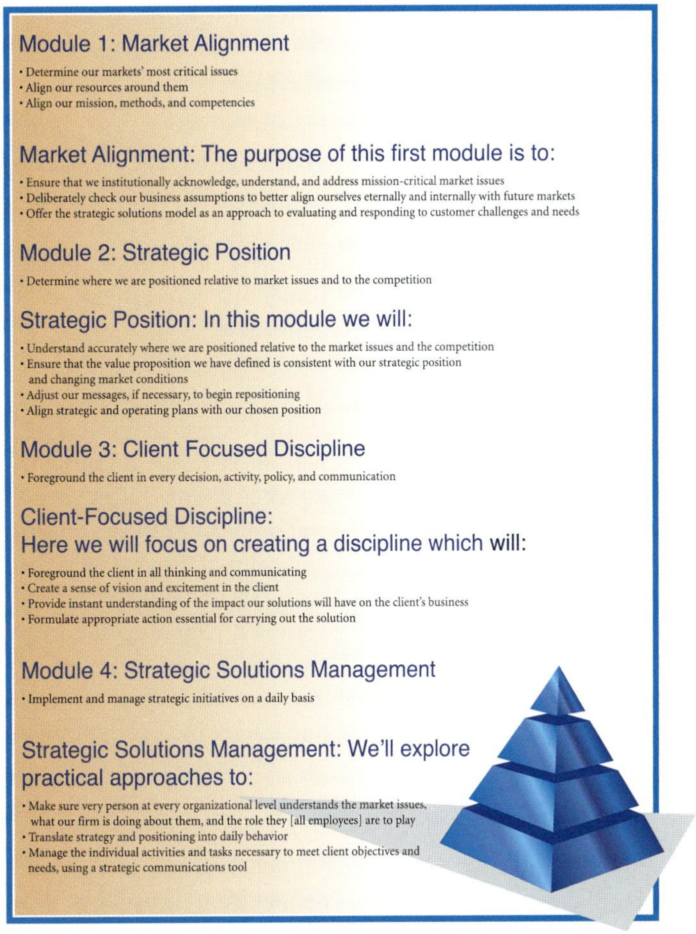

Figure 2. Four Modules on Strategic Selling. *This figure would be unreadable as a slide. It could be the basis for four or five slides if a presenter wanted to present this information graphically.*

Graphics for Presentations

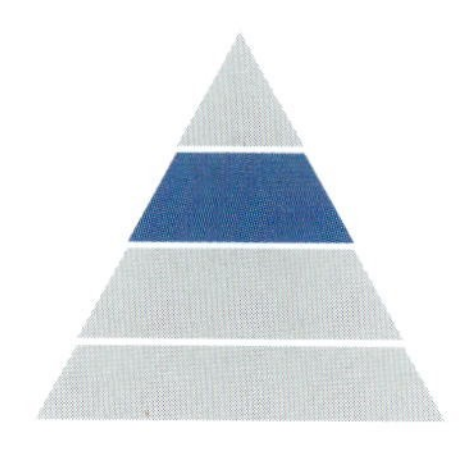

Client-Focused Discipline
• Foreground client in thinking and communicating
• Create vision and exitement in client
• Provide instant understanding of impact our solutions will have on client's business
• Formulate action for carrying out the solution

Figure 3. Client Focus in Strategic Selling (Module 3). *This slide presents only a summary of one of the four modules in figure 2. This focus on essential words and phrases allows for large type, making the figure more suitable for a slide*

For text on a slide, a sans serif typeface is clear and easy for most people to read. If you wish to use a serif font for contrast, use it only for titles. See PAGE LAYOUT for a discussion and examples of both serif and sans serif typefaces. Avoid all caps for titles because they are harder to read; instead, use initial caps and lower case.

When (or if) you use a fancy typeface, such as cursive script, use it sparingly and only for emphasis.

For greater visual emphasis in your graphics, **bold your main points or major headings.**

4. Use graphical special effects only for emphasis.

Today, graphics software features animation, sound effects, built-in clip art, and elaborate fonts. Additionally, infinite libraries of these things are available on the Internet. Use these features minimally and only to draw the attention of the audience to something you really want to emphasize.

These capabilities make it tempting to use technology for technology's sake. Be cautious, because many flashy elements clutter your slides and draw focus from your message. Use animations and special effects sparingly and only for a good reason. If your audience must watch an arrow zigzag across the screen, make sure the zigzag serves a clear purpose. Avoid using more than two font styles in your presentation, even though you have hundreds at your disposal. Generic clip art is abundant but often tacky and can detract from the unified, focused impression you want to make.

Don't overuse frames, boxes, and shadowing, which often just crowd the picture. Finally, fancy animated transitions between slides are more likely to irritate than to charm a business audience.

5. Use color to focus attention, but don't overuse it.

For presentation slides, normally use a light background and dark letters. The high contrast of a dark background and light letters can be powerful if used only occasionally. Some colors just don't work well together; figure 5 is an example of poorly chosen colors.

Use bright colors to focus the eye on important information and dimmer colors for less important information. Don't rely exclusively on color, because some people may wish to print your slides in black and white. Additionally, red-green color blindness affects one of every ten males and a small number of females. Usually, you can combine colors with designs or textures that even a color-blind person can distinguish. Figures 6 and 7 illustrate two uses of color in graphics. A color-blind person would not have trouble interpreting either figure 6 or figure 7.

Use color, but don't overuse it. Color heightens interest and provides emphasis, but too much color may cause a loss of focus on the topic. More than three colors on a slide can make your graphics look busy and cluttered. See COLOR.

E — 108 pt

JWG — 92pt

ARBXE — 72 pt

ZIUVHLJQO — 60 pt

OSKLNYDPRTIE — 48 pt

MQPIXBUWPFCHLOUMV — 36 pt

ZYXWVUTSRQPONMLKJIHGFEDCBA — 24 pt

BADCFEHGJILKNMPOQRTSVUXWZY — 18 pt

LMNOPABCQRSGHIXYZJKDEFUVW — 14 pt

ABCDEFGHIJKLMNOPQRSTUVWXYZ — 12 pt

ABCDEFGHIJKLMNOPQRSTUVWXYZ — 10 pt

ABCDEFGHIJKLMNOPQRSTUVWXYZYXWVUTSRQPONMLKJIHGFEDCBA — 7 pt

Figure 4. Readability Test Sheet. *Make a slide of this page and then use this test slide to determine which font size is best suited to the size of room where you will be making a presentation.*

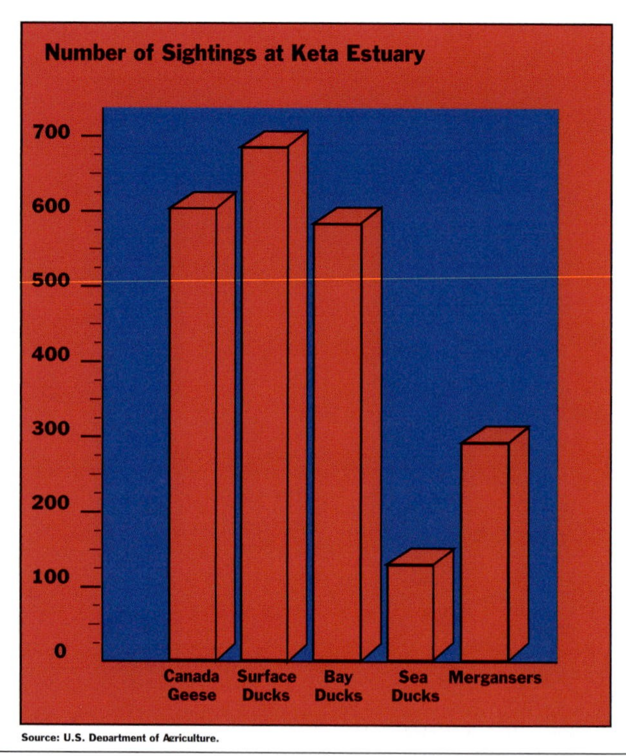

Figure 5. A Dreadful Graphic. *Highly contrastive colors can often be too bright, even glaring. Stay with softer hues and tints. See* *Color.* *The three-dimensional bars are also misleading; for example, do the Canada geese number 600 or 625?*

Single-Family Houses Sold and Sales Price

1990 to 2015

Houses Sold
Millions of houses

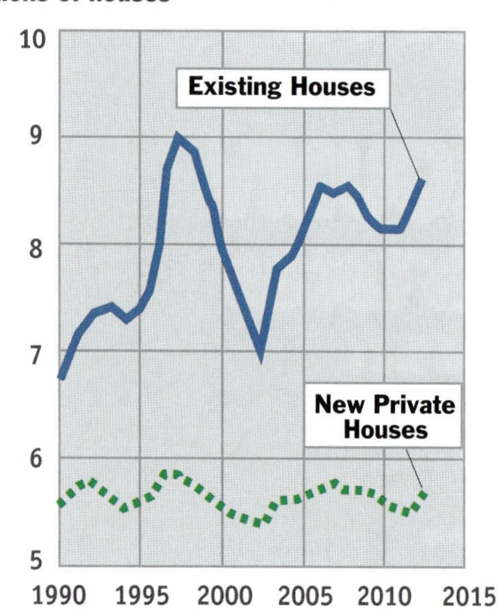

Median Sales Price
Thousands of dollars

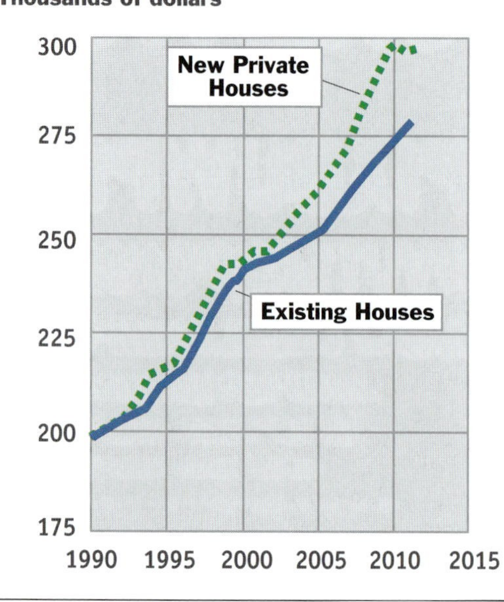

Figure 6. Single-Family Houses Sold and Their Sales Price. *Putting both graphs on the same slide allows for easy comparison and interpretation of the data. In a large room, however, the type used within the graphs would be too small. Two slides would be the better choice.*

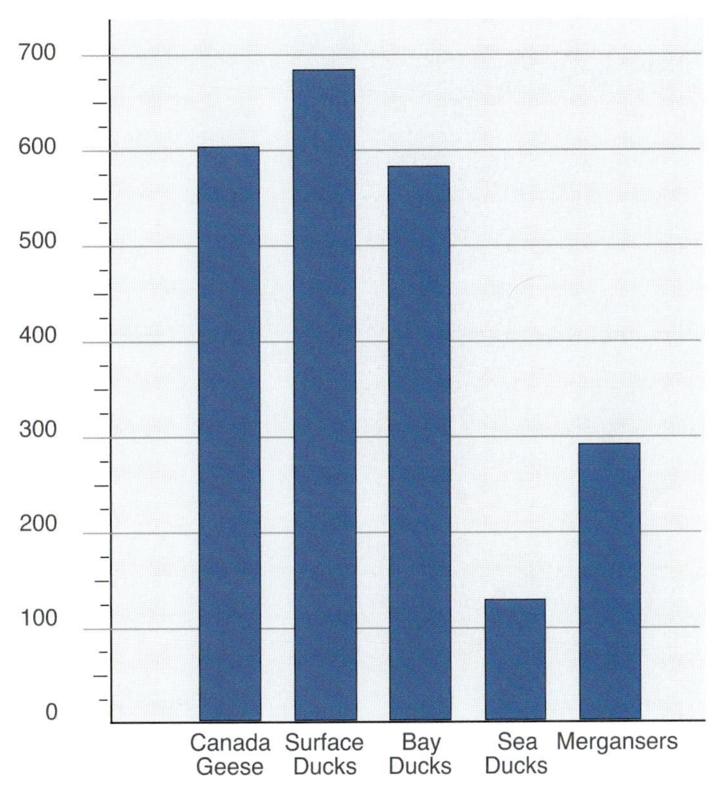

Figure 7. Waterfowl Sightings at Keta Estuary in 2010. *The 2010 annual waterfowl count at the Keta Estuary (Alaska) showed heavy use by some waterfowl, but sea ducks were relatively scarce. Note that this figure and its interpretation would likely require one or more other graphics with data from a baseline year (perhaps 2000) or even earlier (1990). The presenter would need to decide what trend, what conclusion, or what interpretation was most important.*

6. Plan enough time to complete and revise the graphics.

When preparing your own graphics, allow extra time for trial and error adjustments. Also, if you are not familiar with the software, you will experience many delays while producing graphics. See GRAPHICS FOR DOCUMENTS for suggestions about using software to prepare graphics.

Schedule time to review all of your presentation graphics. This is also a good time to do a dry run of your presentation.

7. Use graphics to communicate the structure of your presentation.

Too many slide presentations float by the audience like a stream of debris with no apparent structure. Even a well-organized presentation can seem disorganized and shapeless if the structure of the message is not clearly visible to the audience. If you have three points to make, but viewers never explicitly see the three points displayed, they will be unlikely to grasp those points. See PRESENTATIONS for guidance on how to organize a presentation.

Follow these guidelines for making the structure of your presentation visible and clear:

- Give your presentation a title on the entry slide.
- Create an overview slide to show the outline of your presentation. If you have five points to make and the sequence is important, number the points. If the sequence is unimportant, bullet them.
- In a long presentation, dedicate slides to introducing and summarizing each section so the audience stays oriented to your structure. In a short presentation, use a heading or a change of color to signal a transition from one point to another.
- Keep the section title in a header on each slide so viewers always know where they are in relation to the presentation as a whole.

8. Use graphs, charts, tables, and illustrations instead of text if they are easier to understand.

It's better to show an effective graphic (with highlights) and talk about it than show detailed text and read it. Graphics should be digestible in less than 5 seconds (see rule 3). Simplify, simplify! Leave out data if its loss doesn't detract from the message.

Graphics for Presentations

Titles used on presentation graphics usually appear at the top, as in figures 1, 3, 5, 6, and 7. These same graphics in text would often run into captions at the bottom. See Captions.

Text captions on slides must use larger type if onscreen legibility is a concern

9. Make handouts of your primary presentation graphic.

Always make handouts of your primary graphic showing the main points you want to get across to your audience (See rule 1). Use a primary graphic at the beginning of your presentation to convey your message to people who may not stay for the entire presentation. Then use it later for transitions or to review points. Finally, use it at the end as a summary.

A good practice is to have handouts for **all** graphics, both primary and supporting, that you display during a presentation. If you don't provide handouts, some members of the audience will be frantically trying to take notes or copy graphics. They will often not be following what you are saying.

Another good practice is to make your presentation file available for your audience to download.

Graphs turn numbers into pictures and are useful for showing trends, cycles, cumulative changes, relationships between variables, and distributions. They are not as effective as tables in providing precise data, but readers should be able to extract relatively accurate numerical data from the lines plotted or from the size of the bar or pie segments in bar graphs and pie graphs. Graphs are better than tables if you want to help your readers understand the meaning of your data. See TABLES.

Graphs include **line (or coordinate) graphs, bar graphs,** and **pie (or circle) graphs.** As explained in CHARTS, bar graphs and pie graphs are often called bar charts and pie charts. Don't be bothered by the confusion between terms; the more important issue is to ensure that your graphs (or charts) are well conceived and professional looking.

Line (or coordinate) graphs and bar graphs are normally plotted using grid lines, with a horizontal axis (x-axis/abscissa) and a vertical axis (y-axis/ordinate). Grid lines are usually equally spaced in horizontal and vertical directions and reflect the numerical scales along each axis. Figure 1 is a simple line graph showing the median prices of new single-family houses sold over a 10-year period.

For general information on graphics, see GRAPHICS FOR DOCUMENTS and GRAPHICS FOR PRESENTATIONS. see also COLOR, ILLUSTRATIONS, MAPS, PHOTOGRAPHS, and TABLES.

The most common graphs are **line (or coordinate) graphs, bar graphs,** and **pie (or circle) graphs**. Other types of graphs exist, however, such as graphs using logarithmic scales or from polar coordinates; these other

Graphs

1. Use a line graph, a bar graph, or a pie graph to help readers visualize quickly and easily the major points in a document.
2. Ensure that the visual characteristics of the graph reflect the magnitude and importance of the data being represented.
3. Number each graph sequentially, provide a title, and then add an informative caption to identify the purpose of the graph.
4. Place footnotes and source information below graphs.
5. Use grid lines or tick marks to help readers interpolate data.
6. Orient all labels, numbers, and letters so that they are parallel with the horizontal axis.

Line or Coordinate Graphs

7. Choose and label scales to indicate the quantity, magnitude, and range of each axis.
8. Make your key data lines heavier than axis and grid lines and less important data lines.
9. Use multiple lines, if necessary, to show the relationships between three or more variables, and use different line patterns to depict different variables.
10. Label important values on a data line.

Bar Graphs

11. Clearly label each bar.
12. Make the bars wider than the spaces between them.
13. Use different bar patterns to indicate differences in types of data and label these patterns clearly.
14. Use segmented bars to depict multiple variables or component parts of a whole.

Pie Graphs

15. Identify each sector of the pie and, if appropriate, the percentage it represents.
16. Differentiate adjacent pie sectors by using alternating fill colors or patterns.
17. Group small percentage items under a general label, such as "Other."
18. Use pie-bar combination graphs to show the composition of an important sector of the pie.
19. Use a series of pie graphs to add time as a variable.

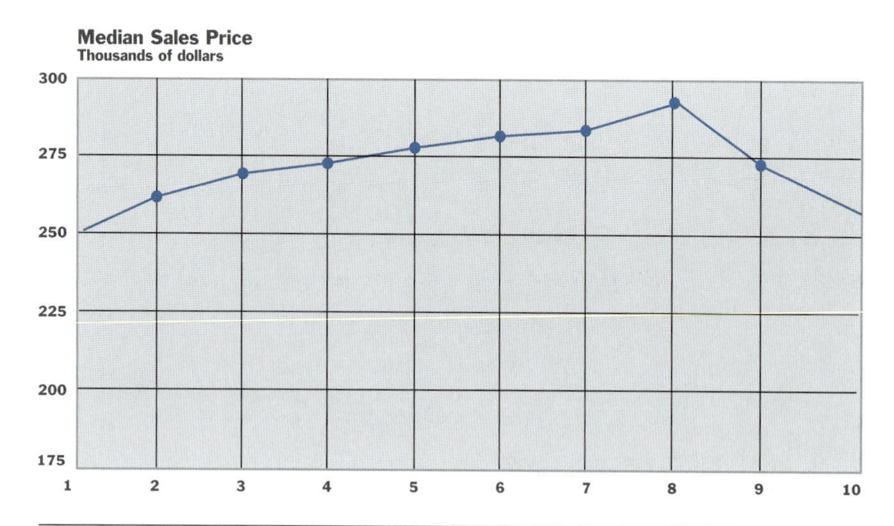

Figure 1. Median Prices of New One-Family Houses. *Over the 10 years of this study period, median sales prices for new homes increased annually until an economic downturn began in year 9.*

types of graphs are beyond the scope of the following discussion, which deals with line or coordinate graphs, bar graphs, and pie graphs.

1. Use a line graph, a bar graph, or a pie graph to help readers visualize quickly and easily the major points in a document.

The most technically complex and even elegant graph fails if it is not simple and easy to read. Keep your graphics simple, even obvious.

Ensure that each graph has a single important point to make or a single relationship to show. Then tell your readers what the relationship is, both in the text and in a caption. See CAPTIONS. Be careful not to confuse readers by assuming that they understand the relationship between variables.

Don't try to do too much with each graph. Figure 2, which is a simple bar graph, shows how revenues and investments in the cell-phone industry changed over a 10-year period. This bar graph has a simple message because it displays only two sets of data, one for 2000 and the second for 2010. In contrast, the original data for this graph comes from Table 1112 in the *Statistical Abstract of the United States 2010*; in this table, data cover nearly all the years of that decade. Also, that table covers several categories of data not presented in figure 2. Figure 2 is designed to focus on only its simplified message, not all the complex year-by-year details.

Complicated graphs are often confusing, even to technically competent readers. Simplify graphs by eliminating everything that does not contribute to the central message of the graph. Do not use more labels, numbers, tick marks, shadings, colors, or grid lines than necessary to do the job. Too much clutter makes graphs difficult to read.

2. Ensure that the visual characteristics of the graph reflect the magnitude and importance of the data being represented.

The value of graphs is their visual impact. Consequently, writers can mislead readers by producing graphs that give more or less prominence to an idea or piece of data the writer wishes to emphasize or deemphasize. See ETHICS.

Distorting scales, bar lengths, or pie slice areas can mislead readers into thinking that something is larger or smaller than it really is. Similarly, using bright colors for insignificant data and dull colors for significant data can confuse readers and lead

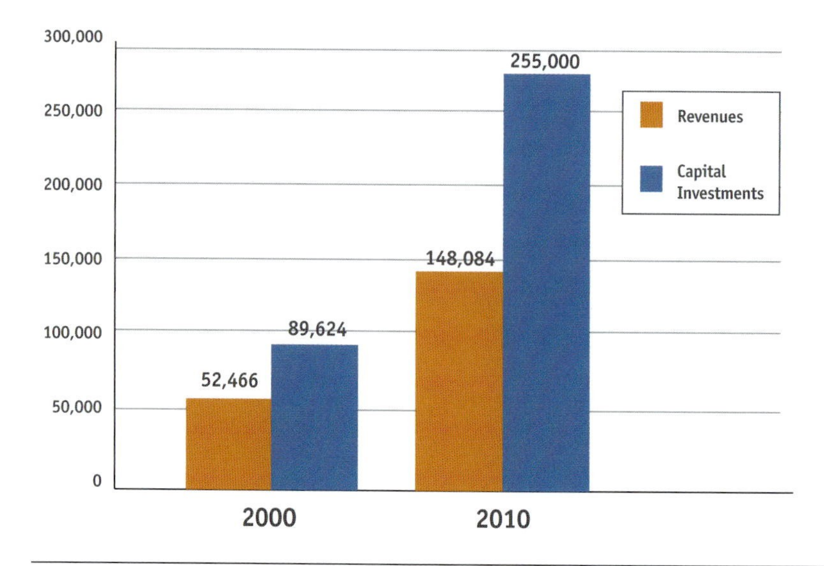

Figure 2. Comparison of Service Revenues and Capital Investment in the U.S. Mobile Phone Sector (in $millions). *While revenues nearly tripled between 2000 and 2010, capital investment hovered at about 40 percent above revenues across the decade. (Based on Table 1112, in the* Statistical Abstract of the United States 2010.*)*

some to think that the insignificant is really significant. To ensure that you have presented a truthful and accurate picture of the situation being depicted, strive to make the visual impression created by the graph consistent with reality. See GRAPHICS FOR DOCUMENTS.

3. Number each graph sequentially, provide a title, and then add an informative caption to identify the purpose of the graph.

Graphs and all other graphics are usually labeled as figures, so the numbering would be *Figure 1, Figure 2*, etc. Or, if the document has chapters, *Figure 1–1, Figure 1–2, Figure 1–3*, where the first number is the chapter number.

Following *Figure* and the number comes the title of the graph:

> **Figure 3–1. Absorption (Percentage) of Drug A Over Time (Minutes).**

Sometimes, however, the title appears above a figure, functioning more like a headline. See figures 1, 3, 4, 6, 7, 8, 11, and 12 for examples of headlines used above figures. Then provide an informative or interpretive caption to guide readers to see the point of the graph:

> **Figure 3–1. Absorption (Percentage) of Drug A vs. Time (Minutes).** *Drug A is very slowly absorbed until it has been in the patient for at least 30 minutes; then absorption is rapid.*

For the graphs used as examples in this discussion, we have varied the presentation to show the different placements of titles and captions. In an actual report or formal document, you would want to choose one placement pattern and use it for all figures. See CAPTIONS and GRAPHICS FOR DOCUMENTS.

4. Place footnotes and source information below graphs.

Footnotes typically explain or clarify the information appearing in the entire graph or in one small part of it. Often footnotes tell what the data apply to, where they came from, or how accurate they are:

> [1]All data are in constant 2010 euros.
>
> [2]For companies with no South Asian assembly plants.
>
> [3]According to the International Union of Geological Scientists. The data come from each nation's database, so some variations will appear because each nation has different reporting standards.

The footnotes for each graph are numbered independently from footnotes in the text and from footnotes in other graphics. Begin with footnote 1 and proceed sequentially for that single graph; for the next graph, start the numbering again. Within the body of the graph, use superscripted footnote numbers. Place the footnote explanations (in numerical order) immediately below the graph and flush with the left margin. Repeat the superscripted footnote number and then provide the appropriate explanation, followed by a period.

If the graph covers more than one page, place the appropriate footnotes with each page. If caption and footnotes appear below a graph, place the footnotes above the caption. If the caption is above the graph, place the footnotes below the graph but ahead of the text.

If footnote numbers would be confusing in the body of the graph, use letters ([a], [b], [c], [d], etc.), asterisks (*, **, ***), or other symbols.

Source information may appear in footnotes if the referenced source provided only that data indicated by the footnote and not the data for the rest of the graph:

> [4]From *Journal of Experimental Biology*, Volume 213 (February 2010).
>
> [5]Source: U.S. Department of the Interior, Bureau of Land Management.

Source information may also appear within parentheses in the caption (regardless of where the caption appears) or within brackets under the caption if the caption appears ahead of the graph:

> **Figure 1. U.S. Aerospace Mergers and Acquisitions, 1990–2010** (U.S. Department of Commerce)
>
> **Figure 1. U.S. Aerospace Mergers and Acquisitions, 1990–2010** [U.S. Department of Commerce]

See FOOTNOTES, CITATIONS, and INTELLECTUAL PROPERTY.

5. Use grid lines or tick marks to help readers interpolate data.

Including grid lines on graphs was the status quo for many years. Now, many graphs appear without grid lines, principally to increase resolution and create a cleaner, more open visual effect (see figure 3). Omitting grid lines is acceptable. However, if you expect readers to extract precise data from graphs, you should not omit grid lines unless you combine the graph with a table. (See figure 6 for an example of a combined graphic.)

If you decide to omit grid lines, retain tick marks to help readers interpret data. Tick marks are short lines on and perpendicular to an axis that indicate the intervals along a scale. Use longer tick marks beside interval labels; use shorter tick marks between labels. Generally, try to use twice as many tick marks as labels (so that you have tick marks at the midpoints between labels). Figure 3 shows the same graph as in figure 1, but in this instance, the grid lines are removed, leaving only tick marks.

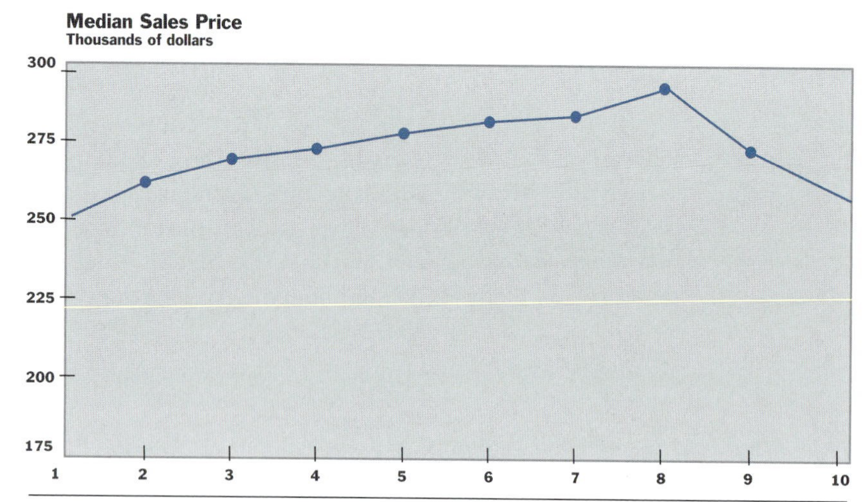

Figure 3. Median Prices of New One-Family Houses. *Over the 10 years of this study period, median sales prices for new homes increased annually until an economic downturn began in year 9.*

Do not use too many tick marks. The more tick marks you use, the more crowded the scale becomes. Do not create finer distinctions than necessary for the data being shown.

6. Orient all labels, numbers, and letters so that they are parallel with the horizontal axis.

Placing all lettering horizontally on the page makes graphs easier to read. The labels in figures 1, 2, 3, and 4 illustrate horizontal lettering. (However, some U.S. Government agencies require that the labeling of the ordinate axis be vertical.)

Overly long vertical labels are an exception to this rule. If your ordinate axis label is too long for horizontal lettering, then turn it vertically with the base of the letters parallel and adjacent to the ordinate axis. Readers should be able to read long vertical labels by turning the page clockwise.

Line or Coordinate Graphs

Line (or coordinate) graphs are normally plotted using grid lines or tick marks, with a horizontal axis (x-axis/abscissa) and a vertical axis (y-axis/ordinate). Grid lines are usually equally spaced in horizontal and vertical directions and reflect the numerical scales along each axis. An exception would be logarithmic graphs or polar coordinates, where the scales are not equally spaced, but these are beyond the scope of the present discussion.

Figure 4 is a typical line graph, with data on U.S. productivity increases from 2000 to 2008. As in most line graphs, figure 4 plots the time (in years) on the horizontal axis and the units to be measured on the vertical axis.

7. Choose and label scales to indicate the quantity, magnitude, and range of each axis.

If the horizontal (x) axis or the vertical (y) axis indicates quantities, magnitudes, and ranges, use scales and grids or tick marks that show the axis minimum and maximum, as well as the numeric intervals. (See the vertical and horizontal scales on figures 1, 2, and 3.)

Label the minimums and maximums, whether they are positive or negative values.

Also label a sequence of intervals along the scale. If the minimum is 0 and the maximum is 1,000, for instance, you might label the scale in steps of 100. These interval labels indicate the scale and allow readers to interpolate data. Do not use so many labels that the scale becomes crowded; however, do not use so few that readers cannot easily interpolate data. Generally, try to leave one or two spaces between interval labels. These labels will correspond to either grid lines or tick marks (see rule 6).

Use axis labels to identify each axis. Placement of these axis labels varies. Sometimes they come under the units on the horizontal axis and to the left of the vertical axis. But as in figures 1, 3, and 4, the label for the vertical axis appears at the top of the graph. Indicate the units of measurement in the axis label or place the units of measurement within parentheses after or below the axis label:

Wheat Shipments, in Metric Tons

Wheat Shipments (Metric Tons)

Wheat Shipments
(Metric Tons)

If time is one variable, plot it on the x axis. See figures 1, 3, and 4.

Scales must increase from bottom to top along the y axis and left to right along the x axis. Scale maximums and minimums must appear on the farthest grid lines along each axis. Scale labels must appear at appropriate intervals to facilitate data interpretation (see rule 8).

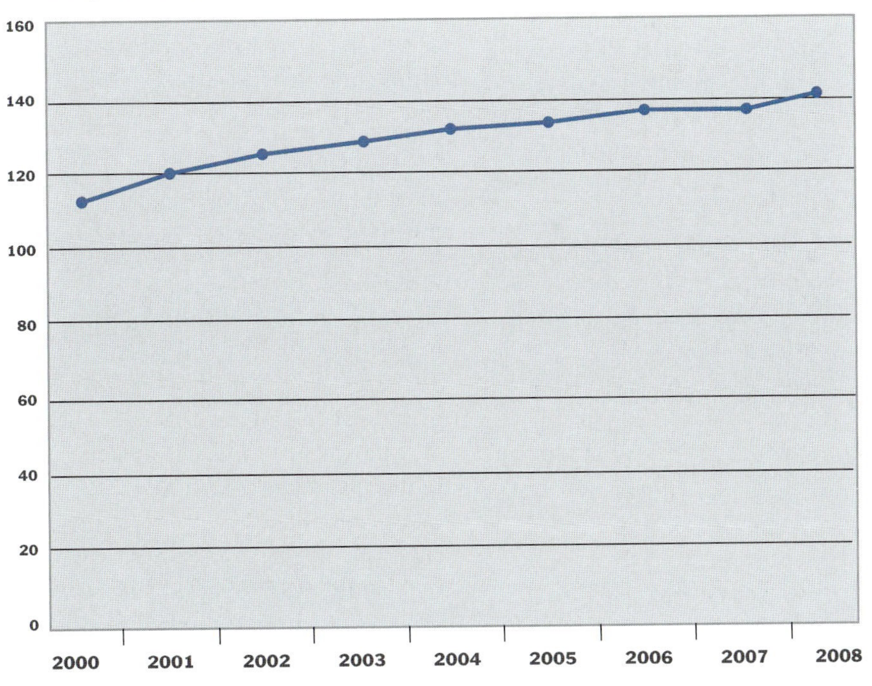

Figure 4. U.S. Business Productivity Rises. *Productivity increased an average 3.25 points per year between 2000 and 2008, from 116 (2000) to 142 (2008). These rapid improvements were probably driven by advances in technology. (Statistical Abstract of the United States 2010, Table 627.)*

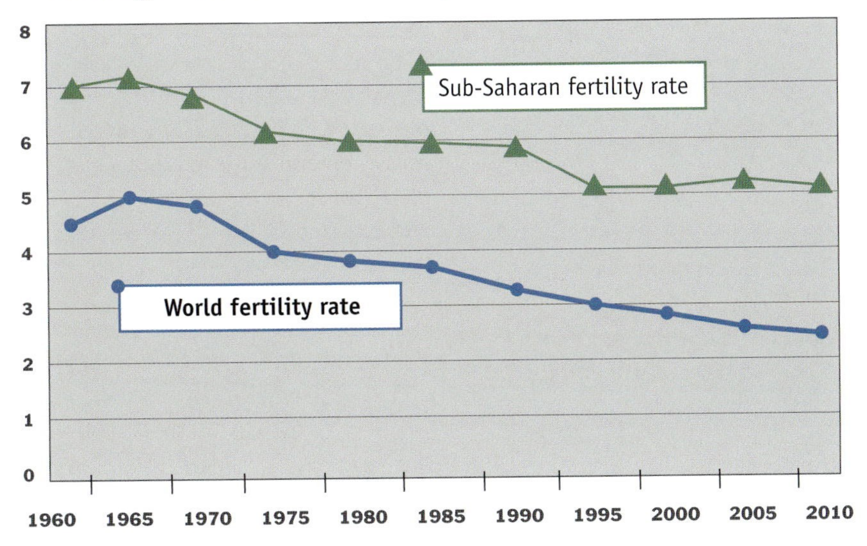

Figure 5. Sub-Saharan and World Fertility Rates. *Although world fertility rates now hover around replacement level, the declining rates in the developing countries of Sub-Saharan Africa have stalled well above that level. (World Bank, World Development Indicators, February 2, 2010.)*

Do not use scales that exaggerate or distort the numerical relationships that actually exist. You can make small and insignificant differences look important by using a minute scale, and you can hide critical differences by using an overly large scale.

Your graphs should reflect the reality of the data being plotted. Therefore, your choice of scale is critically important.

8. Make your key data lines heavier than axis and grid lines and less important data lines.

Axis and grid lines are less important than data lines and must therefore be thinner and lighter. The visual emphasis in the graph must be on the plotted (data) lines, not on the grid lines or axes. Equally, your important data lines must be more emphatic than less important data lines.

Note that in figures 1, 2, 3, and 4, the plotted data lines are darker than the axis or grid lines (if used).

9. Use multiple lines, if necessary, to show the relationships between three or more variables, and use different line patterns to depict different variables.

Lines of different patterns are useful for plotting more than two variables. In figure 5, for example, the graph presents three variables: the year (x axis) and the Sub-Saharan and world fertility rates (y axis).

Line patterns include solid, thin solid, wide solid, thin dashes, wide dashes, small dots, large dots, hollow (two thin lines together), and mixed dots and dashes of varying sizes. If you use different line patterns, label the lines, as in figure 5. Optionally, you can omit the labels within the grid pattern and use an explanatory key to the right or below the graph, as in figure 2.

10. Label important values on a data line.

If you wish to highlight or discuss certain important values on data lines, label them within the grid system. Readers will pay more attention to labeled data values and will not be forced to interpret data values.

In figure 6, for example, the final values of the four regions (the end of the plotted values) are of special interest, so the final values are inserted in the graph. Similarly, other values could be inserted for special years.

Sometimes, for a graph with important data, you may decide to present the table from which the graph came along with the graph. Figure 6 includes the background table for the graph. Note the additional categories (in columns) and the additional values (by year). The graph in figure 6 has the advantage of telling its story at a glance, but the table provides more detail.

Bar Graphs

Bar graphs depict the relationship between two or more variables, one of which is usually time. These graphs typically show how one or more variables change over time. Consequently, bar graphs are useful for depicting trends (see figure 8). Because bar graphs can show multiple variables, they can also depict how several variables change relative to one another over time (see figure 7).

Bar graphs are less useful if the quantities depicted do not differ significantly. And if you expand or distort the axis scales to dramatize slight differences, the bar graphs will look suspicious to alert readers and might damage your credibility.

Bar graphs may be horizontal or vertical. In vertical bar graphs, time is usually plotted along the horizontal x axis (see figure 7). In horizontal bar graphs, time is usually plotted along the vertical y axis.

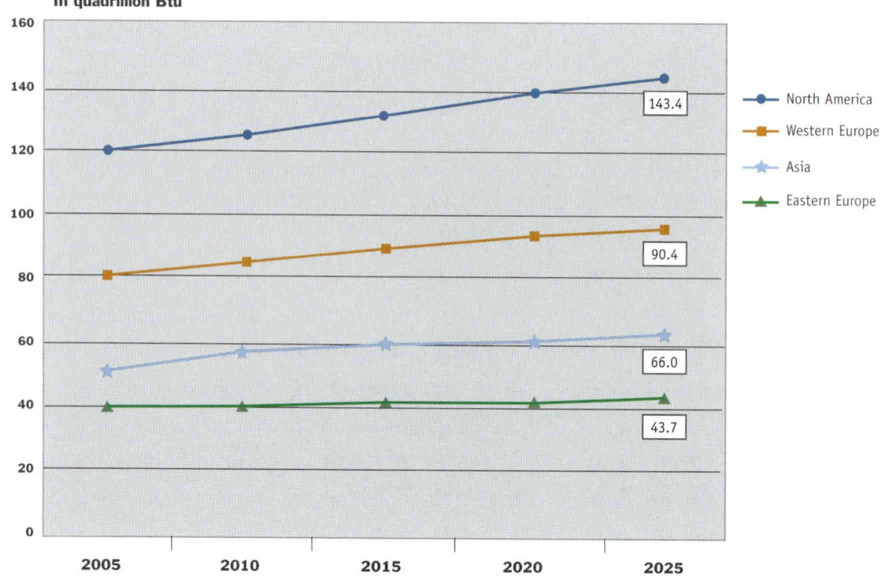

Region	1990	2004	2005	2010	Projections 2015	2020	2025
North America	100.7	120.6	121.3	126.4	132.3	137.8	143.4
Western Europe	70.0	81.0	81.4	83.9	86.8	88.5	90.4
Industrialized Asia	26.8	37.8	38.2	39.3	41.4	42.7	43.7
Eastern Europe & Russia	67.3	49.5	50.7	55.1	59.5	63.3	66.0

Figure 6. Energy Consumption by Industrialized Regions of the Northern Hemisphere. *Cumulative energy consumption across industrialized nations will rise at approximately the same steady rate at least until 2025.* (Statistical Abstract of the United States 2010, *Table 1344.)*

Some writers and illustrators argue that vertical bar graphs are better for showing trends and that horizontal bar graphs are better for comparisons and for showing magnitude changes. Certainly, readers are more used to seeing trends shown along a horizontal axis. However, comparisons and magnitude changes are usually clear in either orientation. Use your judgment.

Bar graphs may be used with other visual forms, such as line or coordinate graphs, maps (see MAPS), or pie graphs (see figure 12).

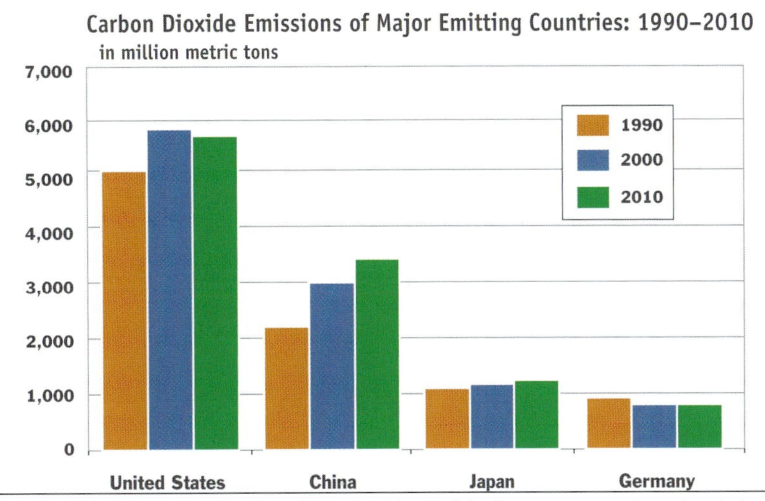

Figure 7. *While carbon dioxide emissions in three of the largest emitters have leveled off, China continues to increase the rate of emissions.* (Statistical Abstract of the United States 2010, *Table 1351; World Resources Institute.)*

11. Clearly label each bar.

Ensure that readers understand what each bar represents. See figures 7 and 8 for examples of two different bar graphs with appropriate labels.

As an example, figure 7 is a vertical bar graph, with countries and time (years) on the x axis and with the volume of carbon dioxide emissions on the y axis. The bars (labeled in the legend) show how emissions varied from year to year. Figure 8 is a horizontal bar graph, showing labeled metals on the y axis, with the volume produced on the x axis.

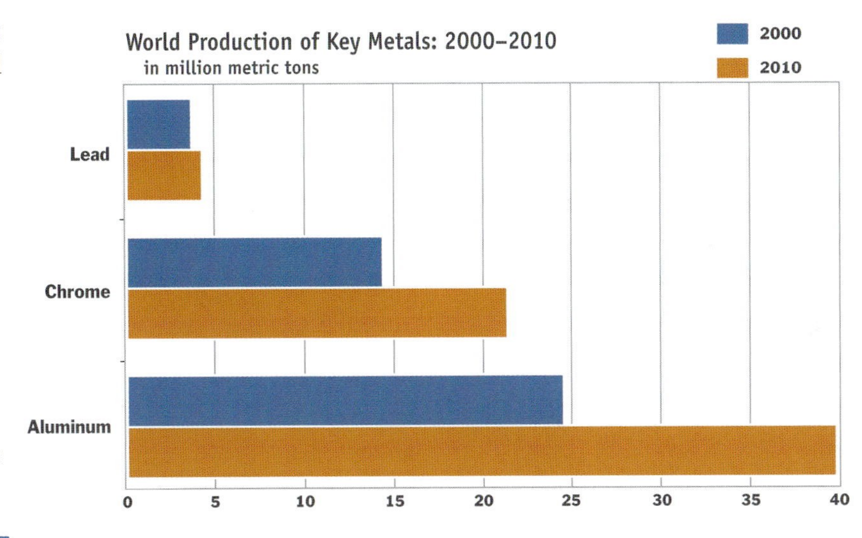

Figure 8. *Aluminum production rates soared by 40 percent between 2000 and 2010.* (Statistical Abstract of the United States 2010, *Table 1341.)*

12. Make the bars wider than the spaces between them.

See figure 7. A bar graph with the spaces wider than the bars would seem "airy." The dominant visual effect in a bar graph should be the bars.

13. Use different bar patterns to indicate differences in types of data and label these patterns clearly.

Bar patterns allow you to distinguish areas, regions, groups, and parts. Patterns also allow you to focus readers' attention on areas of the graph you consider most important. Bar patterns include diagonal lines, cross-hatching, and colors. Remember, however, that colors are a problem if you are using black-and-white photocopies. Whatever pattern of bars you choose, be sure that the pattern is clearly labeled.

14. Use segmented bars to depict multiple variables or component parts of a whole.

Segmented bars allow you to show segments of a total figure.

Figure 9 contrasts household assets with the liabilities for 2010. The shaded segments divide assets and liabilities into their components, which are labeled. In figure 10, the waterfowl sightings for a single year (2009) are presented for three different areas. The different segments in the bar refer to the different areas.

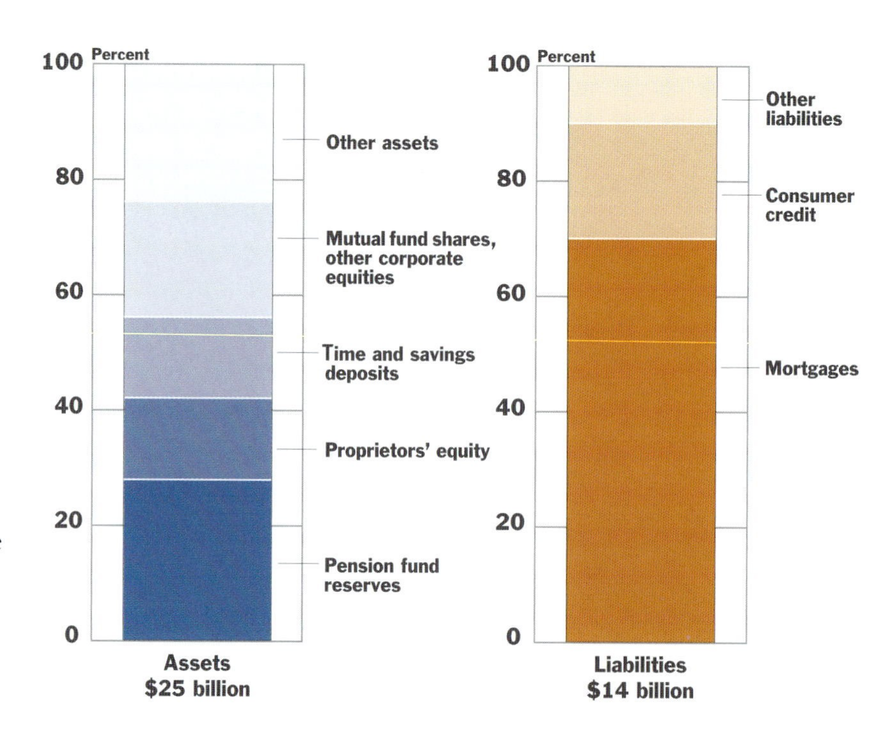

Figure 9. Assets and Liabilities of U.S. Families. *Mortgage obligations represent the overwhelming financial liability of U.S. families. Assets, on the other hand, include a variety of financial fund instruments.* (Statistical Abstract of the United States 2010, *Table 706.)*

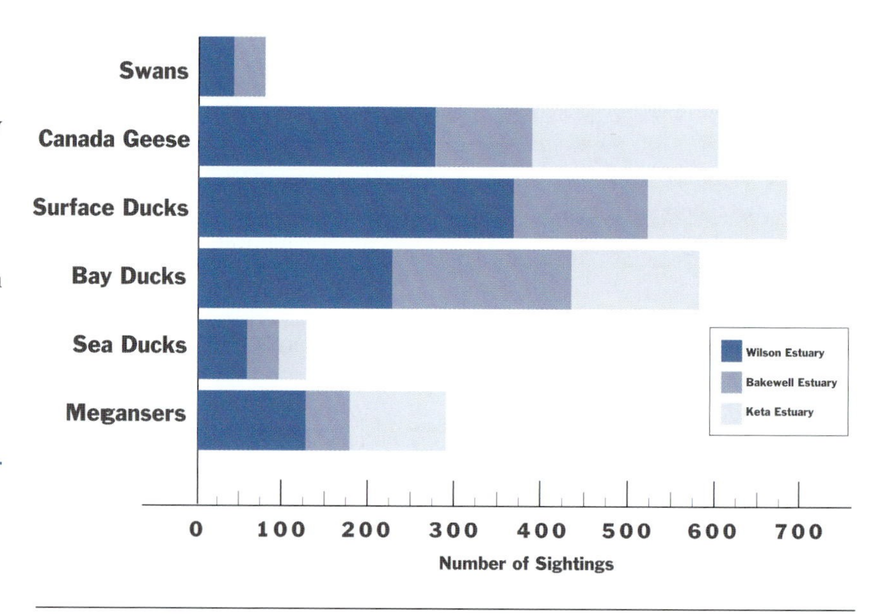

Figure 10. Waterfowl Sightings in 2009 at Three Alaskan Estuaries. *Of the three surveyed estuaries, the Wilson Estuary has the largest and most productive waterfowl population. (Source: U.S. Department of Agriculture.)*

Pie Graphs

Pie graphs are circles (pies) divided into sectors (slices) to show the relationship of parts to a whole. The sectors must add up to 100 percent.

Pie graphs are useful for general comparisons of relative size. However, they are not useful if accuracy is important. Further, pie graphs are not useful for showing a large number of items.

The eye can measure linear distances far more easily than radial distances or areas. Therefore, visual comparisons of bars on a bar graph are much easier to grasp than visual comparisons of sectors in a pie graph. Moreover, readers can usually make more accurate judgments about data relationships expressed in a linear fashion.

So if you need accuracy, use a bar graph or a line graph. If you need to show how parts relate to one another and to a whole—and if precise numbers are not important—use a pie graph.

15. Identify each sector of the pie and, if appropriate, the percentage it represents.

Pie graphs do not have axes and therefore cannot be very precise, so if percentages are important, identify them. As in figure 11, the percentage is more important than the actual numbers, but if a reader wanted to, a simple calculation with the percentages, based on the total

$210 billion, would reveal the actual component numbers.

Always identify each sector of the pie. You can do this by using labels (figures 11 and 12) or fill patterns and an explanation block. For further information on fill patterns, see the preceding discussion of bar graphs.

16. Differentiate adjacent pie sectors by using alternating fill colors or patterns.

To help readers distinguish the sectors, use alternating fill colors or patterns. In figure 11, for example, the various patterns for each slice or segment help distinguish between the segments. These patterns are somewhat redundant (intentionally) with the labels which report the different revenue sources. Reserve the solid (or black) fill pattern for the prominent sectors—the ones you wish to emphasize. Never use the same fill pattern for adjacent sectors.

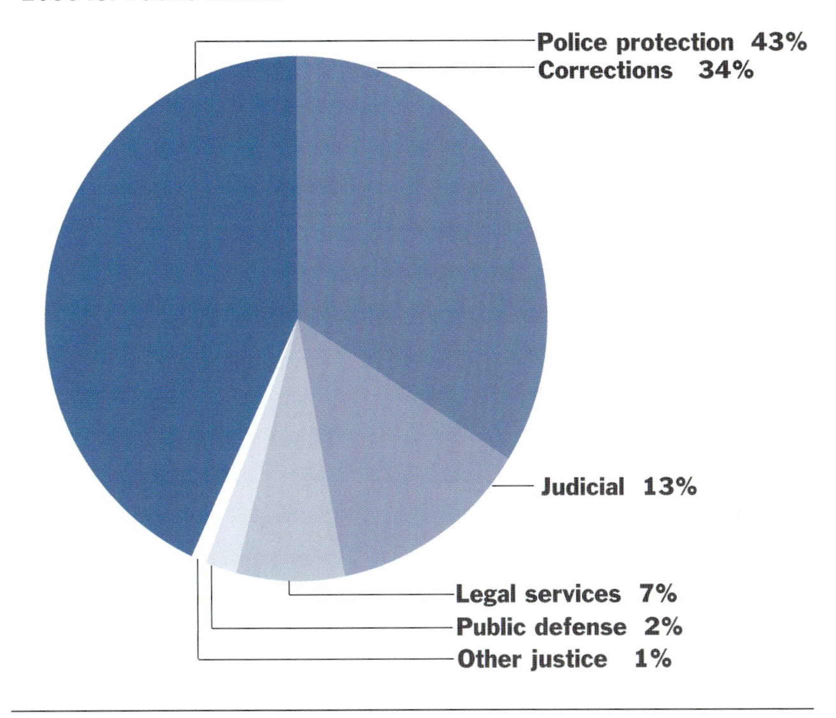

Figure 11. U.S. Criminal Justice System Expenditures. *Despite some reports in the media, expenditures for public defense, which include legal aid services and public defenders, constitute only 2 percent of the money spent nationally on the criminal justice system. (Source: U.S. Bureau of Justice Statistics.)*

17. Group small percentage items under a general label, such as "Other."

Pie graphs should have no more than 8 to 10 sectors, depending on the size of the pie. The larger the pie, the more sectors you can safely divide it into. However, beyond some reasonable number of sectors (10), a pie graph becomes too busy and therefore difficult to read.

If you have a number of small percentage items, you should group them and give them a common label, such as "Other Parts," "Other Exports," or simply "Other" if the context of the graph and the names of the other labels indicate what "Other" refers to.

Sometimes for the "Other" category or another segment of the pie graph, you will want to break out the different components (see rule 18).

18. Use pie-bar combination graphs to show the composition of an important sector of the pie.

Figure 12 shows a combination pie-bar graph, where the largest segment "Sales and gross receipts" would not be very clear to readers who did not know what items went into this category. For such readers, in figure 12, this category is broken down into the adjacent bar graphs, which show the various percentages for the two years.

Each sector of a pie can potentially be expanded into a segmented bar. In this way, you are adding one more variable to the pie graph.

State Government Tax Collections, by Type: 1991 and 2008

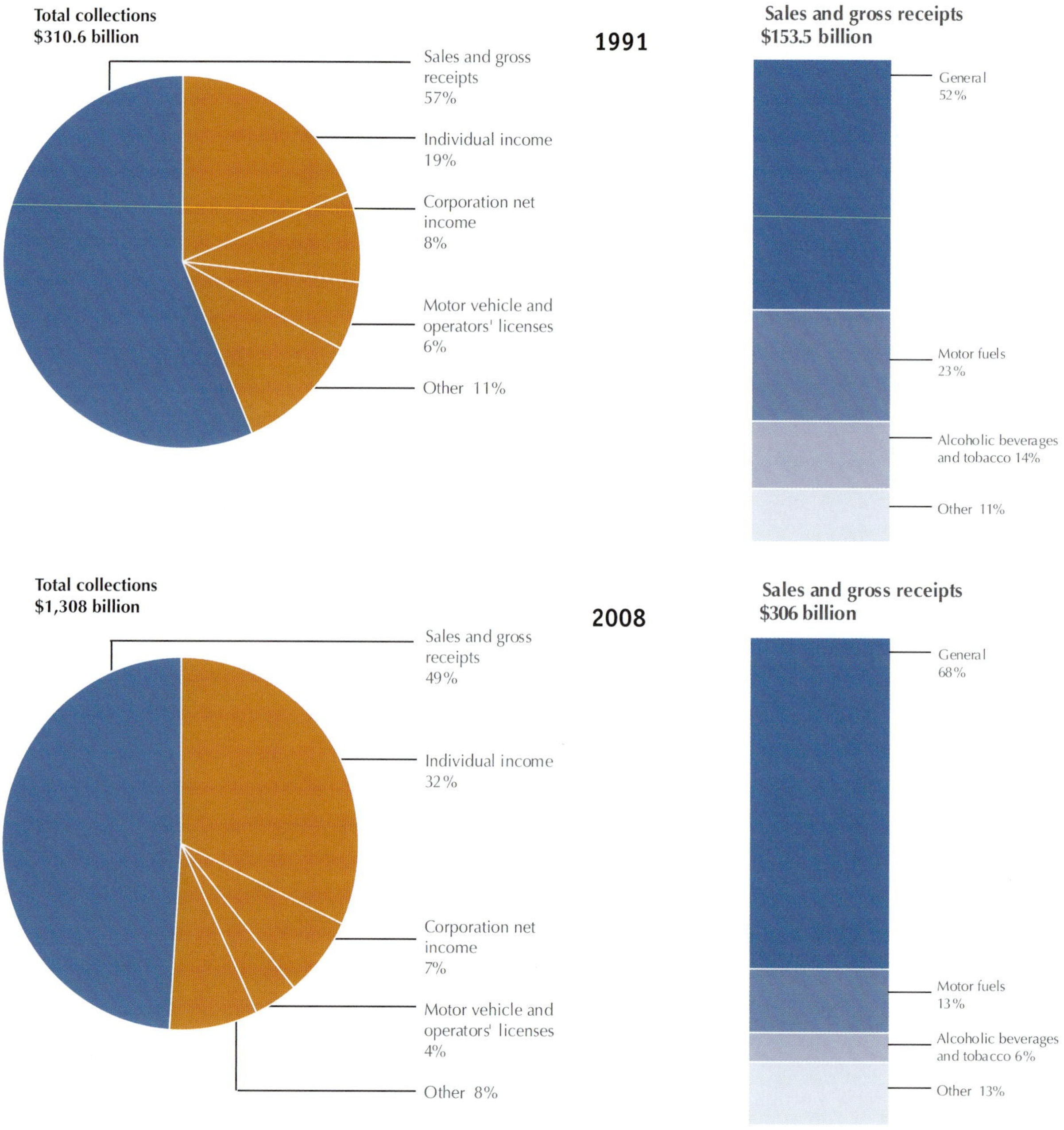

1991

Total collections
$310.6 billion

Sales and gross receipts 57%
Individual income 19%
Corporation net income 8%
Motor vehicle and operators' licenses 6%
Other 11%

Sales and gross receipts
$153.5 billion

General 52%
Motor fuels 23%
Alcoholic beverages and tobacco 14%
Other 11%

2008

Total collections
$1,308 billion

Sales and gross receipts 49%
Individual income 32%
Corporation net income 7%
Motor vehicle and operators' licenses 4%
Other 8%

Sales and gross receipts
$306 billion

General 68%
Motor fuels 13%
Alcoholic beverages and tobacco 6%
Other 13%

Figure 12. *Sales taxes and general consumption taxes (motor fuels, alcohol, and tobacco) still constitute in 2008 the major source of state funds. Taxes on individual income, as a percentage of the total, have increased from 19 percent in 1991 to 32 percent in 2008.*

19. Use a series of pie graphs to add time as a variable.

The two pie graphs in figure 12 are only two of the possible series of pie graphs for tax collections. Depending on the purpose for a document, a writer might choose to present pie graphs for the years 1990, 2000, 2010, and 2020 (estimated). This series would undoubtedly portray the interesting economic adjustments made by state governments over a period of 30 years.

Headings are essential features of any business or technical document. Without headings, documents become a jumble of paragraphs for readers to interpret. Even well-written paragraphs are not easy to read unless they are paired with good headings. See EMPHASIS and PARAGRAPHS.

Well-written headings allow readers to skim and scan a document. Most readers of business documents read to find key information or a missing fact. They do not read a document line by line because their prior knowledge of the subject makes careful reading unnecessary.

Headings include any headline information within a document. The most basic headings appear as labels or signposts before blocks of text. However, headings also include subject lines that introduce a letter or memo and titles of articles or chapters.

Headings

1. Use informative, specific, and inclusive headings in any business or technical document, even those as short as a single page.
2. Design documents by planning and drafting your headings even before you have finished writing the text.
3. Choose different levels of headings to indicate logical divisions and groupings in the text. Avoid using too many levels.
4. To show the level of the heading, use the three types of typographical variations: the *placement* of the heading on the page, the *size* of the type, and the *appearance* of the type.
5. Use a numbering system with your headings if you have more than four levels of headings or if your document is lengthy and you or your readers will need to refer to sections of it by number.
6. Make headings parallel in structure and use the same type of headings at each heading level.
7. Capitalize headings using the same conventions as for titles of books, articles, or other documents.

1. Use informative, specific, and inclusive headings in any business or technical document, even those as short as a single page.

Informative, specific headings allow readers to determine immediately the contents of a section. Unfortunately, many standard headings are neither specific nor informative:

Introduction

Discussion

Results

Better versions of these would be:

Purpose of the Drilling Proposal

Implications of Three Proposed Tests

Valid Data From the Third Test

Informative headings are similar to the captions for graphics. As with headings, good captions are visual signposts for readers who do not want or need to read dense sections of text. See CAPTIONS.

Inclusive headings signal that only material mentioned in the heading will actually be covered in the section. If the heading is *Valid Data from the Third Test*, then no data from the second or first test should appear following the heading.

The subject heading (subject line or title in a letter or memo) is perhaps your best opportunity for communicating your purpose and main point. Again, many subject headings are uninformative:

Town Center Building

Information Management Project

Customer-Satisfaction Survey

Better versions tell readers what is important in the document and what you're asking them to do:

Proposal to rehabilitate the Town Center Building at cost

Recommendation to approve information management project

Invitation to participate in a survey to help us serve you better

Informative subject headings contain key words to signal the purpose of the document:

Proposal to . . .
Recommendation to . . .
Request for . . .
Invitation to . . .
Warning about . . .
Report of . . .
Agenda for . . .
Information about . . .
Notice of . . .
Procedure for . . .

As noted, subject headings should also summarize your main point:

Recommendation to adopt new scrubbers to meet EPA requirements

Proposal to reorganize our division for greater customer focus

2. Design documents by planning and drafting your headings even before you have finished writing the text.

Headings should come early in your writing process so you can write text to fit the heading structure.

For example, when you and colleagues use a prototype or a mock-up to plan a document, that mock-up should include all headings, even if they are provisional. See WRITING AND REVISING.

Writers can see the headings on a blank page and begin to visualize the text that will follow the heading. Writers will be writing to the headings and to the overall structure of the document, as reflected in the headings. This document-design approach will help you avoid costly rewriting and time-consuming editing.

3. Choose different levels of headings to indicate logical divisions and groupings in the text. Avoid using too many levels.

A chapter title, for instance, may be a major (or first-level) heading. Section titles in the chapter may be minor (or second-level) headings. Subsection titles could then be subheadings (third-level headings), and so on.

The level of the heading indicates its logical relation to other headings as well as to the whole. The levels are most apparent in an outline:

3.4 Component Descriptions
- 3.4.1 Gearbox Assembly
- 3.4.2 Brakes
- 3.4.3 Hydraulic Motors
- 3.4.4 AIU Interfaces
  - 3.4.4.1 Controller
  - 3.4.4.2 Encoder
  - 3.4.4.3 Tachometer
- 3.4.5 Servo Valves

In both text and outlines (such as a table of contents), the levels may be indicated by a numbering system (like the one above) or by the placement, size, or appearance of the heading. See OUTLINES and TABLES OF CONTENTS.

How many levels should you use? Dozens are possible, but readers could not comprehend that many levels of subordination. Practically speaking, you should use no more than four or five levels, depending on your readers.

The less technical or less educated your readers, the fewer levels you should use. Experienced and well-educated scientific readers are more accustomed to reading text with multiple levels of subordination. If you suspect that your readers will have trouble remembering the heading levels, use fewer levels.

4. To show the level of the heading, use the three types of typographical variations: the *placement* of the heading on the page, the *size* of the type, and the *appearance* of the type.

Placement variations include centering, placing the heading flush left, indenting, and using a run-in heading (placing the heading on the same line as the text following it):

A Centered Heading

A Flush-Left Heading

An Indented Heading

A Run-in Heading. The text begins following the period (or colon) and one or two spaces.

See PERIODS.
See CAPITALS.

Sometimes a run-in heading has only three spaces between it and the succeeding text and no punctuation:

A Run-In Heading When you do not use punctuation (as in this case), you should make the heading visually distinct from the text in the rest of the paragraph. You might use boldface type or a larger typeface.

Size variations are possible when you can vary the point size of the lettering in the heading. The larger the point size, the greater the level of the heading. A 24-pt heading is on a higher level than a 12-pt heading, etc.:

A 24-pt Head

An 18-pt Headi

A 14-pt Heading

A 12-pt Heading

Appearance variations include ALL CAPITAL LETTERS, underlining, **boldface type**, different typefaces, color, and . See PAGE LAYOUT and EMPHASIS.

Use the following placement and appearance lists to create different levels of headings. The variables are listed in decreasing order of importance:

Placement

1. Centered
2. Flush left
3. Indented
4. Run-in (on the same line as text)

As the preceding list shows, a centered heading is on a higher level than a flush-left heading. A flush-left heading, on the other hand, is on a higher level than an indented heading.

Appearance

1. ALL CAPITALS
2. Underlining
3. **Boldface**
4. *Italics*

The appearance variations may be added singly or in combination. Thus, a heading may be in all capital letters, or it may be underlined, or both. Appearance variations used in combination create higher-level headings than headings with a single appearance feature. So, an underlined, all-capital-letter heading is on a higher level than a heading featuring only all capital letters.

Appearance, size, and placement variations used together allow writers many heading types—and consequently many heading levels. See Page Layout and Word Processing.

5. Use a numbering system with your headings if you have more than four levels of headings or if your document is lengthy and you or your readers will need to refer to sections of it by number.

Report writers normally construct their tables of contents by listing the headings and subheadings in their reports. If you use a numbering system with your headings, include the numbers in the table of contents. See Numbering Systems, Tables of Contents, and Outlines.

6. Make headings parallel in structure and use the same type of headings at each heading level.

Parallel structure means that all headings at the same level have the same basic grammatical structure. So if one heading opens with an *–ing* word, the other headings will also:

this

Developing the Appropriate Tests

Sending Out for Bids

Selecting the Winner

Agreeing on Preliminary Contract Talks

not this

Developing Appropriate Tests

Sending Out for Bids

Selection of the Winner

Preliminary Contract Talks

See Parallelism.

Most headings are declarative. They state or announce a topic:

Facilities in the Local Impact Area

Summerhill Treatment Plant to Close

Complex Traps in Gulf Coast Fields

The Geomorphology of Exploration

An alternative heading form is the question heading:

How Great Is the Avalanche Danger?

Will Surface Water Quality Be Degraded?

What Are the Alternatives?

Should Exxon Proceed With the Project?

Will the Public Accept Our Position?

How Likely Is a Major Bank Default?

As long as you don't overdo them, question headings offer interesting possibilities and can be very effective. Question headings not only announce the topic but also stimulate interest because they pose questions that curious readers will want to see answered.

Question headings are generally more engaging than declarative headings because they seem to speak directly to the reader. However, be careful not to pose obvious or condescending questions:

Doesn't Everyone Know About Anticlines?

Be consistent. If you decide to use question headings, use them for all headings at that heading level. Mixing question and declarative headings at the same level is confusing. However, you can use question headings at a major level and then use declarative headings at subordinate levels, or vice versa.

7. Capitalize headings using the same conventions as for titles of books, articles, other documents:

Three Recommendations for Staff Improvement

A Closing Plea for Sanity

Conclusions About Your Marketing Plan

See Capitals.

Note 1: A few publications, especially ones with an advertising purpose, are beginning to use headings with only the initial word capitalized:

Three recommendations for staff improvement.

Note 2: Subject lines for letters or memos (a type of heading) are increasingly using a capital only on the initial word:

Subject: A closing plea for sanity

This pattern is acceptable, but be consistent within a single document or set of letters.

Hyphens

Hyphenation is one of the trickier aspects of English. There are many rules of hyphenation—including some that apply only in limited circumstances—and all of the rules have exceptions. Below are the most common conventions of hyphen usage. For further discussions of hyphenation, refer to *The Chicago Manual of Style*, *The Gregg Reference Manual*, or the *United States Government Printing Office Style Manual*. See REFERENCES.

Hyphens

1. Hyphenate two or more words that act together to create a new meaning.
2. Hyphenate two or more words that act together to modify another word.
3. Do **not** hyphenate connected words that function as adjectives if they occur **after** the word they modify.
4. Hyphenate compound numbers from twenty-one to ninety-nine and compound adjectives with a numerical first part.
5. Do **not** hyphenate connected words that act as adjectives if the first word ends in *-ly*.
6. Avoid using hyphens with most prefixes.
7. Hyphenate words that must be divided at the end of a line.

Fundamentally, hyphens show a connection. Typically, the connection is between two words or occasionally between a prefix and a word. The connected words (known as compounds) can function as nouns, verbs, or adjectives:

Connected words as nouns

brother-in-law
ex-mayor
follow-up
foot-pound
know-how
run-through
self-consciousness
two-thirds

Connected words as verbs

to blue-pencil
to double-space
to spot-check
to tape-record

Connected words as adjectives

all-around person
black-and-white print
coarse-grained wood
decision-making authority
high-grade ore
high-pressure lines
interest-bearing notes
little-known program
long-range plans
low-lying plains
matter-of-fact approach
off-the-record comment
old-fashioned system
part-time employees
30-fold increase
three-fourths majority
twenty-odd inspections
up-to-date methods
well-known researcher
high-level edit

Unfortunately, not all connected (or compound) nouns, verbs, and adjectives require hyphens. Here are a few of the exceptions:

Connected but unhyphenated nouns

ball of fire
breakdown
fellow employee
goodwill
problem solving
quasi contract
takeoff
trademark
trade name

Connected but unhyphenated verbs

to downgrade
to handpick
to highlight
to proofread
to waterproof

Connected but unhyphenated adjectives

barely known researcher
bright red building
crossbred plants
halfhearted attempts
highly complex task
10 percent increase
twofold increase
unselfconscious person
worldwide problem

As the above examples illustrate, connected words have three possible forms. They can appear as two separate words (*highly motivated*), as one word formed by connecting the two original words with a hyphen (*high-pressure*), and as one word formed by joining the original two words (*highbrow*). See COMPOUND WORDS.

1. Hyphenate two or more words that act together to create a new meaning:

a counterflow plate-fin
the V-space between units
the F-22A airplane
one-half of the annular ring
to double-check the tests

This rule indicates a potential use of the hyphen, not a mandatory one. In some instances the two words become a single word, without a hyphen: *highlight, bumblebee, barrelhead*. In other instances, the two words remain separate: *base line, any one* (one item from a group), *amino acid*. The words sometimes remain separate because combining them would produce strange-looking forms: *aminoacid, beautyshop, breakfastroom*. Because the presence or absence of a hyphen is often a matter of convention, check a current dictionary if you are not sure how the compound word should be written. See CAPITALS for the proper capitalization of hyphenated words in titles.

2. Hyphenate two or more words that act together to modify another word:

brazed-and-welded construction
cross-counterflow unit
engine-to-recuperator mountings
full-scale testing
no-flow heat exchanger
pressure-drop decrease
3-year, multimillion-dollar program
12-foot-wide embayment
up-to-scale modeling
U-tube arrangement
well-documented success

This rule applies only when the connected or compound modifier occurs **before** the word it modifies. See rule 3 below.

3. Do *not* hyphenate connected words that function as adjectives if they occur *after* the word they modify:

The boiler was brazed and welded.
The compartment is 32 feet wide.
The program is well documented.

but

The brazed-and-welded boiler
The 32-foot-wide compartment
The well-documented program

NOTE: An exception to this rule is the use of a hyphen after a verb when the connected words have an inverted order:

They were sun-bathing. (*or* a bathe in the sun)
That transaction was tax-exempt. (*or* exempt from tax)
They were on a fire-watch. (*or* a watch for fire)

4. Hyphenate compound numbers from twenty-one to ninety-nine and compound adjectives with a numerical first part:

thirty-four
eighty-one
five-volume proposal
13-phase plan
24-inch tape
500-amp circuit
4- or 5-year audit cycle

but

22 percent fee

See COMPOUND WORDS and NUMBERS.

5. Do *not* hyphenate connected words that act as adjectives if the first word ends in *–ly*:

highly motivated engineer
poorly conceived design
vastly different approach
completely revised program

NOTE: The words ending in *–ly* are actually adverbs. The *–ly* form indicates the structure of the modifying phrase, so a hyphen is unnecessary.

6. Avoid using hyphens with most prefixes:

counterblow
midpoint
nonperson
progovernment
supercar
undersea

NOTE 1: Hyphens do appear when the prefix precedes a capitalized word:

un-American
mid-August

NOTE 2: Hyphens are sometimes necessary to prevent confusion: *re-treat* (to treat again) versus *retreat* or *un-ionized* versus *unionized*. If you are not sure whether a prefix requires a hyphen, refer to a current dictionary. See COMPOUND WORDS and REFERENCES.

7. Hyphenate words that must be divided at the end of a line.

Words are always divided between syllables, and hyphens should appear at the end of the line where the word division has occurred. Try not to end more than two consecutive lines with hyphens. Try not to divide at the end of the first line or at the end of the last full line in a paragraph. Do not divide the last word on a page.

Hyphens and Technical Terminology

The use of hyphens in technical expressions varies considerably. Technical writers often violate the rules of hyphenation when they believe that the technical expression will be clear:

We will need a high pressure hose.

In this sentence, *high* modifies *pressure*. The sentence refers to a hose that is capable of withstanding high pressures. It is not a pressure hose that happens to be high (off the ground). Yet if we followed the rules of hyphenation strictly, the sentence should be:

We will need a high-pressure hose.

Hyphens are often omitted in technical expressions because the context clarifies the expression. In many cases, however, missing hyphens can cause confusion or a complete lack of comprehension, as in this sentence from an aircraft maintenance manual:

Before removing the retaining pin, refer to the wing gear truck positioning actuator assembly schematic.

Nontechnical (or technical but unknowledgeable) readers can only guess which words are associated with which other words. Does *truck* link with *wing gear*, or does *truck* modify *positioning*? Hyphens would help clarify the modifier relationships:

wing-gear truck-positioning actuator assembly

Proper use of hyphens will not baffle knowledgeable technical readers, and it will help those readers who are not familiar with a technical expression. See ADJECTIVES.

Illustrations

Illustrations, diagrams, and drawings include a wide range of graphics whose purpose is to depict parts, functions, relation-ships, activities, and processes that would be difficult or impossible to describe in text.

Illustrations can help you visualize or conceptualize an idea or a discovery. Rather than merely talk about an idea, try sketching it. A sketch, no matter how rough, can be a valuable tool for you and for others who may be trying to visualize what you are thinking.

The many sketches in Leonardo da Vinci's notebooks are outstanding examples of how illustrations can help in the visualization of concepts. In his notebooks, da Vinci (1452–1519) sketched many images from nature, such as the components of a bird's wing or the muscles in a human arm. Then he used these images to visualize (conceptualize) machines and tools that used the same natural or physical principles.

Figure 1 is an example of a da Vinci sketch. In this instance, da Vinci conceived of a tank-like machine, whose base (in the left image) contained the wheels and gears for providing motion. The right image shows the base covered with a shielded top that would hold men and weapons. The combination of engineering details with the dust clouds suggests rapid movement.

Illustrations

1. Use illustrations, diagrams, and drawings to visualize a system, process, or piece of equipment that would be difficult to describe in text.
2. Use symbols, icons, or other recognizable illustrations to convey messages to readers.
3. Use color to enhance your illustrations, symbols, icons, and logos.
4. Keep illustrations simple, and give each one a perspective that enables readers to understand it.
5. Label each illustration clearly and, if necessary, label the parts of the object shown.
6. In a series of illustrations, make the viewing angle consistent.
7. Ensure that all letters, numbers, and labels are horizontally oriented on the drawing or illustration.
8. If necessary for clarity, remove surrounding detail from illustrations.
9. Use line patterns in an illustration to show how different subsystems interact within a system.
10. If your drawing shows a process, structure the process from top to bottom and left to right.
11. Use special-purpose illustrations when necessary.

Producing good illustrations almost always requires a professional graphic artist. The following discussion does not address the art or mechanics of creating effective illustrations, diagrams, and drawings. Instead, it focuses on how writers should conceive of and use illustrations and what writers can do to assist graphic artists.

Figure 1. Leonardo da Vinci's Sketch of a Tank (Right Image) and Its Working Mechanisms (Left Image). *Although he lived in the fifteenth century, da Vinci's sketches foreshadowed mechanical or engineering discoveries that only appeared centuries later.*

For further information on illustrations and graphics, see Graphics for Documents and Graphics for Presentations. See also Charts, Color, Graphs, Maps, Photographs, and Tables. Always include an informative caption for every illustration. See Captions.

1. Use illustrations, diagrams, and drawings to visualize a system, process, or piece of equipment that would be difficult to describe in text.

Illustrations are very effective at showing views of objects or systems that do not exist (a drawing of a proposed tool), that are abstractions (organizational or functional systems), or that would be impossible to show otherwise (exploded views or cutaways).

Illustrations allow readers to see inside something that is sealed; to see opposite and hidden sides of an object simultaneously; and to see, in close-up, details that would otherwise not be visible. Exploded views like figure 2 allow you to see how components of an assembly fit together.

Like an orange sliced in half, a cutaway or cross section can show hidden detail. Figure 3 shows a cross section of the components of a virus particle. Although drawn to reflect the actual structure of the particle the illustration uses nonrealistic, contrasting colors to make the components more distinguishable. A cutaway can show the internal structure of a mechanism that is normally sealed (see figure 4).

Introduce illustrations in text before they appear. Number them and refer to them by figure number.

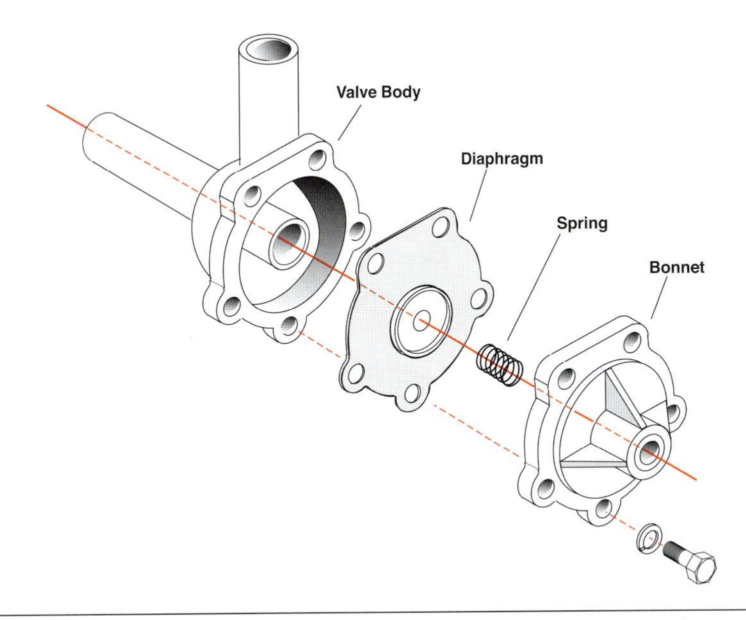

Figure 2. Spring-Actuated Diaphragm. *Remove the bonnet slowly to prevent losing small parts that are under spring compression.*

2. Use symbols, icons, or other recognizable illustrations to convey messages to readers.

Symbols and icons are increasingly valuable communication tools, especially within the international business and technical communities. Figure 5 is an illustration of two of the many symbols that have become international through use if not through governmental agreements. Such symbols usually have an associated color.

This rule contrasts with the suggestion in rule 1 that you use illustrations to conceptualize or visualize your ideas. Actually, as you conceptualize your document or your presentation, you may want to explore your options for using possible symbols or icons. Icons, especially those registered as corporate trademarks or corporate logos, are more and more sophisticated, often with branded colors. See the model documents at the end of the *Style Guide* for examples of various organizational logos. See INTELLECTUAL PROPERTY.

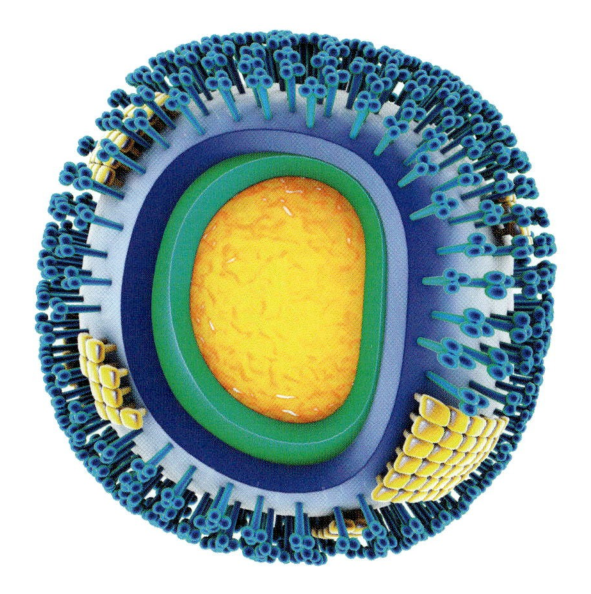

Figure 3. A Cross-Section View of a Virus Particle. *The genetic material (orange) is enclosed by a protein coat (green) and a lipid envelope (blue). The protein spikes on the envelope (blue and yellow) bind to the host cell.*

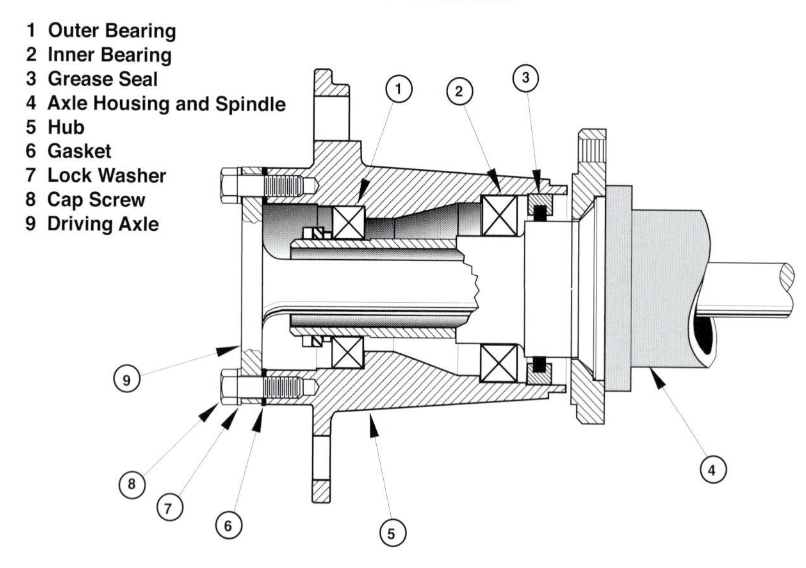

Figure 4. Full-Floating Axle. *Radial and axial thrust loads are carried by the spindle, leaving the driving axle free of all but torque loads.*

3. Use color to enhance your illustrations, symbols, icons, and logos.

Figures 1, 3, 5, 6, 9, and 11 illustrate various ways to use color in illustrations.

Sometimes, you may find that full color is not feasible, based on printing or production costs. An option is to work with two colors. Figure 6 is an example of an illustration using only shades of a single color to represent the different modular furniture. This illustration would be an attractive addition to a document that uses only black text. Of course, the blue-gray color could appear elsewhere in the text beyond this single graphic, most probably in the headings and subheadings or in marginal callouts.

If you decide not to use color, be sure to prepare your illustrations and to design your document so that your message will still be clear and appear professional. In illustrations, for example, you may need to use shading or design patterns instead of color to distinguish between different steps or components of machines.

Always preview sample pages (for documents) or slides (for presentations) to determine if your choice of colors reproduces effectively on paper or onscreen. See Color, Graphics for Documents, and Graphics for Presentations.

4. Keep illustrations simple, and give each one a perspective that enables readers to understand it.

Illustrations and drawings should be focused. That is, they should present a single concept. They should be clean and uncluttered. Everything not pertaining to the single concept should be eliminated. No detail

Figure 5. Standard Symbols. *Symbols such as the two shown are increasingly used in international communication.*

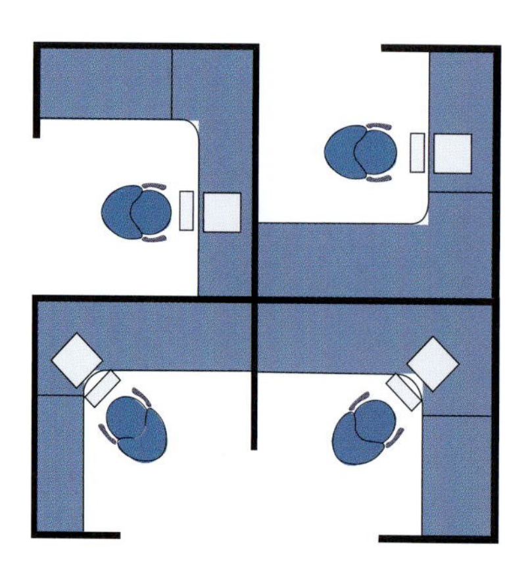

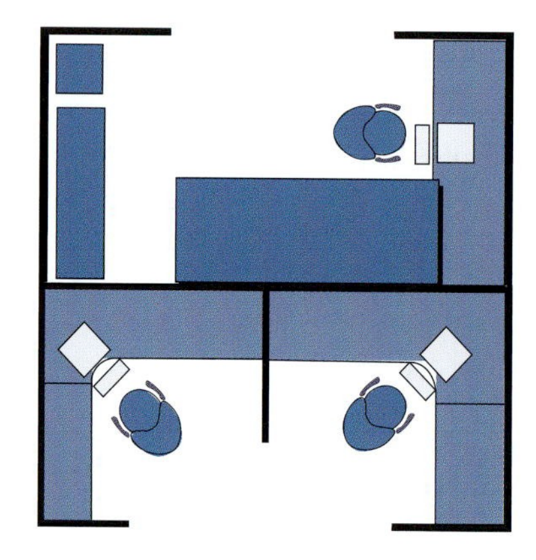

Figure 6. Two Layouts Within a Cubicle Arrangement. *The arrangements shown here, as seen from above, present only two of the many options available.*

should be present that does not contribute to the presentation of that single concept.

As well as being simple, a good illustration has a clear perspective. Illustrations allow you to distort reality, so you must ensure that readers understand the perspective from which the illustration presents its subject. Illustrations almost always show their subjects out of context. Therefore, you might need to establish what the reader is viewing and how that thing relates to other things in its real environment.

Size scales might be necessary if the size and relationship of the object depicted to other things in its environment are not clear. You can also use labels to indicate size, direction, orientation, and nomenclature. If you do not indicate size and distance relationships, readers might not be able to determine the correct proportions of the object shown or its correct orientation in the world outside the illustration or drawing.

5. Label each illustration clearly and, if necessary, label the parts of the object shown.

Figure 2 shows a typical nomenclature illustration. The valve body is shown as one unit because it is not the focus of the illustration. The diaphragm, spring, and bonnet are the reason this illustration exists, so each is labeled separately. Center lines drawn through the axis of each part show how the parts fit together.

Labeling of the significant parts of a drawing is crucial for reader comprehension. You may use word labels, lines, and arrows (figures 2 and 3), or you may use numbers, letters, or symbols in the drawing itself (figure 4) with an explanation block.

6. In a series of illustrations, make the viewing angle consistent.

If you are showing the same object in a series of illustrations, and the point of the series is to show assembly/disassembly steps or operational phases, ensure that readers see the object from the same perspective in each illustration. Changing the perspective is very confusing.

7. Ensure that all letters, numbers, and labels are horizontally oriented on the drawing or illustration.

The text appearing on any part of a drawing or illustration should never be vertically oriented unless the bases of the individual letters or numbers are horizontal, as in this example:

E
X
A
M
P
L
E

The lettering and numbering on an illustration should be oriented so that readers can read it without reorienting the illustration. If you run out of space, use arrows and move the labels away from the busy area of the illustration. If necessary, omit the labels and use letters, numbers, or symbols and an explanation block (see rule 5 above).

8. If necessary for clarity, remove surrounding detail from illustrations.

Figure 7 shows a schematic drawing or illustration in which surrounding but irrelevant detail has been removed. You often see this sort of illustration in subsystem pictorials. Removing the surrounding detail allows readers to focus on the system being shown. The drawing isolates its subject and therefore provides an excellent focus.

9. Use line patterns in an illustration to show how different subsystems interact within a system.

If you are showing how different subsystems fit together and function, you might need to use different line patterns, as in figure 8. The line patterns allow readers to isolate

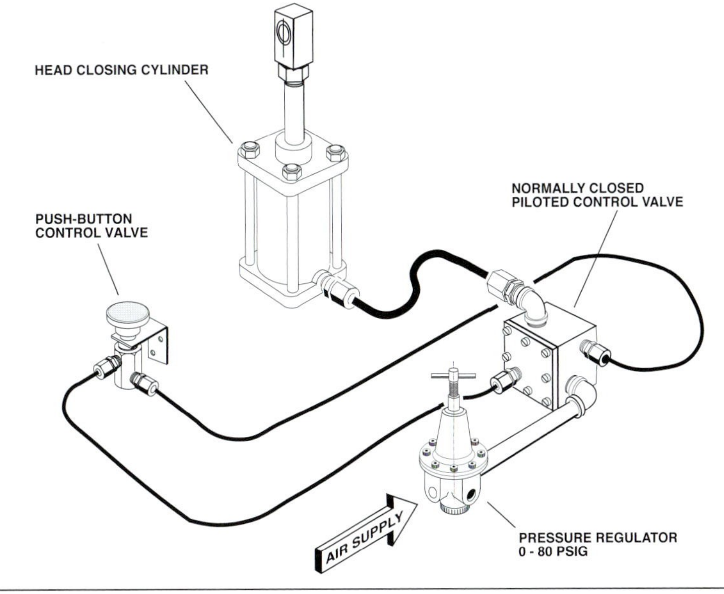

Figure 7. Pneumatic Operating and Control System. *By eliminating electrical controls, potential spark hazards are avoided, allowing operation in hazardous environments.*

subsystems while viewing the whole system. For additional examples of line patterns, see GRAPHS.

If you use line patterns, provide an explanation of what the different patterns represent. For example, in figure 8 the diesel fuel lines and the coolant line use different line patterns.

10. If your drawing shows a process, structure the process from top to bottom and left to right.

Figure 9 is a schematic flow chart. It shows a process and the equipment used in that process. This type of drawing should be oriented from top to bottom and left to right so that readers "read" it as they would read text. In all process drawings, readers will expect the process to start at the left and end at the right. Don't disappoint them. See CHARTS.

11. Use special-purpose illustrations when necessary.

Work closely with graphics specialists when you need special illustrations. The following three examples are only a sample of what illustrations can bring to documents.

Example 1. You might need a diagram of an abstract concept.

Figure 10 is an example of a concept illustrated. The initial step was to sketch many ways of showing the idea that people can contribute more than just their talents at work—they can also contribute what excites them and what their conscience tells them they should contribute in an identifiable need area. Once a sketch was chosen, a graphic artist created the diagram.

Example 2. Illustrations often combine realistic details into a montage designed to capture a theme or an atmosphere. In figure 11, the rich imagery of our hyperlinked world is the main theme. The round

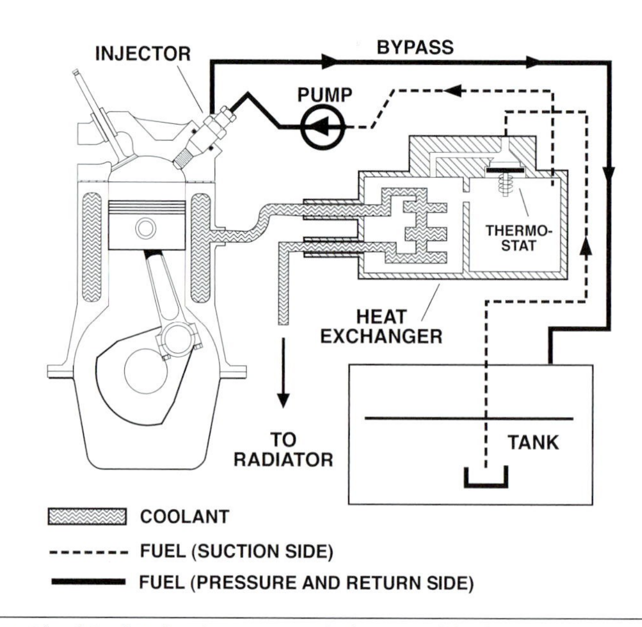

Figure 8. Diesel-Fuel Preheating System. *The burning of the fuel is approximately 10 percent more effective with the fuel preheating system.*

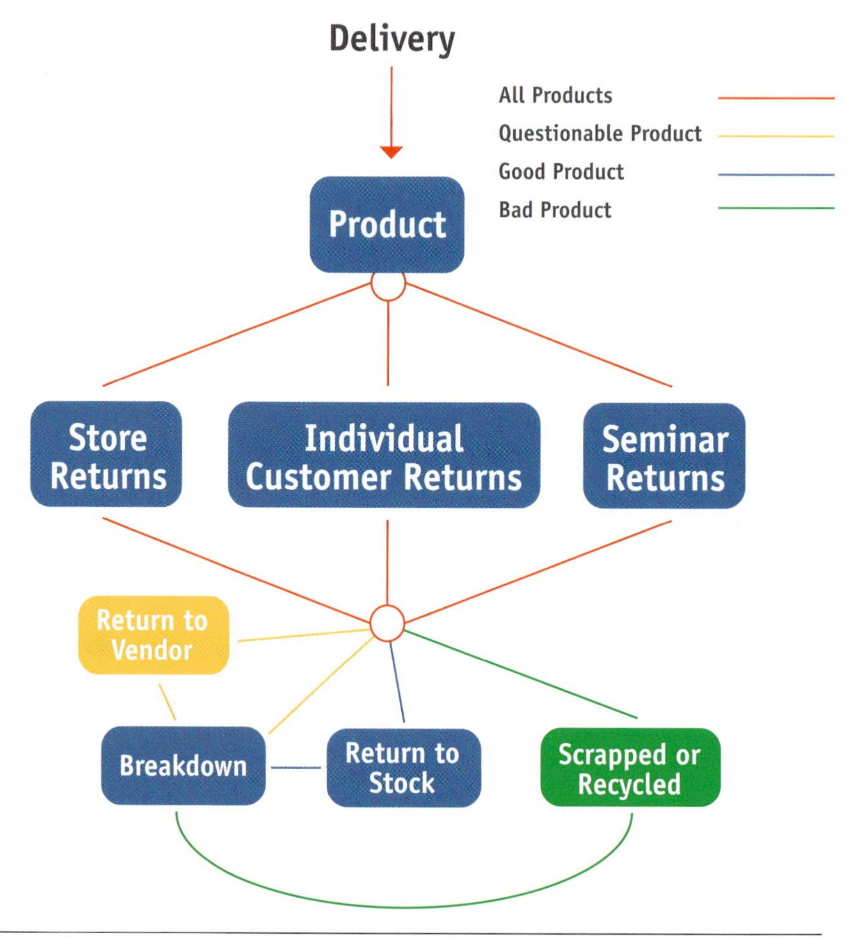

Figure 9. The Returns Process in a Manufacturing Setting. *Products or parts returned fall into one of these categories: good, questionable, or bad. Depending on its category, a part is returned to stock, reworked, or recycled.*

objects echo each other, while the net of optical fiber enmeshes the whole.

Example 3. Illustrations often become part of or combine with other graphics. Figure 12 is a combination between an illustration (the cross section of the dam) and a graph (informal) of the different sediment depths. As a combination graphic, the image does not attempt to be a realistic illustration. Instead, the detailed features about the dam and its operations are ignored (not even labeled). Instead, the labels on the illustration focus on the different elevations for the sediment at selected dates in the future.

A simple bar graph could easily replace this illustration. The bars on the graph could record just how high the sediment would become at a certain date, but the illustration adds a sense of reality to the concept.

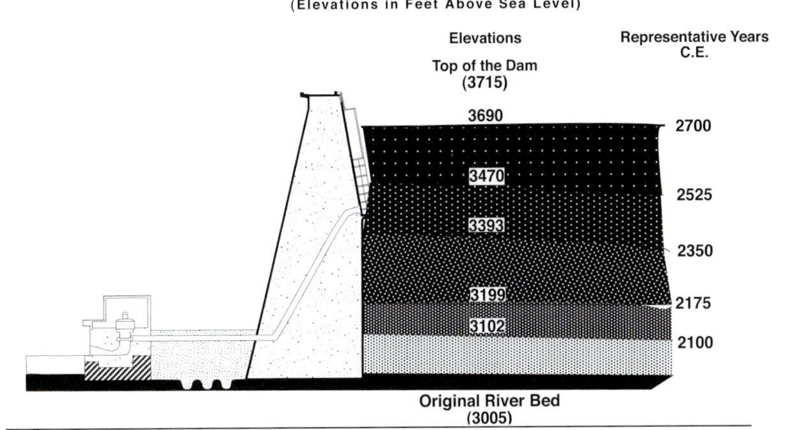

Figure 12. Sediment Buildup Behind Glen Canyon Dam. *Sediment will collect behind Glen Canyon Dam, finally filling in the reservoir (Lake Powell) about 2700 C.E.*

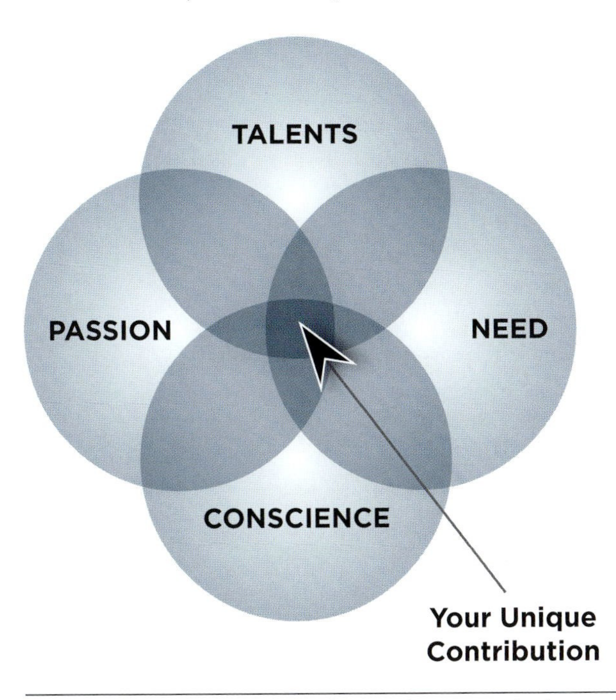

Figure 10. Your Unique Contribution. *You have strengths that can't be found elsewhere—a unique combination of talents, passion, and conscience. Where do your strengths intersect with a compelling market need?*

Figure 11. A Hyperconnected World. *The globe is now like a digital disc, a ball of electronic circuitry connected by fibers of light.*

Indexes

Indexes are alphabetical lists of the subjects within a published (printed) document. For each subject, the index gives a page number (or sometimes a section number) so that readers can easily find where a specific subject is discussed.

Online documents sometimes include an index, especially if the online information is complex and lengthy. See Online Documents.

Published documents usually have indexes unless the document is so short that an index is unnecessary—as, for example, in a brochure or in a marketing summary.

The longer a document, whether published or not, the more helpful an index becomes.

The following rules present the basics of indexing a published (printed) document, but the subject is complex enough that if you want to prepare a detailed index, consult *The Chicago Manual of Style*. See References.

Indexes

1. Provide an index for any document longer than 50 pages and for any document that will have many and frequent readers.
2. Adjust the scope of your index to the needs of your readers.
3. Begin to prepare an index by identifying key words and concepts that readers would likely be interested in locating, given the purpose of the document.
4. Use a computer program to prepare the actual index.
5. In a detailed index, use subheadings and cross-references.
6. Don't forget basic proofreading and cross-checking.

1. Provide an index for any document longer than 50 pages and for any document that will have many and frequent readers.

The 50-page threshold is only a suggestion. Whether you decide to provide an index also depends on how many readers will be using your document. For example, if you are preparing an internal policy manual that will have only 30 or 40 pages, even a manual this short can be confusing when many employees want to look up specific answers to their questions. In such a manual, both a good table of contents and a simple index would be helpful. See Tables of Contents.

For documents with few readers, no index is necessary even if the document exceeds 50 pages.

2. Adjust the scope of your index to the needs of your readers.

Indexes can be fairly simple or very detailed, and they can cover special areas of interest.

A simple index primarily includes key words from the document. Such words are usually those that appear in the title, chapter headings, and subheadings. Such an index might list only two or three topics from each page of a document.

A detailed, professional index can be quite exhaustive, often listing 10, 15, or more topics from each page in a text. Because such indexes are very time-consuming and expensive, you may choose not to prepare them for most unpublished documents. Even published documents sometimes might not need such a detailed index. See the index in this *Style Guide* for an example of a detailed index.

Detailed indexes can also be prepared for special areas of interest. For example, a separate index might list all the people's names in a historical document. Separate indexes sometimes exist for geographical names or for works of art (as in a list of musical titles). Such special indexes usually supplement, not replace, the traditional subject index.

3. Begin to prepare an index by identifying key words and concepts that readers would likely be interested in locating, given the purpose of the document.

Indexing is not a mechanical task. Anyone preparing an index has to have a good grasp of the content and purpose of the document. For this reason, the author of a document is often the best person to prepare the initial or draft index.

Not every fact or every reference automatically appears in an index. For example, if a text mentions the *Mississippi River*, should the index list this reference under *River* or under *Mississippi*? Should the index even list either reference at all?

Answers depend on the purpose of the document. A document discussing all sorts of geographical features might have major index entries on national parks, lakes, rivers, bayous, etc. If the Mississippi itself were merely mentioned in passing as one of a list of well-known U.S. rivers, then it might not appear in the index, and it would surely not be a major (separate) entry in the index.

A document discussing major U.S. rivers, their length, their flow, their flood stages, their pollution, etc., would likely list the Mississippi, the Ohio, and every other river discussed in the document. These individual names would be major entries in the index. Then under each river would appear subheadings:

Mississippi River, 8, 12, 15–18; flood control, 8; pollution on and near, 65, 68–70; source of, 14; tourism on, 48–49

4. Use a computer program to prepare the actual index.

Most word-processing programs have an indexing feature. With this feature, writers can identify key words and phrases either while they are writing the draft or after they have finished the draft. Usually, they code these words and phrases with a computer code that will not appear in the printed copy. The word-processing program can, however, identify the code and will alphabetize the identified words and phrases into major entries and attach the proper page numbers to them.

Such mechanical indexes can be helpful, but they do not replace the many judgments necessary when a full, detailed index is needed. Only a human indexer (or the author of the document) can make such judgments.

5. In a detailed index, use subheadings and cross-references.

Subheadings are important because they reveal how various topics relate to each other. In most cases, subheadings have a grammatical or syntactic relation to the main heading and are presented alphabetically:

Automobiles, 18–20; design of, 43, 48–49; maintenance costs of, 123, 126; maintenance guidelines for, 115–118; optional equipment for, 84–85; resale values of, for retailers, 79; resale values of, for wholesalers, 78

In other instances, the subheadings are merely a list of items:

Fortune 500 companies: Amoco, 45; Exxon, 53; Ford, 17–19; General Dynamics, 115–119; General Motors, 10–11.

As in the preceding examples, the main heading is usually flush with the margin while the succeeding lines are indented. Traditionally, commas separate headings or subheadings from page numbers, and semicolons separate the different subheadings. Subheadings are alphabetized by main words, not prepositions or conjunctions. No final punctuation follows the entry, including all its subheadings.

Inclusion of *See* and *Also see* (or *See also*) references is important when readers are likely to look up a variety of topics.

See references are useful when the indexer has had to choose between several different main headings:

Businesses. *See* Fortune 500 companies

Corporations. *See* Fortune 500 companies

Also see references are usually restricted to added information:

Chevron, 58, 62–64. *Also see* Fortune 500 companies

6. Don't forget basic proofreading and cross-checking.

Nothing destroys reader faith and your credibility more than errors in the index. Users rightly expect to find what you tell them is on a specific page.

Plan, therefore, to allow time for careful proofreading and even cross-checking of items. If possible, go through the entire index verifying that the cited information is indeed on the pages noted. For published indexes, the publisher is usually responsible for this final cross-check of an index for accuracy.

Intellectual Property

Intellectual property includes both ideas and the expression of those ideas. A person's or an organization's intellectual property is potentially protected by the registration of a patent, a trademark, a service mark, or a copyright.

The following discussion summarizes the highlights of current United States copyright rules. It also includes a brief mention of service marks and trademarks. It does not cover patents, which have special and very detailed legal requirements.

For guidance on international copyright regulations, see the website of the World Intellectual Property Organization. This site summarizes the laws of each member state.

If you have questions about copyrights, service marks, trademarks, or patents, you should obtain legal advice from an attorney who specializes in intellectual property.

Intellectual Property

1. Accurately quote and credit the sources for any information you use in your own writing or speaking.
2. Include in anything you produce a written notice of its copyright status and register any commercially valuable work with the U.S. Copyright Office.
3. Identify the source for any photographs or other graphics. Also, do not modify visual creations without considering the rights of the original creator of the visuals.

1. Accurately quote and credit the sources for any information you use in your own writing or speaking.

Abiding by this rule is both ethically and legally important.

Personal ethics require that you give others credit for their ideas, contributions, and originality. You expect no less when someone uses your ideas or words. See Ethics.

Legally, you must carefully quote and cite the source for any published information you choose to use. If you don't, you will violate U.S. copyright laws. You may also be violating non-U.S. copyright laws. Major violations can be costly, both for you and your company. See Quotations and Citations.

Copyrights. Anyone's original work is protected by a copyright, even if the person who creates a work has not chosen to apply formally for copyright (see rule 2). The copyright exists when the writer makes the first original version, even a mere handwritten copy.

Copyright covers the expression of an idea, not the idea itself. You must always credit actual words and phrases that you borrow. Ideas may not be legally copyrightable, but ethically, you should still credit other people's ideas if they were distinctively theirs, not yours. See Quotations.

Copyright protection initially applied only to "writings," but protection now covers videos, paintings, photographs, cartoons, recordings, songs, sculptures, computer programs, Web content, or any other creative product when it is physically recorded. Thus, the manuscript for a song can be copyrighted and the physical recording can be copyrighted, but the actual live performance is not copyrighted unless it is recorded.

Under copyright law, you must obtain permission from the author(s) (or the holder of the copyright if an author has signed over his or her rights to someone else—for instance, a publisher). Often, you will pay a fee if you choose to use copyrighted material in something you wish to publish for profit.

If you obtain permission, you should retain the written permission (usually a brief form letter). In your own work, you should identify the author and the holder of copyright (usually a full bibliographic citation) and include a note like the following:

> Reprinted by permission of
> (the name of the author or the publisher).
>
> *or*
>
> Reproduced courtesy of the
> photographer's heirs. (List the heirs.)

Fair Use. In some instances, especially in academic studies or in critical reviews, you may reproduce without permission excerpts (written or visual) from a copyrighted work. In such cases, you need not obtain written permission before you publish. You must still identify the source for any quotations or excerpts you use.

Fair use applies if you are not using enough of the original to affect its value. The number of words is not the measure of fair use. A quotation of several hundred words from a book might fall under fair use, but a few key words from a short poem might violate fair use. A detail from a single painting might be fair use; the whole painting would likely not be.

Don't rely on the fair use exemption if you are in doubt about the fair use status or if you are going to commercially profit from quoting someone else's copyrighted material. In these cases, write for permission and pay any necessary fees.

Public Domain Works. Works produced by the U.S. government are in the public domain. These works include works produced by individuals working under U.S. government contracts.

Some works also are in the public domain because their copyright has expired. The time limits for a copyright vary, based on different copyright laws. Under the 1976 law, which took effect on January 1, 1978, the general rule is that copyright lasts for the life of the author, plus 50 years. In cases of joint authorship, the copyright lasts 50 years after the death of the last surviving author.

Internet content, such as blog entries, newsletters, ebooks, and music and video recordings fall under the same 1976 law. Additionally, the Digital Millennium Copyright Act of 1998 makes it a crime to circumvent the technologies put in place to prevent unauthorized copying of these works.

You can freely quote, without permission, from public domain works, but you should still include full citations for any materials you are borrowing. Your readers need to know which are your words (and ideas) and which words you are borrowing. See QUOTATIONS and CITATIONS.

Trademarks. Trademarks are names or symbols used by a company to identify a particular product or service. Examples are *Coca-Cola®*, *Mustang®*, and *Chloraseptic®*. In some instances, a trademark can be a graphic symbol and will be called a service mark.

Trademarks are registered with the U.S. government, and their use is prohibited without the approval of the trademark's owner. Trademarks are usually capitalized. See CAPITALS.

As in the preceding examples, trademarks usually have the symbol ® immediately following them. In text, the ® need only appear on the first use. Later references to Coca-Cola would be capitalized, but with no ®.

Before registration, a trademark will have had an attached ™ or SM, to signify the name or symbol as either a trademark or service mark. After registration, the ® replaces the ™. In some instances, both the ® and the ™ will appear next to a trademark.

Other common examples of trademarks include *Kleenex®*, *Jell-O®*, and *Xerox®*. For these three examples, the generic terms would be *facial tissue*, *gelatin*, and *photocopy*. Writers should use the generic terms unless they actually mean, for example, the brand of tissue known as *Kleenex®;* then the name would be capitalized and followed by the ®.

2. Include in anything you produce a written notice of its copyright status and register any commercially valuable work with the U.S. Copyright Office.

Using a written notice applies equally to a few pages as to a full volume. The written notice includes (1) the symbol © or the word *copyright*, (2) the year it originated or was published, and (3) the name of the author (or copyright holder, if they are different people). This notice protects writing or other original works even if they have not been formally registered.

Consider registering any works of particular importance, especially any with commercial value. Your copyright exists even without registration, but registration does fix, for legal purposes, the nature of your work and the date of its publication. Registration is necessary if you have to file suit for copyright infringement, unless your work was published outside the U.S. and its territories.

Fixing the copyright date also starts the legal timing, which under the 1976 copyright law extends copyright protection for the life of the author, plus 50 years. Note that the timing for a copyright is different for works published or copyrighted before 1978, which is when the 1976 law took effect.

Works for Hire. If you are an employee or a paid independent contractor, you will usually be unable to apply for or claim a personal copyright on your work if it is produced for your employer. Whatever you write or produce will be called work for hire. Your employer can and usually will register such work.

3. Identify the source for any photographs or other graphics. Also, do not modify visual creations without considering the rights of the original creator of the visuals.

Photos and computer graphics have copyright protection, just as any other publications do. Sometimes writers borrow a photograph or computer graphic without remembering that a tangible, reproducible image merits the same copyright protection as words.

This same guidance applies, for example, to a cartoon you find online. A one-time use of the cartoon for educational purposes would probably fall under fair use, but ongoing use would not. And even one-time commercial use would not be fair use. As with any copyrighted material, using a cartoon requires you to contact the cartoonist or a publisher for permission, which is usually granted after payment of a one-time fee or a royalty for continued use.

Modifying a graphic does not make it yours! As with the cartoon example, some people provide their own captions or write in new balloon comments. Use of this modified cartoon is still a violation of the cartoonist's copyright.

Similarly, you are not free to modify or adjust a copyrighted photograph or an illustration to suit your own purposes. Even if you have obtained permission, for example, to use a photograph, this permission can include requirements as to how and when you will reproduce the photograph. For example, printing a small black-and-white version of a large color photograph might damage the original intent or artistic value. Sometimes the photographer might even refuse permission based on what you intend to say about the photograph.

Be careful not to "borrow" online images or text that may be copyrighted. Unfortunately, your source may not mark an image with a copyright symbol. Your ignorance of the copyright status does not remove your legal liability.

When you have questions about intellectual property of any kind, consider obtaining legal advice from an attorney who specializes in copyrights, service marks, trademarks, and patents.

International Business English

International Business English is the most common medium for conducting business worldwide. For historical, political, and economic reasons, English is often the only common language among individuals of diverse nationalities and tongues.

Users of English should not assume, however, that they are automatically understood by English speakers from other cultures. The pace, formality, and social context of communication differ among cultures.

Because of the inherent difficulty of cross-cultural communication, even in a common language, business and technical professionals should take pains to use English that is easily understood and discourages misinterpretation.

Following these rules will lead to clarity and a minimum of misunderstanding across cultures.

International Business English

1. Use simple sentences, but do not omit necessary words.
2. Use simple, understandable words, and avoid idiomatic expressions.
3. Be sensitive to the cultural pace of communication.
4. Account for varying levels of formality and personal distance across cultures.
5. Be particularly sensitive to social hierarchies across cultures.

1. Use simple sentences, but do not omit necessary words.

Keep sentences short. Limit yourself to one idea per sentence, and avoid dependent clauses. See SENTENCES.

Often native speakers omit words for the sake of economy, but this practice can confuse nonnative readers:

not this

Send fax this a.m. care of CFO assistant.

this

Please send the fax this morning in care of the assistant to the Chief Financial Officer.

2. Use simple, understandable words, and avoid idiomatic expressions.

Consciously choose words that readers will readily understand. Avoid grammatical complications like verb forms with *–ing* endings. Break up noun strings that contain more than three nouns. See NOUNS.

Avoid using idioms, which are peculiar or specialized expressions you might use every day but which may be unfamiliar to members of the culture you are addressing.

Even common idioms can be a source of confusion. When possible, use synonyms that cannot be misunderstood.

Idioms	*Synonyms*
get ahead	advance/move forward
get someone's goat	annoy/anger someone
get around a place	walk/drive
get at someone	criticize/nag
get away with something	cheat/steal
get up (verb)	stand/rise
get-up (noun)	costume
get-together	meeting/gathering
get-up-and-go	energy
in the right mind	sane
be of one mind	agree
bear in mind	remember
come to mind	remember
slip one's mind	forget
meeting of minds	agreement
prey on one's mind	worry
put up with	tolerate
put down	insult
put over	trick/cheat
put out	offend
put through	connect/get approval
input	enter (into a computer)

English (as do all languages) has thousands and thousands of idioms like the preceding examples. For help interpreting idioms (especially if English is not your native language), refer to the most recent edition of A. S. Hornby's *Oxford Advanced Learner's Dictionary of Current English* (Oxford, England: Oxford University Press). See REFERENCES.

Idioms included in ordinary text are especially hard to spot. In the following example, the words *shutdown*, *thoroughgoing*, *clean-up*, and *re-siting* are all fairly common idioms, but because of these idioms, the original sentence would be hard for a non-English speaker to interpret. Even a native speaker of English would likely prefer to read the revised version. See GOBBLEDYGOOK.

not this

Determining on a site shutdown, we forwarded recommendations for thoroughgoing site clean-up and re-siting of contaminate.

this

We decided to close the site. We also recommended ways to clean the site completely and remove the contaminated soil.

Nonnative speakers should be particularly careful about attempts at using English idioms, as in this sign in a Copenhagen ticket office:

> We take your bags and send them in all directions.

Or, consider the following example from a manual prepared by a car-rental firm:

> When passenger of foot heave in sight tootle the horn. Trumpet him melodiously at first, but if he still obstacles your passage then tootle him with vigor.

International Business Style

Even when communicating in English, business professionals traveling abroad should be aware of the local "communication culture." Although they need not radically change their communicative style, they should be prepared to adapt to differences in three areas: the pace, territoriality, and social hierarchy of the communication culture in which they find themselves.

3. Be sensitive to the cultural pace of communication.

Diverse cultures communicate at different paces. For example, southern Europeans and Asians tend to be more concerned initially with establishing personal relations than with "getting to business." On the other hand, Americans and northern Europeans are more likely to want to go straight to the point and to move quickly through issues.

4. Account for varying levels of formality and personal distance across cultures.

The need for "personal space" or territoriality differs across cultures. The northern European and Asian cultures require a certain formality and distance in language as well as space, whereas Americans and southern Europeans tend to be informal.

North Americans often need more personal space than Latin Americans or southern Europeans do. So a North American can become uncomfortable when a Latin American speaker moves too close, thus violating the North American's space.

5. Be particularly sensitive to social hierarchies across cultures.

Business professionals should be acutely aware of social hierarchy when communicating with members of other cultures. Asians will offer business cards to begin a meeting and expect to see yours so they can gauge your social and professional standing relative to theirs. They are very conscious of rank, and it pays to be deferential and avoid too much familiarity.

Asians and northern Europeans often think of themselves as representatives of their organizations rather than as individuals engaging in business. Southern Europeans and Latin Americans are less concerned with rank or organizational protocol.

Although business professionals abroad need not mirror their hosts' behavior, it makes good sense to be patient, observant, and adaptive to the pace and temperament of the local communication culture.

See British English and Letters for information on dealing with specific cultural differences in documents.

Introductions introduce—as the name suggests. An introduction conducts the reader into a document, usually by establishing the reason for the document's existence; its relation to other documents or projects; and any special circumstances, facts, conditions, or decisions that help the reader understand the body of a document.

The body of a document is more project- or product-oriented—it focuses on the situation or thing being described or analyzed.

The introduction (as well as the conclusion) is more reader-oriented—it orients the reader by providing a context for the reading. In short, the introduction prepares the reader for what will follow.

Introductions

1. Adjust the length and scope of your introduction to meet your readers' expectations and needs.
2. Replace the introduction with an executive summary, especially in business documents.
3. If appropriate, as in a technical or scientific report for publication, prepare a full introduction, using the traditional content.
4. Use a brief introduction in a letter or memo to set the tone, not to delay the main points.

1. Adjust the length and scope of your introduction to meet your readers' expectations and needs.

Ask yourself if your primary readers absolutely must know certain facts or other data before they can understand your document. If they must know such facts, then include them in your introduction. Then adjust the introduction, as discussed below in rules 2, 3, and 4. Otherwise, get right to your major points (usually the *to do* in the document). See LETTERS and MEMOS.

Many short business documents, especially internal ones, may not even require an introduction. Instead, these documents will begin with the writer's proposal or a statement of fact, usually in the subject line. Any background (or setup) information is covered several lines or paragraphs later. See ORGANIZATION.

2. Replace the introduction with an executive summary, especially in business documents.

Introductions are increasingly unnecessary in many business documents. Too often in the past, the introduction was so general that it really didn't help readers.

In contrast, an executive summary contains the essential facts, findings, and interpretations from the later pages in the report. If well done, an executive summary (perhaps only a page or two) will be all that most readers need. Subsequent pages in your document become attachments or file information. See SUMMARIES.

Readers of business reports are often busy managers or supervisors. They are usually aware of the general details of an investigation or project but trust subordinates to evaluate the problem properly and to solve it efficiently. Such readers will become impatient with lengthy digressions and unnecessary explanations, support, and justification. Consequently, they typically want a succinct executive summary. Only if questions arise will they go beyond the summary and read the background information and analysis.

The key feature of a combined introduction and executive summary is a statement of major conclusions and recommendations. A full list of conclusions and recommendations often appears later in the report, but you should never force managers or supervisors to search for them. See REPORTS and SUMMARIES.

3. If appropriate, as in a technical or scientific report for publication, prepare a full introduction, using the traditional content.

Formal reports are likely to have the most developed and structured introductions. See REPORTS. No two introductions are alike, but most include some of the following information:

—The **problem** or **opportunity** prompting the project or investigation

—The **goals** or **purposes** of the project or investigation

—The likely **audience** for the project or investigation

—The **scope** of the project or investigation

—The **sources** of relevant information

—The **methods** used in the project or investigation

—This project's or investigation's **relation** to previous or concurrent projects or investigations

—Any useful working **definitions**

The **problem or opportunity** that the document addresses arises from the historical background. What work was done that stimulated this project? What other work has been done in this or related fields? Often this section will survey relevant literature. See CITATIONS.

If this project is part of a larger or related project, state the relationship briefly. Brevity is important because readers rarely want or need to learn the point-by-point history of a project or related project.

The **purpose** of the project explains why the project was undertaken and what it is expected to achieve. The key objectives might read as follows:

1. To evaluate the groundwater resources of the alluvial aquifer of the Carmel Valley, California, groundwater basin
2. To develop a two-dimensional, digital, groundwater flow model of the aquifer that will aid in the understanding of the geohydrology of the aquifer
3. To identify data inadequacies that might be needed for future studies of this aquifer

The **scope** refers to the limits of the project and the document itself. Provide the scope by stating what the document will cover and what it will not cover.

The **methods** explain how a project was conducted: how the investigator developed the experimental design, constructed or designed the apparatus, collected the data, analyzed the results, and developed the conclusions. If the methods were routine, provide no more than a brief summary. If the methods were unusual or original, explain them thoroughly in the introduction or consider discussing methods in a separate section or subsection of the document.

4. Use a brief introduction in a letter or memo to set the tone, not to delay the main points.

Introductions to letters and memos establish the writer's tone and approach, as well as set the stage for the ideas and supporting details that follow. See LETTERS and MEMOS.

Establishing a Tone

Letters and memorandums are intended to be more personal and less formal than reports. To convey this more human dimension, some writers choose to use personal pronouns: *I, you,* and *we*. Others open with personal remarks or social greetings, much as we often do in personal conversations. Even writers whose purpose is avowedly serious might "break the ice" by calling the reader by name:

> Please let us know, Jim, if our proposal for the replacement pumps begins to answer your needs.

See TONE.

Setting the Stage

Setting the stage might mean no more than a brief phrase: *As we discussed yesterday . . .* or *According to our records . . . .*

In other instances, writers might decide to provide the background for the document before stating the point. This background or setup is quite common in letters conveying bad news, such as enforcement of a financial penalty or a personnel reprimand. See LETTERS.

However, be cautious about spending too much time setting the stage. Almost always, the best strategy is to get to the point quickly. See ORGANIZATION.

Italics

Italics (a slanted typeface) is available only in printed material or on word-processing systems capable of printing italics. In handwritten or typed material, underlining replaces italics. See UNDERLINING.

Some conventions about italics concern only printers. For example, in printed mathematical expressions, letters are italicized and numerals are set in normal type. This distinction is not normally made in handwritten or typed material. The following standard rules apply in most technical and business documents.

Italics

1. Use italics for words or phrases used as examples of language.
2. Use italics for foreign words and phrases that have not yet been absorbed into English.
3. Use italics for titles of books, magazines, newspapers, movies, plays, and other works individually produced or published.
4. Use italics for the names of aircraft, vessels, and spacecraft.
5. Use italics for names of genera, subgenera, species, and subspecies. Names of higher groups (phyla, classes, orders, families, tribes) are not italicized.

1. Use italics for words or phrases used as examples of language.

Words used as examples occur when a writer discusses specific words or phrases as examples of language:

> In all offshore contracts, *consolidation* does not mean what it normally means.
>
> The Anaguae reservoir study was confusing because the author kept referring to the anomalous formations as *anonymous* formations.

NOTE 1: Use italics to emphasize words, phrases, and even letters when discussing them as examples of language. This use of italics (or underlining) sets the words or phrases apart from the other words in the sentence:

> Traditionally, the symbol *M* has been used to mean million. In some disciplines, *M* means thousand, so one indicates million with the symbol *MM*.
>
> One should avoid opening letters with the phrase *In reference to*. Similarly, according to our corporate guidelines, one should never end a letter by saying *Very truly yours*. Both phrases are too wordy.

NOTE 2: Quotation marks sometimes replace italics (underlining) as a way of highlighting words and phrases, especially in handwritten texts:

> In all offshore contracts, "consolidation" does not mean what it normally means.

2. Use italics for foreign words and phrases that have not yet been absorbed into English:

> The initial concept of the United Nations captured a certain *Weltanschauung*.
>
> The *couturier* insisted on keeping the new dress designs secret.
>
> The staple crop in South Africa is *kaffir*, which is a form of sorghum raised for cattle fodder.

NOTE: Some foreign words and phrases have become so common in English that they are not italicized:

> ad hoc
> habeas corpus
> per annum
> rendezvous
> vice versa

Some recent dictionaries indicate if words are still considered foreign, but other dictionaries do not. If your dictionary does not, use your judgment to determine if a word is sufficiently foreign to be italicized. Foreign words usually retain their foreign spellings, pronunciations, and meanings.

3. Use italics for titles of books, magazines, newspapers, movies, plays, and other works individually produced or published:

> To remain current on advances in space technology, we subscribed to *Design Digest*.
>
> The documentary *Before Their Time* showed what is possible when companies wisely invest IR&D funds.

See TITLES.

NOTE 1: Quotation marks sometimes replace italics for titles. For example, newspapers have traditionally preferred quotation marks:

> After some discussion, we decided to order "A Dictionary of Mining, Minerals, and Related Terms" and the "Society of Petroleum Engineers Publication Style Guide." The two volumes should help us prepare articles for the "Journal of Petroleum Technology" and "Petroleum Transactions."

NOTE 2: Sections of these published works are not italicized. So chapters, magazine articles, acts within a play, and editorials in a newspaper require quotation marks, not italics:

> Last week's *Time* had an article entitled "The Roots of International Terrorism."
>
> The final chapter of the annual report is entitled "Prospects for Growth in the 21st Century."

See QUOTATION MARKS.

4. Use italics for the names of aircraft, vessels, and spacecraft:

Discovery
Friendship 6
H.M.S. *Intrepid*
NS *Savannah*
U.S.S. *Iowa*
U.S.S. *Nautilus*

NOTE: In these examples, only the names are italicized, not the abbreviations or numerals associated with the names.

5. Use italics for names of genera, subgenera, species, and subspecies. Names of higher groups (phyla, classes, orders, families, tribes) are not italicized:

the genera *Quercus* and *Liriodendron*
the family Leguminosae

See the *Council of Biology Editors Style Manual* for additional information and examples. See REFERENCES.

Jargon has two meanings: first, using familiar words in unfamiliar ways (using *hot* to mean *crucial* or *exciting*) or, second, using a specialized"shorthand" language unfamiliar to a particular reader or listener (an emergency medical technician shouting "code 99," which is medical jargon for a heart stoppage). Thus, one person's technical or specialized vocabulary becomes another person's jargon. The following discussion focuses on this second, more common meaning of the word jargon.

Every technical discipline needs and has its own vocabulary.

Medical doctors have innumerable special terms, often derived from Latin: *amebic dysentery, uvula, gastric hernia.*

Lawyers also use a number of common terms that have developed special meanings: *property, liability, consideration, conveyance. Consideration*, for instance, means a payment of some kind as a sign of agreement on a contract. This special legal meaning is not obvious to the uninitiated, who might not even know the word has a special meaning. In some contexts, readers might not know whether the ordinary meaning or the technical legal meaning is intended.

Everyone has confronted jargon inappropriate for its intended use. Take for example, this instruction to new owners of a high-definition TV:

> For sync with a single A/V DOOG, use daisy chain connection (optional HDMI and/or optical cable not included) to A/V amplifier and ensure that settings of equipment compatible with HDAV control has not been changed, then set input mode to DOOG and check that DOOG operates.

The writers of this instruction have seriously misread their lay audience. The jargon they use is utterly baffling and likely to irritate rather than to instruct their customers. See STYLE, SCIENTIFIC/TECHNICAL STYLE, and TONE.

Jargon

1. Do not use jargon unless your readers will understand it. If they will not understand—and you must use a term—then define it.

1. Do not use jargon unless your readers will understand it. If they will not understand—and you must use a term—then define it.

The doctor who gives a diagnosis only in medical terminology has failed to communicate with most patients. Similarly, the engineer who speaks only through coordinate graphs and equations will baffle, and perhaps alienate, the general reader who wants an overall sense of the proposed engineering project.

Here are two technical examples, both of which may use jargon unfamiliar to nontechnical readers:

> Wastewater treatment that employs fixed-film biological BOD removals has been shown to be more efficient than was predicted in our pilot studies. This result may be due to product mix, concentration, primary treatment, media type, wall effects, etc.

> Dry rubble stones shall consist of trap rock; granite; gneiss; or other approved hard, durable, tough rock. They should be sound, free from weathered or decomposed pieces, shattered ends, and structural defects, and shall be approved by the Contracting Officer.

In both of these examples, a general or nontechnical reader would encounter unfamiliar terms and abbreviations: *BOD, primary treatment, media type, trap rock, gneiss, shattered ends, structural defects*, etc. To the right reader, these terms have perfectly legitimate meanings, but to the uninitiated reader, the words might be confusing or nonsensical.

Jargon and the Social Sciences

Writers in the social sciences—especially in psychology and sociology—have often been accused of using excessive jargon. Writers in both fields use many common English words with special meanings: *response, learning, training, feeling, concept*, etc.

A psychologist discussing a *tertiary mediated response* is referring to a response coming through an intermediate person and delivered thirdhand. The concept and its expression are valuable in a limited context and to a limited audience. Otherwise, they are meaningless.

Jargon and Gobbledygook

Jargon and gobbledygook are not the same. Gobbledygook is the use of abstract or pompous words and long, convoluted sentences. It is clearly bad writing. Jargon, by contrast, is a specialized vocabulary for a particular technical field and is an often useful shorthand. See GOBBLEDYGOOK.

Key Words

Key words are like flags—they rise above the rest of the text and signify what is most important. In a paragraph, section, or subsection, the key words are those that give the text meaning. Key words impart the central message.

You can deliberately repeat key words and phrases to reinforce your message. Key words ensure that readers who are not reading carefully will still get the point of what you are saying and remember the most important ideas. See REPETITION.

Key Words

1. Use key words to mark key concepts; then repeat key words to reinforce your message.

1. Use key words to mark key concepts; then repeat key words to reinforce your message.

The example below is from a short section on condenser operation. Note how the writer drives home the message by using repeated key words as variations on an important theme:

> This highly effective water-separation process is possible because the condenser design positively prevents two potential **icing** problems: (1) blockage of the low-pressure side by **snow-laden cold** air and (2) **freezing** of condensate on the **cold** metal surfaces. The entering **cold-side** air is below the **freezing** point of water and, although the condenser **heats** it, the outlet **temperature** is still **below 32 degrees F**. Consequently, much of the entrained **snow** is not evaporated and must pass completely through the condenser without blocking the flow passages. Because **cold-side** air **temperatures** are consistently below the **freezing** point, the condenser must be carefully designed so that the metal surface **temperatures** remain above the **freezing** point.

The key words concerning temperature represent one important line of thought in this paragraph. An equally important line of thought concerns the design, mechanics, and operation of the condenser:

> This highly effective water-separation process is possible because the **condenser design** positively prevents two potential icing problems: (1) **blockage** of the low-pressure side by snow-laden cold air and (2) freezing of **condensate** on the cold **metal surfaces**. The **entering** cold-side air is below the freezing point of water and, although the **condenser** heats it, the **outlet** temperature is still below 32 degrees F. Consequently, much of the entrained snow is **not evaporated** and must **pass** completely through the **condenser** without **blocking** the **flow passages**. Because cold-side air temperatures are consistently below the freezing point, the **condenser** must be carefully **designed** so that the **metal surface** temperatures remain above the freezing point.

Note how the two sets of key words work together to create the overall effect and to establish both a primary problem (icing) and a primary need (a condenser design that will prevent it). See PARAGRAPHS and TRANSITIONS.

The key words in the next paragraph provide both a solution and a sharp contrast:

> Figure 2–1 shows how the **condenser design prevents** these **icing** problems. A special **hot** section on the **cold-side** face **prevents ice blockage**. A small **flow** of **hot** air from the compressor outlet **passes** through the **hot** section tubes, **raising** the metal **temperatures above freezing** and allowing the **snow or ice** to be **evaporated** in the main core. The **hot** air then reenters the high-pressure **air-flow** at the turbine inlet.

Headings and Captions

Headings and captions should contain the most important of the text's key words. For example, a heading for the previous paragraphs on condenser design would contain key words:

> **Condenser Design Prevents Icing**

See HEADINGS, CAPTIONS, REPETITION, and EMPHASIS.

Letters remain a key communication tool despite an increased preference by business and technical professionals for email and voice mail. But if you are writing a traditional letter, remember that computers have changed the format and appearance of letters.

Traditional hard-copy letters are now reserved for key, fairly formal communication about agreements or other organizational decisions. The informal and chatty personal letter is an antique, and the traditional written thank-you note is now usually no more than a quick phone call. Many messages conveyed in traditional letters are now emails or text messages.

Today's letters, when written, are less lengthy and rambling than ones written in earlier times because readers want letters that visually emphasize key ideas and that include no irrelevant information.

Letters with business or technical content must be visually effective because readers need to find or retrieve the main points at a glance. Good letters are 100 percent unambiguous. See WRITING AND REVISING and EMPHASIS.

Most readers of today's letters are skim-and-scan readers. They are not interested in a long, rambling journey as the writer explores issues and implications. Such confusing and time-consuming journeys are unfortunately still present in much email; therefore, the guidelines here apply to email as well as to letters. See ELECTRONIC MAIL.

Letters

1. Open letters with an informative and interpretive subject line.
2. Include your main point(s) in the first lines of your text.
3. Design and organize your letters so that key points are immediately visible.
4. Make your letters personal and convincing.
5. Choose direct and simple letter closings.

The following discussion of letters opens with five content rules for effective letter writing, as summarized in the checklist above. Following these rules is a description of letter formats, including the block, modified block, semiblock, and simplified styles.

The final two pages in this discussion explain how to address an envelope.

Effective Letter Writing

The following rules rely on two principles. As explained above, the first and most important is that the message in any letter be 100 percent unambiguous. The second is that a letter be personal and convincing, despite the limitations of written language, versus a dynamic one-on-one oral exchange of information. See TONE and PERSUASION.

1. Open letters with an informative and interpretive subject line.

We recommend that all business and technical letters have a subject line. The only exception would be informal (preferably handwritten) letters, such as a thank-you note or a note of condolence.

Test any letter by covering up all but the subject line. Does the subject line capture the major point(s)? Can you predict exactly what ideas will follow the subject line? If readers can't make an accurate prediction, your letter invites misreading and misunderstanding.

A subject line should be informative and interpretive, giving both the purpose and the key facts related to that purpose. An effective subject will often require six, eight, ten, or more words and will use two or more lines of type. A brief or telegraphic subject line hinders communication.

these

Subject: Authorization to Advertise for an Additional Analyst for Our Marketing Team

Request for Invoicing Turnaround Every 10 Working Days, Not Monthly

not these

Subject: Hiring of an Analyst

NEW INVOICING

As in the preceding examples, subject lines should use boldface, underlining, or other emphasis techniques. See HEADINGS and EMPHASIS. Also, sometimes the lead-in word *Subject* or old-fashioned *Re:* is omitted.

2. Include your main point(s) in the first lines of your text.

If you open your text with main points, your letter will be leading from these main points, not up to main points later in the letter. Readers will see the main points immediately.

The opening lines of text will usually repeat words and phrases from the subject line. This repetition is desirable. See REPETITION.

If your opening lines are not informative, your readers are likely to begin to skim and scan for a main point or supporting ideas. If they don't find anything of importance within 15 or 20 seconds, your letter will go on the bottom of their to-do stack.

> *this*
>
> **Subject: Authorization to Advertise for an Additional Analyst for Our Marketing Team**
>
> Dear Susan:
>
> Please advertise immediately for an additional analyst for the Marketing team. As you noted, our current analyst is unable to provide the information we need in a timely fashion. If possible, let's have the new person hired and trained by June 15.
>
> *not this*
>
> NEW INVOICING
>
> Dear Frank:
>
> As you are aware, we are in an ongoing effort to streamline our organization and cut costs. Please find the enclosed memo documenting this cost-cutting initiative. One of the ways that we project might save costs would be to increase our invoicing cycle so that vendors have less slack time from the time they deliver services to us until they receive a full invoice for these services. We have analyzed the billing cycle and according to this analysis, a two-week interval . . .

NOTE: As in the preceding example of poor content, don't let incidental background information replace a powerful and direct subject line and the crucial opening lines.

Never open a letter with the following common delaying techniques:

—Obvious references that delay or confuse the main point:

> This is in reference to your letter of February 14.
>
> This confirms Mary Evans' telephone conversation with you on June 15 regarding our June 3 meeting with Bilko's attorneys to discuss the acquisition of the Skews property. This is going to be a problem for us if . . .

—Unnecessary and phony social statements:

> Here's hoping that the weather in Tampa is fine and that your family is doing well in the new year.
>
> I hope this letter finds you hale and hearty and that everyone in your family is fine.

—Mechanical enclosure indications:

> Enclosed is . . .
>
> Attached is . . .
>
> Attached herewith . . .
>
> Enclosed please find . . .

—Statements that hint about your topic but not your position on the topic:

> Delaying the division reorganization is an idea of some merit and worth considering. Let's explore some of these notions as we . . .

—Nonessential background information that supports, explains, or illustrates your major points (unless this information legitimately sets up your major points):

> Delaying platform renovation until the 4th Quarter would mean rebudgeting funds that have already been allocated during the 2nd Quarter. I have decided that rebudgeting is not a desirable option at this time.

As in this example, the better version would have led off with the second sentence, which has the actual decision, not the delaying setup information.

NOTE: A document setup is desirable less than 10 percent of the time. Instead, give readers a direct statement in your opening sentence. Then follow up with any necessary background explanations.

Even when you have bad news to convey, you gain little by delaying the message for the purpose of softening the impact. And if you delay too long, you risk annoying the reader, as in the following evasive example:

> Dear Mr. Vikor:
>
> As you can imagine, we received over 100 applications for the position we advertised in the Sunday *Globe*. These responses were from well-qualified people, so our task was doubly hard. We looked for people with a solid work experience and outstanding references. We also wanted someone with demonstrated excellence on the Macintosh computer, which happens to be used throughout our company. Your qualifications, while excellent, did not quite fit all of these requirements.
>
> To conclude, thank you for taking the time to submit an application for employment.
>
> *Better*
>
> Dear Mr. Vikor:
>
> I regret that we can't offer you a job as our corporate accounts receivable clerk. As with many of our applicants, your qualifications were excellent. After a careful review, however, we decided that another applicant's skills more closely matched our company's current needs.
>
> We will keep your resume in our active file of qualified applicants and will call you if we have another opening that could draw on your training and experience in accounting.

3. Design and organize your letters so that key points are immediately visible.

The design of your whole letter should naturally grow out of your strong opening (rules 1 and 2 above). A solid, informative opening often even outlines the supporting points that follow. See ORGANIZATION.

An effective visual design includes headings, lists, graphics, and other emphasis techniques. See EMPHASIS.

Visually design your letters by answering these key questions:

—What are the key points you want to make?

—Which points must the reader remember? Are these the same as your own key points?

—What subheadings and lists should you use?

—Which points come first? Which come next? Should any be moved into attachments?

—How much repetition is useful?

—What tables and other graphics should you include?

—Which letter format is most suitable—block? semiblock? simplified?

—Does the page layout reinforce the message?

A writer's goal is to design and organize a letter so that every reader carries away a 100 percent unambiguous message. Research has shown that the more visual the message or image, the more readable and memorable.

Some years ago, a major corporation surveyed its writers and customers as to the adequacy of the corporation's letters. The survey revealed that nearly two-thirds of the letters required a follow-up call or a revised letter. Messages in the original letters were occasionally missing, often ambiguous, or even deliberately buried.

Answers to the above document-design questions will guide you in choosing subheadings, lists, tables, and other graphics. See EMPHASIS. These and other emphasis techniques are the best way to guarantee that the readers' eyes see exactly what you want the them to see.

For important or lengthy letters, use an early prototype or mock-up to help you visualize your design before spending time writing down everything you want to say. See the discussion of prototypes in WRITING AND REVISING.

4. Make your letters personal and convincing.

A letter, as in earlier handwritten manuscripts, was originally a written record of what would have been one side of a conversation. Oral communication was primary, writing secondary. Writers and scribes even routinely spoke the words as they wrote them.

Writers in these earliest letters assumed they were talking to readers, so oral conventions of presenting information were common. For instance, the closing (or surprise ending) was important, and skillfully building up to major points was the goal. These traits of the oral storyteller worked well around fires in caves, but they do little to help today's business communication.

Today's letters are also less conversational because they rely more on document design, as discussed under rule 3. Document design relies on the readers' visual memories to reinforce the message conveyed by individual words and sentences.

Still, your voice speaks through the words and phrases you choose, even if softly. You must, therefore, choose language that is personal and as convincing as possible. The following techniques will help you make such choices. They will make your letters more credible and effective.

—Use pronouns throughout, especially *you*. Assess your readers' interests and needs, then address these interests by speaking directly to the reader(s):

> One of your priorities included an adjustment of the due date for contract renewal. Your estimate of dates also included associated cost savings. We agree with you.

—Use contractions when they would normally occur in speech. See CONTRACTIONS.

> Once your program has been approved, we'll move ahead to hire the necessary professionals. We're doing this with existing funds because next year's budget has yet to be approved.

—Avoid jargon and gobbledygook, as used in the following bad example. See JARGON and GOBBLEDYGOOK.

> According to the aforementioned agreement, the elements of the items to be discussed are likely of some interest, yet they may well not be greatly and completely significant. These are of some concern because the agenda considered is primarily based on prior assumptions generated and discussed by our committee of professionals.

As in this example of poor and insulting writing, writers of jargon and gobbledygook use language recklessly, without regard for the basic meanings beneath the language. The writer of this example takes 51 words to say nothing. Every phrase is weakened by an empty modifier. Every reader would take away a different, muddled message.

—Prefer the simple, direct sentence and phrasing. See STYLE and SENTENCES.

> The committee met and made its recommendation Monday. They urged a new stock issue by January. This issue would finance an expansion of the Hong Kong plant, which now runs three shifts (about 110 percent of its capacity).

As in this simple and direct example, average sentence length should be low—13 in this instance. Prefer the common word or phrase—for example, *met* instead of *convened* and *plant* instead of *manufacturing facility.*

If you make your writing personal, as discussed in the preceding rules, you will also be convincing. A convincing style arises more from your tone or your implied credibility than from facts and figures. If you keep your language simple and direct, you will gain your readers' respect. See PERSUASION and STYLE.

5. Choose direct and simple letter closings.

Make your closing as simple and direct as you made your opening lines. Don't settle for old-fashioned cliches from traditional letters.

One option is to close your letter with a restatement of your main point:

> I hope you'll make room in your busy schedule for the coordination meeting on June 15 at 2:30 in the Human Resources conference room.

> All of your tests, as outlined above, indicate that we should identify a new vendor for our factory-installed CD player. We hope that you concur and that you'll authorize initial funding by September 22.

> Please call me (Ext. 4455) if you're unable to meet with the Senior VP from King Enterprises. He'll be available in Room 439 all morning April 15. I can set an appointment time, or you can just stop by for a brief introduction.

Another option is the standard reference to questions or problems. If you choose this option, keep it short and be sure to give your phone number:

> If our proposal leaves any questions unanswered, please call either Jane Evans (Ext. 9922) or me (Ext. 8933).

> Please call or write Jack Owens (412-333-2245) if you have any questions about our proposed investment schedule. As our financial advisor, Jack can tell you how we generated the figures in our proposal.

NOTE: Avoid the old-fashioned cliched closing lines:

> Thanking you in advance, I remain,
>
> Yours very truly,

> Should you have any questions, don't hesitate to call.

> Feel free to contact us if you have questions.

Also, as noted below under the discussion of letter formats, avoid the old-fashioned and overdone complimentary closings:

> Very truly yours,
>
> Deepest regards,
>
> Thanking you for everything,
>
> In sincerest and deepest appreciation,

Letter Formats and Styles

Each company or each division within a company should decide which letter format and style is the best for its purposes. Such format and style decisions include how the letter is spaced on the page, the type of letterhead and paper chosen, the typeface used in the text, and even which words and phrases to use or avoid.

Decisions about format and style convey the company's image.

After managers have made such decisions, writers can start with a given format (style sheet), so the writer's options properly focus on content questions. See PAGE LAYOUT for a sample style sheet.

Letter Formats

Two major types of letter formats are possible: the traditional ones and the simplified one.

Traditional formats include the block, modified block, and semiblock. See the illustrations in figures 1, 2, and 3 for the way these traditional formats would appear on a page. Traditional formats remain popular in most business contexts because they are what many readers expect to see. Traditional formats rarely draw on all the potential format options available in today's software programs.

The **simplified format** is less popular, but still very useful. As figure 4 shows, the simplified letter omits the traditional salutation (e.g., *Dear Ellen*) and the complimentary

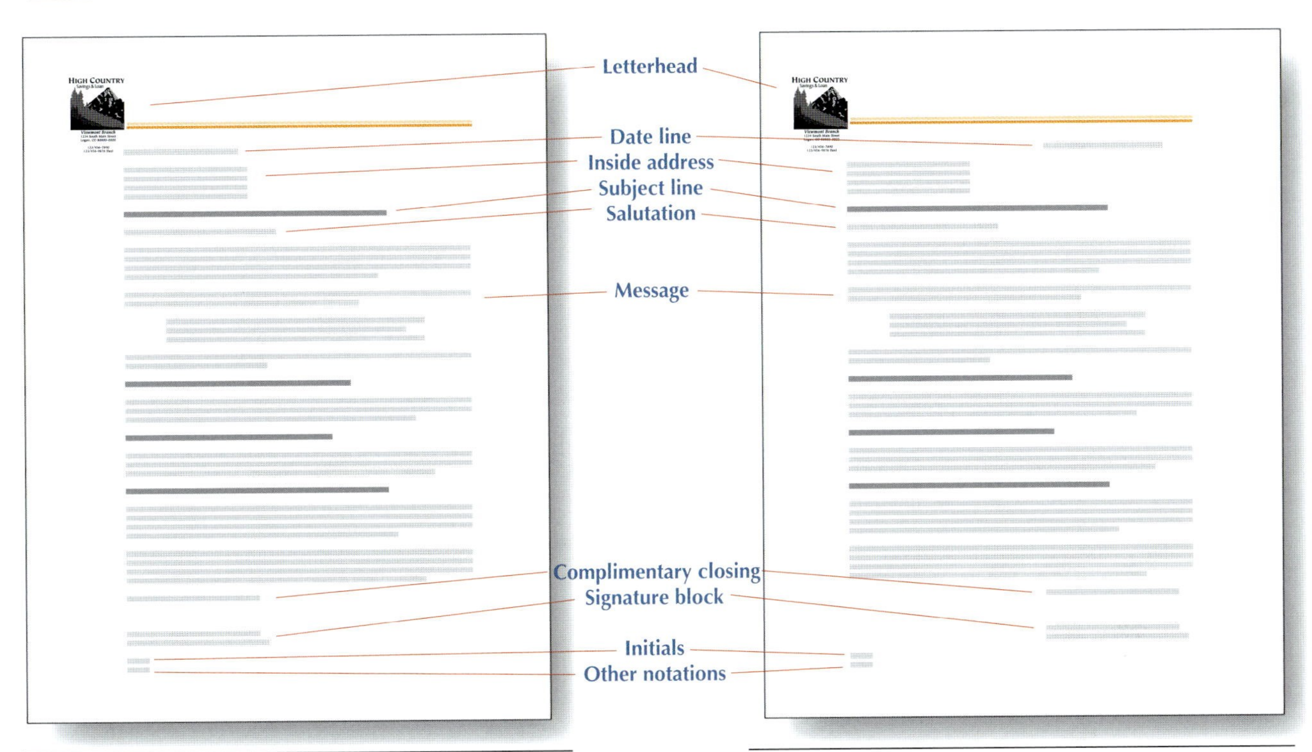

Figure 1. Block Style. *All parts of a block letter are flush left.*

Figure 2. Modified Block Style. *This letter is the same as the block style except that the date line and the complimentary closing are positioned to the right of the center line.*

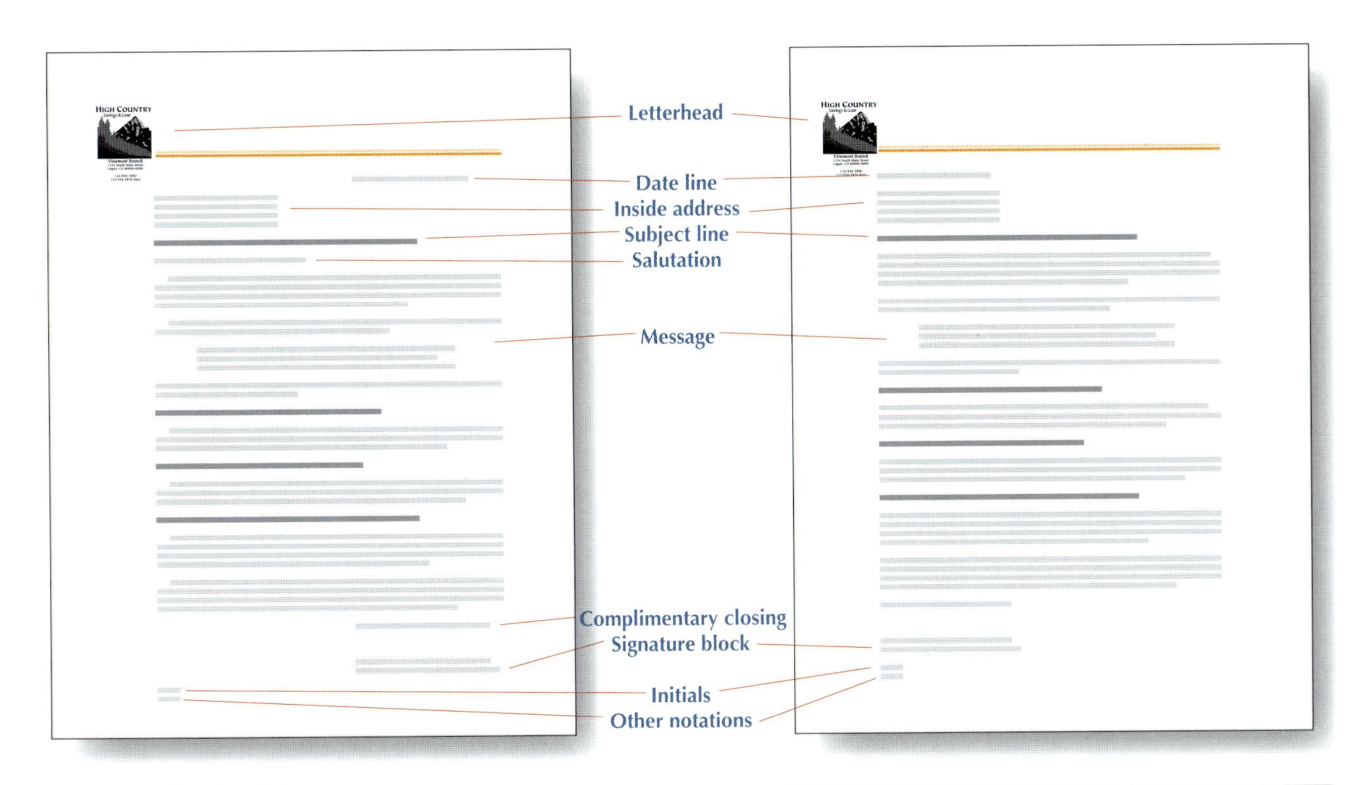

Figure 3. Semiblock Style. *The semiblock and the modified block are the same except that the semiblock has indented first lines of paragraphs.*

Figure 4. Simplified Style. *This style omits both the salutation and the complimentary closing. The rest of the letter is flush left like the block style.*

closing (e.g., *Sincerely,*). A simplified letter looks almost like an internal memo except that it still appears on the company's letterhead.

Simplified letters eliminate most of stock phrases typical of traditional letters, such as *Dear* ______, or *Yours sincerely*. Although outdated, such phrases are reassuring to readers, so many companies choose not to use the simplified format. They want to maintain a more formal, traditional image.

Punctuation Styles for Letters

In the open style, the writer omits all nonessential punctuation, including the colon after the salutation and a comma after the complimentary closing. See PUNCTUATION. The simplified letter reflects the open style because it omits both the salutation and the complimentary closing.

In the standard style, the writer uses minimal punctuation in the letter but does include a colon (or a comma) after the salutation and a comma after the complimentary closing. The standard style is almost always used with the traditional formats.

The closed (or close) style has all but vanished from business letters written in the United States, although some European firms still use it. In the closed style, writers retain all of the punctuation marks used in standard punctuation and add others:

—A period after the date

—Commas after each line of the address, with a period after the final line:

Mr. Edwin Jones,
Acme Office Supplies,
2211 Elm Street,
Dayton, Ohio 65432.

—Commas after each line in the signature block, with a period after the last line:

Elaine P. Hardesty

Elaine P. Hardesty,
Senior Financial Analyst,
Investments Department.

Margins and Spacing

The left and right margins should be roughly equal, but the exact spacing depends on the length of the letter. In long letters, the margins are usually at least 1¼ inches wide. In short letters, the margins can be wider; in extremely short letters, wide margins combine with double or triple spacing of the text.

Your goal should be to center the letter, so you might have to reset the standard margins in your word-processing software.

Features (Parts) of Letters

Traditional formats retain the bolded items in the following list of letter features. The nonbolded items are optional for either traditional or simplified letters. As noted above, the simplified format omits both the salutation and the complimentary closing.

Letterhead/return address

Date line

Reference line

Special notations

Inside address

Attention line

Salutation (Optional for simplified letter)

Subject line (Recommended for all business letters, but optional for informal notes)

Text

Headings for continuation pages

Complimentary closing (Optional for simplified letter)

Signature block

Reference initials

Enclosure initials

Copies and blind copy notations

Postscript

The following pages survey each of these features of typical business letters. The survey presents information in a three-column table.

- Column 1 identifies and defines the feature.
- Column 2 presents actual examples.
- Column 3 adds notes and warnings about the use and abuse of each feature.

Explanation	Examples	Notes
Letterhead/Return Address All business letters should have a letterhead or a return address with the same information as in a letterhead: —Logo (optional) —Full legal name of organization or person —Full address—including post office box, suite number, street, city, state, and full ZIP code —Area code and telephone number(s) —Fax or modem instructions —Email or website address For a letterhead, the above items are variously arranged. Sometimes, they all appear at the top of printed letterhead paper. Letterhead paper often has an address or other details as a footer, with the logo and company title reserved for the header. See PAGE LAYOUT.	High Country Savings & Loan 1234 South Main Street Logan, CO 80000-0000 555/456-7890 www.highcountrysl.com	NOTE 1: Your letterhead is one of the first things a customer sees produced by you. Make it a quality product by hiring a professional artist to help design the page. Then have the letterhead paper printed on high-quality letter bond paper. NOTE 2: Occasionally, departments, branch offices, or even company officers will have their own letterhead, which can include names and titles of managers, building numbers, and other identifiers. NOTE 3: In handwritten or typed personal letters, replace the letterhead with the writer's legal address. By tradition, the writer's name is not at the top of the page, only the address is.
Date Line The date line is either flush left (as in the block letter and in the simplified letter) or positioned several spaces to the right of the center line (as in the modified block and semiblock letters). See figures 1, 2, 3, and 4 for an illustration of these placements. The date line should never extend into the right or left margins. The date line usually appears two or three lines below the letterhead, but you can leave more spaces if your letter is very short. If you do not have letterhead stationary, then the date will come below the return address, as in the second example.	March 15, 2011 *or* 15 March 2011 ——— In letters written without a printed letterhead: 1729 Jackson Avenue Sparta, AL 12345-6789 March 15, 2011	Note 1: The optional format (15 March 2011) is common in letters written in the U.S. military services. This optional format requires no commas. See COMMAS. NOTE 2: Write March 15, 2011, not March 15th, 2011. Even though numbers in dates, when read, have the suffix indicating that they are ordinals, the standard U.S. practice is not to use this suffix in writing. British practice uses the suffix: June 7th 2011. Note the omission of the comma in the British version. NOTE 3: Do not use abbreviations for the month in general business letters. But U.S. military services often do use abbreviations in their correspondence—for example, 15 MAR 11. NOTE 4: Do not use the shortened version with numbers and slash marks: 5/6/11. These shortened versions are often ambiguous—for example, this abbreviation could mean either May 6, 2011 (U.S. style), or 5 June 2011 (British or European style). See NUMBERS.

Letters

Explanation	Examples	Notes
Reference Line or Block Reference lines appear beneath the date line, usually two lines down. Reference lines are typically aligned with the date, either flush left or several spaces right of the center line, depending on the chosen format for the letter. Reference lines are optional, but use them whenever you are responding to prior contacts. Mentioning documents or information in the reference lines will eliminate the need for you to include such references in the opening lines of the text of the letter. As an option, writers sometimes include the reference lines below the inside address, either flush left or centered. These reference blocks often begin with RE: and may run several lines. In the style used in the Department of Defense, the references are listed by number or letter so that the letter can easily cite the references as they are discussed. When letters run several pages long, the reference lines (if short) often appear on all continued pages as part of the continuation headings on these pages.	March 15, 2010 Invoice SX-4567A *or* 15 March 2010 Phone conversation ——— 15 March 2010 ***References*** A. DoD Directive 5202.43-1 B. SECNAVINST 452.1 C. COMCINCPAC Ltr 85-00064-5, dtd 23 April 2009 D. DoD Manual 34.3 ——— Reference E suggests that formulation of a new European policy is imminent. However, reference F cites an EEC memo stating that economic goals set last year would not be changed until 1995. Header for continuation pages: Page 4 Lynn J. Flaherty March 15, 2010 Invoice SX-4567A	NOTE: Reference lines include both written or published material and informal contacts, such as phone calls or meetings. These lines can serve, therefore, to record what went on earlier during these informal contacts. You might even want to include in your reference line the names of the attendees at the meeting. Directors' Meeting, 5 November 2009. Attendees: S. Lynne, J. Swan, and F. Trenner.
Special Notations Special notations appear between the date (and the optional reference lines) and the inside address. These notations usually appear two lines above the inside address and are usually typed in all capitals. If two or more notations apply to a letter, the additional one(s) appear directly below the first one, with no lines between the notations. Such notations should also appear on the envelope, above and between the address and the stamps. Do not place them below the address block on the envelope because they will interfere with the U.S. Postal Service's optical scanner.	SPECIAL DELIVERY REGISTERED MAIL CERTIFIED MAIL CONFIDENTIAL PERSONAL	

Explanation	Examples	Notes

Explanation

Inside Address

The inside address includes the following:

—The addressee's courtesy title (Mr., Mrs., Ms., Miss, or Dr.)—Optional
—The addressee's name
—The addressee's title—Optional if the name is sufficient
—The name of the organization
—The street address and the post office box number
—The city and state
—The ZIP Code (+4 if known)

The inside address usually appears three or four lines below the date, but the spacing will vary depending upon the length of the letter. For very short letters, the inside address can be six or eight lines below the date.

The format and content in the inside address should match as closely as possible the information in the receiver's own correspondence (assuming you have copies of it). You should spell the company's name as they do; you should use their exact spelling for the address; and you should use the person's courtesy and business titles if they appear on the person's correspondence.

Courtesy titles are a complex issue. Sometimes even the sex of the person you are writing to is unknown—for example, letters written to you from C. L. Saxon. And if you do happen to know that C. L. Saxon is a woman, you may not have any information about whether she prefers Mrs. or Ms. See BIAS-FREE LANGUAGE.

When in doubt, use merely the person's name as you received it—C. L. Saxon, as in the preceding example, or Carolyn L. Saxon, if you know the person's given name. As an option, call C. L. Saxon's office and ask for information about titles and, if appropriate, the correct address.

Courtesy titles for national political figures, religious dignitaries, royalty, or foreign heads of state are all different and unpredictable. If you are writing to one

Examples

Ms. Louise H. Johnson
Director of Human Resources
The Locklear Company, Inc.
First National Building
456 SECOND STREET SUITE 408
HOUSTON TX 12345-6789

Dr. Edwin B. Roberts
Chief, Psychiatric Services
Saint Benedict's Hospital
PO BOX 67
NORTH MEDFORD OR 76543-4321

C. L. Saxon
District Engineer
U.S. Corps of Engineers
879 RIVER STREET
COLUMBUS OH 99876-5432

Attention C. L. Saxon
District Engineer's Office
U. S. Corps of Engineers
879 RIVER STREET
COLUMBUS OH 99876

Notes

NOTE 1: The last two lines of the inside address are the most important, especially if you print the address on the envelope from the same computer file as the letter. The U.S. Postal Service asks that the last two lines be printed in capitals and with no punctuation. We recommend using the same choices in the inside address.

NOTE 2: To assist the optical scanning process, use clear sharp letters, preferably all capitals. Do not use italic or slanted script or other unusual fonts. Sans serif fonts are best. See PAGE LAYOUT. Also, use dark high-contrast print because the scanner will have trouble with light colored text or text from a poor-quality dot-matrix printer. Do not use any punctuation in the last line.

No special notations or corporate logos should appear in the address box or below it. These will confuse the scanner.

NOTE 3: Formats for Canadian and other foreign addresses are different from U.S. addresses. The U.S. Postal Service recommends that only the country name in capitals appear at the bottom of the address block. Also, include any foreign delivery zone information if known.

Explanation	Examples	Notes
Inside Address (continued) of these categories, check a standard secretarial reference, such as the *Gregg Reference Manual* or Susan Jaderstrom, et al., *Complete Office Handbook.* The two most critical lines in the inside address are the last two: the street address and the city, state, and ZIP Code + 4. The U.S. Postal Service uses these in its optical scanning system. So these lines in the inside address should match those in the address on the envelope. If you intend to print both the inside address and the envelope address from the same computer file, the format of the envelope address should take precedence. For example, the U.S. Postal Service prefers capitals, especially for the critical last two lines. See the example of V. H. Crenshaw's address.	V. H. Crenshaw Bolter Machine Tools, Inc. 4571 CHESTNUT DRIVE IDAHO FALLS ID 99543-2278 *or* V. H. CRENSHAW BOLTER MACHINE TOOLS, INC 4571 CHESTNUT DRIVE IDAHO FALLS ID 99543-2278 ——— MR. THOMAS CLARK 117 RUSSELL PLACE LONDON W1P6HQ ENGLAND ——— Josephine Andrews First Option Software Corp, Ltd. 129 Western Province Boulevard Edmonton, Alberta T5J2H7 CANADA	NOTE 4: Use the approved U.S. Postal Service abbreviations. We give only those for states and territories. These two-character abbreviations are mandatory for envelopes. For consistency, more and more organizations are also using these same abbreviations for inside addresses. We recommend their use in inside addresses. Within the text of letters, however, you would use *Alabama* instead of *AL.* NOTE 5: Omit the comma between city and state in the last line of addresses prepared to match U.S. Post Office envelope standards.

Approved U.S. Postal Service Abbreviations:

State	Abbr.	State	Abbr.
Alabama	AL	Montana	MT
Alaska	AK	Nebraska	NE
Arizona	AZ	Nevada	NV
Arkansas	AR	New Hampshire	NH
California	CA	New Jersey	NJ
Colorado	CO	New Mexico	NM
Connecticut	CT	New York	NY
Delaware	DE	North Carolina	NC
District of Columbia	DC	North Dakota	ND
Florida	FL	Ohio	OH
Georgia	GA	Oklahoma	OK
Guam	GU	Oregon	OR
Hawaii	HI	Pennsylvania	PA
Idaho	ID	Puerto Rico	PR
Illinois	IL	Rhode Island	RI
Indiana	IN	South Carolina	SC
Iowa	IA	South Dakota	SD
Kansas	KS	Tennessee	TN
Kentucky	KY	Texas	TX
Louisiana	LA	Utah	UT
Maine	ME	Vermont	VT
Maryland	MD	Virgin Islands	VI
Massachusetts	MA	Virginia	VA
Michigan	MI	Washington	WA
Minnesota	MN	West Virginia	WV
Mississippi	MS	Wisconsin	WI
Missouri	MO	Wyoming	WY

Explanation	Examples	Notes
Attention Line An attention line was formerly used when the inside address did not contain either the name of an individual or a department. The attention line allowed writers to identify a person or department that would review the letter as soon as it arrived. In traditional letters, the attention line came between the inside address and the salutation. Instead of this traditional placement, the U.S. Postal Service prefers the attention line, if included on envelopes, to be above the street address and the city and state. This preference means that the traditional attention line, which in letters appeared under the inside address, may be still used, but you need to remember to move it up above the address when you print the envelope.	In letters: H. Allen and Associates, Inc. Valley Bank Building, Suite 3409 408 Pico Boulevard LONG BEACH CA 55387-2298 Attention Georgia Banks ——— Denver Regional Office Midland Oil and Gas Company 4509 Western Avenue DENVER CO 96499-2235 Attention EXPLORATION DEPARTMENT On the envelope to these letters: Attention Georgia Banks H. Allen and Associates, Inc. Valley Bank Building, Suite 3409 408 Pico Boulevard LONG BEACH CA 55387-2298 ——— Attention EXPLORATION DEPARTMENT Denver Regional Office Midland Oil and Gas Company 4509 Western Avenue DENVER CO 96499-2235	
Salutation The salutation has traditionally begun with the greeting *Dear* followed by the courtesy title and name of the addressee. See the discussion of courtesy titles under the discussion on inside address. When you know the person's preferred title, then use it along with his or her family name. If you don't know the addressee's gender or preference as to a title, the safest choice is to use no title, merely the person's name: Dear A. L. Jones: Dear Abigail L. Jones: In many cases, the use of no title is the best choice even when gender is not an issue: Dear Clarence Johns:	When the title is known: Dear Mr. Neal: Dear Frank: (*or* Dear Frank,) Dear Mrs. Skoal: Dear Miss Anderson: Dear Ms. Branck: Dear Cheryl: (*or* Dear Cheryl,) Dear Professor Maloney: Dear President Crouch: Dear Ms. Dearden and Mr. Wong: Dear Mrs. Anderson and Ms. Blaine:	NOTE 1: Some formal notes in the past used a formal salutation: My dear Mr. Devon: My dear Susan: These salutations sound too stiff to be acceptable today. Avoid them. You should also avoid the previously acceptable salutations using French titles: Dear Messrs. Franks and Harris (*for two men*) Dear Mesdames Long and Minor (*for two women*) Even very educated readers in the Unites States would have difficulty pronouncing these French forms and would likely consider the writer odd.

Letters

Explanation	Examples	Notes
Salutation (continued) In cases where you don't know who the addressee is or whether the addressee is a company or department, the choice of salutation is a problem. Do not choose the traditional *Gentlemen*, which ignores the women who may receive the letter. The option of *Ladies and Gentlemen* is almost as poor. Some people now object to the word *Ladies* as suggesting an out-of-date image of women. What is politically or socially acceptable is increasingly unclear! See BIAS-FREE LANGUAGE. We recommend omitting the salutation from a letter if you have any doubts about the correct salutation. If you follow this recommendation, you will be following the basic format and arrangement of the simplified letter. In a simplified letter you'll have the inside address, a subject line, and then the text begins. The salutation is unnecessary. In the open-punctuation style, no punctuation follows the salutation. In standard punctuation, use a colon (for formal letters) or a comma (for informal letters) after the salutation.	When a title is uncertain or unknown: Dear G. B. Harrison: Dear Sydney Carton: Dear Georgie Cohan: When the identity of the reader(s) is unknown: Omit the salutation, as in the simplified letter format.	
Subject Line We recommend using subject lines in all business and technical letters. The only exception would be informal letters and notes, often handwritten. The subject line (or lines, in many instances) should be informative and interpretive. As noted at the beginning of this entry (rule 1), test your letter by covering up the text and reading only the subject line. If you can't predict the message and content of the letter, you've failed to capitalize on the potential of the subject line. Traditionally, subject lines, if used, appeared two lines below the salutation and two lines above the first line of text.	*these* Recommend That We Replace All Macintosh Pluses With Quadra 900 Models Three Phases in Merger of Personnel and Functions With Compunet, Inc. Approval Requested for New Marketing Task Teams Organized Around Market Niches Instead of Regional Areas *not these* Replacement Computers Our Merger Plans Marketing Teams	NOTE 1: Many subject lines, if adequate, will run six, eight, or ten words and will require two or three lines of text. This length is often surprising to traditional writers, who occasionally assumed that a subject line was brief, even telegraphic. NOTE 2: Most subject lines should be capitalized like titles of books. See CAPITALS. This practice occasionally changes, especially when the writer makes the subject line a full sentence. In this case, only the first word and any proper nouns are capitalized: We recommend that Acme Industries take a one-time loss of $5.5 million to cover back payments to the employee pension plan.

Explanation	Examples	Notes
Subject Line (continued) In a few instances, writers prefer placing the subject line before the salutation. Either practice is acceptable, but just be consistent. Highlight the subject line by using boldface, all capitals, or underlining. Also, you need not use the word *Subject* and a colon; the role of the subject line is clear without this lead-in word.		
Text The text (or body) begins two lines below the subject line, or if there is no subject line, below the salutation. In the simplified style, the text begins three lines below the subject line, and no salutation is used. The text of most letters is single-spaced, with double spaces between paragraphs. If a letter is very short, it may be printed with double spacing instead of single spacing. Use a full range of emphasis techniques to highlight key information in the text. These include boldface, lists, headings, single-sentence paragraphs, and graphics. See EMPHASIS. Also see rule 3 above at the beginning of this entry on letters. Indentation of paragraphs is optional. In the block, modified, and simplified styles, do not indent paragraphs. In the semiblock style, indent paragraphs, usually five spaces. Some organizations prefer up to 10 spaces. Reversed indentation (sometimes called "hanging indentation") is a format option, especially in letters with lists. Use this reversed indentation with any of the letter styles. A hanging-indented paragraph begins flush with the left margin, but subsequent lines in the paragraph are indented, usually three to five spaces.	*not this* *this*	

Explanation	Examples	Notes
Headings for Continuation Pages Continuation pages are necessary if you have more than three lines of text left over from the preceding page. With fewer than three lines left over, adjust the spacing on the first page to accommodate the additional lines. The continuation page opens with a header that contains the name of the person receiving the letter, a page number, and (optionally) the date.	Page 2 R. E. Blessing May 15, 2011 R. E. Blessing Page 2 May 15, 2011	NOTE: Most word processing programs now allow for easy creation of a header, so the items in the continuation page can be and often are spaced across the top of the page, not lined up flush with the left margin. Either design of the continuation page is acceptable. See PAGE LAYOUT.
Complimentary Closing Traditional letters require a complimentary closing followed by the writer's name and signature. Simplified letters omit the closing while retaining the writer's name and signature. Alignment of the complimentary closing varies. In the block style, the complimentary closing is flush with the left margin. In modified and semiblock styles, the complimentary closing appears right of center. Choice of a closing signals how formal you intend your letter to be. In most business letters, you can use the simple closings of *Sincerely* or *Sincerely yours*. Another less common option would be *Thank you*, which moves away from the more traditional ones with *Sincerely*.	Use these in most business letters: Sincerely, Sincerely yours, Or in less formal letters: Thank you, Best wishes, Regards, Cordially, Best regards, Or in highly formal letters (to dignitaries or governmental officials): Respectfully yours, Respectfully, Yours sincerely,	NOTE 1: As in all of the examples, the first letter of the first word in the closing is capitalized. Later words in the closing are lowercase. NOTE 2: A comma follows the closing if you are using standard punctuation. If you choose open punctuation, no punctuation is needed.

Explanation	Examples	Notes
Signature Block The signature block follows the complimentary closing. In the simplified style, the signature block appears four or five lines below the last line of the text. Alignment of the signature block varies with letter formats. In the block and simplified styles, the signature block is flush with the left margin. In the modified and semiblock style, the signature block is right of center. The contents of the signature block are as follows: • Company name (optional) • Handwritten signature, either full name or only the given name • Full typed name of writer • Title of the writer (optional)	Yours sincerely, D&L Drilling Equipment *Dwight G. Edwards* Dwight G. Edwards Sales Manager Sincerely, *Jane G. Nostromo* Jane G. Nostromo Chief District Engineer Best wishes, *Howard* Howard G. Balaock, PhD Personnel Manager Engineering Division Sincerely, *Elaine Raddison* Elaine Raddison Treasurer *or* (Mrs.) Elaine Raddison Or rarely today Sincerely yours, *Elaine Raddison* Mrs. Thomas Raddison Treasurer --- Grace Babbitt, M.D. *not* Dr. Grace Babbitt, M.D.	NOTE 1: The company name appears in the signature block only when the letter represents a company policy, position, or decision, especially in legal matters. Even with the company name, the signature block also contains the name and title of the person signing for the company. NOTE 2: Sign formal and official letters with your full legal name. In informal letters, you need to sign only your first name. NOTE 3: Do not include courtesy titles such as *Mr., Mrs., Miss, Ms.*, or *Dr.* in a signature block, either in the typed name or in the handwritten signature. But see note 4 below for an exception. NOTE 4: Women's signatures generally include only the woman's given and family names, with no courtesy titles. These titles, if the woman chooses to include them, would appear with the typed name, either with or without parentheses. Most women today do not choose to note their marital status. After all, men never have indicated their marital status when signing letters. NOTE 5: Academic titles or professional degrees are not part of the handwritten signature. If you use them, they are typed with the name and usually come after the name. NOTE 6: If you are signing for a colleague or your manager, you have a choice. One way is to sign his or her name and add your initials to the handwritten signature. The other way is to sign your own name and then include the word *for* in the typed title. Sincerely, *James Westwood* For Frank Proctor, P.E. District Engineer

Letters

Explanation	Examples	Notes
Reference Initials The reference initials appear two lines below the last line of the signature block and are always flush with the left margin. These initials consist of the secretary's or clerk's initials, along with (optionally) the writer's initials.	The secretary's initials alone: goj Writer's plus the secretary's: HFG:goj *or* HFG/goj Initials of the person signing, plus the writer's, plus the secretary's: KR/HFG/goj *or* KR:HFG:goj	
Enclosure Notation Enclosure notations remind readers that one or more documents were enclosed with the letter. Such notations usually come directly under the reference initials. Enclosure notations differ greatly in their forms, as the examples show. We recommend that you list the actual enclosures under the word *Enclosure* (whichever form you choose). A list of items, unless so long as to be unwieldy, helps readers check to see what should be attached. Sometimes separate items are listed as being sent separately. This listing helps readers check to see what they should be receiving, either with the letter or in a separate mailing.	Enclosure Enclosures (4) 4 Enclosures Enclosures 4 Enc. With content information: Enclosures 1. Invoice 547/33 2. File 54A-R33 3. Map 54M With separate mailings: Enclosures 1. Invoice 547/33 2. File 54A-R33 Sent separately 3. Map 54M	

Explanation	Examples	Notes
Copy Notations Copy notations show the distribution of the letter. Originally copy notations were abbreviated *cc*, which stood for carbon copies. With the decline of carbon paper use, *cc* became associated with "courtesy copy." The recent trend is to use merely a single *c*; and, as in metric abbreviations, the single *c* stands for both singular and plural forms of copy. Whichever abbreviation you choose to use (*c* or *cc*), either with or without a colon, you should list the people who are receiving copies. This distribution list is usually important for organizational reasons. In some cases, you may not want to list the distribution on the original copy, but merely on the copies to be distributed. In this instance, writers use the abbreviation *bc* for blind copies. As with *c*, a list of names follows the *bc*.	c T. E. Damer Alicia Johnstone *or* c: Mel Barnes L. H. Strachan V. Sacloft *or* Copy to Claudie Mannes --- Copies to George Siebert A. S. Viceroy Uhlan Yost --- bc L. H. Strachan V. Sacloft *or* bc: L. H. Strachan V. Sacloft	
Postscript Postscripts are additions to a letter after it has been printed. These were handy format options in earlier times when retyping the whole letter was time-consuming and an annoyance. With today's computer files, postscripts are almost never necessary because insertions and corrections can be made to the letter file and a new copy printed in a few seconds. If you use a postscript, start it flush with the left margin and use the abbreviation *PS* without any punctuation. We recommend avoiding postscripts unless you have a last-minute handwritten reminder you wish to add to your letter.		

Letters

Envelopes

Addressing tips on the website of the U.S. Postal Service provide the best guidance on how to address letters so they will move smoothly through the postal system.

Addressee's Name and Address

Figure 5 comes from the U.S. Postal Service website. As the figure shows, you should put all necessary information in the address block and use capital letters so that the address can be optically scanned.

The processing of letters begins with the optical scanning of the address. Next, a machine prints a bar code on each letter specifying exactly where a letter is to go. This bar code then controls where the letter is sent.

Poor-quality printing, inaccurate information, or other problems can prevent optical scanning, thus delaying your letter.

As noted above in the discussion of the inside address for letters, the format requirements of the envelope are making changes in how we expect the information to appear inside the letter.

For instance, we recommend using the same format for the city, state, and ZIP code in both the inside address and on the envelope. Using the same format will help prevent problems, especially if you are printing mailing labels from the same file where you have the inside address.

So the city, state, and ZIP Code always will appear like this in letters and on envelopes:

BOUNTIFUL VA 84010

As in figure 5, the other lines on the envelope should be in capitals. Whether you choose to use all capitals for all lines in your inside address is a matter of personal preference—the choice is yours.

Sender's Return Address

Place the sender's name and correct address, including the ZIP Code, in the upper left corner of the envelope two or three lines from the top and five spaces from the left edge.

Be careful to print the return address so the lines in it do not enter the block reserved for the addressee's name and address.

Special Notations

Special mailing notations (*Special Delivery, Certified Mail, Registered Mail*) come beneath the stamp in the upper right corner. These should not extend into the area reserved for the name and address of the addressee.

Similarly, position miscellaneous notations (*Personal, Confidential, Please Forward,* and *Hold for Arrival*) above the name and address of the addressee. These should not extend into the area reserved for the name and address.

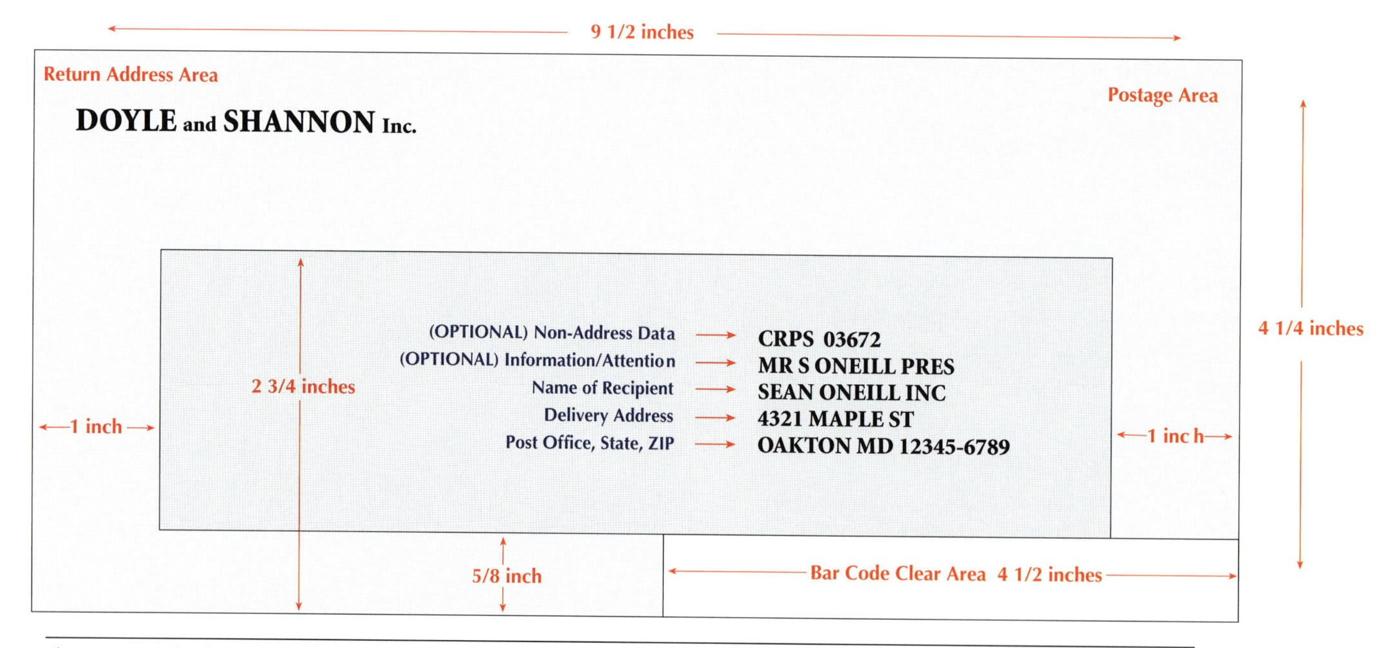

Figure 5. Standard U.S. Post Office Envelope Format. *Use this format for all envelopes. Note the use of capitals, especially in the last line.*

Note 1: Use only U.S. Postal Service abbreviations. Earlier in this discussion of the inside address, we provided the approved U.S. Postal Service list of abbreviations for states and territories. In addition, the U.S. Postal Service can provide an extensive list of other abbreviations, including shortened forms of cities with longer names as well as abbreviations for common terms like *road* or *university.*

Note 2: Carefully fold letters before inserting them into envelopes. The two common methods of folding letters are as follows:

Folds for Long Business Envelopes (No. 10)

—Fold the bottom third of the letter up and crease. Next, fold the top third of the letter down and crease. (Caution: The top fold should not come far enough down to bend or crease the third of the paper folded up from the bottom.) See figure 6.

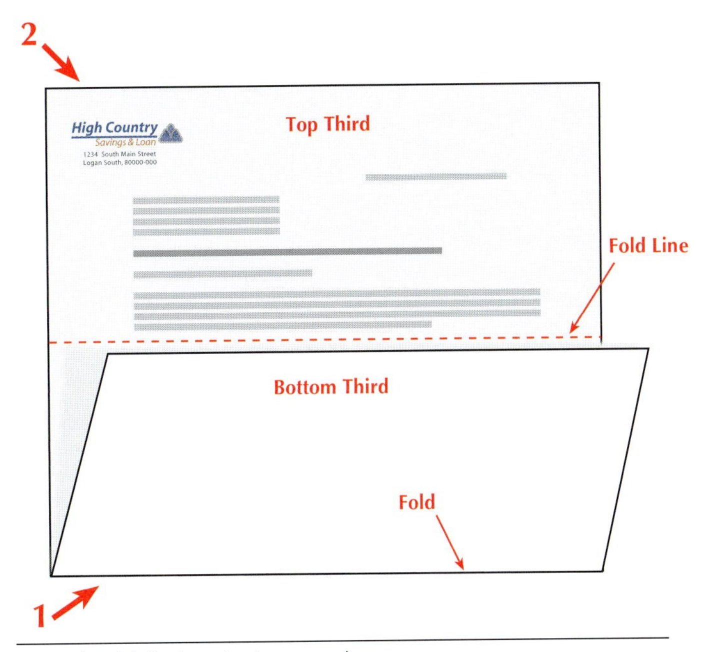

Figure 6. *Folds for long business envelopes.*

Folds for Regular Business Letters (No. 6 3/4)

—Fold horizontally almost in half, with about one-half inch of the top of the paper visible above the folded portion.

—Then fold the paper in the vertical direction. This time, fold the paper into thirds: the right third over the middle third, and the left third over the other thirds. (If folded properly, the upper left corner of the letter is on the top of all of the folds.) See figure 7.

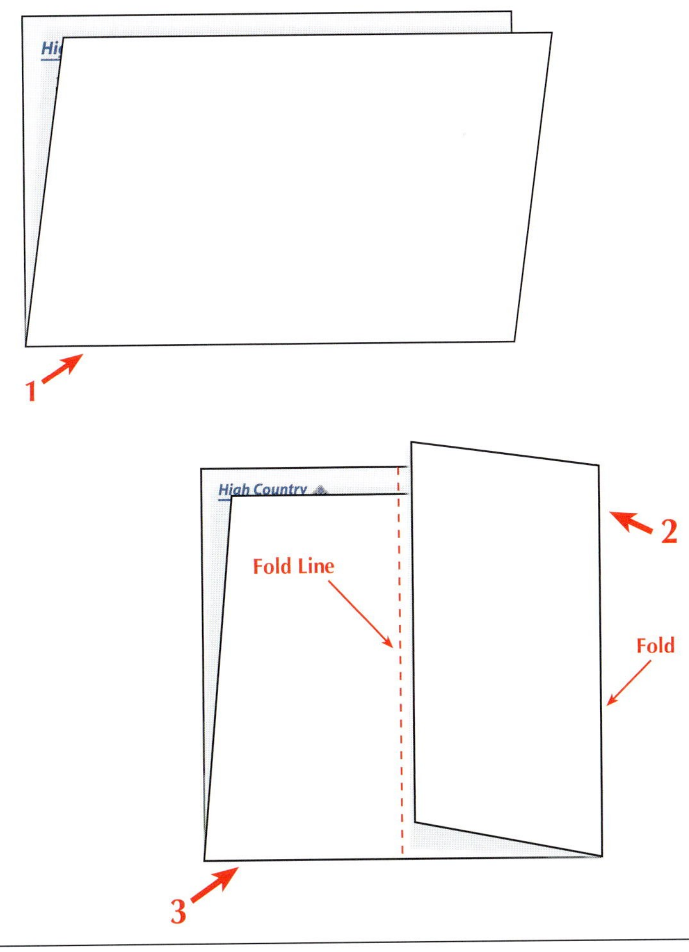

Figure 7. *Folds for regular business envelopes.*

Lists

Lists are increasingly important as a major aid for readers in interpreting and remembering information.

Lists, headings, graphics, and other emphasis tools are now essential in business and technical writing. These tools have now replaced the long, academic paragraphs of earlier times. See EMPHASIS and PARAGRAPHS.

Lists help readers in two ways. First, a reader can easily scan a list to survey the points mentioned. Careful reading of every word and phrase is often unnecessary. Second, a reader can remember more information longer if it appears in a list. The reason is that the visual image of a page reinforces the verbal message.

The most effective lists are **displayed lists**, where items are listed vertically.

1. Listed item a
2. Listed item b
3. Listed item c

Paragraph lists, where items are embedded within a paragraph, are less visually obvious to readers and less memorable.

Lists

Displayed Lists

1. Use a displayed list for a long series of items and for any series you wish to emphasize.
2. Use numbers, letters, bullets, or dashes to identify each item in a displayed list.
3. Capitalize the first word of each item in a displayed list.
4. Use bullets or dashes to identify each item in a displayed list when the list contains items of equal importance and those items will not have to be referred to by number or letter.
5. Use a colon to introduce a list when the sentence preceding the list contains such anticipatory words or phrases as *the following, as follows, thus,* and *these.*
6. Do ***not*** end the introductory sentence with a colon if the sentence is lengthy and the anticipatory word or phrase occurs very early in the sentence or if another sentence comes between the introductory sentence and the list.
7. End items in a displayed list with periods if one or more of the items is a complete sentence.

Paragraph Lists

8. Use a list within a paragraph whenever the list is short (fewer than six items) and you do not wish to emphasize the list.
9. Use numbers or letters to identify each item in a paragraph series. Enclose the number or letter within parentheses.
10. Capitalize the first word of each item in a paragraph list only if each item is a complete sentence or if an item begins with a proper noun.
11. In paragraph lists, do not precede the list with a colon if the list follows a preposition or a verb.

Lists Within Lists

12. Whenever one list occurs inside another list, use numbers for the outer list and letters for the inner list.

Lists and Parallelism

13. Ensure that items in lists are parallel in structure. Begin each item with the same type of word (noun, verb, adjective, etc.).

Displayed Lists

Displayed lists are now an essential tool in business and technical writing. A well-written displayed list can and often should replace a traditional paragraph. See PARAGRAPHS.

Use displayed lists for the following information, especially in longer documents:

- Preview of topics
- Summary of reasons
- Assumptions
- Steps in a procedure
- Findings
- Recommendations
- Conclusions
- Methods surveyed
- Sources of information

Often the items in a displayed list will repeat points made elsewhere in a document. For example, all of the preceding types of information can be briefly listed, but then, they will usually require more detailed explanations. See REPETITION.

Lists

1. Use a displayed list for a long series of items and for any series you wish to emphasize:

Five collective protection countermeasures were identified:

1. Simple activated-carbon absorption filters
2. Regenerative filters
3. Closed-loop or recirculation environmental control systems
4. Pyrolytic destruction of agents
5. Corona discharge and other molecular disruption techniques

2. Use numbers, letters, bullets, or dashes to identify each item in a displayed list.

Use numbers or letters whenever the list is lengthy, whenever the text must refer to items in the list, or whenever the items are listed in decreasing order of importance. The numbers or letters should not be enclosed by parentheses, but they should be followed by a period:

a. Definition of systems and subsystems

b. Progressive apportionment of figure-of-merit requirements to subsystems

c. Progressive definition of functional requirements for subsystems to meet mission requirements

d. Definition of subsystem design and interface constraints as dictated by the chosen deployment strategy

See NUMBERING SYSTEMS.

3. Capitalize the first word of each item in a displayed list:

- Item a
- Item b
- Item c

NOTE: The exception to this rule occurs whenever the listed items complete the thought begun in the introductory sentence (see rule 7, note 2).

4. Use bullets or dashes to identify each item in a displayed list when the list contains items of equal importance and those items will not have to be referred to by number or letter:

We selected these means of accomplishing the scope of work for the following reasons:

- They respond to the program needs as defined in the RFP.
- They reflect the approach to project definition steps that experience indicates is effective in other isotope separation projects.
- They reflect our experience in assisting R&D personnel in transferring the requisite technology to documentation rapidly enough to support tight project schedules.

5. Use a colon to introduce a list when the sentence preceding the list contains such anticipatory words or phrases as *the following, as follows, thus,* and *these*:

The 1553 interface will be programmed to respond to the following mode commands:

a. Synchronize (without data word).
b. Synchronize (with data word).
c. Transmit status word.
d. Reset terminal.

NOTE: A recent trend is to omit anticipatory words and even break the syntax of the lead-in sentence:

As written, the 1553 interface will allow users to:

a. Synchronize
b. . . . .

Some editors would object to this use of the colon to break the syntax. For consistency in this *Style Guide*, we have avoided such uses of the colon. See rule 11.

6. Do *not* end the introductory sentence with a colon if the sentence is lengthy and the anticipatory word or phrase occurs very early in the sentence or if another sentence comes between the introductory sentence and the list:

The following steps are required to process the message received by the bit processor. Note that the sequence parallels both subsystem interface architectures.

1. Recognize a valid RT enable flag.
2. Read the 16-bit word from the bit processor.
3. Check the T/R bit.
4. Decode the subaddress field where incoming data will be stored.

7. End items in a displayed list with periods if one or more of the items is a complete sentence:

Western Aeronautics has a wide range of related experience:

- We built and tested crash-proof recorders for the T41 and T7G engine data analyzers.
- We designed, developed, tested, and manufactured the Central Air Flight Data Computer (CAFDC) systems for the Air Force.
- We produced the full range of flight data computers with 18 different designs flown on 35 different aircraft.

NOTE 1: The above listed items are clearly sentences, so periods are necessary. Sometimes, however, items may not be clearly sentences:

—Convene.
—Caucus.
—Vote.
—Adjourn.

Each of these words is an imperative sentence and should require a period (see SENTENCES), but their shortness has led some editors to prefer no periods. Items without periods also reflect a trend in advertising and in media documents to use lists without final punctuation.

Note 2: Sometimes a displayed list completes the sentence begun with the introductory statement. In this instance, even though the listed items are not sentences, the final one will require a period. Also, note that such lists usually do not follow a colon and that they use commas or semicolons after all but the last item:

You may gain access to the Human Resources Department database by

1. clicking on the HRD logo on your screen,
2. logging on by entering your name and department code, and
3. entering your password and the last four digits of your phone number.

Lists with continued punctuation (as in this example) are now rare, probably because writers, influenced by advertising, are using colons, capitals, and spacing to make lists more visually emphatic.

Paragraph Lists

8. Use a list within a paragraph whenever the list is short (fewer than six items) and you do not wish to emphasize the list:

Five collective protection countermeasures were identified: (1) simple activated-carbon absorption filters, (2) regenerative filters, (3) closed-loop or recirculation environmental control systems, (4) pyrolytic destruction of agents, and (5) corona discharge and other molecular disruption techniques.

9. Use numbers or letters to identify each item in a paragraph series. Enclose the number or letter within parentheses:

The HCF memory is in three sections: (1) program, (2) nonvolatile RAM, and (3) scratch-pad RAM.

or

The HCF memory is in three sections: (a) program, (b) nonvolatile RAM, and (c) scratch-pad RAM.

10. Capitalize the first word of each item in a paragraph list only if each item is a complete sentence or if an item begins with a proper noun:

We propose that the qualification program include (1) documentation of tests conducted on similar equipment, (2) service history of similar equipment, and (3) Bell Laboratories' testing of the equipment.

11. In paragraph lists, do not precede the list with a colon if the list follows a preposition or a verb:

The 1553 interface will be programmed to respond to (a) synchronize (without data word), (b) synchronize (with data word), (c) transmit status word, and (d) reset terminal.

Note 1: If this list becomes a displayed list, however, use the colon even though the list follows a verb:

The 1553 interface commands are:

a. Synchronize (without data word).
b. Synchronize (with data word).
c. Transmit status word.
d. Reset terminal.

Note 2: This example (where a colon follows a verb introducing a list) is more and more widely accepted. However, some editors would still revise the lead-in sentence to read as follows:

The 1553 interface has these commands:

In this *Style Guide*, for instance, we have avoided introducing displayed lists with sentences that end with a verb or preposition. See rule 5.

Lists Within Lists

12. Whenever one list occurs inside another list, use numbers for the outer list and letters for the inner list:

1. The physical characteristics of the regenerator include (a) ferritic stainless steel construction, compatible with a moist, coastal salt-air environment; (b) an internally insulated turbine exhaust duct; (c) a horizontal configuration; and (d) high-performance rectangular fins.
2. The performance data include (a) 88 percent thermal effectiveness, (b) over 4,600 hours of operating time, and (c) no evidence of corrosion or fouling.

Note: For a third level of nested lists, use lowercase Roman numerals (*i*, *ii*, *iii*, etc.). Also use caution. Lists within lists within lists become confusing and irritating.

Lists and Parallelism

13. Ensure that items in lists are parallel in structure. Begin each item with the same type of word (noun, verb, adjective, etc.):

This programming language interface allows the simulation model programs to access the database in the following ways:

1. Retrieve records with specific key values.
2. Retrieve records in database sequences.
3. Insert records into the database.
4. Delete records from the database.
5. Modify and replace records in the database.

Each item begins not only with a verb, but also with the same kind of verb. The list would **not** be parallel if the verb or sentence forms were changed:

This programming language interface will allow the simulation model programs to access the database in the following ways:

1. Retrieve records with specific key values.
2. Records in data base sequences can also be retrieved.
3. Insertion of records into the database.
4. Deleting records from the database.
5. Modification and replacement of records in the databasea.

Lists, whether in paragraph or display form, must always be parallel. The items listed must be consistent in form and structure. See Parallelism.

Managing information is a major productivity challenge for business and technical professionals. The amount of data that pours into your world can be staggering. Email, social networks, text messages, RSS feeds, blog updates—this deluge of data adds exponentially to the already overwhelming volume of input from older media such as television, the telephone, and print. Add to this picture the real-time streaming of commentary from countless sources, and coping with it all seems hopeless, much less the prospect of staying current.

The principles of information management are, nevertheless, basic and simple. To begin with, data is not information. A data point becomes information to you only when it becomes relevant to your purposes. While data is abundant, time and resources are limited; so you must have productive strategies for categorizing, accessing, and analyzing information. The following rules will help you manage the flow of information that is so critical to your professional success.

Managing Information

Categorizing Information

1. Create space for important information.
2. Assign information to the proper category.

Accessing Information

3. Provide for easy and rapid retrieval of information.
4. Avoid overloading others with information.

Analyzing Information

5. Review your purpose for collecting and analyzing the information.
6. Question the reliability of the data.
7. Use appropriate methods for interpreting data.

Categorizing Information

1. Create space for important information.

Not all information is created equal. Email messages, for example, stack up in your inbox without regard to priority. As a result, you might spend a lot of time on a minor request that appears at the top of the screen and have no time left for an important request that appears at the bottom.

Ensure that your most important information takes priority by creating space for it. The most accessible space on your office bookshelf should be reserved for the books you need to refer to most. For the same reason, an accessible space on your computer should be reserved for the most important information that comes to you.

Importance depends on relevance. To manage information effectively, you must first decide what your most important purposes are. What are your organization's top-priority goals? What is your team's key initiative right now? What is your most critical project? Create folders for these most important priorities and keep them on your desktop—the most accessible place.

2. Assign information to the proper category.

All data that comes to you in the workplace can be categorized under three headings: tasks, appointments, and reference documents.

When you open your email inbox, delete any message that isn't relevant to your priorities (see rule 1). Then sort through the relevant messages and assign them to a category. Click and drag requests for action to your task list. Meeting requests go to your calendar. Documents useful for later reference go to your notes or to the relevant folders on your computer or in your filing cabinet.

Alternatively, you can insert notes and reference documents directly into your task list or calendar. For example, if someone emails you the agenda for a meeting, you can click and drag it into the meeting notice on your calendar. If someone emails you an article relevant to a scientific experiment you are doing, you can add reading the article to your task list and insert the article itself into the task.

Of course, you can also create protocols for automatically deleting or filing incoming messages. For example, you can create a "boss" folder and a rule that directs all emails from your boss into that folder. You can set up rules for routing emails by content or keyword as well. To supplement automatic spam filters, your own rules can block unwanted messages.

Managing Information

Accessing Information

3. Provide for easy and rapid retrieval of information.

For you, the ability to retrieve information is obviously important; for your organization, it can be critical. Regulatory authorities and auditors often require access to background information. From a legal and operational point of view, you should be able to produce a documentation trail for key decisions.

If your filing system is effective, retrieving information should be easy. If you can't locate an item, the "search" or "find" function of your computer can help. One major pitfall, however, is that a search might produce hundreds of documents, and it becomes time-consuming to find exactly what you seek.

If you want to avoid searching through stacks of files for what you need, tag each document with keywords specific to that file when you create it. To retrieve that exact file, you enter the keywords you associate with it. Tagging can even eliminate the need to file a document in a folder.

4. Avoid overloading others with information.

Today the sheer quantity of information weighs down most organizations. You can help manage the flow of information. Your organizational email system is not a social network; don't add to the overload with unnecessary copying on messages or irrelevant forwards. Write concise, action-oriented emails. Be clear about what you want.

If information overload becomes acute, you might meet with others and develop a set of shared protocols to relieve the problem. For example, you might establish a set time for checking emails each day. You might adopt a rule for signaling action desired on an email or you might stop the practice of attaching weighty files to email messages. For guidance, see Electronic Mail.

Analyzing Information

Recall that information is the interpretation and analysis of data (facts, opinions, ideas). In this sense, information is an intellectual product. Unanalyzed data is not an intellectual product and serves no one well. Since analysis always comes in the form of writing or presentation, good writing is in part defined by the quality of analysis.

The job of most knowledge workers consists of analyzing information and making decisions based on analysis. Whether it's hiring a new supplier, comparing field studies for an environmental assessment, decreasing cycle time in a key process, or prescribing the right medication for a patient, job success depends on the quality of information analysis.

Knowledge workers have always used their judgment based on the information available. In our world, however, the quantity, diversity, and short life cycle of information have made analysis enormously more challenging. The following rules can help.

5. Review your purpose for collecting and analyzing the information.

Review the questions that led you to collect the information. What provocative question are you attempting to answer? It's easy to lose sight of your purpose once you get into the thick of the data. Many reports are useless because they fail to meet the needs of the audience, even if the information is high-quality and the analysis first-rate.

6. Question the reliability of the data.

Information is only as good as the accuracy of the data behind it, so an analytical thinker always questions the reliability of the data. Are the findings current? Are the financial statements truly representative? Has bias crept into the study design? Are there gaps in logic? Is this white paper just marketing propaganda? Are the sources for these figures reputable and trustworthy?

7. Use appropriate methods for interpreting data.

The purpose for data analysis is to arrive at a decision or conclusion. The better the analysis, the better the decision. As you analyze the data, ask yourself what story it tells you. You are looking for meaningful patterns, trends, and unexpected developments that you can turn into a story to tell decision makers.

For example, in analyzing the financial statements of a business, you might find healthy revenues and profit margins, but also a downward trend in cash on hand. What's the story here? Is the firm using cash intelligently to invest in the future, or are they spending recklessly? Obviously, the story you tell influences any resulting investment decisions.

Methods for interpreting data can be specific to a profession. For example, the methods a lawyer uses to interpret evidence in a legal case are quite different from the methods used by an engineer to interpret time-to-failure data on an electronic component. You should understand and strictly comply with the interpretive standards of your field.

Some interpretive methods are generally useful regardless of the field. Gap analysis is particularly helpful in developing rational solutions to business problems that arise, and is used widely by quality-improvement professionals, auditors, engineers, and scientific researchers. To do gap analysis, ask yourself these questions:

- What is the current condition (according to the most reliable data)?
- What is the desired condition (based on established criteria, plan, or best practice)?
- What is causing the gap between the current and the desired condition (root causes, both proximate and remote)?
- What is the effect of the gap (in terms of lost time, money, quality, etc.)? Answers to this question will tell you whether the gap is worth addressing.
- What do you recommend to close the gap? (Recommendations should address not only the gap itself but also the causes of the gap.)

The result of gap analysis should be a compelling story to motivate action.

The range of interpretive methods is wide, from statistical analysis with its strict protocols to deconstructive analysis, which assumes that any conclusion that can be drawn is inadequate and self-contradictory. As a knowledge worker, you must be aware of and able to defend the interpretive method you use because your audience will inevitably question it. For more guidance, see Thinking Strategies.

Maps

Maps integrate geographic data with other kinds of data to help business and technical professionals solve problems. Geographic maps can show geologic features, economic trends, demographics, housing developments, manufacturing sites, wells, crops, and how to get to grandmother's house.

Not all maps are geographic—that is, Earth-based creations. Maps exist of the Milky Way and of distant galaxies. Medical researchers map the areas within the brain or the genes within a chromosome. Metallurgists prepare maps of the atomic structure of a crystal. These and other nongeographic maps are beyond the scope of the following discussion.

The principles discussed below and the sample maps all focus on geographic maps. By analogy, however, these principles might apply to nongeographic maps, especially those included in technical or business documents.

Geographic maps today rely on increasingly sophisticated GIS (geographic information system) technologies. GIS is a computer application that generates maps using information from a database. GIS starts with an actual map or a plot of a geographic area or site. Next, database information is combined with the map features. A database can include information about all sorts of site characteristics or features:

—Income distribution of a population

—Retail density

—Rivers, brooks, lakes, and ponds

—Roads, trails, and cow paths

—Water, gas, and electrical lines

—Legal, political, or zoning boundaries

—Distributions of plants and animals

—Elevations above or below sea level

—Buildings, monuments, and cultural sites

—Census data by age, employment, sex, race, etc.

—Any set of data tied to land locations

For ongoing guidance on map design, see the online Map Book Gallery of ESRI (Environmental Systems Research Institute), which is continually updated. Many open-source GIS software programs are on the market, notably GRASS GIS (Geographic Resources Analysis Support System). For general information on graphics, including maps, see GRAPHICS FOR DOCUMENTS and GRAPHICS FOR PRESENTATIONS. For information on captions for maps, see CAPTIONS. For listing a number of maps in a document, see TABLES OF CONTENTS.

Maps

1. Choose maps that will help you achieve the purpose of your document.
2. Use GIS capabilities creatively to generate insights and ease decision making.
3. Label and number each map, explaining its features and including an informative caption.
4. Use inset location maps to orient readers to the location of the site recorded on a map.
5. Include latitude and longitude references if relevant, and always orient north toward the top of the page.
6. Use maps to connect key business or technical data with geographical distributions.

1. Choose maps that will help you achieve the purpose of your document.

Like any graphic, a map is only useful if it helps you make your point. Too often, maps are included only to add "visual interest" to a document. Figure 1, for example, was found on the first page of an analytical report intended for prospective investors in bonds issued by the state of Tennessee. The map clearly serves no purpose other than to decorate the page. It also represents a missed opportunity: the writers could have included a map of economic activity in Tennessee that would help investors decide whether to buy the state bonds.

Choose the proper scale and decide on the amount of detailed

Figure 1. A Missed Opportunity. *This map appeared on the first page of a bond-rating report on the state of Tennessee. A map showing economic features of the state would have been more useful to potential investors than this decorative map.*

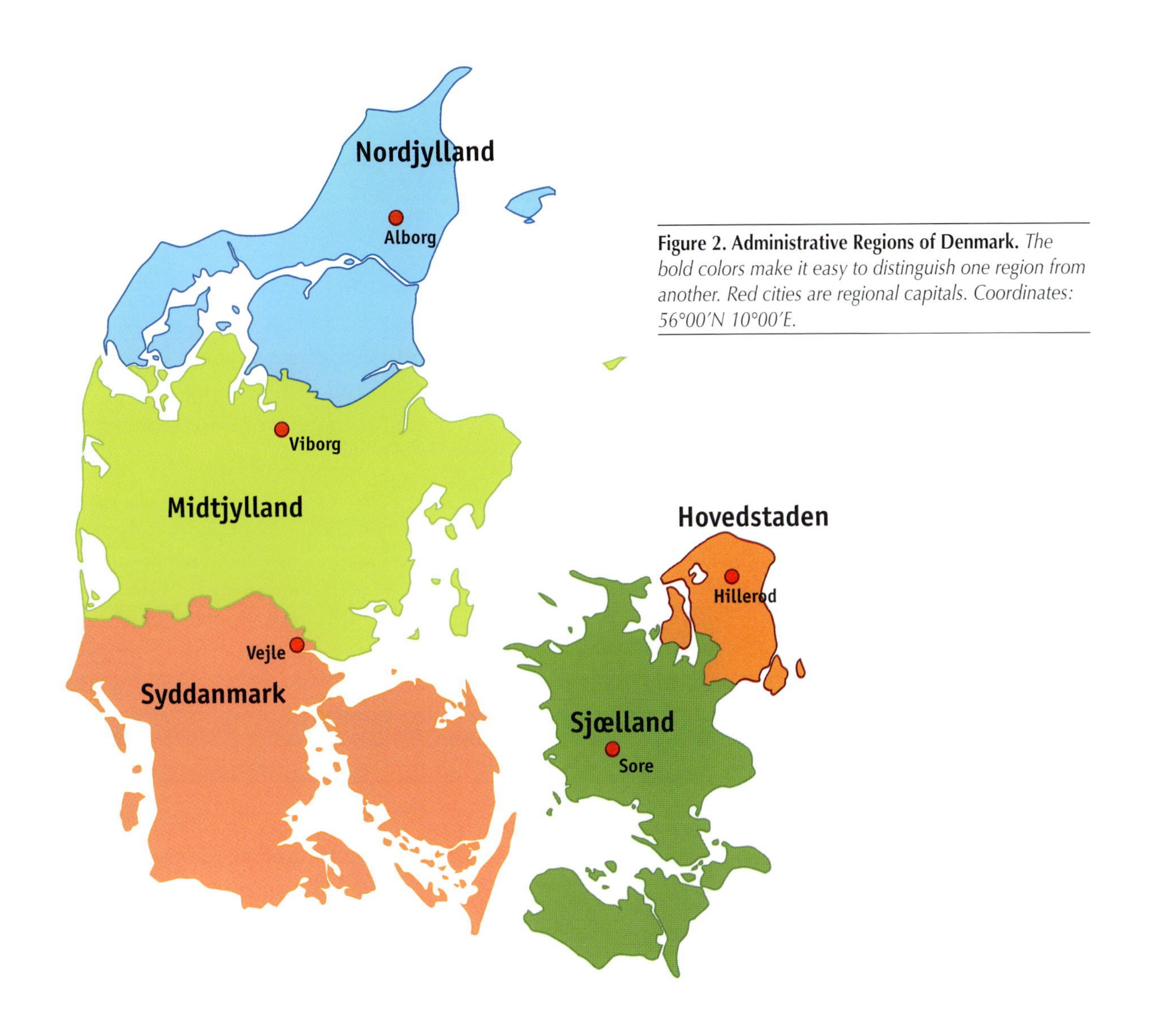

Figure 2. Administrative Regions of Denmark. *The bold colors make it easy to distinguish one region from another. Red cities are regional capitals. Coordinates: 56°00′N 10°00′E.*

information to include based on your purpose. The wrong scale can make map features too large or too small to be useful to readers of your document. Similarly, too many extraneous features on a map become mere clutter.

Without careful planning, the details recorded on a map may be excessive. In figure 2, for instance, the location of each regional capital might be irrelevant to a report on the prospects of expanding a business in Denmark. On the other hand, if the business needs to work closely with administrative officials in each region, the map becomes very relevant.

Design maps so that they will have good readability at the scale you select. Remember that changing the scale by reducing the size on a slide or photocopy can produce maps that are unreadable. For example, the letter "o" usually becomes a dot when the map is reduced by a factor of two or more. Remember, also, that maps photocopied several times lose much of their contrast and clarity. With GIS, you can test legibility and usability at various scales, then leave out or include data sets to improve legibility.

Plan the kinds of maps to use in the earliest phases of project and document planning, not when you are preparing the final draft of your document. Include map specifications and standard GIS parameters in the document style sheet. The earlier you plan your maps, the better their quality and relevance will be. See Page Layout.

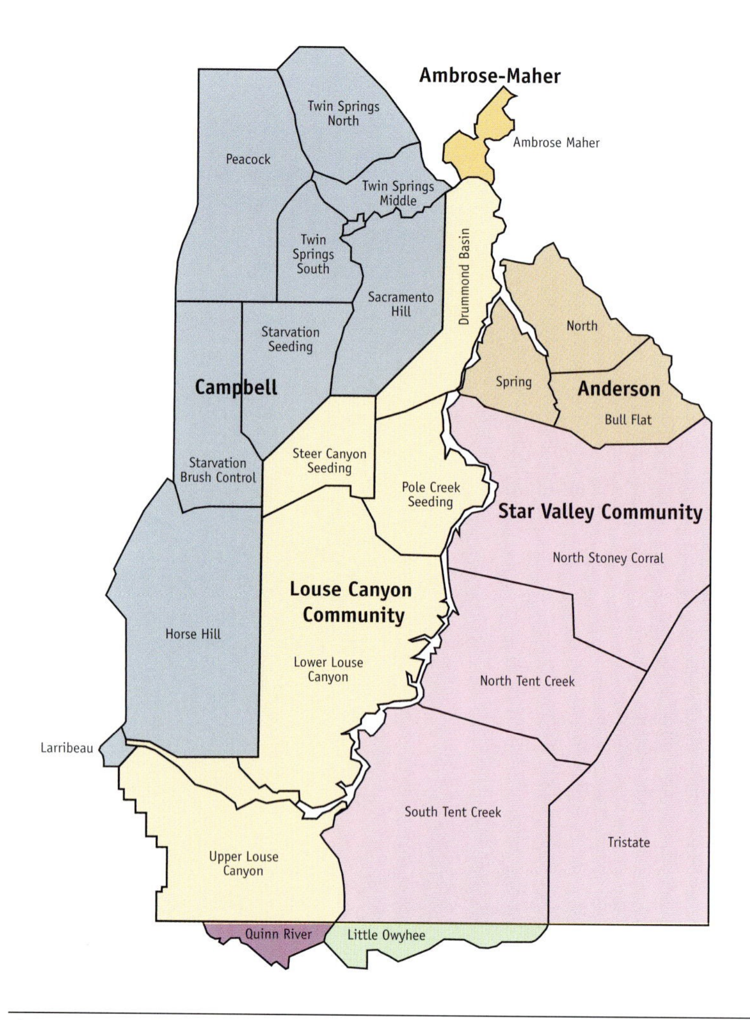

Figure 3. Grazing Allotments and Pastures, Star Valley Region, Oregon. *Colored areas indicate ranching-community licenses. Coordinates at center: 42.2035°N 117.244°W Source: U.S. Bureau of Land Management.*

2. Use GIS capabilities creatively to generate insights and ease decision making.

Maps have always been important in decision making, beginning when hunter-gatherers drew in the mud to indicate where a herd of bison was feeding. Until GIS, maps primarily captured a static view of a geographic site and its geographic features. With GIS maps, you can create varying views to help users discover insights and make decisions. One map can be changed to serve differing purposes, audiences, and media. For example, a bank might superimpose a map of high-income neighborhoods over a map of its own branch locations to determine where to close old branches and open new ones.

To illustrate, figures 3 and 4 are different versions of one map viewed according to two different data sets. Figure 3 shows grazing parcels allotted to ranchers. Figure 4 shows points where the densest population of sage grouse may be found. By comparing the two maps, decision makers can come up with ideas for balancing the needs of the ranchers and the birds. You can even project future conditions by asking "What if?" about the existing baseline data.

GIS can map different future scenarios so that their effects can be assessed. What if the land-management authority chose to zone cattle grazing away from the sage-grouse population? What would be the impact on the birds if new grazing allotments were granted? In such cases, GIS maps can be a powerful tool for conceptualizing and analyzing options.

NOTE: With GIS, bad data are like the proverbial rotten apple. One error in the database can ruin all the dependent analyses. Each extrapolation from the data will magnify the errors.

Be sure to include complete citations to any sources of data you are using. See CITATIONS.

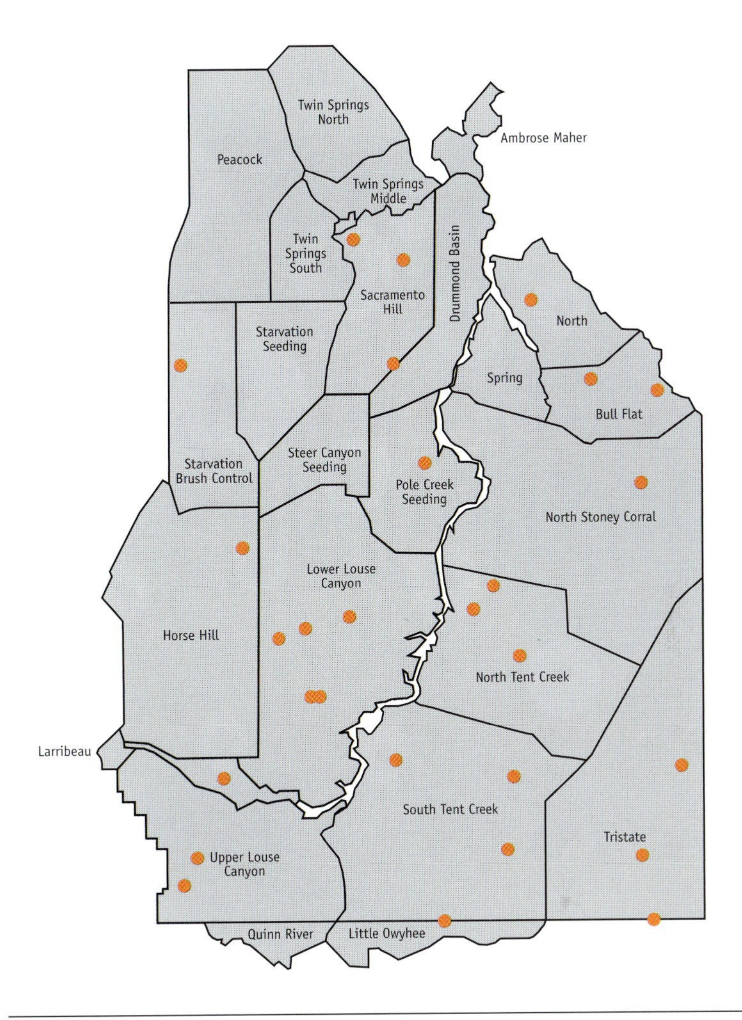

Figure 4. Sage-Grouse Populations, Star Valley Region, Oregon. *Each dot represents a nesting community of at least 50 birds. Coordinates at center: 42.2035°N 117.244°W Source: U.S. Bureau of Land Management.*

3. Label and number each map, explaining its features and including an informative caption.

Because maps are a type of figure, label and number your maps figure 1, figure 2, etc. If you are using many maps in your document, you might label them as maps, not figures. In this case, you would also have a separate list of maps as well as a list of figures following the table of contents. See TABLES OF CONTENTS.

The name of the map should indicate the principal feature(s) of the map (see figures 2 through 6). A map name should be brief but accurate. Accuracy is more important than brevity.

Leave out "Map of" as part of the name on the map. But in a list of figures, add *Map of* or use *Map* in parentheses to help readers tell which figures are maps.

Include an explanation block (legend, key) to ensure that readers understand all of the symbols, numbers, letters, patterns, and colors on your maps (see figure 6).

Orient the letters, numbers, and labels in your legend horizontally on the page. Place the explanation where it does not interfere with items on the map. Keep the explanation block small enough that it does not dominate the map. Include the scale in the explanation block. Your GIS program will do these things for you, but make sure your maps follow these guidelines.

Include an informative caption (following the name of the map) that states the key message you want the

Unemployment Rates in Metropolitan France

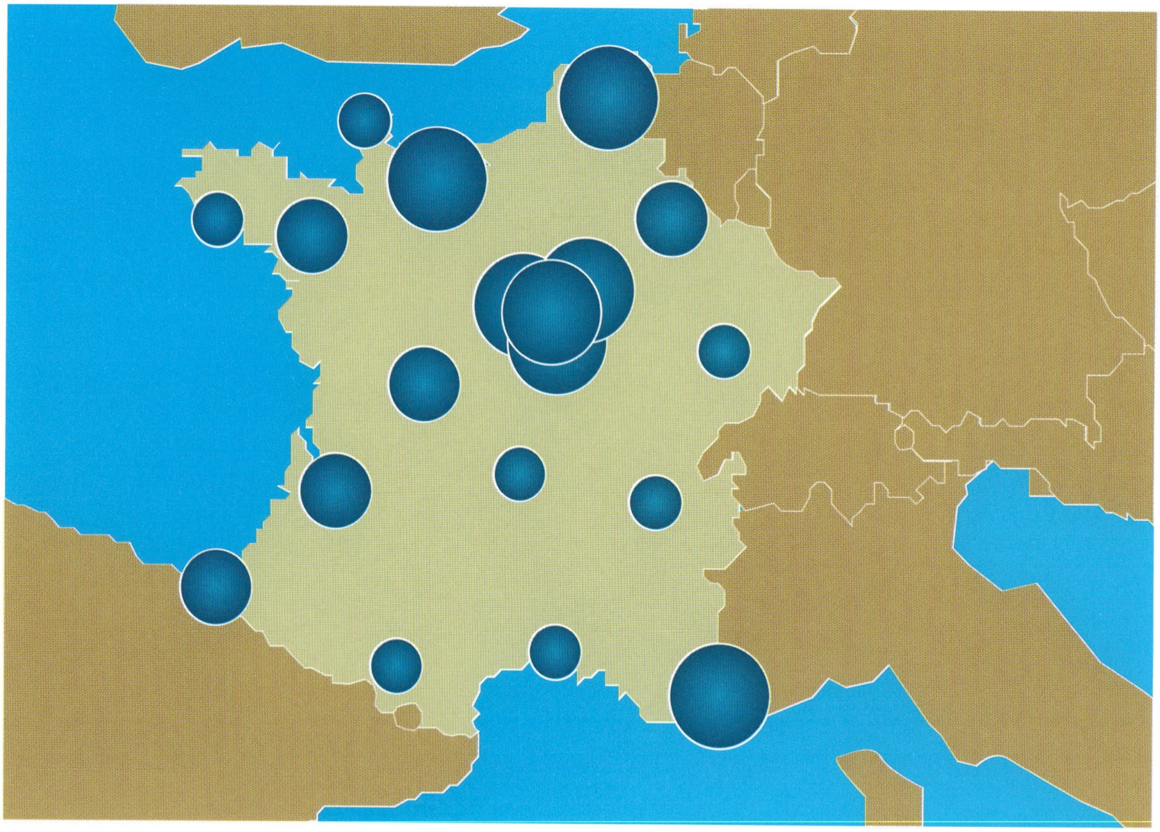

Figure 5. Poor Map Design. *Users mouse over the circles in this interactive map to see the "rate" of unemployment in each region. Actually, they see unemployment numbers, not rates, so the title is inaccurate. Circles are best used to show quantity rather than rate. The poor design is not helped by the lack of an informative caption or a legend.*

reader to learn from studying the map (see the informative captions on figures 2, 4, and 6). See CAPTIONS. The caption and the accompanying text will usually repeat each other so that the map (or figure) can stand alone.

Be sure to include the source for any of your maps. Usually, the source follows the caption. See CITATIONS and BIBLIOGRAPHIES.

4. Use inset location maps to orient readers to the location of the site recorded on a map.

Use inset maps either to establish the geographical location and perspective or to provide greater detail for some site or feature. The insets in figure 6 show Alaska, Hawaii, and Puerto Rico as non-contiguous parts of the United States.

Use inset maps only when necessary. Figures 3 and 4 do not include inset maps because the natural users of these maps will know that the Star Valley region is in Oregon.

5. Include latitude and longitude references if relevant, and always orient north toward the top of the page.

Unless you are using a map of Antarctica, you should orient the map so that north is toward the

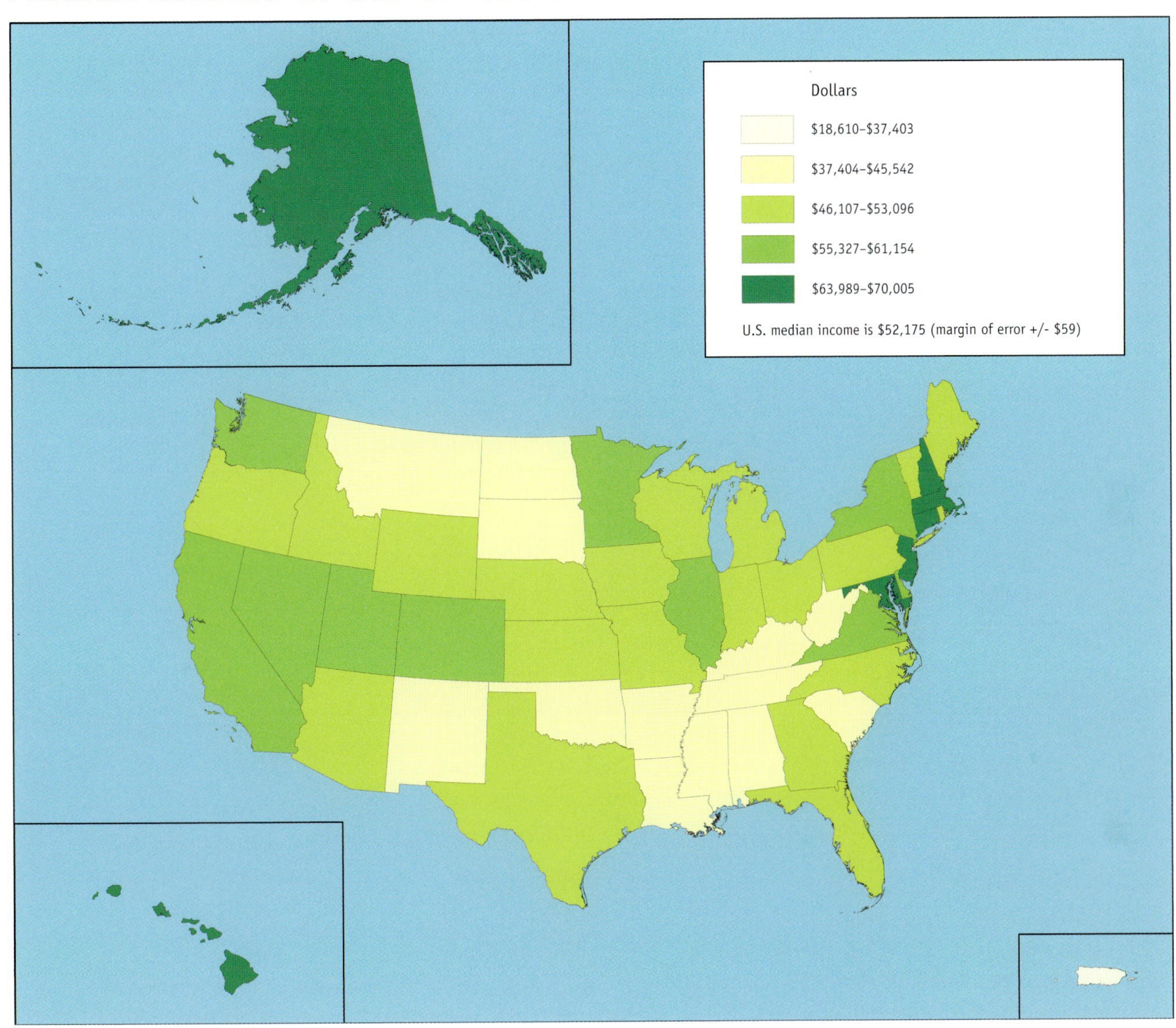

Figure 6. A Well-Designed Map. *To find the median income for a state, users click on the state in this interactive map. The movement from light to dark shades enables users to grasp instantly the differences in income from state to state and region to region. The caption is informative and the legend simple: Median Income in the United States. The coastal Northeast remains the wealthiest region, with the Great Lakes and the Far West regions gaining rapidly.*
Source: U.S. Census Bureau, 2006–2008 American Community Survey.

top of the page and south is toward the bottom. In any case, include a north arrow unless the orientation is obvious.

A map of Antarctica (yes, we were serious) should be oriented with the South Pole toward the top of the page. Everything in Antarctica is upside down. Even in maps of Antarctica, an exception is often made if a map includes any part of another continent. Then, north is usually at the top.

If precise location is important, a latitude and longitude reference is better than a north arrow. GPS technology on the Internet can instantly provide latitude and longitude for any point on land or sea.

6. Use maps to connect key business or technical data with geographical distributions.

Data often can be interpreted or divided along geographical divisions. Economists compile income data into totals for each state or for major metropolitan areas. Climatologists speak of regions when they profile data on rainfall, wind-chill factors, or soil moisture. Criminologists survey their statistics by summing up trends across a region, within a province, or even within a neighborhood.

Figure 5 is an example of data presented from a geographical perspective. In this interactive map, users mouse over the circles to find the unemployment rate in a particular region of France.

Figure 5 is, however, a poorly designed graphic. The first problem with the figure is its reliance on circles to represent rates, as circles are more useful for communicating comparative quantities. The lesson: select the best graphic feature for the job. Use a dot to represent a city on a small-scale map. Use lines for highways. Use an arrow to pinpoint a feature. Use a line graph rather than a map to illustrate differences in rate.

The second problem with the figure is more subtle. The circles representing unemployment turn out to be proportional to the population centers they represent. The user might conclude that there is more unemployment in the Paris region simply because Paris is the largest city in France. Make sure your map communicates the message you want it to communicate.

Figure 6 avoids the problems with figure 5. Figure 6 tells a simple story in a picture of contrasts that are easy to understand. There are no fancy but misleading graphics. While colors are used, the graphic does not rely solely on color. A black-and-white copy of the figure would still communicate its message because the colors are in differing shades. See COLOR.

Mathematical Notations

Mathematical notations vary from one journal to another and from one publisher to another. So before preparing manuscripts for professional submission, determine which style the editor or publisher prefers. Adhering to the preferred style is especially important if the manuscript will be typeset because setting mathematical symbols is often the most costly phase of typesetting.

The following principles of mathematical notations apply to most publications.

Mathematical Notations

1. Be consistent in writing mathematical signs, symbols, and units.
2. Use displayed (separate-line) expressions for lengthy equations or for special equations.
3. Use expressions within the text when the equations are minor or routine, when they are short, and when they are not important enough to highlight.
4. Do not punctuate displayed expressions that continue a sentence in the text.
5. Divide displayed expressions before the equals sign or the sign of operation when the expression extends beyond one line.
6. For reference, number displayed expressions in parentheses to the right of the expression.
7. For grouped expressions, place parentheses inside brackets inside braces inside parentheses.

1. Be consistent in writing mathematical signs, symbols, and units.

Because the conventions for writing mathematical signs, symbols, and units vary, you must establish a consistent methodology and adhere to it. If you are writing the Greek letter *sigma* (Σ) for instance, you can use either σ or ς for lowercase sigma. Either form is acceptable, but be consistent in the one you choose.

A related principle is to be consistent in your choice of units from one equation to the next. If in one equation a figure is expressed in meters, then related equations should express equivalent figures in meters, not yards. See Signs and Symbols.

Note 1: Consider keeping a list of the conventions you prefer for writing signs, symbols, and units. Referring to the list as necessary will help you maintain consistency throughout your text. See Editing and Proofreading.

Note 2: Style guides and editors are not consistent as to how to space mathematical symbols within equations or expressions. Some authorities suggest no space between the common operations and numerals:

2+4

b-c

6÷3

axb

The authroitative *Chicago Manual of Style* recommends using narrow spaces around these operations as well as the equals sign (=), greater than (>), less than (<), and other operational signs:

2 + 4

b - c

6 ÷ 3

a x b

2 + 4 = 6

9 > 3

3 < 9

Software programs designed to present mathematical expressions will often provide automatic spacing. If so, follow that pattern used in your software unless you have reasons to change the spacing. The key is to decide on your spacing pattern and then be consistent throughout your text.

2. Use displayed (separate-line) expressions for lengthy equations or for special equations.

Displayed expressions are separated from the text and are usually centered on the page with two or more spaces above and below the equation. Major equations and those equations too lengthy or too complicated to place within the text should be displayed. You can also display equations you wish to emphasize:

The initial form of the equation was

$$\int_a^b f(x)\,dx = \frac{b-a}{6}\left[f(a)+4f\left(\frac{a+b}{2}\right)+f(b)\right]$$

Mathematical Notations

3. Use expressions within the text when the equations are minor or routine, when they are short, and when they are not important enough to highlight.

Expressions within text are more difficult for readers to see. However, in many circumstances, the writer does not need to highlight an equation by displaying it. Expressions written within text must be simple enough for readers to comprehend easily:

> The expression $1/(x+y)$ becomes increasingly smaller as the values of x and y increase.

When you convert displayed expressions into textual expressions, add parentheses or brackets as necessary to clarify the mathematical relationships:

$$\frac{3}{x-y}\,ab \quad \text{becomes} \quad \left[3/(x-y)\right]ab$$

$$6+\frac{x+y}{3}+18 \quad \text{becomes} \quad 6+(x+y)/3+18$$

Similarly, when converting a textual expression into a displayed expression, remove unnecessary parentheses and brackets:

$$1/(x/y) \quad \text{becomes} \quad \frac{1}{x/y}$$

$$(a+3)/(b+6) \quad \text{becomes} \quad \frac{a+3}{b+6}$$

4. Do not punctuate displayed expressions that continue a sentence in the text:

> The revised equation is
>
> $$y = A+B\,(x-x_1) + C(x-x_1)^2$$

5. Divide displayed expressions before the equals sign or the sign of operation when the expression extends beyond one line:

$$\int_a^b f(x)\,dx = \frac{1}{3}\Big[y_o + y_n + 4(y_1 + y_2 + \ldots + y_{n-1})$$

$$+2(y_2 + y_4 + \ldots + y_{n-2})\Big]\,\Delta\chi$$

Note 1: The opposite is true for expressions within text:

> After conversion, the alternate version has $\pi\left(y_{i-1}+y_i\right)\left(\Delta x_i^2+\Delta y_i^2\right)=$ $p\left(y_{i-1}+y_i\right)\left[1+\left(Dy_i/Dx_i\right)^2\right]^H Dx_i$

Note 2: Do not divide short expressions. If possible, avoid dividing any expressions.

6. For reference, number displayed expressions in parentheses to the right of the expression:

$$V\,\overset{5}{2}\pi\int_a^b xf(x)\,dx \qquad (15)$$

Note: Unless you refer to the expressions elsewhere in the text, numbering of displayed expressions is unnecessary.

7. For grouped expressions, place parentheses inside brackets inside braces inside parentheses:

$$x = a\left(b\left\{c+\left[d+2(e+7)\right]\right\}\right)$$

Note: Sometimes these symbols appear by themselves or the sequence does not apply:

> Sets
>
> $$\{a, b, c\}$$
>
> Expressions with functions
>
> $$f\left(g(x)\right)$$
>
> Expressions with upper and lower limits
>
> $$\left[\tan\theta\right]_o^\pi$$

See Brackets.

Meetings are one of the most expensive methods of gathering or disseminating information. Therefore, the best advice we might offer is: Don't hold a meeting!

Granted, meetings are sometimes necessary—and even the best way to achieve a given purpose. But if you can achieve the same purpose by another less costly and time-consuming means, don't hold a meeting.

Follow these guidelines when conducting a meeting or a teleconference. Some of the guidelines apply to those who conduct meetings, others to participants. But since most people often find themselves in both roles, we have not maintained that distinction in the rules that follow.

Meetings Management

1. Hold meetings only when necessary.
2. Make a meeting plan or agenda.
3. Define the purpose of your meeting.
4. Invite only those people who need to be there and who have something to contribute.
5. Start and end your meetings on time.
6. Fulfill your responsibility (either as a meeting leader or participant) to keep the meeting on track, moving toward the achievement of the purpose.
7. Conduct your meetings with appropriate "people skills."
8. Identify and follow up on action items.
9. Hold effective teleconferences.

1. Hold meetings only when necessary.

Meetings cost money, and the cost goes up incrementally with every person who attends. We suggest, therefore, that if you can achieve the same goal by a telephone call, an email message, or a couple of informal chats in the hall, then do so.

Meetings, like bad habits, tend to be part of our daily routines: "Today is Monday—time for the management meeting again." Before holding any meeting, we should always ask ourselves: Why are we holding this meeting? Is it necessary? Is there a better way of achieving what we want to accomplish?

On the other hand, short, highly focused meetings can achieve a great deal, particularly when it's crucial to track progress on a project or gain rich input on a decision that must be made.

2. Make a meeting plan or agenda.

Unfortunately, too many meetings are held with little or no advance thought or preparation. The result is wasted time, unjustified costs, and frustration.

An effective meeting plan is not just a list of topics to discuss. Your plan should answer such questions as:

- What do I want to accomplish with this meeting?
- Whom should I invite and what will be their contributions?
- What methods will we use to accomplish the purpose(s) of the meeting?
- What preparations are necessary to ensure that the meeting will run smoothly and effectively?
- Is the meeting worth the cost?

Unless you consider these and other relevant questions, your meetings will not be as effective as they might otherwise be.

3. Define the purpose of your meeting.

Before you can achieve any purpose, you must know what that purpose is. Writing that purpose down on the whiteboard or on the meeting request will focus your thinking about what you really want to achieve. A written purpose also gives you a means of keeping the meeting on track.

One of the best ways of defining your purpose is to write a purpose statement and distribute it with the meeting request. A purpose statement is more than a topic and looks like a sentence. Each agenda item should also be phrased as a sentence and not just as an empty topic. (See figure 1.)

Also ask, "What do I want the participants to know going into the meeting?" This might require some additional effort in preparing and distributing information before the meeting so participants will come prepared to discuss items intelligently.

NOT THIS	THIS
Budget Meeting	Resolve outstanding budget issues in R&D and product management.
Cash Flow	Brainstorm ideas for improving cash flow this quarter.
Market Share Report	Hear latest market-share data and generate recommendations for increasing share.

Figure 1. Weak Versus Strong Agenda Items. *Write agenda items as clear objectives, not as empty topics.*

Gather information on controversial topics and distribute it in advance. This will enable participants to think through the issues beforehand to save time in the meeting.

If you don't have time to prepare, start your meeting by making a quick agenda. Ask others for input to the agenda.

Whatever your purpose, it should be made clear at (or before) the very outset of the meeting. It provides the focus for a truly productive meeting.

4. Invite only those people who need to be there and who have something to contribute.

How often have you sat in a meeting and wondered, "Why have I been asked to attend?" Inviting people who have nothing to contribute or who have no business being at the meeting is one of the greatest hidden business expenses.

By inviting people unnecessarily, you not only incur additional expense, but also increase the level of inefficiency and frustration.

The role of any participant depends on the purpose of the meeting. If someone has certain information, knowledge, skill, or the power to make decisions regarding the purpose of the meeting, he or she should be invited to attend. If people have nothing to contribute toward achieving the purpose, why invite them?

Once you determine that someone has a contribution to make to the meeting, include that expectation in the meeting request. If possible, let the person know how much time will be scheduled for that contribution.

5. Start and end your meetings on time.

If you begin on time, even though not everyone has arrived—and if you continue that pattern—your chances for prompt arrivals in the future will be greatly improved.

Waiting for latecomers to arrive before starting meetings is one of the greatest hindrances to productive meetings. It is also very costly.

Why penalize those who are on time by having them wait for those who come late? Build a reputation in your organization as one who starts (and ends) meetings on time.

6. Fulfill your responsibility (either as a meeting leader or participant) to keep the meeting on track, moving toward the achievement of the purpose.

As a leader or as a participant, you have a responsibility to assist the group to achieve its purpose. Everyone in the meeting should be there because they have important contributions to make.

Diverting the attention of the group to irrelevant topics, monopolizing the time, and wasting time on unnecessary details or explanations are expensive ways of ruining a meeting.

7. Conduct your meeting with appropriate "people skills."

People skills are simply common acts of courtesy, kindness, and tact.

Ask yourself as you determine the purpose of your meeting, "What do I want the participants to feel?"

Seldom do you achieve your purpose if you instill negative

feelings among the participants. Therefore, treating every participant with respect and kindness can go a long way in helping you reach your desired purpose.

Some people can be "meeting wreckers" or "meeting robbers." Dealing tactfully and calmly with such individuals can sometimes be a challenge.

The first deterrent is the meeting plan or agenda itself. If you as a meeting leader have distributed it beforehand and have defined roles and determined just how much time each person has in the meeting, chances are that those people will be less apt to cause problems than if their role in the meeting had not been clearly defined.

The second deterrent is to tactfully yet firmly refer to the purpose of the meeting. A gentle (or sometimes more forceful) reminder about the purpose of the meeting also tends to keep annoying interruptions from becoming major hurdles.

8. Identify and follow up on action items.

Without follow-up, a meeting may be just a waste of time.

You hold a meeting to achieve a certain purpose. Often that purpose is something that can only be achieved through the implementation of the decisions made during the meeting. Unless someone implements those steps, the purpose of the meeting may never be accomplished.

You will also want to establish some method for letting the meeting participants know about the progress and completion of the assigned tasks. If appropriate, distribute minutes that are concise and objective, with the emphasis on decisions and items for follow-up. See MODEL DOCUMENTS, "Minutes."

9. Hold effective teleconferences.

Teleconferences are now routine as video, audio, and Internet communication becomes better and cheaper. The rules above apply to teleconferences as well as to face-to-face meetings. However, teleconferences present some unique challenges. Physical remoteness makes it harder for people to participate and to "read" one another's reactions. People only "half listen" or check out to do other tasks. To facilitate a teleconference effectively, keep the following suggestions in mind.

Rotate opportunities to speak. Use a "round robin" approach to poll people one by one for their opinions. Assign different people to facilitate different parts of the meeting. Invite comments from those who aren't actively participating. If participants know they can be called on at any moment, they are more likely to stay engaged.

Say your name when you make a comment and address other people by name. This might seem awkward at first, but you will all be clearer about who is speaking and to whom.

Announce it if you leave the meeting. Then don't "lurk" if people believe you are no longer on the line. If you return to the meeting, announce it.

Memos

Memos are essentially letters written to persons within a writer's organization. Hence, they are often called "interoffice correspondence." As noted below, the principles of good letter writing apply equally well to memos. See LETTERS.

Memos is a shortened version of *memoranda* or *memorandums*. Most writers now prefer the short form, but either of the long forms is correct.

Email is increasingly replacing printed memos. Email is much faster and more immediate in its impact. However, a drawback of email is that it does not automatically provide a hard-copy record of important information (unless the writer or the recipient of the email message prints a hard copy). Organizations should have a clear policy on what information to retain in printed form versus information merely in electronic files. See ELECTRONIC MAIL, MANAGING INFORMATION, and ETHICS.

Memos can include any type of organizational information—from brief notices of meetings to full analyses of data. In some companies, for example, memos record internal technical or scientific reports. So no clear distinction exists between memos and reports, except perhaps in their format. A memo uses a memo heading (*To*, *From*, etc.) while an external report will often have a title and an abstract on the cover page. See REPORTS.

Effective Memo Writing

These rules apply to any memo, whether you email it or print it. Also, because memos are essentially letters that stay within an organization, the principles of good letter writing apply equally to memos:

Memos

1. Open memos with an informative and interpretive subject line.
2. Include your main point(s) in the first lines of your text.
3. Design and organize your memos so key points are immediately visible.
4. Make your memos personal and convincing.
5. Choose a direct and simple memo closing.

1. Open memos with an informative and interpretive subject line.

See LETTERS.

2. Include your main point(s) in the first lines of your text.

3. Design and organize your memos so key points are immediately visible.

4. Make your memos personal and convincing.

5. Choose a direct and simple memo closing.

For a thorough discussion of these five principles, see LETTERS. See also ORGANIZATION, EMPHASIS, REPETITION, KEY WORDS, and PARAGRAPHS.

Memo Format

Memo format varies considerably from organization to organization. However, memos often have these components: **heading, body, signature line, reference initials, attachment notation,** and **courtesy copy notation.** See figure 1 for an illustration of these components.

Heading

Memo headings, whether printed or typed, usually contain these elements:

To:
From:
Subject:
Date:

The order of these elements, their spacing and punctuation, and their placement on the page vary considerably. In printed memo forms, the heading elements often do not have colons. Typed headings usually have colons.

Some memos open with *To* and then give the subject line. Others place the date after the *To* line. Still others arrange the items in two parallel lists:

To	From
Department	Department
Subject	Date

Some memos omit the *From* line, opting instead for a typed name and signature at the end of the memo.

Two optional elements of the heading are a distribution list and a reference line or block.

The distribution list can appear in a box that follows or includes *To*. If used in the heading, the distribution list replaces the *cc* (courtesy copy) list at the end of the memo.

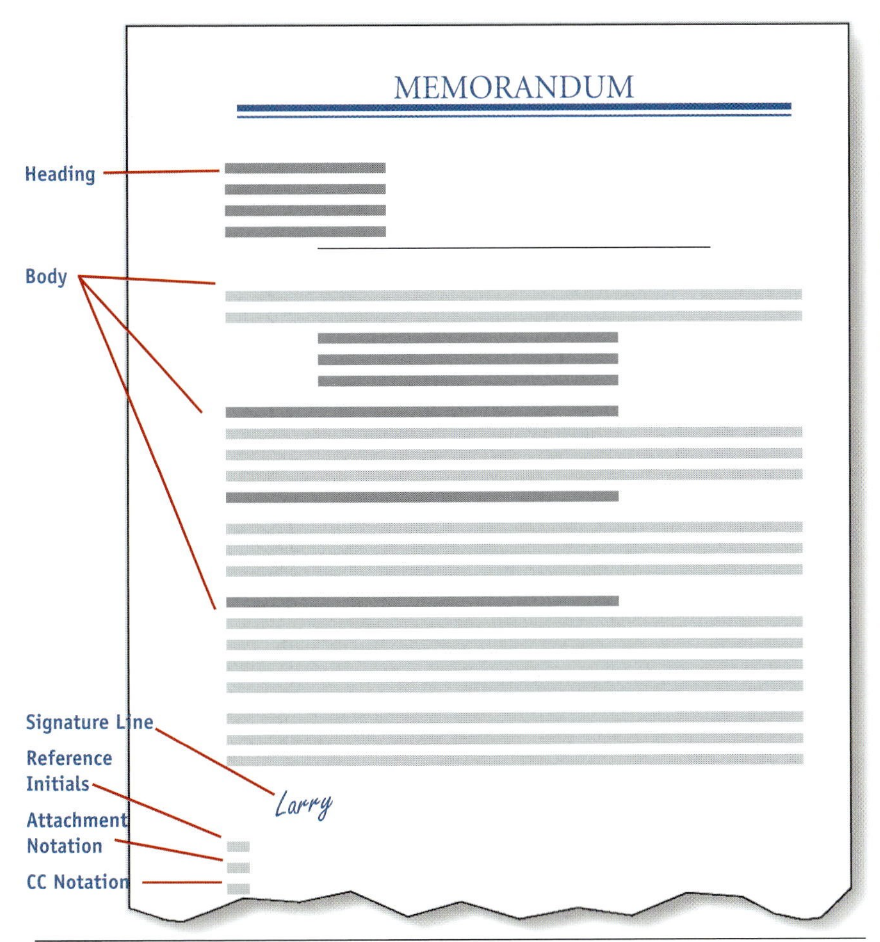

Figure 1. Sample Memo Format. *Most word-processing programs have automated formats for memos.*

Short reference lines or blocks can appear two lines below the date if the date appears by itself just right of the center of the page. Extensive references need to have their own lines, usually before or after the *Subject* line:

1. F. H. Howell, "Testing of the Wing Plate Assembly," May 18, 2010.
2. J. K. Jameson, "Design Options in the Wing Plate Assembly," March 22, 2010.

Numbering references helps writers refer to the references later in the text:

> Reference 1 notes that all wing plate assemblies have passed inspection this year. However, reference 2 indicates that design modifications must be undertaken to improve reliability.

The names of both the sender and the receiver do not require courtesy titles (*Mr., Mrs., Ms.,* or *Miss*), but *Dr.* is sometimes used. Names should be as complete as possible even if the sender and receiver are close friends. Long after the memo has been filed, future readers will probably not know whom Hank or Sue refers to, and the names could be important.

The subject line should be as specific as possible (see HEADINGS):

this

Subject: Recommendation to Test Two Methods of Lowering Salt Content

not this

Subject: Salt Content Tests

this

Subject: The Sales Decline in the Northern Region

not this

Subject: Northern Region Sales

Body

Paragraphs in the body of the memo are usually single-spaced with a double space between them. These paragraphs may or may not be indented (5 to 10 spaces). Both the indented and block forms are correct and usually acceptable; however, your company may have a preferred style.

Headings and lists are important devices, especially when a memo is more than a page or two long. See HEADINGS and LISTS.

Continuation Pages

Use a heading on pages beyond the initial page (with its memo format). Usually, continuation pages have the name of the receiver(s), the date, and a page number. The most common arrangement is to have the name of the receiver and the date flush left, with the page number centered.

Signature Line

Traditional memos have no signature line because the author's name appears after *From* in the heading. However, you can attach an automated signature to any memo you email.

Reference Initials

Reference initials in memos usually contain only the typist's initials. These initials appear either two lines below the signature line or two lines below the final line of text (if the memo has no signature line). The initials are usually in lowercase letters:

jtk

Memos

If the reference initials also contain the author's initials, then they would precede those of the typist and would follow one of these forms:

GLK/jtk GLK:jtk glk:jtk

When someone other than the sender writes the memo, the sender's initials come first, then the writer's initials, and then the typist's initials:

GLK/TER/jtk GLK:TER:jtk

NOTE: Reference initials are unnecessary if the writer prepares and then prints the memo with no help from anyone else.

Attachment Notation

Attachment notations are not very common except in hard-copy memos. If used, they appear on the line immediately below the reference initials:

GLK/jtk
Attachments (3)

In some very technical hard-copy memos with a number of attachments (such as maps or charts), the attachments may be listed at the bottom of the memo following the attachment notation.

Courtesy Copy Notation

In a hard-copy memo, *cc* or *c* (courtesy copy or carbon copy) notations appear two lines below the reference or attachment notations. Increasingly, the single *c* is used rather than *cc*—much as symbols in the metric system. The form varies:

c
c:
cc:
Copy to
Copies to

Memos sent to a large number of readers often have a distribution list instead of a courtesy copy list. The word *Distribution* appears in the heading following *To. Distribution* also appears instead of *c* or *cc* in the courtesy copy notation, and following *Distribution* is a list of the names and (if appropriate) departments of those people who should receive copies of the memo.

Only occasionally do *bcc* (blind courtesy copy) lists appear on memos. Blind courtesy copy lists appear only on the copies and not on the original memo.

If used, the *bcc* list appears two lines below the *cc* list. See SPACING.

Metric units of measurement are now used worldwide by scientists in the physical and biological sciences. The most precise version of the metric system is the International System of Units or SI (from *Système international d'unités*).

Despite its widespread acceptance, SI has not been adopted by all U.S. firms and government agencies. Retooling to metric standards has been a slow and costly process, and redrafting existing maps and design layouts has been unfeasible.

Nevertheless, SI is accepted internationally and uses unambiguous symbols. It is, therefore, the preferred system of measurement for all sciences and many areas of engineering.

Metrics

1. Use base units with their SI symbols.
2. Do not capitalize or italicize SI symbols, except those derived from proper names (e.g., A and K) and except for L for liter. The symbols do not change in the plural and are never followed by a period.
3. Use a point (period) as the decimal marker, and use spaces to separate long numbers into easily readable groups of three.
4. Some SI-derived units have special names.
5. Some SI-derived units have no special names.
6. Use the table of prefixes to form the names and symbols of multiples and submultiples of SI units.
7. Some non-SI units are still permissible within SI.
8. Avoid certain metric units that have been replaced by SI units.
9. Use table 6 to convert SI units into the common units of measure still widely used in the United States.

1. Use the following base units and their SI symbols:

length .... meter (m)
mass .... kilogram (kg)
time .... second (s)
current .... ampere (A)
thermodynamic temperature .... kelvin (K)
amount of substance .... mole (mol)
luminous intensity .... candela (cd)

NOTE: Volume (liter) is not considered an SI base unit because it can be derived from length (dm^3). See table 4 for the listing for volume.

2. Do not capitalize or italicize SI symbols, except those derived from proper names (e.g., A and K) and except for L for liter. The symbols do not change in the plural and are never followed by a period:

46 m (*not* 46m)	6 K
1 kg	6 kg
15 s	22.5 cd

NOTE 1: Use the metric abbreviations only when the metric unit follows a number. If the metric unit appears without a number, spell it out:

We measured 2 kg of salt.

but

We had several kilograms of salt.

NOTE 2: Wherever possible, choose SI units so that the numerical values will be between 0.1 and 1000:

54 m (*not* 54 000 mm)
3.6 mm (*not* 0.0036 m)

3. Use a point (period) as the decimal marker, and use spaces to separate long numbers into easily readable groups of three:

45 671.378 34
0.634 701

NOTE 1: When only four numbers appear on one side of the decimal, the space is optional but not preferred:

5.7634 *or* 5.763 4
8764 *or* 8 764

NOTE 2: In some countries outside the United States, writers use a comma as the decimal marker. If you are writing for a non-U.S. journal or publisher, you may need to use a comma as the decimal marker:

5,763
0,634
3,1415

See DECIMALS and PERIODS.

4. Some SI-derived units have special names.

See table 1 on the next page.

Table 1. SI-Derived Units With Special Names

Quantity	Name	Symbol	Derived From
absorbed dose of ionizing radiation	gray	Gy	J/kg
activity of radionuclides	becquerel	Bq	s^{-1}
electric capacitance	farad	F	C/V
electric charge	coulomb	C	A·s
electric conductance	siemens	S	A/V
electric potential, potential difference, electromotive force	volt	V	W/A
electric resistance	ohm	Ω	V/A
energy, work, quantity of heat	joule	J	N·m
force	newton	N	$m \cdot kg/s^2$
frequency	hertz	Hz	s^{-1}
illuminance	lux	lx	lm/m^2
inductance	henry	H	Wb/A
luminous flux	lumen	lm	cd·sr
magnetic flux	weber	Wb	V·s
magnetic flux density	tesla	T	Wb/m^2
plane angle	radian	rad	m/m
power, radiant flux	watt	W	J/s
pressure, stress	pascal	Pa	N/m^2
solid angle	steradian	sr	m^2/m^2

5. Some SI-derived units have no special names.

See table 2.

6. Use the table of prefixes to form the names and symbols of multiples and submultiples of SI units.

See table 3.

NOTE 1: Without using a space or a hyphen, attach the prefixes directly to the SI base unit: kilogram, millisecond, gigameter, etc. Similarly, the abbreviations for the prefixes attach directly to the abbreviation for the SI units: cm, Mg, mK, etc.

NOTE 2: Do not use two or more of the prefixes to make compounds of the SI units. Write ns (nanosecond), not mμs (millimicrosecond).

NOTE 3: Although kilogram is the base unit for mass, the prefixes are added to gram (g), not kilogram (kg).

7. Some non-SI units are still permissible within SI.

See table 4.

8. Avoid certain metric units that have been replaced by SI units.

See table 5.

9. Use table 6 to convert SI units into the common units of measure still widely used in the United States.

For an excellent reference, see Theodore Wildi, *Metric Units and Conversion Charts*, Second Edition (New York: IEEE Press, 1995).

Table 2. SI-Derived Units With No Special Names

Quantity	Description	Expressed in SI Units
acceleration		
—linear	meter per second squared	m/s^2
—angular	radian per second squared	rad/s^2
area	square meter	m^2
concentration (of amount of substance)	mole per cubic meter	mol/m^3
current density	ampere per square meter	A/m^2
density, mass density	kilogram per cubic meter	kg/m^3
dynamic viscosity	pascal second	Pa·s
electric charge density	coulomb per cubic meter	C/m^3
electric field strength	volt per meter	V/m
energy density	joule per cubic meter	J/m^3
heat capacity, entropy	joule per kelvin	J/K
heat flux density, irradiance	watt per square meter	W/m^2
luminance	candela per square meter	cd/m^2
magnetic field strength	ampere per meter	A/m
molar energy	joule per mole	J/mol
molar entropy, molar heat capacity	joule per mole kelvin	J/(mol·K)
moment of force	newton meter	N·m
permeability	henry per meter	H/m
permittivity	farad per meter	F/m
specific energy	joule per kilogram	J/kg
specific heat capacity, specific entropy	joule per kilogram kelvin	J/(kg·K)
specific volume	cubic meter per kilogram	m^3/kg
speed		
—linear	meter per second	m/s
—angular	radian per second	rad/s
surface density of charge, flux density	coulomb per square meter	C/m^2
surface tension	newton per meter	N/m
thermal conductivity	watt per meter kelvin	W/(m·K)
volume	cubic meter	m^3
wave number	1 per meter	m^{-1}

Metrics

Table 3. Prefixes and Their Symbols for SI Units

Multiplying Factor			Prefix	Symbol
1 000 000 000 000 000 000	=	10^{18}	exa	E
1 000 000 000 000 000	=	10^{15}	peta	P
1 000 000 000 000	=	10^{12}	tera	T
1 000 000 000	=	10^{9}	giga	G
1 000 000	=	10^{6}	mega	M
1 000	=	10^{3}	kilo	k
100	=	10^{2}	hecto	h
10	=	10^{1}	deca (deka)	da
0.1	=	10^{-1}	deci	d
0.01	=	10^{-2}	centi	c
0.001	=	10^{-3}	milli	m
0.000 001	=	10^{-6}	micro	μ
0.000 000 001	=	10^{-9}	nano	n
0.000 000 000 001	=	10^{-12}	pico	p
0.000 000 000 000 001	=	10^{-15}	femto	f
0.000 000 000 000 000 001	=	10^{-18}	atto	a

Table 4. Non-SI Units Permissible within SI

Quantity	Name	Symbol	Definition
area	hectare	ha	1 ha = 1 hm^2 = 10 000 m^2
mass	ton, tonne	t	1 t = 1000 kg = 1 Mg
plane angle	degree	°	1° = (π/180) rad
	minute	′	1′ = (π/10 800) rad
	second	″	1″= (π/648 000) rad
temperature	degree Celsius	°C	0°C = 273.15 K However, for temperature intervals 1°C = 1 K
time	minute	min	1 min = 60 s
	hour	h	1 h = 3600 s
	day	d	1 d = 86 400 s
	year	a	
volume	liter	l or L	1 l = 1 dm^3

Table 5. Metric Units Replaced by SI Units

Quantity	Metric Unit	Symbol	Definition
absorbed dose of ionizing radiation	rad	rad	1 rad = 10 mGy = 10 mJ/kg
activity	curie	Ci	1 Ci = 37 GBq = 37 ns^{-1}
area	are	a	1 a = 100 m^2
	barn	b	1 b = 100 fm^2
conductance	mho	mho	1 mho = 1 S
energy	calorie	cal	1 cal = 4.1868 J
	erg	erg	1 erg = 0.1 μj
force	kilogram-force	kgf	1 kgf = 9.806 65 N
	kilopound	kp	1 kp = 9.806 65 N
	dyne	dyn	1 dyn = 10 μN
illuminance	phot	ph	1 ph = 10 klx
length	angstrom	Å	1 Å = 0.1 nm
	micron	μ	1 μ = 1 μm
	fermi	fm	1 fermi = 1 femtometer = 1 fm
	X unit	—	1 X unit = 100.2 fm
luminance	stilb	sb	1 sb = 1 cd/cm^2
magnetic field strength	oersted	Oe	1 Oe corresponds to $\frac{1000}{4\pi}$ A/m
magnetic flux	maxwell	Mx	1 Mx corresponds to 0.01 μWb
magnetic flux density	gauss	Gs, G	1 Gs corresponds to 0.1 mT
magnetic induction	gamma	γ	1 γ = 1 nT
mass	metric carat	—	1 metric carat = 200 mg
	gamma	γ	1 γ = 1 μg
pressure	torr	torr, Torr	1 torr = 1.333 22 x 10^2 Pa
viscosity			
—dynamic	poise	P	1 P = 1 dyn·s/cm^2 = 0.1 Pa·s
—kinematic	stokes	St	1 St = 1 cm^2/s
volume	stere	st	1 st = 1 m^3
	lambda	λ	1 λ = 1 μl = 1 mm^3

Metrics

Table 6. Metric Values and Their Equivalents

Length

Myriameter (obs.)	10,000 meters	6.2137 miles
Kilometer	1,000 meters	0.62137 mile
Hectometer	100 meters	328 feet 1 inch
Dekameter	10 meters	393.7 inches
Meter	1 meter	39.37 inches
Decimeter	0.1 meter	3.937 inches
Centimeter	0.01 meter	0.3937 inch
Millimeter	0.001 meter	0.0394 inch

Area

Hectare	10,000 square meters	2.471 acres
Are	100 square meters	119.6 square yards
Centiare	1 square meter	1,550 square inches

Weight

Unit	Number of Grams	Volume of Water Corresponding to Weight	Avoirdupois Weight of Water
Metric ton, millier, or tonneau	1,000,000	1 cubic meter	2,204.6 pounds
Kilogram or kilo	1,000	1 liter	2.2046 pounds
Hectogram	100	1 deciliter	3.5274 ounces
Dekagram	10	10 cubic centimeters	0.3527 ounce
Gram	1	1 cubic centimeter	15.432 grains
Decigram	.1	0.1 cubic centimeter	1.5432 grains
Centigram	.01	10 cubic millimeters	0.1543 grain
Milligram	.001	1 cubic millimeter	0.0154 grain

Capacity

Name	Number of Liters	Metric Cubic Measure	United States Measure	British Measure
Kiloliter or stere	1,000	1 cubic meter	1.308 cubic yards	1.308 cubic yards
Hectoliter	100	0.1 cubic meter	2.838 bushels 26.417 gallons	2.75 bushels 22.00 gallons
Dekaliter	10	10 cubic decimeters	1.135 pecks 2.6417 gallons	8.80 quarts 2.200 gallons
Liter	1	1 cubic decimeter	0.908 dry quart 1.0567 liquid quarts	0.880 quart
Deciliter	.1	0.1 cubic decimeter	6.1023 cubic inches 0.845 gill	0.704 gill
Centiliter	.01	10 cubic centimeters	0.6102 cubic inch 0.338 fluid ounce	0.352 fluid ounce
Milliliter	.001	1 cubic centimeter	0.061 cubic inch 0.271 fluid dram	0.284 fluid dram

Common Measures and Their Metric Equivalents

Common Measure	Equivalent
Inch	2.54 centimeters
Foot	0.3048 meter
Yard	0.9144 meter
Rod	5.029 meters
Mile	1.6093 kilometers
Square inch	6.452 square centimeters
Square foot	0.0929 square meter
Square yard	0.836 square meter
Square rod	25.29 square meters
Acre	0.4047 hectare
Square mile	259 hectares
Cubic inch	16.39 cubic centimeters
Cubic foot	0.0283 cubic meter
Cubic yard	0.7646 cubic meter
Cord	3.625 steres
Liquid quart, United States	0.9463 liter
Dry quart, United States	1.101 liters
Quart, imperial	1.136 liters
Gallon, United States	3.785 liters
Gallon, imperial	4.546 liters
Peck, United States	8.810 liters
Peck, imperial	9.092 liters
Bushel, United States	35.24 liters
Bushel, imperial	36.37 liters
Ounce, avoirdupois	28.35 grams
Pound, avoirdupois	0.4536 kilogram
Ton, long	1.0160 metric tons
Ton, short	0.9072 metric ton
Grain	0.0648 gram
Ounce, troy	31.103 grams
Pound, troy	0.3732 kilogram

Modifiers are words or groups of words that describe or limit other words. Modifiers include adjectives, adverbs, prepositional phrases, nouns used as adjectives, and clauses that function as adjectives or adverbs:

> The **entire** proposal had **excellent** graphics. (*adjectives*)
>
> The manager **eventually** explained the reasons for his disapproval. (*adverb*)
>
> The pump **next to the intake line** was serviced last month. (*prepositional phrase*)
>
> The Sky Aviation proposal, **which scored second in technical merit**, had some interesting innovations. (*adjectival clause*)
>
> The ventilation fan was replaced **because its peak circulation volume fell short of our needs**. (*adverbial clause*)

See ADJECTIVES, ADVERBS, NOUNS, PREPOSITIONS, and CONJUNCTIONS.

Modifiers

1. Ensure that modifiers, particularly those expressing action, have a clear noun to modify.
2. Ensure that modifiers appear either next to or as close as possible to the word or words modified.

Writers and editors usually depend on their ears to tell them where a modifier should appear in a sentence. Essentially, however, modifiers should be as close as possible to the words they modify. If they aren't, readers might misinterpret the sentence. The most common sentence problems associated with modifiers result from dangling or misplaced modifiers.

Dangling

Modifiers dangle if they do not seem to be related to anything in the sentence or if they are not placed near enough to the words they modify to seem attached to those words. Modifiers dangle when they float, unattached, in a sentence.

Dangling modifiers can be adjectives, adverbs, prepositional phrases, infinitive verbs, appositives, or clauses. Quite often, dangling modifiers are participial phrases, usually beginning with a present participle (such as *knowing*):

> Knowing that standard 3/4-inch pipe was too small, the specifications included provisions for larger pipe.

The phrase beginning with *knowing* seems to modify the noun *specifications*, but clearly, specifications cannot know anything. The phrase must modify a human being to make logical sense, but no humans are mentioned in the sentence, so the modifier dangles.

Whenever you open a sentence with an action stated with an *–ing* verb (present participle) or *–ed* verb (past participle) and do not follow it with the name of the person doing the action, you will have a dangling modifier (sometimes called a dangling participle):

> *this*
>
> After discussing interest-rate trends, we decided to refinance our present loan.
>
> *not this dangling modifier*
>
> After discussing interest-rate trends, the decision was made to refinance our present loan.

> *this*
>
> Having analyzed the technical problems, she recommended routing the feed-forward signal through a broadband transmitter.
>
> *or this*
>
> An analysis of the technical problems led researchers to suggest routing the feed-forward signal through a broadband transmitter.
>
> *not this dangling modifier*
>
> Having analyzed the technical problems, the recommendation was to route the feed-forward signal through a broadband transmitter.

1. Ensure that modifiers, particularly those expressing action, have a clear noun to modify:

> *this*
>
> Having missed our connecting flight, we discovered that no later flights were going to Albuquerque.
>
> *not this dangling modifier*
>
> Having missed our connecting flight, no flights later that day were going to Albuquerque. (*Who missed the flight?*)

> *this*
>
> While reviewing the figures, we discovered many errors.
>
> *not this dangling modifier*
>
> While reviewing the figures, many errors became apparent. (*Who reviewed the figures?*)
>
> *nor this dangling modifier*
>
> Many errors became apparent while reviewing the figures.

Modifiers

NOTE 1: Dangling modifiers do not necessarily introduce the sentence; they can appear anywhere:

> The report was inaccurate, comparing it with the prior ones. (*Who compared it?*)

NOTE 2: Some introductory participles (usually ending in *–ing*) have become so common that they do not require clear words to modify:

> Considering your reluctance, you should not represent us before the Texas Railway Commission.
>
> Judging from the revised figures, the report will never be approved.

Misplaced Modifiers

Modifiers are misplaced when they do not appear in their customary place in a sentence. Readers often misread sentences in which the modifiers are misplaced:

this

The Human Resources manager told Hughes that the company no longer needed him.

not this misplaced modifier

Hughes was told that he was no longer needed by the Human Resources manager.

---

this

The factory reopened on March 1 after a 3-month lockout of employees.

not this

The factory reopened after a 3-month lockout of employees on March 1.

In both of these examples, the writer might not have intended what the "better" versions say, but the "better" versions are much clearer. Your goal as a writer should be to write so that you cannot be misunderstood. One way to achieve this goal is to ensure that modifiers appear where they should.

2. Ensure that modifiers appear either next to or as close as possible to the word or words modified:

this

The manager was interested only in production data.

not this misplaced modifier

The manager only was interested in production data. (*Does* only *modify* manager *or* data*?*)

See ADVERBS.

The book on the shelf with all the samples is our only copy. (*Does the book or the shelf contain the samples?*)

The report on osteoarthritic outpatients that our team studied was as up to date as possible. (*Did the team study the report or the outpatients? The clause* that our team studied *should appear immediately after* report *or immediately after* outpatients.)

---

this

A computer program for estimating gradients has been written for the mainframe IBM in the Production Department.

not this

A computer program has been written for calculating estimates of the gradients on the mainframe IBM in the Production Department.

---

this

We are separately shipping the faulty circuit board.

or this

The circuit board that failed is being shipped separately.

not this

We are shipping the circuit board that failed separately. (*Was it okay until you shipped it separately?*)

NOTE 1: If a modifier refers to two nouns, it should appear with the first noun mentioned:

this

The land is rocky on the west side of the allotment and somewhat less rocky on the east side.

not this

The land is rocky on the west side and somewhat less rocky on the east side of the allotment.

NOTE 2: Unmodified nouns might need an article (*a, an,* or *the*) or an adjective to clarify their meaning:

this

The secretary and the treasurer attended our meeting.

not this

The secretary and treasurer attended our meeting. (*Is* the secretary and treasurer *one person or are they two people?*)

Nouns

Nouns signify persons, places, things, and ideas. Even more important, nouns are the main words in a variety of noun phrases:

a **bottle**
the comprehensive **report**
a slowly changing **pattern**
some **tomatoes** for lunch
the young **engineer** who works next door

Noun phrases, in turn, become key building blocks in the English sentence. Within a sentence, a noun phrase can be a subject, an object, or a complement:

The proposed electrical changes will be expensive. *(subject)*

The engineer designed **two holding ponds**. *(object)*

The applicant was **the person who was busily filling out forms**. *(complement)*

Noun phrases can also complete a prepositional phrase by becoming the object of the preposition:

near **the fuel storage tank**
beyond **the property line**
at **the amount we requested**

Nouns

1. Distinguish between count and non-count nouns.
2. Use collective nouns and the names of companies as either singular or plural.
3. Distinguish between common and proper nouns.

1. Distinguish between count and non-count nouns.

Count nouns can be counted individually—*people, posts, ducks, pencils*. Non-count nouns cannot be counted individually—*honey, gasoline, air, hospitality*. See Articles.

Count nouns can be either singular or plural:

a bottle/two bottles
every desk/six desks
neither proposal/three proposals
each pump/30 pumps
either ox/five oxen

Non-count nouns are collective in meaning, not singular or plural:

furniture/some furniture
meat/most meat
wheat/less wheat
hospitality/more hospitality
warmth/some warmth

See Plurals and Agreement.

Count and non-count nouns accept different modifying words:

count nouns

fewer animals (*not* less animals)
three fewer pumps (*not* less pumps)
fewer gallons of gasoline (*not* less gallons)

non-count nouns

greater warmth (*not* seven warmths)
less gasoline (*not* fewer gasolines)
more hospitality (*not* three hospitalities)

Native speakers of English usually choose the proper modifying words unconsciously. Only occasionally do they make mistakes: *The zoo had less animals than we expected.* Because animals can be counted, the proper modifier is *fewer*: *The zoo had fewer animals than we expected.*

Many nouns can belong to either type, but their meanings change:

She's had many odd experiences. *(count)*
This job requires experience. *(non-count)*
The talks will take place in Cairo. *(count)*
He dislikes idle talk. *(non-count)*

2. Use collective nouns and the names of companies as either singular or plural.

Collective nouns are nouns that signify groups of people or things: *staff, team, family, committee, majority, crew, squad*, etc.:

The committee has met, and it has rejected the amendment. *(singular)*

The committee have met, and they have rejected the amendment. *(plural)*

The majority has made its viewpoint clear to the candidate. *(singular)*

The majority have made their viewpoint clear to the candidate. *(plural)*

Company names are similar to collective nouns because both can be either singular or plural:

Sky Aviation has submitted its proposal. *(singular)*

Sky Aviation had their annual report sent to all stockholders. *(plural)*

See Agreement and Articles.

3. Distinguish between common and proper nouns:

common nouns

a company	the professor
three lines	the avenue
some paper	a river
an idea	our dentist

proper nouns

Acme Glass Company
Professor Thomas Miles
Second Avenue
the Mississippi River
Dr. John Wray

Note: Proper nouns are capitalized, and common nouns are not. For more information, see Capitals, Titles, and Adjectives.

Numbering Systems

Numbering systems are used with outlines, tables of contents, and headings to display a document's organization and allow readers easy access to parts of the document. The two basic numbering systems are the traditional outline system and the decimal system. See Outlines and Tables of Contents.

Numbering Systems

1. Use the decimal system for most outlines, especially if your outline will be the basis for headings in a longer document.
2. Use no more than four or five numbered subheadings in documents.

Traditional System

Traditional outlines use the following numbering and lettering conventions:

```
I.
  A.
    1.
      a.
        (1)
          (a)
            1)
              a)
```

Decimal System

In the decimal system, successive dots or periods indicate levels of subordination:

```
1.
    1.1
            1.1.1
            1.1.2
            1.1.3
                    1.1.3.1
                    1.1.3.2
2.
    2.1
    2.2
            2.2.1
            2.2.2
```

Note: The title "Decimal System" is a handy label, but it is not quite accurate. This numbering system does not use mathematical decimals.

To prevent confusion with traditional decimals, some publications use a raised dot: 1·1·1 or 2·2

1. Use the decimal system for most outlines, especially if your outline will be the basis for headings in a longer document.

The decimal system is preferable in lengthy documents with a multitude of numbered subsections. For example, a reader seeing heading number 6.3.5 knows that the heading is in chapter 6, subsection 3, and additional subsection 5. Traditional outlines would only have a 5, which is not a clear indicator of chapter or subsection.

2. Use no more than four or five numbered subheadings in documents.

Even four or five levels is often confusing. Additional subheadings can be unnumbered and with format options:

Soil Types. In the Lornack Allotment, most soils are . . .

Numbering Systems and Punctuation

In the traditional system, a period or a single right parentheses always follows the outline number or letter:

```
I. First-order heading
  A. Second-order heading
    1. Third-order heading
      a. Fourth-order heading
        (1) Fifth-order heading
          (a) Sixth-order heading
            1) Seventh-order
               heading
              a) Eighth-order
                 heading
```

In the decimal system, two or more spaces follow the number. No punctuation is used after the number:

```
1.   First-order heading
     1.1 Second-order heading
     1.1.1 Third-order heading
     1.1.2.1 Fourth-order heading
     1.1.2.1.1 Fifth-order heading
     1.1.2.1.1.1 Sixth-order heading
```

or

```
1.   First-order heading
     1.1 Second-order heading
      1.1.1 Third-order heading
       1.1.2.1 Fourth-order heading
        1.1.2.1.1 Fifth-order heading
         1.1.2.1.1.1 Sixth-order heading
```

See Headings and Lists.

Numbers can be written out or can appear as figures, depending on the size of the number, what it stands for, and how exact it is. The stylistic conventions for number usage vary, so you will find conflicting suggestions from one dictionary or style guide to the next. The recommendations that follow are based on the current standard practice for technical and scientific writing.

Numbers

1. Use numerical figures for any number expressing time, measurement, or money.
2. Write out numbers if they are below 10. If they are 10 or above, use figures.
3. Write out numbers that begin a sentence.
4. Rewrite sentences beginning with a very large number.
5. Use figures to express approximations that are based on experience, evidence, or both.
6. Write out approximations that are obvious exaggerations for effect.
7. Use a combination of letters and figures for very large round numbers (1 million or greater).
8. Be consistent.
9. Use figures for quantities containing both whole numbers and fractions.
10. Always use figures for percentages and decimal fractions.
11. Always use figures for dates.
12. Form the plural of a number expressed as a figure by adding a lowercase *–s*.
13. Use a comma to separate groups of three digits

1. Use numerical figures for any number expressing time of day, measurement, or money:

3 a.m.
$15
45 ft
1 in.
8 cm
34.17 m

Measurement includes length, weight, volume, velocity, and other units of measurement.

Figures are easier to read and are remembered more accurately and longer than their spelled-out versions. Rules 2 through 12, which follow, present instances where rule 1 may not apply, however.

2. Write out numbers if they are below 10. If they are 10 or above, use figures:

five systems
15 systems
three mission capabilities
14 mission capabilities
two technicians
22 technicians

Note 1: Regardless of the number's size, use figures if they are followed by a unit of measurement (see rule 1):

5 pounds
2 yards
1 kilometer

Note 2: In nontechnical writing, writers often write out numbers less than 100—for example, *thirty-five, seventy-one, eighty-nine*. Note the hyphens in these written-out forms. See Hyphens. Writing out numbers less than 100 avoids overemphasizing double-digit numbers in nontechnical documents, which typically contain few numbers.

3. Write out numbers that begin a sentence:

Thirty-three patients were involved in the double-blind tests.

Twelve inches from the centerline are two slots for plate fins.

Four years ago, we initiated an IR&D study of argon-atmosphere braze furnaces.

The last two examples do not use figures even though they are followed by units of measurement. Rule 3 overrules rule 1.

4. Rewrite sentences beginning with a very large number:

not this

363 times a second the oscillator receives a signal from the bit generator.

nor this

Three hundred sixty-three times a second the oscillator receives a signal from the bit generator.

this

Every second, the oscillator receives 363 signals from the bit generator.

5. Use figures to express approximations that are based on experience, evidence, or both:

about 3,000 samples
approximately 60 applicants
roughly 2 cubic feet per second
over 3 million orders this quarter

Numbers

NOTE: While some editors would retain the written out forms in the above examples, others (notably the *U.S. Government Printing Office Style Manual)* prefer numerals with such words as *nearly, about, around,* and *approximately.* Use your judgment. Figures convey a greater sense of precision than words. Thus, figures may seem to contradict the idea of approximating.

6. Write out approximations that are obvious exaggerations for effect:

That computer is not worth two cents.

The boss received a thousand telephone calls today.

His mother told him a million times to clean up that mess.

7. Use a combination of letters and figures for very large round numbers (1 million or greater):

We have invested over $45 million on laser research in the last 5 years.

Our annual IR&D budget exceeds $16 million.

8. Be consistent.

Treat numbers of the same type equally within a sentence, paragraph, or section. However, **never** begin a sentence with a figure:

this

Unit A will require 5 outlets; Unit B, 17 outlets; Unit C, 9 outlets; and Unit D, 14 outlets.

not this

Unit A will require five outlets; Unit B, 17 outlets; Unit C, nine outlets; and Unit D, 14 outlets.

---

this

Seven of the stations carry 39 spare controllers. The other 14 stations carry only 8 spares.

not this

7 of the stations carry 39 spare controllers. The other 14 stations carry only eight spares.

nor this

Seven of the stations carry thirty-nine spare controllers. The other fourteen stations carry only eight spares.

nor this

Seven of the stations carry 39 spare controllers. The other fourteen stations carry only 8 spares.

The sentence cannot begin with a figure, so *seven* must be written out. The *14 stations* uses figures because 14 is greater than 9; so the two references to *stations* cannot be consistent. The number *39* is too large to write out, so both of the numbers referring to spare controllers are written as figures, although *8* is less than *10.*

9. Use figures for quantities containing both whole numbers and fractions:

The proposal calls for 8H- by 11-inch paper.

See FRACTIONS.

10. Always use figures for percentages and decimal fractions:

The rectangular fins are 0.07 in. high.

The maximum core diameter is 2.54 mm.

The tests require an 8 percent solution.

NOTE: In the last example, *8%* would also be acceptable, although many style guides prefer that writers use the percent sign only in tables and visual aids. In accounting and other financial documents, the percent sign is common in text. See DECIMALS and SIGNS AND SYMBOLS.

11. Always use figures for dates:

June 14, 2010
14 June 2010
the 14th of June 2010
June 2010

NOTE: If you use the preferred U.S. style (month-day-year, as shown in the first example above), always separate the day and year with a comma. The second example shows the alternate style: day-month-year, with no punctuation.

The third example is wordy but still acceptable in some contexts, such as in legal documents.

If you write only month and year (as in the last example above), use no punctuation. Separating the month and the year is unnecessary. See COMMAS, PUNCTUATION, and LETTERS.

12. Form the plural of a number expressed as a figure by adding a lowercase *–s:*

before the 1970s
temperatures well into the 100s
the 5s represent actual strikes

NOTE: Plurals of numbers written out are formed like the plurals of other words:

in the twenties
groups of threes or fours

See PLURALS.

13. Use a comma to separate groups of three digits:

55,344,500
10,001
9,999
678

NOTE 1: In some technical fields, the preferred style is to omit the comma separating digits in numbers only four digits long:

5600
9999

NOTE 2: A practice outside the United States is to use a space instead of a comma to separate groups of three digits:

7 143
98 072.1
1 742 600 503

See METRICS.

Online Documentation

Online documentation includes any information designed to appear on an electronic screen, from a stadium-wide flat screen to a pocket-size mobile phone screen. Designing how each screen will appear and deciding how multiple screens will link with each other are key design issues.

Online Documentation

1. Decide on the best online medium for your message.
2. Survey or profile potential online users as to how they will likely use the online documentation.
3. Create a map of your entire document and its major sections.
4. Design your screens so that content items are clear, consistent, and predictable.
5. Break your content into manageable and readable chunks.
6. Use graphics, but don't overdo them.
7. Prepare an online index for documents that are complex and lengthy.
8. Use a writing style that suits online users.
9. Keep screen size in mind.

Designing for a screen is a visual problem. Elsewhere in this *Style Guide*, we emphasize that document design (visualization) is a key step in the writing process. Design is even more important when information is to appear on a computer monitor. Screen designers have to choose icons, fonts, margins, frames, colors, and other visual tools. Many of these tools are similar to those used in the design of a page of printed text. See PAGE LAYOUT.

Electronic screens are not the same as sheets of paper or a printed document. Although screen resolution is improving all the time, block of text on a screen is often more difficult to read than the same text printed on a page. Also, online documents exist in cyberspace. Unlike the pages of a book, Web pages are not sequentially numbered and can be accessed in any order.

Furthermore, an online document is rarely truly finished and might need constant updating. In the case of open-ended websites such as Wikipedia, virtually anyone can comment on or even change the content at will. More and more, online documents are actually conversations. With the explosion of social media, this trend will increase. See ELECTRONIC MAIL.

Every day, more business and technical professionals are designing online information in the form of websites, blogs, and videos. The following rules are only general suggestions for designing these media. If you need more detailed guidance, you will find a wealth of useful "dos and don'ts" in a quick search of the Web.

1. Decide on the best online medium for your message.

Business, consumer, and technical users of the Internet are usually looking for quick solutions, and the faster your message fills their need, the better. Regardless of how beautiful or flashy, your online document fails if users can't get what they want from it. Choose your medium and design based on user needs.

For example, users learning a complex process will benefit from a carefully designed website rich with illustrations. On the other hand, a user who wants to do a simple task will prefer a short video demonstration. An academic research audience will expect abundant links to sources, while a shopper looking for a certain pair of shoes just wants to see a picture with a "buy" button next to it.

2. Survey or profile potential online users as to how they will likely use the online documentation.

A survey means that you go directly to users and ask them questions such as those listed below. A profile, in contrast, means that you use existing information to compile probable answers to the same list of questions.

Whether you use an actual survey or existing information, you should have answers to the following questions before you begin to design online documentation.

—Who are the users? Their job titles? Their experience? Their knowledge?

—What hardware will be available for each user? What applications? What connectivity?

—When and how often are users likely to want to access the information?

—What information are they most interested in finding?

—What biases or assumptions will they bring to the task of finding information?

—What provisions are there for updating the information?

—Will users want to print the information? Do you want them to be able to do that?

Answers to these questions will enable you to design a usable document. Without these answers, you will probably end up frustrating users. For example, one of the most common complaints of website users is their inability to compare the prices of a product or service among suppliers. You must carefully weigh the possibility of driving frustrated customers away against the disadvantage of revealing your prices to competitors.

3. Create a map of your entire document and its major sections.

Too often designers spend more effort on individual Web pages than on the design of the whole website. Another trap is to get hung up on colors, font sizes, and other details before getting clear on bigger issues: the structure of the website, the main sections, titles, and relationship to other websites. Developing an overall map of your site or document should be one of your earliest tasks (following your survey of users, as described in rule 2).

Usually, the map consists of a "wireframe," a prototype that contains all the key features of the website, including headings, menus, graphics, and application interfaces.

Figure 1. Two Wireframe Examples. *The designers of a mobile-phone career-planning app conceived these wireframes as alternatives for a site map.*

Prototyping saves time and money because it allows for early reviews of the website concept. Circulating the wireframe for review brings to the surface issues you could not by yourself anticipate. Get reviews from as many interested parties as possible, including prospective users of the website. See WRITING AND REVISING for guidelines on how to prototype a document and conduct reviews.

Reviewers of the wireframes in figure 1, for example, will raise issues about the advantages of one alternative design over another. They will test the concepts with potential users. They might come up with a third alternative and then test it in the same way.

As you design the wireframe, keep in mind that users want to find things with as few clicks as possible. This is a guiding principle for online documentation: the more clicks, the more likely users will abandon your site.

Visually display your chosen option in a site map that will make navigation simple and quick. Include an overall navigation menu bar on each screen of an online document. Figure 1 shows wireframe examples of site maps and navigation bars.

As the main purpose of designing a wireframe is to get input from reviewers, include explanatory information in the wireframe where necessary. In figure 2, the designers insert callouts to explain the features of the wireframe.

4. Design your screens so that content items are clear, consistent, and predictable.

Clarity arises from making sure that you don't overload a screen or Web page with too much information and too much text. Beyond a certain point, users will be overwhelmed

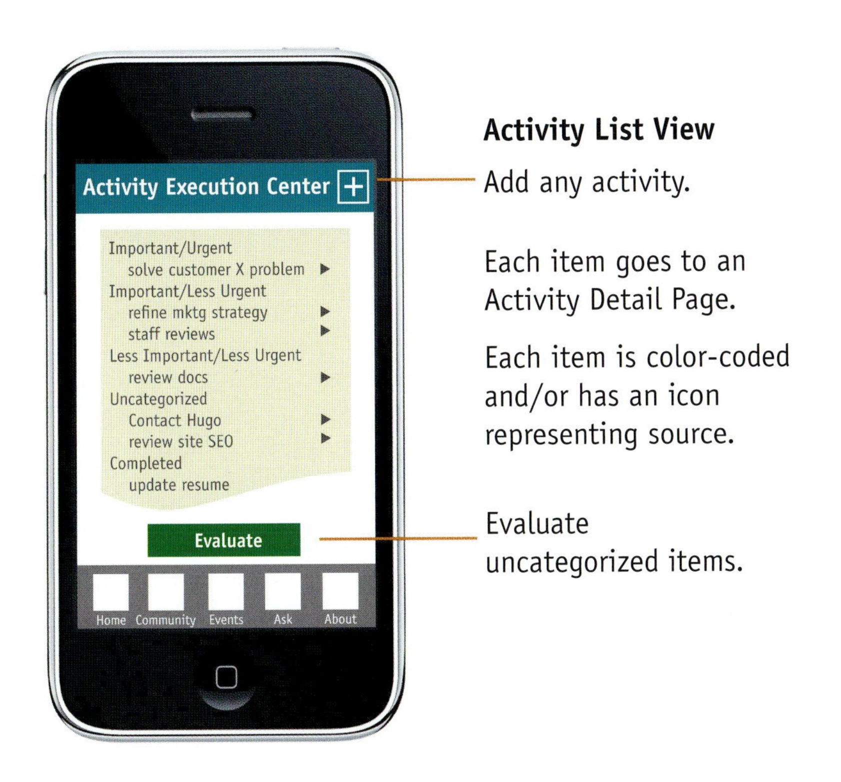

Figure 2. Wireframe With Explanatory Callouts. *The text on the side explains for reviewers the purpose or function of the app features.*

if you try to pack too much information into a single screen.

Font size should be no smaller than 9 or 10 points, and be sure to choose a proportionally spaced typeface (and one that is not too ornate so as to be difficult to read). For maximum ease of use, column width should generally not exceed 600–800 pixels. This will avoid forcing users to read wide swaths of text or to scroll sideways. Typically, single spacing is too tight for onscreen text, so choose a wider leading (spacing between lines). Be sure to leave an extra line between paragraphs and provide generous margins so that text doesn't crowd the edges of your screen. See Page Layout.

Once you have arrived at a screen design, show it to users. Ask them if they find it readable and accessible and then make adjustments. This survey, actually a mini usability test, should be part of any documentation process (either online or printed).

Consistency means that each screen has the same items in the same place and in the same order. Perhaps you have developed a standard set of icons. These should appear throughout in the same format and in the same spatial position on the screen.

For example, assume that you want to provide automatic safety warnings within an online maintenance manual. These warnings should always appear on the screen with the same format (icons, font size, typeface, color, margins, etc.).

If you are clear and consistent, you will be predictable. The more predictable your document, the more usable (readable) it becomes.

Predictability also means that your users will not get lost in cyberspace. Users should be able to tell from each screen exactly where in the program they are. A predictable screen design will have enough road signs (menu bars, icons, headings, cross references) so that users are comfortable navigating through the program and can quickly start over again if they choose.

5. Break your content into manageable and readable chunks.

Make your text as schematic and as visually accessible as possible. This rule applies whether you are adapting printed text or writing an online document from scratch.

Traditional text is too dense and difficult to read online. It has long paragraphs and too few headings, lists, and tables. See figure 3 for an example of how traditional text can be revised to become more manageable and readable. See Paragraphs and Tables.

Textual chunks should be designed to fit within the screen format you have developed (see rule 3). The format of the online document should govern how you decide to chunk or format text. The most efficient way to create new text is to write to fit the format, not to write text and then have to rewrite it to fit the format. See Emphasis for a list of techniques for emphasizing information for maximum usability. Most of the techniques mentioned apply equally well to online documents as to printed documents.

6. Use graphics, but don't overdo them.

Graphics aid users (readers) to interpret and remember key information. So plan to use graphics creatively to capture key information, to aid navigation, and to highlight major features.

Employees who are on travel assignments should be classified for reporting in the following ways. Those who are on assignments for at least one month but not more than six months are classified as Extended Travel. Temporary Change of Station applies to those offsite for six months but for less than one year, while those gone for between one and two years should be categorized as on Indefinite Assignment.

Classify employees on travel assignments in these ways:

Classification	For Employees on Assignment
1. Extended Travel	At least one month but not more than six months
2. Temporary Change of Station	Six months but less than one year
3. Indefinite Assignment	At least one year but not more than two years

Figure 3. Contrast Between Two Screens. *The original (top) screen comes directly from the paper copy with no editing. The revision (bottom) is more readable, whether on paper or on a screen.*

Remember, however, that graphics will not save content that is not otherwise solidly conceived and well prepared. Also, graphics are time-consuming in two ways: they take time to create, and they often use a lot of memory.

Use flashy, memory-intensive graphics only when absolutely necessary to your purpose. Too often users have to wait for large images to download, only to find that the images are irrelevant. Make sure you compress memory-heavy graphics to avoid slowing the download.

Avoid elaborate backgrounds, frames, Flash introductions, and busy animations that delay and distract business or technical users. Google.com has been a traditional standard for simple, highly usable website design.

Online documents can include any of the common graphics: charts, graphs, illustrations, maps, photographs, and tables (see the separate discussions of these graphics). As with graphics for documents and for presentations, graphics for an online document need to be well designed and appropriate. See Graphics for Documents and Graphics for Presentations.

Color is one feature of graphics to use sparingly in an online version. Some colored features are now conventional; for example, links generally appear in blue text and change color when they have been clicked. But too many colors can be a distraction. Contrast helps—light text on a dark background is easier to read than on a light background. Avoid color combinations that make text illegible, such as red text on a blue background. Also, different monitors convey colors differently, and some users may be color blind. If you do use color, add a redundant feature that does not depend on color alone. For example, if you decide to use a red message box, you might add a separate icon or a distinctive border that does not depend on color to convey its message. A link in blue text might also be underlined. For more guidance, see Color.

7. Prepare an online index for documents that are complex and lengthy.

The terms "complex" and "lengthy" are necessarily subjective, but you should consider preparing an online index if you know that users of your online documentation will often need to search for information. Also, your survey or profile of users (as discussed in rule 2) should indicate that they are not fully knowledgeable about your topic, so they will need assistance searching for key concepts or distinctions. One of the most common user complaints is difficulty in finding exactly what they want without wading through a vast sea of possibilities. An index can help.

The design features of online indexes are still developing and evolving, but two general observations are important starting points if you intend to prepare an online index.

First, an online index is not like a screen copy of the index for a printed document. In a document, the index topics are usually keyed to pages, but pages don't exist online and individual screens are not numbered in a traditional fashion. Also, the hierarchical arrangement within the traditional index will not match the sequence of screens in an online document.

Second, a traditional search tool does not replace a separately designed online index. In a search box, the user types a word or a phrase and then sees all uses of the phrase in the text. For example, a user of the online *Style Guide* who searches for the term *organization* will likely see 200 different contexts using that word. Such a lengthy list is not very helpful and is time-consuming to scan.

An online index is a separate subsection of your online document. Users, having accessed the index, can then either visually scroll through the index or, more efficiently, use the search tool on only the index to discover related information, synonyms, or other helpful links between sections of the document.

As an example of how the online index might work, a document on environmental issues might contain numerous uses of the following related terms (actually partial synonyms):

impacts
effects
consequences
irreversible/irretrievable commitments of resources
short-term uses vs. long-term productivity

In the online index, these terms would be linked so that users unfamiliar with one or more of these would be alerted to search the text for the other phrases, each of which is related to the simple terms *impacts* or *effects*. These links parallel, of course, the *See* and *Also See* functions of a traditional index, but what is different here is that the links help users navigate back and forth from the index to the document itself. See INDEXES.

8. Use a writing style that suits online users.

The tone of the Internet is far more informal than the tone of traditional business and technical communication. The Web audience is relatively young, tech-savvy, and casual about everything from business relationships to dress—and that includes language. The Web is a conversation, and it can be quirky.

Traditionally, a tip for website designers might sound like this:

"Users should not be required to scroll horizontally to access information."

In the language of the Internet, the same tip might sound like this:

"Don't make me (ack!) scroll sideways."

For online documentation, choose a tone of voice that fits the mood and purposes of your readers, but in general, you should lean toward a casual tone. You can create such a tone by using simpler sentences, shorter and more concrete words, and contractions.

Also, keep in mind that your Internet audience is potentially enormous and unpredictable. The Internet makes it possible for nonspecialists to access highly specialized documents—an ordinary patient, for example, can now study the clinical trial reports related to the medication she's taking, and she might want to!

Thus, you should adopt a plain style wherever possible to accommodate the needs of a wide range of users.

Plain language instead of jargon will also bring you more readers because people search the Internet using *their* words, not yours. See GOBBLEDYGOOK.

Finally, much weak online documentation is just text from old printed sources dropped into the Internet. It's usually better to adapt this kind of text for the medium or even start over.

9. Keep screen size in mind.

Screens now come in all sizes. There are giant flat-screen monitors, compact little notebook screens, and handheld devices. The rules above generally apply to any screen size, but you'll want to adjust your design for very large or very small screens.

Keep in mind that very large screens, when maximized, will require users to read wide columns of text. The effect on the reader will be like watching a tennis match. It's usually better to fix column width if readers will use large screens.

Designing for small mobile screens presents a different challenge. A document for a small screen should be simple, with few graphics and features. Break large pages of content into small chunks accessible through a menu of links. Enlarge links and search boxes to make them clearly visible. A page of text for a mobile device should probably not exceed 500 words.

Organization

Organization is the key writing principle. If you organize your documents well, you almost surely will have successful documents—even if you violate other writing principles. But if your documents are poorly organized, nothing can save them.

The ideas presented in a document should be structured in a natural but emphatic sequence that conveys the most important information to readers at the most critical times.

The principles of organization differ slightly from document to document, depending on the type of document, the readers, the content, and the writer's purpose. Nevertheless, logic and common sense dictate that a well-organized document must have certain features:

- The document should announce its organizational scheme and then stick to it.
- The ideas in the document must be clear and sensible, and comprehensible, given the readers.
- The document should conform to the readers' sense of what the most important points are and of how those points are arranged.

Letters, memos, and reports differ somewhat in their organizational patterns, mostly because their readers differ. See LETTERS, MEMOS, and REPORTS.

Readers of letters are typically outside the company or agency sending the letter. Their relationship to the writer is therefore more distant, and consequently more formal, than the relationship between the writer and others within the writer's company. See LETTERS.

Organization

1. Organize information according to your readers' needs.
2. Group similar ideas.
3. Place your most important ideas first.
4. Keep your setups short.
5. List items in descending order of importance.
6. In most business or technical documents, preview your most important ideas and your major content areas, and review (summarize) major points at the end of sections.
7. Discuss items in the same order in which you introduce them.
8. Use headings, transitions, key words, and paragraph openings to provide cues to the document's organization.
9. Use the Document Planner when you need to design a document quickly and efficiently.

Readers of emails, on the other hand, are typically from within the writer's company or agency and share various assumptions, experiences, and knowledge—all of which tend to make emails less formal than letters.

The distance and formality between writer and reader affect organization in several ways. The greater the distance, the more the need to set up (introduce and perhaps explain) the ideas in the document. The greater the distance, the greater the need to substantiate information that might be subject to differing interpretations. The more formal the document, the more the writer must consider format traditions and reader expectations in organizing material.

Reports, technical or otherwise, often have prescribed organizations. Scientific-report organization is based on a long tradition in the sciences. The organization of such reports is strictly prescribed, and writers have very few options in varying that organization. Technical (but nonscientific) reports offer somewhat more latitude, but some companies still have strict guidelines on organizing technical reports.

Within the limitations imposed by tradition, logic, and audience, writers must carefully consider how to arrange their ideas and supporting data so a document serves its purpose and satisfies the readers' needs.

The first eight rules listed below suggest how you can accomplish these tasks. Rule 9 introduces a basic organizational template, a document planner that shows how the preceding eight rules would apply to the design of documents. See LETTERS, MEMOS, REPORTS, and ELECTRONIC MAIL.

1. Organize information according to your readers' needs.

How you organize information depends on your readers. You might organize the same information differently for different readers, depending on their needs and your purpose in writing to them. Here, for instance, is the text of an email written to the test director of a laboratory:

> We request the following tests on the dry field cement samples that we shipped on July 20 to Mr. J. F. Springer of your laboratory:
>
> - Thickening time
> - Rheology
> - High temperature/high pressure fluid loss
> - 12- and 24-hr compressive strength
>
> Davidson-Warner, a cementing company, has been using this cement in our Mt. Hogan Field. On July 17, they experienced a cementing failure while setting a string of 3½-in. casing at 11,323 ft in our Hogan BB-62 well. They pumped 688 barrels of cement and 78 barrels of displacement fluid before halting displacement when the pressure increased to 5,000 psi.
>
> To facilitate your testing, we have attached pertinent well logs, cement data, and a copy of Davidson-Warner's laboratory blend test results. Please submit your findings to me at your earliest convenience.

This email begins, appropriately enough, with a request. The writer wants something of the reader. Establishing what the writer wants makes sense as an opening statement. The specific details concerning the cementing failure do not appear until the middle paragraph because this particular reader will not need to know this information except as background for conducting the tests. The details of the cementing failure are less important than a list of the tests the writer is requesting.

However, if the document had been written to the production engineer who will now be responsible for this well, it might have begun like this:

> The Hogan BB-62 is currently shut in because of a cementing failure that occurred on July 17. The regional office would like us to return this well to production by July 28.
>
> On July 15, this well was shut in to allow Davidson-Warner to set a new string of 3½-in. casing from 10,500 ft to 11,890 ft. While setting the string at 11,332 ft, they halted displacement when the pressure increased to 5,000 psi. Before stopping, they had pumped 688 barrels of cement and 78 barrels of displacement fluid. They left approximately 35 barrels of cement in the casing (with a cement top at 8,992 ft).
>
> Wiley Laboratories has been asked to test dry field samples of the cement. In the meantime, AGF Cement has been contracted to finish setting the string. They will be onsite no later than July 25. You should plan to be present.
>
> Mt. Hogan Field production figures are down 4.3 percent in July, primarily due to this cementing failure. The regional production manager has asked that we resume full production by July 28. If you need assistance, call me at 555-6666.

This email is written from supervisor to subordinate. Its tone is obviously different (more forceful, more directive) than the letter written to the laboratory. The organization of ideas is also very different.

The email to the engineer begins with a statement of fact (a setup), followed by a deadline. As in the first email, the details of the cementing failure appear in the middle, but in this second example, the details lead to an amplification of the implied directive that appears in the opening paragraph. The email closes with a compelling reason for action (production figures down) and a reminder of the deadline.

As you organize a document, always consider what information your readers need from you. In the examples above, the test director at Wiley Laboratories will not care that Davidson-Warner left 35 barrels of cement in the casing. The engineer will not care that the dry field samples were shipped to Mr. Springer. Each document above reflects those concerns that its readers will care most about.

2. Group similar ideas.

Separating similar ideas creates chaos. In the examples above, the details concerning the cementing failure appear in the same place. If they had been scattered, the effect could have been baffling for readers:

> The Hogan BB-62 is currently shut in because of a cementing failure that occurred on July 17. Wiley Laboratories has been asked to test dry field samples of the cement.
>
> On July 15, this well was shut in to allow Davidson-Warner to set a new string of 3½-in. casing from 10,500 ft to 11,890 ft. Please try to return this well to production by July 28. AGF Cement has been contracted to finish setting the string. Before stopping, Davidson-Warner had pumped 688 barrels of cement and 78 barrels of displacement fluid. AGF Cement will be onsite no later than July 25.

3. Place your most important ideas first.

A frequent problem with business and technical writing is the tendency to lead **to**, rather than **from**, major ideas. Many writers believe that they have to build their case, that skeptical readers will not agree with their conclusions unless they first demonstrate how they arrived at those conclusions. This tendency results in documents that are unemphatic, difficult to follow, and filled with unnecessary detail.

Organization

The strongest part of a document is its beginning. Readers typically pay more attention at the beginning because they are discovering what the document is about. The beginning, then, is the most emphatic part of the document by virtue of its position. Because the beginning is so strong, you should begin with the most important ideas in the document—then support those ideas by presenting your evidence afterward.

The Scientific Format. Many of those writers who tend to lead down to their major ideas have been schooled in the scientific method. According to the scientific method, one presents the facts, observations, and data that lead to and support a conclusion. The strength of this method is that it presents a series of steps that culminates in an **inevitable** conclusion. Therefore, the steps are as important as the conclusion.

In some scientific reports (notably those written from one scientist to another), an organizational scheme based on the scientific method is desirable:

Abstract
Summary
Introduction
Materials and Methods
Results and Discussion
  Fact 1
  Fact 2
  Fact 3
  Fact 4
  (therefore)
Conclusions
Recommendations (optional)
Summary (optional)

This format is acceptable only if readers will be as interested in the process of arriving at the conclusions as they are in the conclusions themselves. When readers are more interested in the conclusions, follow the managerial format.

The Managerial Format. You should follow the managerial format in all documents except scientific documents written for scientific peers.

The managerial format is the reverse of the scientific format. Managers (and most other nonscientific readers) are far more interested in the conclusions than they are in the steps leading to them. This is not to say that these readers will not want to see the conclusions supported—only that they will want the conclusions before the results and discussion:

Summary/Executive Summary
Introduction
Conclusions (and Recommendations)
(because of)
  Fact 1
  Fact 2
  Fact 3
  Fact 4
Results and Discussion

Having the conclusions early in the report facilitates reading because the reader is given a perspective from which to understand the facts and data being presented. Furthermore, busy managers often know the background and tests that have led to the conclusions. See REPORTS.

NOTE 1: The principle of emphasis through placement extends to all documents and all sections of documents. Your most important ideas should appear at the beginning of your documents and of individual sections. The most important idea in most paragraphs should appear in the opening sentence. The most important words in a sentence typically come at the beginning of the sentence. See PARAGRAPHS and SENTENCES.

NOTE 2: A corollary to note 1 is that you should always subordinate detail. Place it in the middle of sentences, paragraphs, sections, and documents. Detail includes data, explanation, elaboration, description, analyses, results, etc.

NOTE 3: In lengthy documents, begin **and** end with important ideas.

The longer a document becomes, the more crucial this rule is. Readers of long passages need to be introduced to the subject, learn the most important points early, receive the supporting detail and explanation, and then have it all wrapped up in a tidy closing statement that reiterates the important points.

An adage regarding oral presentations (but applicable to writing) is that you should "tell 'em what you're gonna tell 'em, tell 'em, and then tell 'em what you told 'em." See REPETITION, REPORTS, and EMPHASIS.

4. Keep your setups short.

Sometimes you cannot begin by stating your most important idea because the reader either will not understand it or will not accept it. If such is the case, you need to set up the most important idea by providing introductory information meant either to inform readers or to persuade them.

A fundamental of organization in business and technical writing is to keep your setups short. Do not delay your major ideas any longer than necessary.

When you give people positive information, you should give them the positive information right away. They want to hear it, and hearing it will make them more receptive toward you and the rest of the information you provide.

However, when you give readers negative information, giving them the negative information first might put them off, and they will not be receptive to what follows. Moreover, they might become antagonistic toward you.

Therefore, you should say no to readers only after you have set them up for it. Be careful, however, not to delay the "no" too long. Keep your setups short, as in the following example:

> I have been asked to reply to your request for additional compensation following approval of your Engineering Change Order dated March 3.
>
> As you know, a Health Department inspector ordered the design changes, and our contract states that all design changes required for safety reasons are warranted under the contractor's bond. Therefore, additional compensation would be inappropriate at this time.

The first sentence sets the stage. The second provides brief rationale for the decision. The third states the decision. The two-sentence setup in this example makes the decision more palatable than if the writer had begun by saying: "*We will not be providing the additional compensation you requested.*" See INTRODUCTIONS.

5. List items in descending order of importance.

Readers typically assume that information in lists appears in descending order of importance: most important listed item first, least important item last. Numbering and lettering systems reinforce this assumption. We all know that being number 1 is better than being number 6. We know from school that an A is better than an F. Rightly or wrongly, we assume a natural ranking of items. Therefore, writers should list items in descending order of importance.

If you wish to create a list in which items are equally important, use bullets or dashes instead of numbers or letters, and state that the listed items are equal. See LISTS.

6. In most business or technical documents, preview your most important ideas and your major content areas, and review (summarize) major points at the end of sections.

In most business and technical documents, you must establish the structural framework of the documents. Opening previews and concluding reviews are essential if you want readers to grasp your major points. See INTRODUCTIONS, SUMMARIES, REPORTS, and REPETITION.

Even mechanical or routine features of a document can help readers understand its content. For example, a detailed, quite specific table of contents can almost be a summary of the document. In some special types of documents, the table of contents is supplemented by a matrix outlining where in the document each requirement or issue is addressed. Such a matrix is especially helpful in proposals, where the writer must respond to every one of the client's requirements. See TABLES OF CONTENTS.

Summaries and introductions are ideal devices for previewing content, but you can also preview content in opening paragraphs. Generally, however, when the preview refers to itself as a preview, it is obtrusive. Your preview should sound natural and should be unobtrusive:

this

> The Hamerling Study (March–October 2009) found that predators have played only a minor role in the recent population decline of the cutthroat trout. Far more serious impacts on this species are (1) a degraded watershed, (2) temperature increases, and (3) deforestation.
>
> Together, these environmental changes have reshaped the cutthroat trout's habitat, perhaps beyond the species' ability to adapt.

not this

> This report discusses the results of the Hamerling Study (March–October 2009), which found that predators have played only a minor role in the recent population decline of the cutthroat trout. The first section concerns the quality of the watershed, which has declined significantly since 1965.
>
> Following that section is a discussion of the role of climate changes, particularly a 2-degree increase in temperature throughout the study area. In section 3, the report notes the effect of deforestation in one part of the study area. In its concluding section, the report discusses the combined impact of watershed degradation, climate changes, and deforestation. As the report notes, these changes have reshaped the cutthroat trout's habitat, perhaps beyond the species' ability to adapt.

The first (preferred) version amounts to a summary of the report. It could actually appear in the summary, become part of an abstract, or open the introduction. It could even appear in all three places.

7. Discuss items in the same order in which you introduce them.

When you introduce items, you should discuss them in the same order later. Saying that you are going to talk about A, B, and C, but then beginning with B violates the readers' sense of order. Follow these examples:

> The three greatest influences on cutthroat trout population are *a degraded watershed, temperature changes,* and *deforestation.*
>
> *The watershed* has been declining in quality since 1965 when . . .
>
> *Temperature changes* over the last 5 years have resulted in a 2-degree . . .
>
> *Deforestation* through the study area has also affected . . .

> *this*
>
> The acquisition improved our *cash flow* while providing significant tax advantages and allowing us to capitalize expenses. Prior to the takeover, we had negative *cash flow* on several . . .
>
> *not this*
>
> The acquisition improved our *cash flow* while providing significant tax advantages and allowing us to capitalize expenses. Prior to the takeover, our *expenses* were not capitalized. . . .

This second example demonstrates a subtle but important use of organization. The writer introduces three ideas: cash flow, taxes, and expenses. To be consistent with the order in which these ideas were introduced, the writer must follow the introductory statement with *cash flow,* not *expenses,* as occurs in the final version.

8. Use headings, transitions, key words, and paragraph openings to provide cues to the document's organization.

Throughout documents, you should signal organizational shifts or changes in direction by using headings, transitions, repeated key words, and opening or closing statements in paragraphs.

Headings are especially useful when you need to signal abrupt changes in direction, such as the transition from one topic to another (unrelated) topic. If the shifts are too radical, you cannot easily indicate them in text.

Transitions and repeated key words provide for smoother changes in direction and are useful between sentences and paragraphs, as in the example below. Note how the bolded words indicate organizational patterns and shifts in direction:

> The coal seam trends northwesterly for approximately 9,500 meters before pinching out on a fault line. **However**, seismic evidence suggests that **another** seam of coal extends from a point 75 meters downdip of the pinchout. This **second** seam appears to trend northerly for another 5,000 meters. **Together**, these seams represent a sizeable reserve of recoverable coal, **but** initiating mining operations will still be extremely **difficult**.
>
> The biggest **difficulty** is landowner resistance to strip mining. . . .

See HEADINGS, KEY WORDS, PARAGRAPHS, and TRANSITIONS.

9. Use the Document Planner™ when you need to design a document quickly and efficiently.

Use the Document Planner, as shown on the next page, to organize and design documents of all types and lengths. The Document Planner can help you plan letters, memos, and short reports (up to 8 to 10 pages long). For these shorter documents, you would fill out only a single Document Planner.

For longer documents—business plans, economic studies, technical reports, audit reports, etc.—you could potentially use the Document Planner more than once. First, use the Document Planner to plan the whole document. Next, use a second Document Planner to design each major chapter or subsection. So if the whole document has four chapters, you would fill out four additional Document Planners, one for each of the four chapters.

The following paragraphs explain what you should record on a Document Planner.

1. Purpose. State the main point for the document. This written statement should focus on what you want readers to do and to know.

The to-do statement is especially important, and most effective letters or memos should have a clearly identified action (a to-do statement):

> ***To-Do Statements***
>
> I want the Financial VP to approve the funding request for the Oatmark Building.
>
> Jaclyn (my boss) should authorize the hiring of two new sales representatives.
>
> Contemporary Architects, Inc., should design a new entry atrium for our headquarters.
>
> Financial Analysts, Inc., should prepare a new staffing plan to reflect changes in our global market.

Occasionally, however, a letter or memo will not identify an action; instead, it will present information that readers should know. Such a document should begin with a to-know statement:

> ***To-Know Statements***
>
> Our Executive Board wants to know why the Plum Creek Assembly Plant is less efficient than the Albany Plant.
>
> Our CEO needs to know the steps for acquiring a controlling interest in Balkan Corporation, Inc. (He has already decided to acquire Balkan.)
>
> The manager of marketing wants an explanation of the minimum marketing requirements for HiGro (a new liquid for growing hair).

Title/Subject Line. Write a subject line to reflect your initial to-do or

Purpose: State what you want readers to do and to know.	Title/Subject Line: Opening paragraph:
Preview: List your key points	• • • •
Details: Expand your key points in order.	
Review: Restate your purpose	

Figure 1. Sample Document Planner. *Use this organizer to plan whole documents or sections of documents. The four boxes guide you to use a managerial organization for documents.*

to-know statement. For example, here are two subject lines based on two of the preceding to-do and to-know statements:

> Title/Subject: Authorization to Hire Two New Sales Representatives
>
> Title/Subject: Steps for Acquiring a Controlling Interest in Balkan Corporation, Inc.

These subject lines are likely to be very similar to the phrasing you have written in step 1. This repetition is deliberate and desirable. Also, make your subject line long enough so that readers can grasp your content and understand how you plan to approach the content. See LETTERS and MEMOS.

2. Preview. Rule 6 suggests that writers preview information early (usually toward the top of the first page of a document). This preview helps readers know (and remember) what points you will be discussing in the rest of the letter, memo, or report.

Do not overlook the need to preview your points even though you are writing a short letter or memo. These shorter documents can still profit from a preview list of points, as in this example:

> The Plum Creek Plant is less efficient than the Albany Plant for three reasons:
>
> 1. Qualified vendors for components are not located as near to Plum Creek as they are to Albany.
> 2. The rejection rate for components at Plum Creek is double the rate at Albany.
> 3. Plum Creek is assembling a model with newly engineered components, not with the well-tested components used at Albany.

This Plum Creek document might be only a page or two long, depending upon what the readers will need to know. Or you might decide to prepare a major report with 30 or more pages of data and analysis. In either case, you should preview your main points.

3. Details. Provide as much detail about your previewed points (listed above in step 2) as necessary. Sometimes you can discuss major points in a few sentences. In other instances, you may need several pages of explanation. For shorter, more routine documents, you might need to fill out only a single Document Planner, both front and back. For longer documents, you might use a new organizer for each major topic discussed.

4. Review. Even a one-page letter can profit from a review of major points (perhaps only a reminder that a meeting is being scheduled). If appropriate, end the review with your phone number and an offer to answer additional questions. See LETTERS.

Outlines

Outlines are convenient tools for the schematic organization of material. See ORGANIZATION.

Preliminary or draft outlines help writers determine early in the writing process, usually before the document is written, whether the content is logical and complete. Preliminary outlines do not have to be neat or accurately numbered.

Final outlines (which usually form the table of contents) display the overall structure of the content. Final outlines may or may not be numbered. If they are numbered, they typically use either the traditional format (I/A/1/a, etc.) or the decimal format (1.0/1.1/1.1.1, etc.). Your word-processing program contains automated outlining features.

Traditional Outlines

Traditional outlines are those using the following numbering and lettering system:

TITLE

I. First-level division

    A. Second-level division

        1. Third-level division

            a. Fourth-level division

                (1) Fifth-level division

                    (a) Sixth-level division

II. First-level division

    A. Second-level division

    B. Second-level division

NOTE 1: Some writers and editors prefer *a)* instead of *a.* to indicate a fourth-level division.

NOTE 2: Because Roman numerals vary in length, they are customarily aligned according to the period, not the length of the numeral:

  I.
 II.
III.

Outlines

1. Use a numbering system when you need to cross-reference sections of your outline (and the resulting document).
2. Use decimal numbering when your document is more than a few pages long and has several levels of subsections.
3. Avoid numbering systems (either traditional or decimal) with more than four or five levels of subordination.
4. If possible, design your outline so that each subdivision has at least two points.
5. Use an outline to check the logical consistency and basic organization of a piece of writing.

Decimal Outlines

TITLE

1. First-level division

    1.1 Second-level division

        1.1.1 Third-level division

            1.1.1.1 Fourth-level division

    1.2 Second-level division

        1.2.1 Third-level division

            1.2.1.1 Fourth-level division

2. First-level division

NOTE 1: Numbers in decimal outlines do not require a period following them. Instead, two or three spaces visually set off the text from the decimal.

NOTE 2: A variation of the decimal format uses hundreds and tens. This format is not widely used, perhaps because it is less flexible than the decimal and traditional formats:

TITLE

100 First-level division

    110 Second-level division

        111 Third-level division

        112 Third-level division

    120 Second-level division

200 First-level division

1. Use a numbering system when you need to cross-reference sections of your outline (and the resulting document).

A well-designed numbering system helps readers track ideas and remember them longer. Also, a numbering system allows for cross-referencing between sections and subsections in longer technical and business documents. See NUMBERING SYSTEMS.

Occasionally, writers choose not to use a numbering system. Instead, they use a system of headings, where the placement, the type size, and the appearance of the headings tell readers which is a main heading and which is a subheading. See HEADINGS.

If you decide not to use a numbering system in a document, you will need to use page numbers to reference information. This decision often means that you cannot insert references until you are nearly ready to publish the final document. This decision may delay your project.

2. Use decimal numbering when your document is more than a few pages long and has several levels of subsections.

Decimal numbering has one main advantage over the traditional system. Readers can tell from the numbering exactly where they are in a document. See Numbering Systems.

For example, if readers open to the middle of a document, they might find the number *5.3.7* in the margin before a subheading. They instantly know that they are in Chapter 5, in its third subdivision, and in the seventh topic within the subdivision.

By contrast, the same page with traditional numbering would have *7.* in the margin. Readers cannot know which chapter or which subsection the *7* falls under unless, of course, the document has specific information in a header or footer. See Page Layout.

3. Avoid numbering systems (either traditional or decimal) with more than four or five levels of subordination.

As noted above in rule 2, the traditional system might have *a)* in the margin, but this sixth-level heading is of little use to a reader who usually can't remember the five levels above the *a)*.

Numerical numbering is also difficult to interpret when the numbers string out into six or seven divisions: *5.3.7.3.2.5.* In this case, consider using numbers for the first three or four levels. Then change over to format options; for example, the fifth-level could be an unnumbered run-in heading:

Discount rates. Studies of the discount rate structure include . . .

4. If possible, design your outline so that each subdivision has at least two points.

If a subdivision has only one subpoint in it, then the subpoint should become the subdivision heading:

this

3. Overhead rates
4. Labor issues

not this

3. Cost analysis
    a. Overhead rates
4. Labor issues

In the second example above, the subdivision for *cost analysis* has but a single point. If the cost analysis consists of nothing more than overhead rates, why list cost analysis as an activity? The first example properly recognizes that the cost analysis is nothing more than a determination of overhead rates.

5. Use an outline to check the logical consistency and basic organization of a piece of writing.

If an outline is not parallel and is not logical, then the document based on the outline is likely to be chaotic. See Organization and Tables of Contents.

Page Layout

Page layout uses visual or graphic-design techniques to enhance and arrange information on a page.

A good page layout ensures that language, graphics, and colors combine on a page to promote clear communication. Readers of the page will find it pleasing and easy to read even though they may not be conscious of all the page layout techniques. See Color, Emphasis, Graphics For Documents, and Writing and Revising.

Just because computers now allow everybody to design and produce documents, not all do-it-yourself page layouts are successful. Some documents are cluttered and disorganized. Others are well designed but poorly written.

A high-quality, professional document still might require the help of outside professionals (graphics specialists and writers) when the outcome is crucial. These professionals will work with you to prepare a page layout for all your writers to use.

Page Layout

1. Develop styles for your documents.
2. Select a page shape (orientation) and design features that will clearly and effectively communicate to your readers.
3. Set your margins and borders so that you have enough white space to make the page attractive and readable.
4. Add headers and footers that will help your readers know exactly where they are as they are reading your document.
5. Use more than one column to enhance readability.
6. Choose a font size and style that complements your page design.
7. Establish a consistent system of headings and lists and stick with it throughout your document.
8. Choose and place graphics for maximum impact and increased readability.
9. Avoid overloading pages with too much text and too many layout features.
10. Don't forget the basics just because you have a fancy design.

1. Develop styles for your documents.

Styles are page layout choices such as the kind of margins, headers, and fonts you plan to use. These choices are essential, particularly when many people will be working on the document. If several writers and designers are contributing to a single document, they should all agree on the styles before they begin writing.

Even a writer working alone will profit from choosing styles before beginning to write the text.

Of course, your computer program will already contain default styles that might satisfy your purpose. But for various reasons, you might want to change those style choices. For example, you might need to give a branded look to a slide presentation or a report, using the company logo and approved fonts and colors. If so, you can create a template or style sheet that will apply those styles automatically.

A good template or style sheet shows and tells writers exactly what the page layout or format will be; how to break up text using space or headings; how long (roughly) paragraphs should be; and what sorts and sizes of graphics will work best. In short, a good style sheet helps ensure that what the writer produces will fit the chosen style and format.

Because it already contains all the format options you want, the template file becomes the starting point whenever you start a new document. You will call up the template, write some text, and then save the file using another file name.

As figure 1 shows, a template or style sheet includes fonts, type sizes, and any other information you or someone else would need to use to prepare the document. Fonts, font sizes, and other format options are discussed in the following rules.

2. Select a page shape (orientation) and design features that will clearly and effectively communicate to your readers.

A normal business letter and most printed pages are longer than they are wide. This shape is called a vertical or portrait format. This *Style Guide* uses a portrait format for its pages.

However, if you expect to use many charts and graphs or if you will present the message on a video screen or computer monitor, you might choose a horizontal or landscape format. In a horizontal format, pages are wider than they are long.

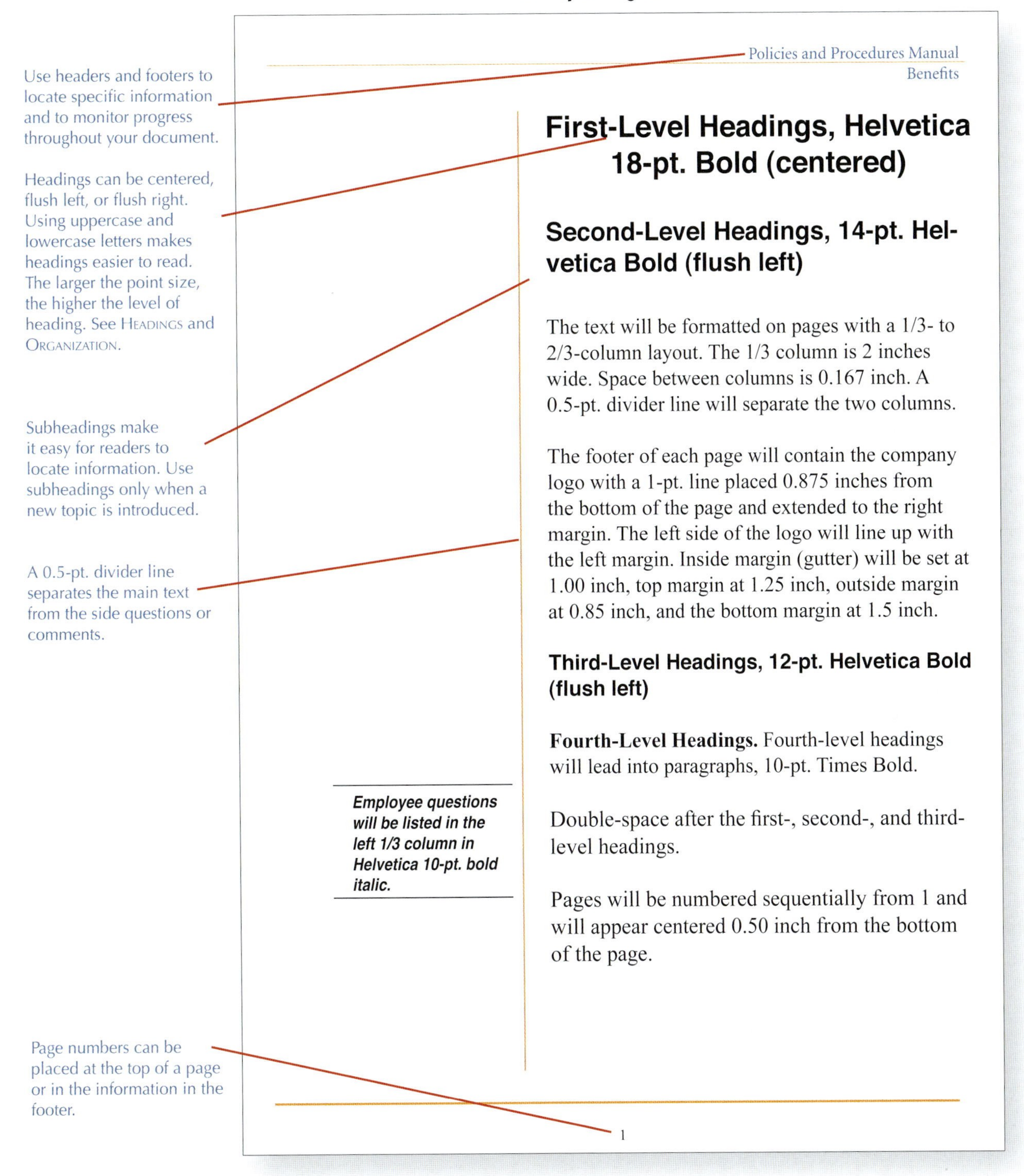

Sample Style Sheet
Use headers and footers to locate specific information and to monitor progress throughout your document.
Headings can be centered, flush left, or flush right. Using uppercase and lowercase letters makes headings easier to read. The larger the point size, the higher the level of heading. See HEADINGS and ORGANIZATION.
Subheadings make it easy for readers to locate information. Use subheadings only when a new topic is introduced.
A 0.5-pt. divider line separates the main text from the side questions or comments.
Page numbers can be placed at the top of a page or in the information in the footer.
Policies and Procedures Manual
Benefits
First-Level Headings, Helvetica 18-pt. Bold (centered)
Second-Level Headings, 14-pt. Helvetica Bold (flush left)
The text will be formatted on pages with a 1/3- to 2/3-column layout. The 1/3 column is 2 inches wide. Space between columns is 0.167 inch. A 0.5-pt. divider line will separate the two columns.
The footer of each page will contain the company logo with a 1-pt. line placed 0.875 inches from the bottom of the page and extended to the right margin. The left side of the logo will line up with the left margin. Inside margin (gutter) will be set at 1.00 inch, top margin at 1.25 inch, outside margin at 0.85 inch, and the bottom margin at 1.5 inch.
Third-Level Headings, 12-pt. Helvetica Bold (flush left)
Fourth-Level Headings. Fourth-level headings will lead into paragraphs, 10-pt. Times Bold.
Double-space after the first-, second-, and third-level headings.
Pages will be numbered sequentially from 1 and will appear centered 0.50 inch from the bottom of the page.
Employee questions will be listed in the left 1/3 column in Helvetica 10-pt. bold italic.
1

Figure 1. Sample Template for a Vertical or Portrait Format. *A good template or style sheet provides very clear and specific instruction to contributors so that all contributors' material is consistent.*

Sample Style Sheet (continued)

Policies and Procedures Manual
Benefits

Times 10-pt. will be used for all normal text. We will use boldface for emphasis. Text will be unjustified.

1. **Whenever possible, we will use lists.**
2. **All lists will be Times 10-pt. bold.**

- **Where numbering is not crucial, lists will begin with bullets.**

√ **We may also use some checklists.**

NOTE: The tab for lists will be set at 0.25 inch before the number, bullet, or check. Right margins will be indented 0.25 inch also.

Special statements will be in 10-pt. Helvetica italic and placed in a 10 percent screened 0.5-pt. line box.

Visuals will be numbered sequentially throughout the response, beginning with figure 1.

Visuals will appear in 0.5-pt. line boxes in the right 2/3 column. Text in visuals will be 8-pt. Helvetica Captions will have a boldface figure number and title followed by an action caption, 10-pt. Times italic.

Figure 1. Insurance Benefits. *The medical insurance costs for 2008–9 reflect a 35 percent increase over the previous year.*

2

Displayed lists emphasize a series of important items. See LISTS.

Boxes and screens highlight and emphasize text and visuals. See EMPHASIS and GRAPHICS FOR DOCUMENTS.

Next to headings, captions are the most-read part of your document. Use action captions whenever possible. Keep them short and to the point. See CAPTIONS.

After you choose a page shape, you need to choose the design features you will use on that page. Design features include the following options:

—Margins

—Size

—Borders and divider lines

—Indention

—Line spacing or leading

—White space

—Headers and footers

—Columns

—Font styles and sizes

—Page color (if any)

—Headings

—Graphics

—Callouts

—Text wrapping

Figure 1 illustrates these design options as presented in a portrait format. Each of these options is addressed in separate rules.

Figure 2 shows these same design options, but in a landscape format.

3. Set your margins and borders so that you have enough white space to make the page attractive and readable.

Each page and each section of a page has margins—that is, the white space separating pieces of text from each other or from the sides of the page.

Your computer has default margins, but these might be too narrow or too wide. So try different margins to see how the page will look with these margins.

Setting margins for printed letters can be a problem, especially if the text is very brief. The overall rule is to center a letter on the page so that the page looks balanced. The left and right margins will be roughly equal, and if the printed letter is very short, the text should be dropped down so that white space at the top roughly equals the white space at the bottom. See LETTERS.

Borders and divider lines can help frame your text and can be as simple as a line around the entire text on a

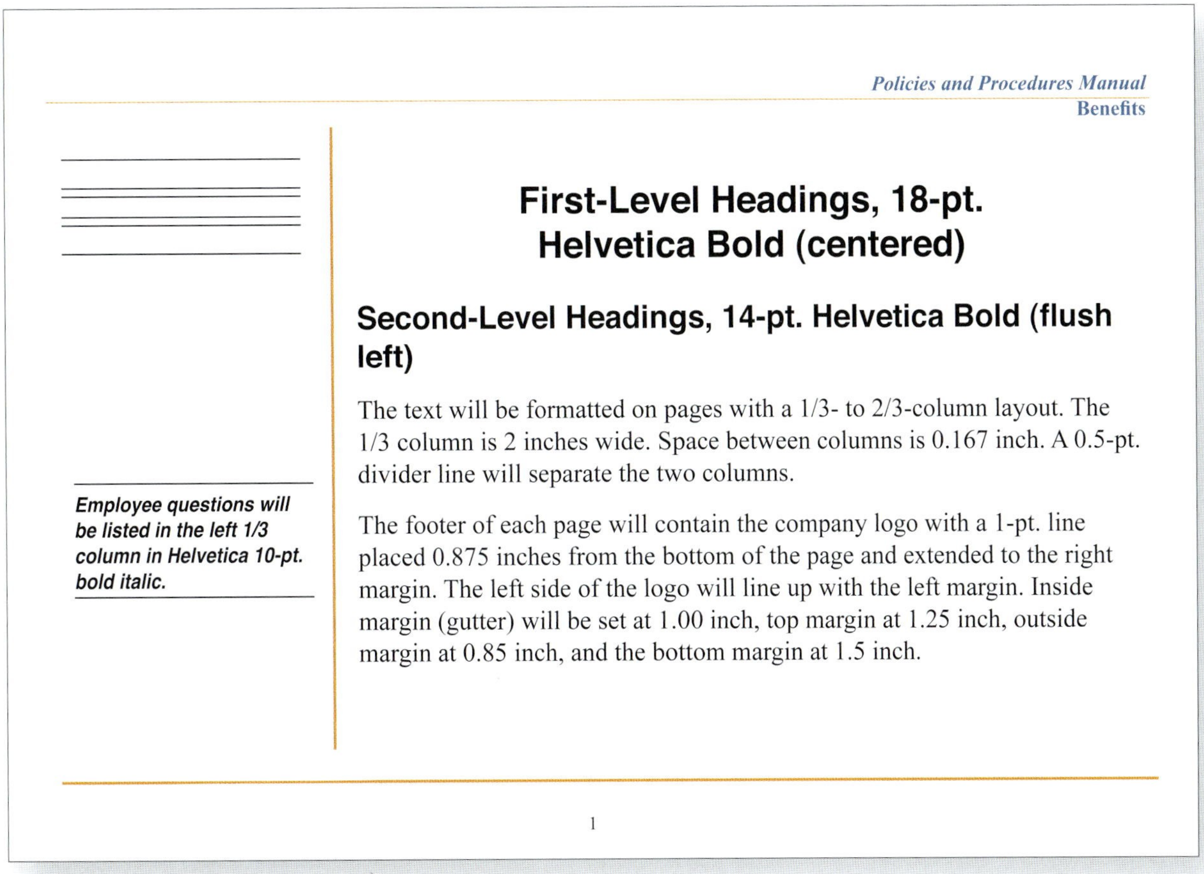

Policies and Procedures Manual
Benefits

First-Level Headings, 18-pt. Helvetica Bold (centered)

Second-Level Headings, 14-pt. Helvetica Bold (flush left)

The text will be formatted on pages with a 1/3- to 2/3-column layout. The 1/3 column is 2 inches wide. Space between columns is 0.167 inch. A 0.5-pt. divider line will separate the two columns.

Employee questions will be listed in the left 1/3 column in Helvetica 10-pt. bold italic.

The footer of each page will contain the company logo with a 1-pt. line placed 0.875 inches from the bottom of the page and extended to the right margin. The left side of the logo will line up with the left margin. Inside margin (gutter) will be set at 1.00 inch, top margin at 1.25 inch, outside margin at 0.85 inch, and the bottom margin at 1.5 inch.

1

Figure 2. Sample Template for a Horizontal or Landscape Format. *This sample is the opening section of the same template or style sheet from the previous pages, presented in a horizontal or landscape format.*

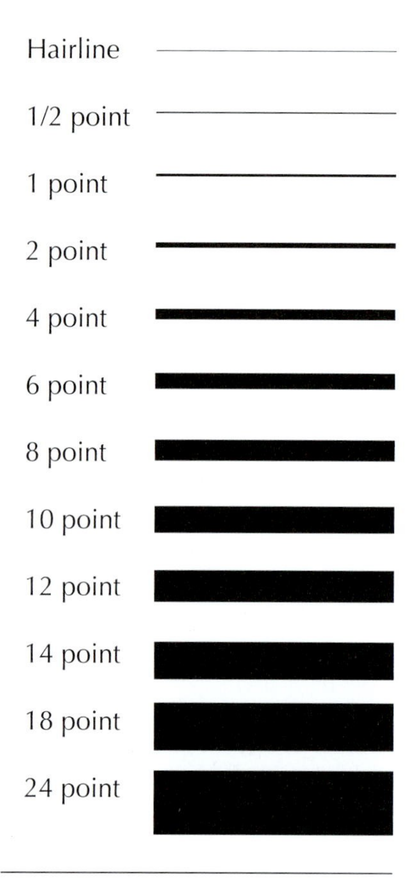

Figure 3. Sample Rules. *Divider lines or rules, of differing widths, are labeled with their point size.*

page. They can also be complex—for example, a line of asterisks or even a curving design. Divider lines are lines used to divide or set off a column or a section of text.

Printers call these lines rules. Rules can be very fine (called hairlines and 1 point wide) or quite heavy (perhaps 12 point). See figure 3 for a sample of several common divider lines, with their point size indicated.

Remember that you want your page to look simple. Plenty of white space can contribute to the simplicity of your page. Remember, also, that too many borders and divider lines can ruin an effective page. Use them sparingly and only when absolutely necessary.

4. Add headers and footers that will help your readers know exactly where they are as they are reading your document.

Headers and footers are like road signs. They tell readers which section, subsection, and page they are reading. Most published books, whether printed or online, have headers and footers. Notice the header and footer on this page.

You will help orient your readers if you include headers and footers in reports, presentations, and other common business documents.

Headers usually have a page number and the number and title of a chapter or subsection in a lengthy chapter. Footers often have the name of a company or an organization, as well as an organizational logo or some other symbol, such as a trademark. See INTELLECTUAL PROPERTY.

5. Use more than one column to enhance readability.

Two or even three columns on a page can make your text seem to be more professional and attractive than a single column. Multiple columns have shorter lines of text, which make them more readable than text that moves all the way across a page. The eye can read a narrow column about 50 characters wide (six or seven words) faster than lines as wide as a full page (about 75 characters).

Two or three columns are basic to most formal publishing situations, but four or even five columns can be useful on occasion. A lengthy list of rivers, cities, or foods would go very nicely in four or even five columns. Four or five columns to the page would rarely be used for normal text, but might be used in a document printed in horizontal (landscape) format.

6. Choose a font size and style that complements your page design.

A font includes both the size and style of letters, numbers, and symbols. You probably have a wide selection of fonts available to you. Use a plain font when quick comprehension is needed. A fancy or display font should only be used when fast comprehension is not important or to make a point—STOP—for example. The examples in the following discussion will suggest some of the options you have for either plain or fancy fonts.

Choose a font size and style and then stick to your choices. Too many font choices on the same page become busy-looking or cluttered.

The names of fonts—such as Arial or Times New Roman—come from old printing fonts, which had to be set up character by character in the days before typesetting machines and computers. See figure 4 for examples of these and other common fonts.

Arial
Avant Garde
Bookman
Century School Book
Futura
Garamond
Helvetica
Helvetica Condensed
Optima
Minion
Palatino
Souvenir
Times
Times New Roman

Figure 4. Typical Type Fonts. *Many popular proportionally spaced typeface fonts are based on old printing font-families. Although these faces are all shown in 12-pt. size, some look smaller because the descenders on j, p, q, and y are part of the height of a font.*

Sometimes two fonts look almost alike but have different names. There are differences in some characters; look particularly at *a*, *e*, *f*, *g*, and *q*. These differences are important because artists design fonts, and these designs are often copyrighted. See INTELLECTUAL PROPERTY.

Manual typewriters generally had mono-spaced type, unlike the proportional fonts used today. Courier font is a reproduction of a traditional typewriter font (see figure 5).

We recommend that you avoid using mono-spaced fonts that look old-fashioned and stiff today.

Courier
Geneva
Chicago
Letter Gothic
Prestige Elite

Figure 5. Typical Mono-Spaced Type Fonts. *Nonproportionally-spaced fonts are mono-spaced, meaning all letters are about the same width.*

Serif and **sans serif** differ in their shape or design.

A **serif** font has extra lines or hooks on letters, which makes them more decorative. In a serif font, the letters have extra finishing lines—for example in figure 4, Garamond is a serif font, as illustrated in the lines at the base of the *a* or the *d* and the hooks on the *G*.

A **sans-serif** font does not have the finishing lines of the serif fonts. For example, in figure 4, Avant Garde is a sans-serif font. Note that both the *A* and *a* are plain in appearance and have no finishing feet or hooks.

In figures 4 and 5, the serif fonts are Bookman, Courier, Century School Book, Garamond, Minion, Palatino, Prestige Elite, Souvenir, Times, and Times New Roman. The sans-serif fonts are Arial, Avant Garde, Chicago, Futura, Geneva, Helvetica, Helvetica Condensed, Letter Gothic, and Optima.

Different styles of letters exist for both serif and sans-serif fonts. The common optional styles include bold, extra bold, condensed or narrow, italic, and outline (see figure 6).

Times
Times Italic
Times Bold
Times Bold Italic
TIMES SMALL CAPS
Times Outline
Times Shadow

Figure 6. Times Roman Font Shown in Several Forms. *Italic and bold forms of any font are common; use other forms sparingly and only for compelling reasons.*

Font size is measured in points—that is, from the tallest ascender (on *b*, *d*, *l*, or *h*) to the deepest descender (on *g*, *p*, *q*, or *y*). One inch (2.54 centimeters) is 72 points. See figure 7 for examples of different font sizes.

36 pt. Ti
24 pt. Times
18 pt. Times
14 pt. Times
12 pt. Times
10 pt. Times
9 pt. Times
6 pt. Times

Figure 7. Times Font Shown in 6 pt. Through 36 pt. *Most programs permit font-size changes in fractions of a single point size.*

7. Establish a consistent system of headings and lists and stick with it throughout your document.

Design a system of headings and lists that allows readers to scan quickly for key ideas. A consistent, well-designed set of informative headlines helps readers grasp the structure of your document. Include headings in business letters and memos, especially if a letter or memo is longer than a single page.

Set up a system of headings that signal the different sections and subsections of your text. See figure 1 for examples of possible heading levels. Also see HEADINGS and LISTS.

Your program will enable you to select heading styles quickly and easily to signal various levels. As in figure 8, your decisions about headings will include the fonts to use, their size, and any other distinguishing features.

FIRST-LEVEL HEADINGS

Second-Level Headings

Third-Level Headings

Fourth-Level Headings. Text . . .

Figure 8. *Typical headings use various font options—for example all capitals, different sizes, boldface, and even italics.*

A very common pattern is to use sans-serif fonts for headings and a serif font for the text. This is the practice followed in this *Style Guide*. See rule 6 for a definition of *serif* and *sans-serif*.

You might choose to number headings—for example, 2., 2.1, 2.1.1, etc.—especially in a complex technical document.

Lists also allow for a number of options. At the simplest, decide when and where you will number lists, use bullets, or use some other symbol such as checkboxes or dashes. Also, you need to decide how you intend to indent or frame lists using white space. See figure 1 and LISTS.

8. Choose and place graphics for maximum impact and increased readability.

A well-chosen, high-quality graphic is worth a thousand words. And it is ten times more memorable than mere text. So plan for and prepare high-quality graphics.

Your template or style sheet (see rule 1) should plan for graphics and the placement of captions as a part of graphics. See GRAPHICS FOR DOCUMENTS, GRAPHICS FOR PRESENTATIONS, and CAPTIONS.

Decide what your main point is before you decide what graphics you want to include. If your goal is to present a trend or a contrast between two different methods, you might want to use a bar graph or a line graph showing two trends.

In other cases, if you want to describe a piece of unfamiliar equipment, you should consider either a photograph or an illustration (maybe a schematic or outline version of the equipment).

Finally, you might want to use a flow chart or other tree diagram to show the steps in a process or to present who is serving on different committees.

Your goal is to make your graphics fit your main point or document purpose.

Placing graphics is often a subjective skill, but the following suggestions are generally true:

—Mention a graphic in the text before it appears.

—Make a graphic and its caption largely independent from the text.

—Make a graphic large enough to be readable and don't expect readers to turn the page to read the caption or notes.

—If a graphic is too complex or cluttered, break it into two graphics or consider making it an attachment to your document.

9. Avoid overloading pages with too much text and too many layout features.

Writers often cram too much writing onto a page, trying to stay within page limits or adhering to custom, an arbitrary format, or old myths about how pages ought to look.

Writers may even ignore the appearance of the document because they believe that page layout is someone else's concern. They think their readers will "figure it out." People who write mostly email messages develop this habit, which works against them when preparing more formal documents or presentations.

Later they (or a boss or an editor) have to go through the text shortening a discussion, changing an introduction, and adjusting all the graphics. Time and money are wasted because much of the text must be rewritten to meet readers' needs.

10. Don't forget the basics just because you have a fancy design.

Final proofreading and style checking are essential. Errors creep into the best of documents. For instance, changing the form of a subheading in the middle of a chapter may also require changes in many other places: the introduction to the chapter, the table of contents for the document, the index, the summary of key points in the chapter, and others. Unlike grammar and spelling mistakes, style mistakes will go unnoticed by your computer.

Most programs tag text (attaching invisible programming characters) for aggregation into indexes or tables of contents. See INDEXES and TABLES OF CONTENTS.

However, some other changes may not occur automatically. With computers, the ease of changing something can cause some problems. Moving text often results in leftover letters, odd punctuation, or spacing errors.

Paragraphs were the mainstay of writing a generation ago. Many excellent documents broke their content into logical sections by using only paragraphs. Headings and lists were unknown or infrequent.

The traditional paragraph, especially in its more lengthy versions, is becoming less common in today's business or technical writing. Interestingly, academic writing still relies on and composition teachers still teach the structure and form of the long paragraph.

The first eight rules in the box to the right apply to situations where you want to write your paragraphs with a traditional logical structure. Rule 9 suggests an alternative to the traditional paragraph.

In many types of documents, however, you will find that headings, lists, graphics, and other emphasis techniques can and should replace traditional paragraphs. See figure 1 for an example of how emphasis techniques can easily replace a traditional paragraph. See Emphasis, Headings, and Lists.

Paragraphs

1. Limit paragraphs to a single topic or major idea.
2. Do not allow paragraphs to become too long.
3. Vary the length of your paragraphs.
4. Ensure that the opening sentence of every primary paragraph accurately reflects the content of that paragraph and any following secondary paragraphs.
5. Organize paragraphs logically.
6. Use key words and other devices to ensure that paragraphs are coherent.
7. Emphasize the important ideas within a paragraph.
8. Provide transitions between paragraphs.
9. If appropriate, break up or replace paragraphs with lists.

As figure 1 illustrates, a traditional paragraph is often difficult to read. The typical reader cannot easily scan a traditional paragraph because an assumption behind the traditional paragraph is that a reader must read it line by line and word by word.

In contrast, the schematic or emphatic paragraph in figure 1 assists readers by visually segmenting the points into chunks. Scanning schematic paragraphs is easy.

1. Limit paragraphs to a single topic or major idea.

Ensure that your paragraphs focus on a single topic or idea. When you go to a new topic, start a new paragraph. If your paragraph on a single topic becomes too long, start a new paragraph at a logical point and have two (or more) paragraphs dealing with the same topic. When such is the case, you normally focus each paragraph on a subtopic related to the overall topic.

Traditional Paragraph	Schematic or Emphatic Paragraph
The engineering design team needs a full-time graphics specialist assigned to be in the team's work area. Current graphics support does provide high-quality graphics, but their delivery times are not as rapid as we would like. In some cases a requested graphic takes over a week before coming back to the design team for review and revision. Some graphics require several review loops, so some graphics may require as much as a month before they are ready for publication in the design team's documentation procedures. Having a full-time graphics person would lessen the delays and would allow the team members easily to share their design concepts with a graphics specialist.	**Recommendation: Full-Time Graphics Support for the Team** Having a full-time graphics specialist at the team site would provide these advantages: • **Shorten turnaround time** for routine graphics. Current times average nearly a week for requested graphics. • **Decrease review loops** by working with team members during the conceptual phase. Current review loops often require as much as a month. • **Allow immediate corrections and adjustments** while the graphic specialist is creating a graphic. This collaboration would also enhance conceptualization.

Figure 1. Two Contrastive Versions of the Same Content. *The traditional paragraph is 112 words long and follows the traditional pattern of a topic sentence, supporting details, and a concluding sentence. The schematic or emphatic version (78 words) covers many of the same ideas, but it is much more readable.*

Paragraphs

2. Do not allow paragraphs to become too long.

Quantifying paragraph length is difficult, but in business and technical writing, paragraphs exceeding 100 to 125 words should be rare. Most paragraphs will consist of three to six sentences. If a single-spaced paragraph goes beyond one-third of a page, it is probably too long. A double-spaced paragraph should not exceed half a page in length.

The document's format should influence paragraph length. If a document has narrow columns (two or three to the page), then paragraphs should be shorter, perhaps on the average of no more than 50 words. If a document uses a full-page format (one column), then average paragraph length can reach 125 words.

Length is therefore a function of appearance and visual relief. Almost all readers have difficulty with dense pages of print, no matter how well-written and logically organized the text may be. Remember that paragraphs are visual devices meant to make reading easier, so keep them shorter rather than longer.

3. Vary the length of your paragraphs.

A document containing paragraphs of uniform length would be dull and difficult to read. For the sake of variety and to stimulate reader interest, you should vary the length of your paragraphs, especially in documents longer than one page.

The length of successive paragraphs will, of course, depend on content. The logic of the material will dictate, at least to some extent, where paragraphs can logically begin and end, but you still have a great deal of latitude.

A particularly involved point may require lengthy explanation and two or three examples. If so, you might state the point, explain it in one or two paragraphs, and then make each example a separate paragraph. Dividing the topic in this fashion is necessary in a double- or triple-column page format.

Your paragraph stating the main point could be relatively short. Short paragraphs usually draw attention to themselves, so they are useful for stating major ideas. The explanatory paragraph should be much longer. The paragraphs providing the examples should vary in length, with the most important example appearing in the longest paragraph.

Are single-sentence paragraphs acceptable?

Yes. A common misconception about paragraphing is that single-sentence paragraphs somehow violate a principle of writing. In fact, single-sentence paragraphs are very emphatic, especially if they are surrounded by longer paragraphs. You should take care not to use single-sentence paragraphs too often, however. See EMPHASIS.

4. Ensure that the opening sentence of every primary paragraph accurately reflects the content of that paragraph and any following secondary paragraphs.

Primary paragraphs introduce an idea. Secondary paragraphs develop and support that idea. All primary paragraphs should have an opening sentence that introduces the content of the primary and any following secondary paragraphs. The opening sentence is called a topic sentence.

The opening sentence should establish a key word or phrase that indicates the paragraph's topic. If your paragraph will focus on health problems, then the opening sentence should contain at least two key words: *health* and *problems.* These key words help establish the paragraph's viewpoint, which is often called its thesis. The three examples below each have a topic sentence that announces the thesis of the paragraph:

> Software sales in the Eastern Region have substantially increased. From Q1 to Q2, the Sales Division conducted four sales campaigns that . . .

> MOGO's May reservoir study indicates that remaining recoverable reserves exceed previous estimates by over 66 percent. Seismic data gathered in conjunction with the study . . .

> The Packaging Department examined the problem and recommends replacing our standard cardboard containers with molded plastic wrap. The plastic is applied from a hot roller after the cases . . .

Secondary paragraphs do not begin with topic sentences. In a technical report discussing a series of tests, for instance, the results section of the report might have several paragraphs opening as follows:

> Test 1, series 1, involved decreasing eluants by 0.4 cm^3/hr and noting pH changes occurring as the solution was heated to 250 degrees F . . .

> During test 1, series 2, eluants were removed altogether, and the solution was subjected to pressure variations during heating . . .

These paragraphs develop and support the thesis that was established within a topic sentence in a previous primary paragraph.

As readers, we should expect the first secondary paragraph to focus entirely on test 1, series 1. Any information in that paragraph that is not related to test 1, series 1, does not belong there. Similarly, the second secondary paragraph should focus on test 1, series 2.

Open every primary paragraph with a topic sentence that states the thesis of that paragraph and any following secondary paragraphs. See ORGANIZATION.

5. Organize paragraphs logically.

The structure of the ideas within a paragraph should be logical. Furthermore, the paragraph structure should be obvious to readers.

Sometimes this structure follows a classic organizational pattern: chronological, whole to parts, problem to solution, cause to effect, most important to least important, general to specific, and so on. Sometimes the structure follows some logic that is inherent to the subject. A paragraph on drilling-rig problems, for instance, might be organized according to a series of problems that relate to each other in some way that uninformed readers would not perceive.

The following paragraph is paraphrased from Charles Darwin's *Origin of Species*. It demonstrates a classic organizational pattern: general to specific. Note that the opening sentence is a topic sentence and that the three succeeding sentences substantiate Darwin's thesis. The boldfaced key words indicate one of the most common methods of achieving coherence: repeating key words. The key words in the Darwin paragraph form a clear link between sentences.

> Without exception, every species naturally **reproduces** at so high a **rate** that, if not destroyed, the earth would soon be covered by the progeny of a single pair. Even **slow-breeding** man has **doubled** in 25 years, and at this **rate**, in less than 1,000 years, there would literally not be standing room for his progeny. Linneaus has calculated that if an annual plant **produced** only two seeds—and no plant is so **unproductive**—and their seedlings next year **produced** two, and so on, then in 20 years there should be 1,000,000 plants. The elephant is reckoned the slowest **breeder** of all animals, and I have taken some pains to estimate its minimum **reproductive rate**; it will be safest to assume that it begins **breeding** when 30 years old, and goes on **breeding** till 90 years old, bringing forth six young in the interval, and surviving till 100 years old; if this be so, after 750 years there would be nearly 19,000,000 elephants alive, descended from the first pair.

6. Use key words and other devices to ensure that paragraphs are coherent.

Coherence refers to the cohesiveness of a paragraph's sentences. In a coherent paragraph, the sentences seem to "stick together"—they all clearly belong in the paragraph and are logically connected to one another.

As you read the paragraph below, note the lack of coherence:

> A great number of apparatus are available today for field work in the broad sense, including gas-chromatographs, as well as infrared, electrochemical, and other analyzers. In connection with the early prediction of possible pollutants and the assessment of natural discharge prior to the development of geothermal resources, however, hydrogen sulfide and volatiles such as ammonia, mercury, and arsenic are the major concern. Under the conditions prevailing before industrial development, preliminary evaluation and prediction of the discharge of such chemicals depends to a sizeable extent on water analyses. Surveying mercury content in air might deserve consideration; however, it is not discussed here, as mercury determination in soil is likely to be a valid substitute for it.

The opening sentence to this paragraph suggests that the paragraph will discuss the apparatus available for field work, particularly those apparatus listed. However, this equipment is never again discussed. The second, third, and fourth sentences seem loosely connected, but the paragraph never "gels"; it never seems to be focused on a single topic. In short, the paragraph is incoherent. See PARAGRAPHS.

You can achieve coherence by opening with a topic sentence, by using a clear organizational scheme, by repeating key words, by using transitional words (such as *however, furthermore, consequently, next, then, additionally,* etc.), and by using pronouns to link sentences to the major idea or theme of the paragraph. See ORGANIZATION, KEY WORDS, TRANSITIONS, and PRONOUNS.

7. Emphasize the important ideas within a paragraph.

As figure 1 illustrated, one way to emphasize important ideas is to use a schematic or emphatic approach to the content. Lists, headings, and graphics are all common tools for emphasizing important ideas. See EMPHASIS.

In one sense, extensive use of such emphasis techniques makes many of the traditional rules for writing paragraphs unnecessary.

In traditional paragraphs, however, important ideas are emphasized by being placed in the initial or topic sentence. The initial sentence is visually and logically the most important sentence in a paragraph, so use it to capture the most important ideas.

Paragraphs

Traditionally, the final sentence in a paragraph was the second most emphatic position. This viewpoint assumed that readers would take time to read the final sentence. In most technical or business writing, however, the final sentence either drops out (because it is repetitive) or it moves up to become part of the initial sentence.

8. Provide transitions between paragraphs.

In most well-written documents, the information flows from paragraph to paragraph. To achieve this effect, you must provide smooth transitions between paragraphs.

Writers can set up transitions by previewing content. If you announce, for instance, that you will be discussing five topics and then list those topics, you have set up a progression that the reader will expect. As you move from topic to topic, the transitions will be automatic:

> The first topic concerns . . .
>
> Likewise, the second topic . . .
>
> The third topic . . .
>
> However, the fourth topic . . .
>
> Finally, the fifth topic . . .

As you can see, these paragraph openings also use some transitional words to make the transition, but merely moving to and announcing the next topic is sufficient.

Sometimes you can create the transition between paragraphs by using key words to connect the closing sentence of one paragraph and the opening sentence of the next:

> . . . because deep **salt domes** usually occur as a result of normal **faulting.**
>
> Thrust (reverse) **faults**, on the other hand, are normally responsible for piercement **salt domes** . . .

In this excerpt, the first paragraph closes with a key word (*faulting*), a variation of which is repeated in the opening sentence of the next paragraph. The first paragraph concerns deep salt domes; the second concerns piercement salt domes. Repeating a variation of *faulting* helps make the transition.

In this example, the opening sentence of the succeeding paragraph makes the transition. However, the transition could also be made by the closing sentence of the preceding paragraph:

> . . . because deep **salt domes** usually occur as a result of normal **faulting.** Thrust (reverse) **faults**, on the other hand, are normally responsible for **piercement salt domes.**
>
> **Piercement domes**, which are common along the Texas and Louisiana Gulf Coast, are produced from traps caused when complex **faulting** forces a salt core upward through overlying sediments . . . .

In some documents, the information cannot easily flow from paragraph to paragraph because the paragraph topics are too disjointed. When such is the case, use headings, lists, and numbering systems to indicate the transition from one topic to another:

> . . . Reef-producing areas might or might not be obvious from overlying sediments.
>
> **Piercement Domes**
>
> Piercement domes, which are common along the Texas and Louisiana Gulf Coast, produce from traps caused when complex faulting forces a salt core upward through overlying sediments . . . .

See Headings, Lists, Numbering Systems, and Transitions.

9. If appropriate, break up or replace paragraphs with lists.

If a paragraph contains or consists of a long series of items, consider replacing the paragraph with a displayed list. Lists are more emphatic than paragraphs, so if you want to emphasize the series of items, display it. See Lists.

A Final Case Study in Paragraphing

Some writers dump everything they can think of about a topic into a single paragraph. Then when a paragraph becomes long enough (by whatever standard), these writers pause, indent, and start another paragraph. This sort of paragraphing is neither logical nor effective, as the following example illustrates:

> Oxides of nitrogen include nitrogen dioxide (NO_2) and nitric oxide (NO). NO_2 is a pungent gas that causes nose and eye irritation and pulmonary discomfort. NO is converted to NO_2 by atmospheric chemical reaction. Both NO and NO_2 participate in photochemical reactions leading to smog. Sulfur dioxide (SO_2) is a colorless and pungent gas that causes irritation to the respiratory tract and eyes and causes bronchoconstriction at high concentrations. Hydrocarbons react with NO or NO_2 and sunlight to form photochemical oxidants or smog. Health effects include irritation of the eye, nose, and throat. Extended periods of high levels of oxidants produce headaches and cause difficulty in breathing in patients suffering from emphysema.

What did the writer of this paragraph want to accomplish? Is the first sentence on NO and NO_2 an accurate reflection of the rest of the content? How do the other facts and points in the paragraph fit together? Can readers see a definite pattern or structure to the facts?

A Traditional Rewrite

You could rewrite the paragraph using these traditional remedies:

- Shorten the paragraph and focus on only one topic.
- State the topic in the opening sentence.
- Supply organizational cues in the opening sentence and, as appropriate, in later sentences.

By applying these remedies, we can improve the paragraph as follows:

> Nitrogen oxides, hydrocarbons, and sulfur dioxide—these constituents of smog can cause health problems. Nitrogen dioxide (NO_2) is a pungent gas that causes nose and eye irritation and pulmonary discomfort. Hydrocarbons that react with NO_2 or with nitric oxide (NO) and sunlight form photochemical oxidants that can irritate the eyes, nose, and throat. Extended exposure to high levels of oxidants can produce headaches and cause persons with emphysema to have trouble breathing. Sulfur dioxide (SO_2) is a colorless and pungent gas that irritates the eyes and respiratory tract and, at high concentrations, can cause bronchoconstriction.

A Nontraditional Rewrite

You might decide to transform the paragraph into a graphic, especially if you decide that the information merits high emphasis. Figure 2 is a possible version of such a nontraditional rewrite. See Organization and Graphics for Documents.

Components of Smog	Health Effects
Nitrogen dioxide (NO_2)	Eye irritation and pulmonary discomfort
Photochemical oxidants (Hydrocarbons + NO_2 or NO)	Eye, nose, and throat irritation
Sulfur dioxide (SO_2)	Eye and respiratory irritation (especially bronchoconstriction)

Figure 2. Major Components of Smog and Their Health Effects. *All three components are colorless, yet pungent and irritating, especially in high concentrations.*

Parallelism

Parallelism

1. Ensure that two or more parts of speech behaving similarly in a sentence, or coordinated (connected) in some way, are parallel in construction.
2. Make items in lists parallel.

Parallelism is essentially a convention of sentence construction. The principle behind it is that similar ideas should be expressed in a similar fashion, thereby demonstrating their similarity and making reading easier. The following sentence is not parallel:

> The analysis will include organizing, dividing, and assessment of turnaround functions.

The sentence verb *include* is followed by three key words: *organiz*ING, *divid*ING, and *assess*MENT. These three words appear in series. They are equal in purpose and use in the sentence. Therefore, they should have the same grammatical form:

> The analysis will include organizing, dividing, and assessing turnaround functions.

1. Ensure that two or more parts of speech behaving similarly in a sentence, or coordinated (connected) in some way, are parallel in construction.

Parallelism applies not only to verbs, but also to nouns, adjectives, phrases, and every other part of a sentence:

> The Interface Team will be responsible **for integrating** the functional units developed by the QA Team and **for executing** the model test matrix.

> Applying **abstraction**, **partition**, and **projection** to the system development process results in the traditional top-down view of the software engineering process.

> Figure 2.2–1 shows the documentation relationships: **where things happen, why things happen,** and **how things can be changed.**

> Multilevel training was necessary to meet the needs of **managers, designers**, and **programmers**.

> A final report was prepared **describing** the case-study process and **referencing** the documents containing the code.

NOTE: Often a repeated *who, which,* or *that* will signal a sentence with parallelism:

> The departmental white paper specified **who** would head the new project team and **who** would be the team members.

> *not*

> The departmental white paper specified the head of the new project team and who would be team members.

2. Make items in lists parallel.

Parallelism is especially important in lists. A list, whether displayed vertically on the page or embedded within a paragraph, is a series. To make it parallel, each item should be constructed similarly and should begin with the same kind of word (noun, verb, etc.):

> This file will include the following items:
>
> 1. Problem headings
> 2. Database specifications
> 3. Reporting intervals
> 4. Restart options
> 5. Report macros

The following list is also parallel (each item completes the sentence started by the introductory statement). Note that each item begins with the same kind of verb:

> The study concluded that the ATAC fighter must:
>
> 1. Have a long-range, high-payload capability.
> 2. Be flexible in mission and payload design.
> 3. Be survivable against A-A and S-A threats.
> 4. Be maneuverable in the F-15/F-16 class.

NOTE: This example has several variations. Many editors, including the authors of this *Style Guide*, would prefer a different lead-in sentence (one with a complete grammatical structure):

> A study concluded that the ATAC fighter must have these features:

See COLONS, CONJUNCTIONS, and LISTS.

Parentheses are used to insert material into a sentence. Dashes—which are more emphatic than parentheses—are also used to insert material. Commas, which are also used to insert material, are less emphatic than either dashes or parentheses.

Using parentheses appropriately and effectively is an art. The following rules will help you develop this art.

Parentheses

1. Parentheses enclose explanatory sentences within a paragraph.
2. Parentheses enclose references, examples, ideas, and citations that are not part of the main thought of a sentence.
3. Parentheses enclose numbers in a paragraph list.
4. Parentheses enclose acronyms, abbreviations, definitions, and figures that have been written out.

1. Parentheses enclose explanatory sentences within a paragraph:

> Only the total systems approach can deal with the trade-off in performance between the weapon and the aircraft platform. Existing beyond-visual-range air-to-air missiles are inhibited, for instance, by the lack of an effective IFF system. The total systems approach, with its full range of analysis tools, may be the only acceptable means of evaluating trade-offs prior to the detail design phase. (The discussion of IFF design under Targeting Systems on p. 89 reveals how we solved the problem cited above.)

2. Parentheses enclose references, examples, ideas, and citations that are not part of the main thought of a sentence:

> Our Level 6 analysis (see figure 9.4) illustrates how a single multi-mission destroyer can contribute to task force operations.

> Our design accounts for all environmental factors that may affect sensitivity (smoke, terrain, weather, and physical damage).

> Affordability (cited in the RFP as a primary concern) was the guiding principle behind our application of new technologies.

> Our previous state-of-the-art survey (conducted over a 3-month period in 2007) suggested that RDF SOPs were not current.

> The most recent research (Smithson 2008) revealed pollution problems from nearby gasoline storage tanks.

See CITATIONS.

Parentheses, Commas, and Dashes

Commas and dashes also enclose explanatory ideas. Commas are less emphatic than parentheses; dashes are more emphatic. Note how emphasis progressively increases in the following examples:

> Cost analyses using both parametric and detail O&S cost methodologies helped us determine the right support systems.

> Cost analyses, using both parametric and detail O&S cost methodologies, helped us determine the right support systems.

> Cost analyses (using both parametric and detail O&S cost methodologies) helped us determine the right support systems.

> Cost analyses—using both parametric and detail O&S cost methodologies—helped us determine the right support systems.

See COMMAS and DASHES.

3. Parentheses enclose numbers in a paragraph list:

> The operational characteristics we will discuss below are (1) manning, (2) training, and (3) providing required support.

See LISTS.

4. Parentheses enclose acronyms, abbreviations, definitions, and figures that have been written out:

> The CARP (Capital Area Renovation Project) is adequately funded as long as the contractor trims costs by using off-the-shelf materials wherever possible.

> United's South Fork Mine can deliver over 20,000 dwt (deadweight tons) of ore every month.

> Artesian water (water naturally confined in the ground under pressure) is the primary source for the city's culinary use.

> By the project deadline date, Northrop will deliver fifty (50) centrifugal pump assemblies to the San Diego facility.

See ABBREVIATIONS and ACRONYMS.

NOTE 1: As in the first example above, we recommend using the acronym followed by its explanation in parentheses. Some editors prefer the reverse:

> The Capital Area Renovation Project (CARP) is adequately funded as long as the contractor trims costs by using off-the-shelf materials wherever possible.

NOTE 2: The practice of writing out numbers and enclosing the figure in parentheses is not necessary except in legal, contractual, or requisition documents. Do it only when you need to protect against unauthorized alteration of numbers in a document. See NUMBERS.

Parentheses

Parentheses and Brackets

Brackets are, in effect, parentheses. They enclose incidental or explanatory words and phrases within parentheses or within quoted material:

> The environment and activities of opposing forces may change the capabilities of a particular sensor (see appendix 4, Battlefield Adaptability Requirements, for a fuller discussion of RGS [Remote Ground Sensing] and ground-based sensor limitations).

> Your original letter stated: "Our onsite project coordinator [Walt Petersen] will be responsible for maintaining the Schedule of Deliverables."

See BRACKETS.

Parentheses and Periods

If the entire sentence is enclosed by parentheses, the period at the end of the sentence goes inside the closing parentheses:

> (See appendix 2 for the complete test results.)

If only part of a sentence is enclosed by parentheses and the closing parenthesis occurs at the end of the sentence, the period goes outside the closing parenthesis:

> Hydrostatic and thermostatic monitors ensure system equilibrium (see figure 5–15 for monitor locations).

See PERIODS.

Parentheses and Question Marks

Question marks come inside parentheses when they are part of the parenthetical (added) information:

this

> The project deadline (April 1?) is never stated in the Statement of Work.

not this

> The project deadline (April 1)? is never stated in the Statement of Work.

See QUESTION MARKS.

Periods primarily indicate a break or a full stop in text. At the end of a spoken sentence (signaled by a period), the voice drops and the speaker takes a breath.

Periods

1. Use periods following statements, commands, indirect questions, and questions intended as suggestions.
2. Use periods following numerals or letters marking a list, but periods need not follow the items listed unless they are full sentences.
3. Use periods (decimal points) to separate integers from decimals.
4. Use a period with run-in headings, but not with displayed headings.
5. Use periods following some abbreviations.

1. Use periods following statements, commands, indirect questions, and questions intended as suggestions:

Statements

The workover plan was finished.

Tomorrow we will visit the mine site.

Mr. Smythe owes OP&L $75.

Commands

Stop working on the project now.

Please help us tomorrow.

Redesign the pump housing to accommodate the larger intake pipe.

Indirect questions

I wonder how he managed the project.

Jane Greer asked whether we would approve the budget.

Questions intended as suggestions

Will you please return the forms by a week from Monday.

Would you let me know if you have any questions.

2. Use periods following numerals or letters marking a list, but periods need not follow the items listed unless they are full sentences (see rule 1 above):

a. A larger pump
b. An extra ventilation fan
c. A heavy-duty circuit breaker

1. The cost is 50 percent greater than was budgeted.
2. Materials were not equal to those specified.
3. Installation procedures were violated.

Note: You may need a period to end a list that continues the syntax established in its lead-in sentence:

We tested the procedure by
1. increasing the flow,
2. decreasing the temperature, and
3. contaminating the water.

This pattern of continued syntax and punctuation is much rarer than it used to be. See Lists.

3. Use periods (decimal points) to separate integers from decimals:

4.567
327.5
1,456.25

Note 1: In tables, the decimal points (periods) are aligned:

4.567
327.5
1,456.25

Note 2: In some foreign countries, a comma separates integers from decimals, and spaces separate groups of three numerals in longer numerals:

4,567
56 764,534 45

See Metrics and Numbers.

4. Use a period with run-in headings, but not with displayed headings:

Two Options. The first option is to discontinue the testing until safety procedures are developed. The second option . . .

This same heading would not be followed by a period if it appeared on its own line:

Two Options

The first option is to discontinue the testing until safety procedures are developed. The second option . . .

Note: You can also use run-in headings with dashes, colons, or no punctuation. But be consistent. Once you've established a pattern, use it throughout a document. See Headings.

5. Use periods following some abbreviations:

10 a.m.	6 p.m.
C.E. 1910	225 B.C.E.
U.S.A.	Mr./Mrs./Ms.
e.g.	S. Pugh
Dr. Will Lange	U.K.
i.e.	

Note: Many abbreviations and most acronyms no longer require periods, especially names of fraternal organizations, government agencies, corporations, and colleges and universities:

BPOE	BLM
DOE	GM
UCLA	LSU
HUD	FERC
NASA	

See Abbreviations, Acronyms, and Parentheses.

Persuasion

Persuasion relies on any techniques used by a writer or speaker to gain agreement or to support an idea. The most obvious persuasive technique is an appeal based on a logical rationale and on solid reasons, but the most effective technique is really the one that also enlists the reader's or listener's emotional support.

Successful persuasion is 50 percent emotion and only 50 percent logic and facts. Without the emotion and the subjectivity, logic and facts just don't convince.

The following rules use examples primarily from written documents. The rules, however, would apply equally well to oral presentations.

The following rules also refer to customers rather than readers. A customer is someone involved in a process or an activity and someone to negotiate with. World-of-work documents presume that readers are like customers—that is, these readers have something they need to learn, and your task is to service their needs.

Some documents are all persuasion—for example, a sales letter or a proposal to hire more employees. Other documents may not be obviously persuasive, yet they usually have a persuasive intent. A scientific study or a financial audit report may be primarily factual, but their writers still intend that readers understand and act upon the value and reliability of the information presented. Therefore, the scientific study and the audit report are persuasive documents, whatever their other purposes might be.

Persuasion

1. Assess your customers' needs and objectives.
2. Define your own role and objectives in relation to those of your customers.
3. Collaborate with customers to generate persuasive solutions and benefits.
4. Design a message that speaks clearly, effectively, and persuasively to your customers.
5. Maintain a credible position so customers have reasons for agreeing with you and your ideas.

1. Assess your customers' needs and objectives.

The customers' needs and objectives are important starting points because persuasion must be a two-way process. The old model of the salesperson with a pitch is gone. Today's persuasive writing is based on problem solving and consultation, not pitching a canned presentation to skeptical customers.

Initially, list your customers, both internal and external. Next, profile each of them by asking the following questions:

—What are their jobs and professional responsibilities?

—What level and type of decisions can they make?

—Whom do they work for and whom do they supervise?

—What problems or questions are they working on now?

—What do they already know about your services or your ideas?

—How likely are they to use the information you can provide?

Your primary goal in asking these questions is to determine what your customers' needs are.

Example 1—Recommendation Document. Your customer needs to obtain a more reliable source for computer network service. Your questions reveal just how important this need is to the customer, and you discover that the customer values a solution enough to spend money hiring a new employee or a new vendor. You have to decide what to recommend to the customer.

Example 2—Research Report. Your customers are readers of your report on a new surgical procedure. As you profile these customers, you determine how they intend to use your report and how likely they are to agree with your views. Based on what you determine about their needs, you can either adjust your existing report or, perhaps, add to it.

As in both of these examples, you must identify your customers' needs before you can begin to design persuasive documents.

2. Define your own role and objectives in relation to those of your customers.

As two-way communication, effective persuasive writing requires you to know your own role and objectives.

The following examples expand on the two examples introduced under rule 1 above.

Example 1—Recommendation Document. You decide that you would propose that the customer hire a new network computer specialist rather than obtain a new vendor. You might have a choice of roles, however. If your role is as an advisor, then your approach is more factual and neutral than it would be if you would profit from the decisions being made. Your second role might be as an advocate for a friend or family member. With this role, you move from neutrality to a more aggressive presentation of your points. Your changing roles control the slant of your persuasive documents.

Example 2—Research Report. If you discover that readers of your report are likely to be very skeptical about your new surgical procedure, you can adjust your research report to make it more convincing. Perhaps you add more detailed explanations of the prior literature related to your procedure.

At times, your role in relation to customers may lead you to ask questions where your honesty will be tested. For instance, do you need to admit to the customer that you made a mistake or that your equipment won't meet the written specifications? Such questions and problems are essentially ethical issues. See ETHICS.

Hesitation over an ethical question is perhaps the surest way to lose your persuasive credibility. See rule 5 below.

Unless you establish your own role, your objectives and, if appropriate, your ethical stance, you cannot prepare persuasive documentation.

3. Collaborate with customers to generate persuasive solutions and benefits.

This rule means that you continue to work with your customers, validating your initial perceptions and checking up on your progress. You never want your customers to feel neglected.

The following two examples build on details already introduced in the prior rules.

Example 1—Recommendation Document. As you work to list the pros and cons of the hiring of a new computer network specialist, you discover that you still have some unresolved questions about the problem. How frequent and severe are the network problems? Who is currently handling them and what does this service now cost? You need the data before you can complete your recommendations.

Your goal with these or similar questions is to refine your ideas so you can specify the benefits related to your proposal.

Example 2—Research Report. You transmit a working copy of your research report to a colleague who will be a sharp critical reviewer. Your intent is to find out if your approach and the details in your draft are on target. This review—essentially a usability test—is a crucial step if you want to guarantee that your report has the most persuasive impact possible. This review would parallel the peer review often required prior to publication in the scientific arena.

See the reviewing section of WRITING AND REVISING.

4. Design a message that speaks clearly, effectively, and persuasively to your customers.

Make your message customer-centered. An effective and a persuasive message will have the customer in the foreground. You and your interests remain in the background.

Test yourself by looking at a recent persuasive letter or sales document: The *you's* in your message should outnumber two to one the *I's* and *we's*.

The two letters in figure 1 illustrate what we mean by customer-centered writing. The "Before" letter is typical of many sales letters. Its persuasive power is lost because the writer can't stop talking about her own product and her own interests.

The "After" letter moves the customer to center stage. It also illustrates the following three writing principles:

1. **Be as clear as possible.**
2. **Organize your information with your customers' needs in mind.**
3. **Choose effective and forceful examples.**

Be as clear as possible. An undefined term or an unclear explanation allows customers to dismiss your ideas on the trivial grounds of their not being able to understand you. So do everything you can (or need to) to make sure that you have been clear. Be especially careful to define all technical terms and technical assumptions. Use page layout, boldface, headings, lists, and other format options to make your message as unambiguous and as effective as possible. See EMPHASIS.

Organize your information with your customers' needs in mind. Organization means that you have considered both the logic of your points and the arrangement of these points for the maximum effectiveness on your customers. See ORGANIZATION.

Example 1—Recommendation Document. You have two options. One is to open with your recommendation and state it both in the subject line and in your opening lines of the document. This option is the strongest and clearest way to begin your document.

A second option would be to lead up to your recommendation by giving some background—perhaps the scope of your investigation or some details about similar problems you've analyzed. This option delays your recommendation for a few lines or a paragraph or two. You would use this option only if you felt that your customers needed to be prepared for your recommendation. This opening is not as forceful as the opening in option 1, and it will likely not be as clear.

Example 2—Research Report. Often you have no discretion as to how to organize the body of your report. The major headings are standard as they are in many scientific reports: Introduction, Methods, Materials, Results, Conclusions, and Recommendations. Still, you have some choices to make. For example, you may precede the report with a summary, an abstract, or a transmittal letter. If so, you need to decide how direct and specific you want or need to be. This decision will rely on your view of the customers' point of view about your proposed surgical procedure.

Choose effective and forceful examples. Persuasive writing requires an interplay between general principles and good examples. Choose examples

BEFORE

Ann Devi
Vice President, Sales
Ringer-Helvetica LLC

Subject: Oblako 9

This letter is a follow-up to a conversation we had some time ago about switching your customer-relations management program to Oblako 9. As you may not be aware, Oblako is now up and running in the EMEA region. Our Oblako 9 is a mature product tested in the IT market for the last 2 years.

Oblako 9 is actually a whole set of capabilities. It enables visual processing of channel sales efforts, interactive/collaborative functionality that bridges fluidly across departments and divisions, and the ability to track multichannel opportunities. Additionally, the analytics of Oblako 9 are second to none, providing you the ability to assess your multifaceted sales process. Oblako, the Russian word for "cloud," is accessed via a secure Internet connection, thus eliminating the capital expense of purchasing hardware and maintaining it.

We have a great product and I could go on, but I really wante to remind you that Oblako has opened a direct office in your area. I hope you'll visit us to see for yourself the full-featured capabilities of Oblako 9. I have attached a customer survey fo you to fill out.

Hoping all is well with you. If you have any questions, please not hesitate to call.

AFTER

Ann Devi
Vice President, Sales
Ringer-Helvetica LLC

Subject: **Invitation to Discuss Improved Customer-Relationship Management**

Ann, in our discussion, you identified three challenges we can help you meet. Please consider a visit from us to explore these challenges further.

Lowering the cost of maintaining your current CRM system

Your in-house CRM system requires constant updating and maintenance at a cost of €77.000 per year, in addition to the licensing fees you pay. A secure Internet application like Oblako 9 can eliminate that maintenance outlay overnight.

Reducing sales cycle time

Your current sales-cycle time is about 36 weeks. We have typically seen Oblako 9 users cut 15 to 20 percent from the time between first contact and close. You could see sales closing anywhere from 6 to 8 weeks faster.

Increasing sales-force adoption rate

Your sales representatives struggle with the complexity of your current system, resulting in only a 50 percent adoption rate. By contrast, more than 80 percent of Oblako 9 users fully adopt the program within 3 months of installation because it's easy to use and significantly improves deal management.

Our local representative, Tessa Jorgen, will ring you to arrange a meeting.

Sincerely,

Figure 1. Before and After Versions of the Same Letter. *The "Before" version is writer centered and not well organized. The "After" version focuses on the reader (customer) and uses a variety of emphasis techniques.* *See Emphasis.*

that match the backgrounds of your customers. If your customers have a limited technical background, choose nontechnical examples.

Example 1—Recommendation Document. You might choose to develop a scenario with associated times for troubleshooting a network service problem. This hypothetical example would demonstrate the cost savings to your customers.

Example 2—Research Report. You may have limited discretion to change the information you include. Certain figures, data, and analysis procedures would likely be required for the discussion of any new surgical procedure. You might decide, however, to present before-and-after pictures of the typical patient. You make this decision based on your estimate of your customers' needs for a visualization of the benefits of your procedure.

As examples 1 and 2 show, no rules exist about which examples or details would be most effective and most desirable. You can only choose appropriate ones if you have profiled your customers. For example, if your customers were raised in the 1980s, you can refer to incidents during the presidencies of Ronald Reagan and George H. W. Bush. For customers a generation younger, you might use allusions to later presidents.

5. Maintain a credible position so customers have reasons for agreeing with you and your ideas.

You will be establishing and maintaining your credentials from your first discussion with your customers to the writing of your final document. Your persuasive case stands or falls depending upon your customers' perception of your credibility.

Do the customers see you as honest? knowledgeable? professional? a good listener? sympathetic? In other words, should the customers trust what you have to say or to recommend?

From a personal standpoint, honesty and trust are the basis of any persuasive relationship. Without these virtues, you will have no success mustering reasons or developing a logical case. No questions about your ethics should ever arise. If they do, you will have damaged your own persuasive case. See ETHICS.

Don't rely on purely personal facts about yourself or your organization to help maintain your credibility. You can occasionally refer to your background or your prior business experiences. But these references will not buy you any credibility if you can't, for instance, produce a professionally appearing document or sound analyses of the problems at hand.

From what the clients see in your documents, you also establish your persuasive credibility by providing the following features or information:

—Careful, well-designed format—one that is professional and excellent

—Clear, persuasive organization and examples (see rule 4 above)

—Technically sound, valid analyses of data and examples

—No slips in spelling or punctuation

Photographs

Photographs convey realism and authenticity. In the past, photos were especially persuasive because readers trusted this realism. Today photos are still persuasive, but readers now know that a photograph often does not portray reality. Digital enhancements and virtual-reality techniques mean that the images in a photo or on a monitor may never have existed except in someone's imagination.

Choose a photograph if you want to show your readers what is (or, perhaps, what could be). Choose a diagram or an illustration if you want to convey a concept, not realism. See ILLUSTRATIONS.

The following discussion presents a few principles for selecting and placing photographs in your documents and presentations. Techniques of how to take a photograph are beyond the scope of this discussion. Several of the rules, however, mention principles that would help you take effective photographs.

For related information about graphics, see GRAPHICS FOR DOCUMENTS and GRAPHICS FOR PRESENTATIONS. See also COLOR, ILLUSTRATIONS, and PAGE LAYOUT.

Photographs

1. Before choosing (or taking) a photograph, visualize its role by deciding exactly what your message is.
2. Survey different sources of photographs so that you have a rich array of photos to choose from.
3. Remove or adjust the features in a photograph to complement the purpose of your document or presentation.
4. Choose a photo with a simple yet interesting composition.
5. Select photographs with lighting that enhances the subject and reinforces the message.
6. Choose a photograph with an angle that best shows your subject and reinforces your purpose.
7. Use color photographs to enhance the effectiveness of your document or presentation.
8. Establish the size, scale, and orientation of the objects in a photograph.

1. Before choosing (or taking) a photograph, visualize its role by deciding exactly what your message is.

A document or a presentation should have a clear, forceful message. Graphics and any photographs you choose must support this message.

Identifying your readers (viewers) is the first step in deciding what your message should be. What do you want readers to do, to know, to feel, and to remember? You must strive to view each photograph as your readers would view it.

Assume, for instance, that you are preparing a brochure to promote a hospital. You have done your homework by talking to medical doctors, nursing supervisors, hospital directors, and others in health care. As they explained their needs, they said their highest goal was to center everything they do on the health and welfare of the patient.

Figure 1 (actually two pages with a single photo) shows how the marketing challenge was solved. The right side of figure 1 is the cover for the brochure. The left side of figure 1 is the back cover of the brochure, so the two halves are continuous if a reader opens up the brochure and looks at both halves at the same time.

The photograph in this figure shows two doctors talking while a patient with a leg brace and crutches moves toward them. Only the patient appears in color and front center, communicating his central importance to the hospital.

Figure 1 illustrates the rich options you have for creating layouts and presentations that are visually effective. In this figure, the underlying photograph supports the overlaid text and the two panels flow together to convey a single message: high-quality patient care.

As in the preceding health care example, each photograph (image) you choose must capture your readers' interest and convey the message you want to send. Choosing the right photograph should be an early step in the design of an effective document. See EMPHASIS and WRITING AND REVISING.

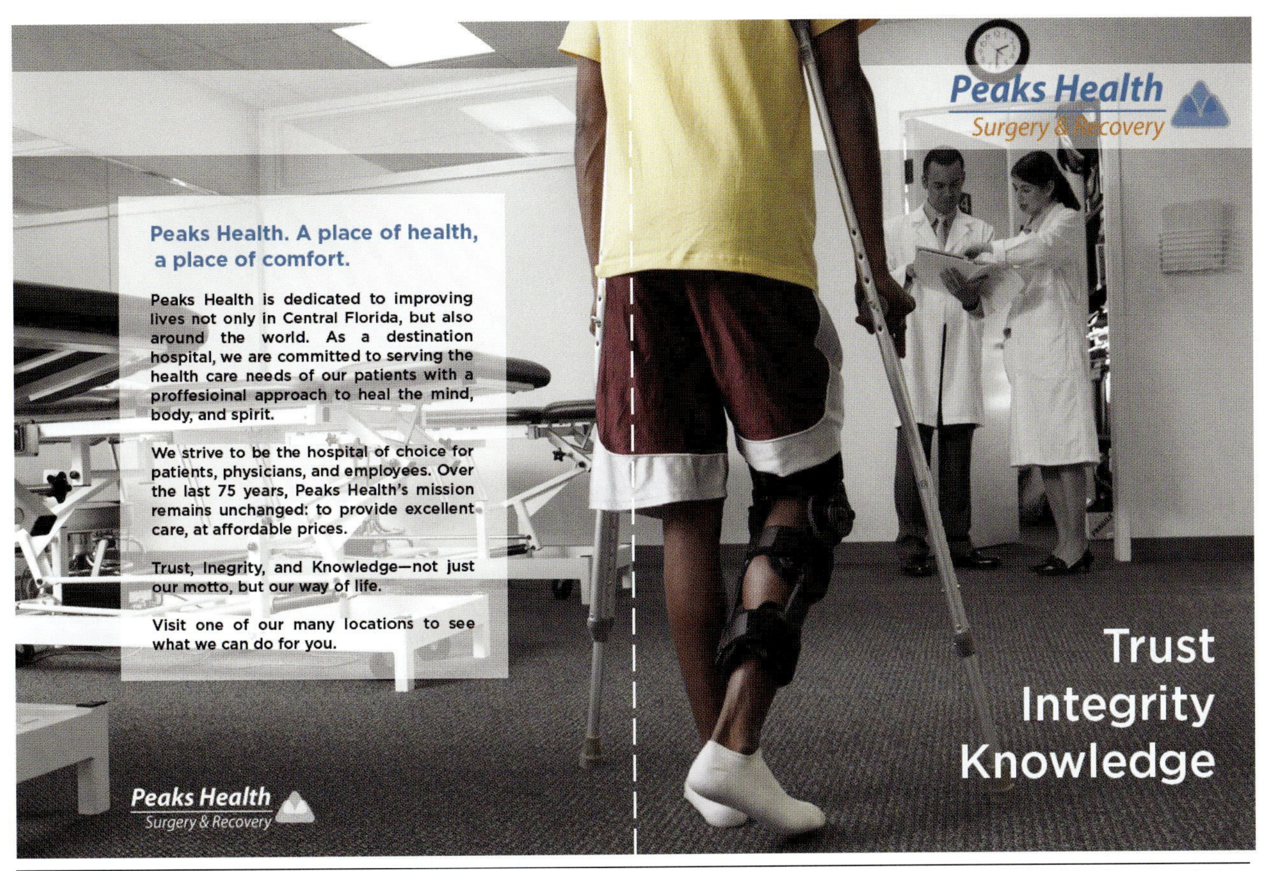

Peaks Health
Surgery & Recovery
Peaks Health. A place of health, a place of comfort.
Peaks Health is dedicated to improving lives not only in Central Florida, but also around the world. As a destination hospital, we are committed to serving the health care needs of our patients with a proffesioinal approach to heal the mind, body, and spirit.
We strive to be the hospital of choice for patients, physicians, and employees. Over the last 75 years, Peaks Health's mission remains unchanged: to provide excellent care, at affordable prices.
Trust, Inegrity, and Knowledge—not just our motto, but our way of life.
Visit one of our many locations to see what we can do for you.
Peaks Health
Surgery & Recovery
Trust
Integrity
Knowledge

Figure 1. The Front and Back Covers for a Brochure. *The horizontal photograph links the front cover (the right side of the figure) with the back cover (the left side). As in this example, the creative use of a photograph enhances the message in the text.*

2. Survey different sources of photographs so that you have a rich array of photos to choose from.

Professional-quality photographs are widely available and easy to copy and paste into a document.

A convenient option is to access online collections of stock photographs. These collections often contain thousands of pictures categorized by theme—such as business architecture or coins and money.

Remember, however, that almost all photographs are protected by copyright, as are other graphics. You may not use someone else's photograph without permission and, often, the payment of a fee.

Some websites contain photos that are free for any use, but they are often of low quality.

Also, when you request permission, you should ask if you are restricted from changing the photo to suit your purposes. Some photographers will not grant you permission to print a black-and-white version of a color photograph. Other photographers want to know what size of reproduction you intend to make. See Intellectual Property.

If you decide to shoot your own photographs, be sure to get a signed release from anyone appearing in your photos. Such a release allows you to publish and distribute the photos.

3. Remove or adjust the features in a photograph to complement the purpose of your document or presentation.

This rule expands on rule 1 because, after you have identified a document's message, you may find that no available photograph conveys your exact message.

Tools for choosing or adjusting a picture are common and are increasingly easy and inexpensive to use. Once a picture is electronically downloaded or scanned into a computer file, you can adjust the light, the texture, the color, or even add or crop images to fit the intended purpose.

Photographs

The high angle of the photo in figure 2 dramatizes the setting of the ruins of Petra in Jordan. Figure 3, a cropped and enlarged version of figure 2, emphasizes the scale of the monument by bringing the human figures closer.

Figure 4 is an example of a computer-enhanced photo. In this photograph, images relating to the digital world are imposed on the photo of a monitor. Viewers can easily see that this photograph is not realistic even though some of its components began as realistic images.

If you are taking a picture for a document, you should eliminate distracting details before you even shoot the picture. You might move a person or subject so that the background is not distracting. Or you can adjust your camera angle or direction of the shot so that a distracting image is not part of the picture. Figure 5 is an example of a picture with a distracting image. The people standing to the left of the red arch are likely a distraction to most viewers, so you would want to remove the people either by changing the camera angle (while the picture is being taken) or by cropping the picture, as in figure 3. Of course, if your purpose is to illustrate the scale of the arch, the people should stay.

Whether you are using someone else's photographs or taking your own pictures, you are responsible for the final images your readers will see. Choosing the proper image is just one step in the design of a document. See EMPHASIS and GRAPHICS FOR DOCUMENTS.

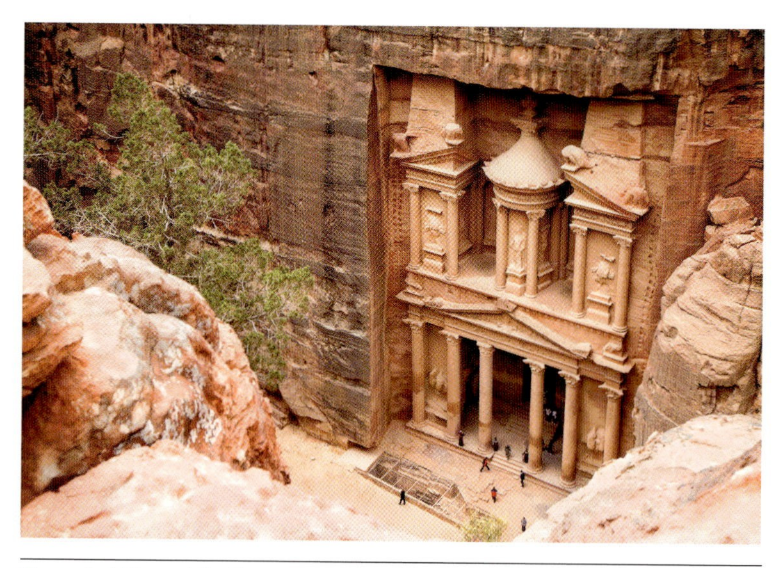

Figure 2. A View of the Treasury at Petra, Jordan. *This high-angle photo captures the entire front side of the historic ruin. The dramatic size of the ruin is, however, diminished because it is so small. An optional format would be to enlarge the area directly around the ruin, cutting out the foreground area. See figure 3 for an example of this optional format.*

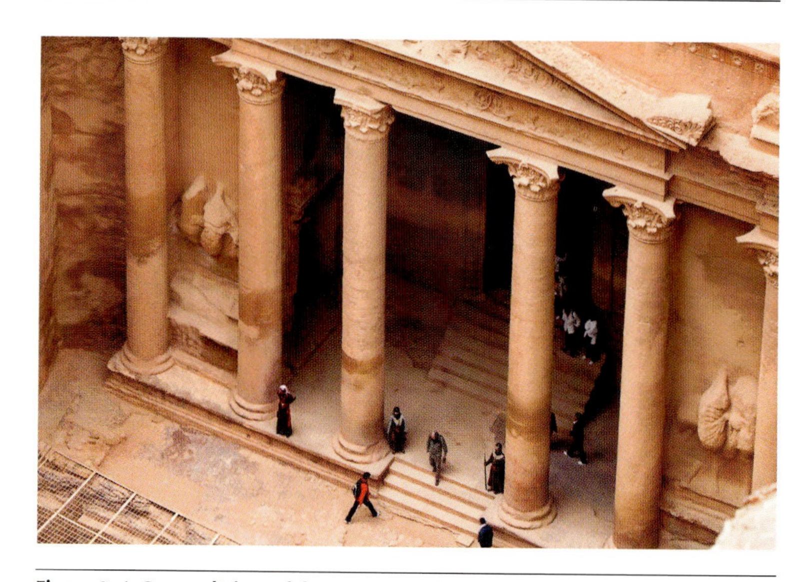

Figure 3. A Cropped View of the Treasury at Petra. *In this photograph, the emphasis is directly on the massive, awe-inspiring size of the ruin. The human figures show its massive scale.*

4. Choose a photo with a simple yet interesting composition.

Ensure that your subject is the principal point of interest in the photo, but a subject in the center of a photo often results in a static, uninteresting picture. Instead, choose a photograph where the main image is a bit off-balance or off-center.

This off-center arrangement—called the rule of thirds—is a widely accepted principle. Imagine that the picture plane is divided into thirds, vertically and horizontally. See figure 6 for a photograph with a superimposed grid of thirds. This grid shows in its four intersections where the primary points of emphasis or focus would usually be.

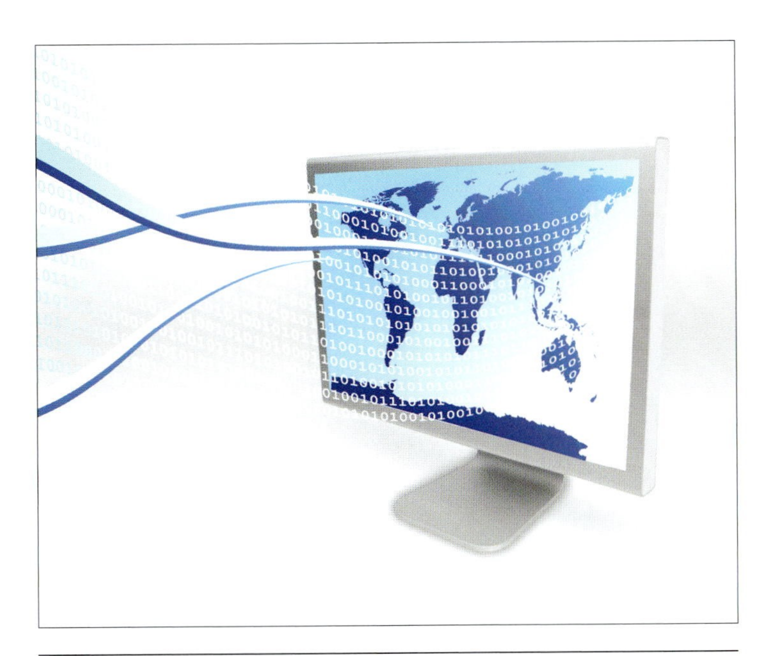

Figure 4. The Digital World. *This combination of illustration and photograph communicates both symbolism and reality.*

In this picture, the giant pylon lies on the vertical line marking the left-hand third of the picture. Another more distant pylon recedes into the lower third of the image.

Your goal is to choose a photo that has a balanced and interesting composition. Experimenting with the balance between different points of interest is why photographic professionals often take dozens of shots of the same scene, but from slightly different angles or with variations in the lighting.

Figure 7 has a balanced and interesting composition. The staircase with its converging spiral and ascending steps conveys a strong sense of depth and power. The composition is not quite centered, and the upper and lower handrails would line up horizontally with the left and right grids in the rule of thirds.

5. Select photographs with lighting that enhances the subject and reinforces the message.

Photographs, either in color or monochrome, record light in two dimensions, but we see in three dimensions. In a photograph, the illusion of three dimensions is created by the reflection of light at various angles off the subject(s). Proper lighting also produces shadows in a photograph, helping to create images and make the photograph more interesting.

Look at photographs with a critical eye. Is the existing light sufficient to provide depth and dimension? Light direction, for instance, is important in how we perceive an object.

Direct light from behind the camera makes colors bright but may reduce detail and texture in the background. Sometimes, objects may seem flat. In Figure 8, the direct light from behind the camera throws the emphasis on the tiger, and the overall impression is more two-dimensional than three-dimensional.

Side light emphasizes shape, texture, and detail, but it also can accentuate only those features you want readers to focus on. In figure 9, the light coming from the left of the picture highlights dramatically the glass vial and the gloved hand holding the vial. Notice that the side lighting gives

Figure 5. Delicate Arch in Arches National Park. *This arch, a favorite with tourists, appears on many professional postcards. People usually don't appear in such photos even though the people do help viewers realize how large the arch is. (Photo courtesy of Bill Grubbs)*

Figure 6. A Rule-of-Thirds Grid Overlaying a Photograph. *Professional photographers often choose an off-center or off-balance view of a scene. In this case, the near pylon is left of center, allowing a view of more distant pylons and thus the scale of the electrical transmission line. Another, less effective version of this picture might have moved the near pylon into the vertical center, with the other pylons out of the picture.*

Figure 7. A Balanced and Formal View of Ascending Steps. *The photograph of these steps effectively uses both the composition (arrangement) of the architectural details and the light coming down from a high angle. As rule 4 says, choose photographs with an impressive, interesting composition.*

the ice and the other vials a jewel-like texture. Also, as in this figure, side lighting often fades out any background details, most of which would likely be irrelevant to the point of this photograph.

Back light is especially dramatic. A subject lit only with a back light is a silhouette. Added side, overhead, or direct light can create a striking photo. This can be the most difficult lighting to do well because often the photo will be over- or under-exposed. Figure 10 uses back lighting to convert the two standing figures into silhouettes. The interaction between the two figures is the focus of the photograph, as expressed in their body postures and the extended hand of the figure on the left. Notice that the photograph does not include any distracting details, such as their facial expressions or their clothing.

6. Choose a photograph with an angle that best shows your subject and reinforces your purpose.

The angle from which a photograph was taken is often not particularly obvious, but it still affects how viewers interpret the picture.

A **normal** straight-on, frontal shot is often the choice for routine photos.

Figure 8. A Tiger in Direct Light (from Behind the Camera). *Notice that the light tends to flatten the image of the tiger. The tiger's face, which is offset to the left, illustrates the rule of thirds, as discussed above in rule 4.*

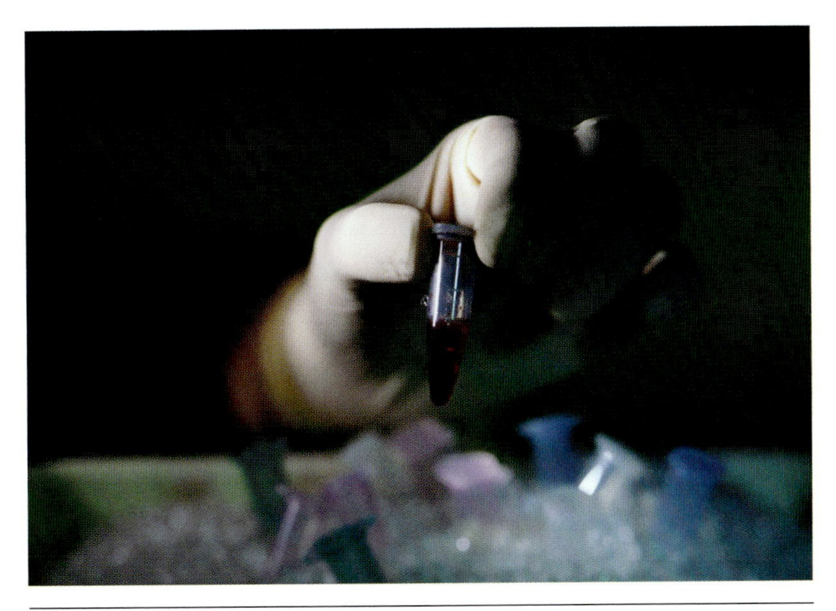

Figure 9. Side Light on a Hand Holding a Vial. The light comes from the left of the picture and highlights the hand and its interaction with the vial. Notice that background details vanish into darkness because the side light is falling on the vial and the ice.

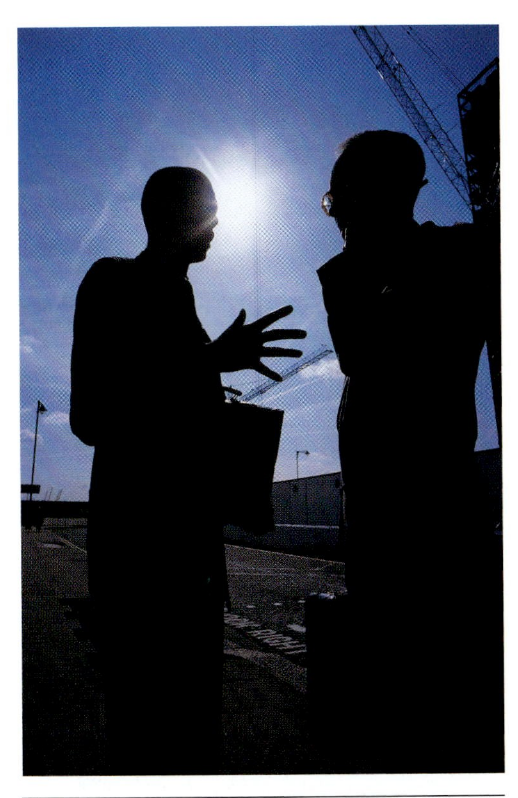

Figure 10. Back Light of a Business Conversation. *The back lighting makes the images of the men dramatic, even energetic. Small details, such as the facial features of the man facing the camera, are not the purpose or goal of the photograph.*

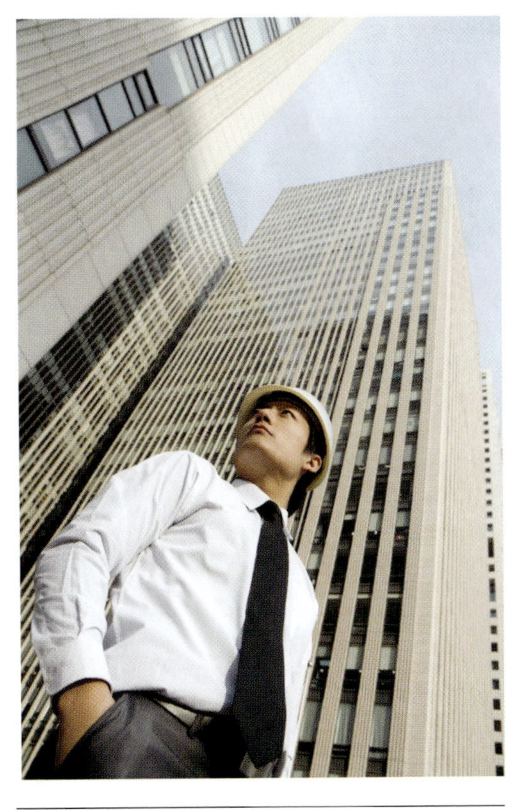

Figure 11. A Low-Angle Photograph of an Engineer. *The extremely low angle makes the figure seem unrealistically tall, yet also powerful and in command of a situation. This message may well have been the photographer's goal in this instance. In other cases, however, this picture might not be acceptable precisely because it is unrealistic.*

Figure 12. Operator of Assembly-Line Equipment. *Both the raised angle and the foregrounding of the equipment de-emphasize the role of the man.*

This angle (or lack of an angle) would be appropriate for a control panel or for a static display.

In many other contexts, however, different angles would produce more interesting effects.

A picture taken from a low camera angle accents the subject and makes it more impressive, perhaps even dominant. Too obvious an angle, however, will make the subject seem distorted, so be cautious in choosing low-angle shots. Figure 11 is an example of an extremely low-angle shot. Notice that it makes the man seem to be unrealistically tall. As with any such shot, this lack of realism was likely the photographer's goal, especially if the photographer wanted to show a commanding figure, with implications of control. The man is also looking into the distance, as if to signal vision and determination. As rule 1 suggests, you (and a photographer) must choose photographs to fit your intended message.

A picture taken from a high camera angle makes human subjects more abstract and appear less human. The image of the man in figure 12 is overshadowed by the mechanisms around him; the raised angle further de-emphasizes the man.

7. Use color photographs to enhance the effectiveness of your document or presentation.

Color photographs provide more information than a black-and-white (monochrome) picture. Colors convey tints, hues, and textures that no monochrome can hope to convey. As figure 13 (the inset color version and its monochrome version) shows, a color photograph has richer details and a sharpness of focus. If you need to print in only one or two colors, either take monochrome pictures or pictures that have a full range of contrast and color shades. See COLOR.

Reproduction quality remains an issue, whether you are using color or monochrome photos. Some printers make photographic images fuzzy or indistinct. In other cases,

Figure 13. A Monochrome Version of a Color Photo. *As the small inset photo shows, the original photograph was in color. The monochrome print is still effective even though it has fewer contrasts between light and dark areas and between different hues or shades of color. (Photo courtesy of Bill Grubbs)*

the final colors may not be true to the original image. So be sure to print a trial version of a photograph. Validate that the quality is what you or your client expects.

Sometimes a photograph can be the background for a page of text. Figure 1 is an example of this technique. If you decide to use a photographic background, experiment with the color balance of the picture. Sometimes you might shift the main color toward blue or even fade it out toward a neutral tan or even white, rather than a high-contrast black-and-white.

Figure 14. Close-Up of a Computer Circuit Board. *The American one-cent coin shows how tiny the circuitry is.*

8. Establish the size, scale, and orientation of the objects in a photograph.

A photograph can be puzzling to viewers if the size, scale, or orientation of an image is not immediately obvious.

When the size of an image is not obvious, include rulers, a human hand, or common objects in the photograph to make the scale obvious. As options, use a ballpoint pen, a key, or a coin next to the main image. This technique is illustrated in figure 14, where the size of the circuit board would have been uncertain without the accompanying figure of an American one-cent coin.

For similar reasons, geographical photographs may also be puzzling. Often viewers would need to know from which direction a picture was taken. The freeway interchange in figure 15, for instance, could run either north and south or east and west. If the orientation is important, then, as in figure 15, insert an arrow showing which direction would be north.

Figure 15. Freeway Interchange. *The superimposed arrow in the lower right corner shows the north/south orientation of the interstate in the photo.*

Plurals

Plurals of most nouns and pronouns are signaled by their spelling. Such changes are no problem if they follow the regular pattern: an *–s* or an *–es* added to the singular form makes the plural form:

report + s = reports
book + s = books
church + es = churches
tax + es = taxes

Problems arise when the plural does not follow the regular pattern:

mouse	mice
datum	data
chassis	chassis
fungus	fungi
matrix	matrices
I	we
he, she, it	they

The following discussion covers these and other irregular plurals. The best guide, however, is a good recent dictionary. See References, Nouns, and Spelling.

Plurals

1. Check a list of irregular forms to determine the correct plural, especially those plurals borrowed from Latin or other languages.
2. Add the plural ending (usually *–s* or *–es*) to the most significant word in compound terms.
3. Add an *–es* to form the plural of nouns ending in *–o* with a preceding consonant.
4. Add an *–s* or sometimes an apostrophe plus an *–s* to form the coined plurals of abbreviations, titles, figures, letters, and symbols.
5. Plurals of pronouns, when they exist, are very irregular.

1. Check a list of irregular forms to determine the correct plural, especially those plurals borrowed from Latin or other languages:

addendum, addenda
agendum, agenda
alga, algae
alumnus, alumni (*masc. or mixed*)
alumna, alumnae (*fem.*)
antenna, antennas (antennae, *zoology*)
appendix, appendixes (*or* appendices)
axis, axes
basis, bases
cactus, cactuses
calix, calices
cicatrix, cicatrices
Co., Cos.
coccus, cocci
consortium, consortia
crisis, crises
criterion, criteria
curriculum, curriculums (*or* curricula)
datum, data
desideratum, desiderata
ellipsis, ellipses
equilibrium, equilibriums (equilibria, *scientific*)
erratum, errata
executrix, executrices
focus, focuses
folium, folia
formula, formulas
fungus, fungi
genus, genera
gladiolus (*singular and plural*)
helix, helices
hypothesis, hypotheses
index, indexes (indices, *scientific*)
lacuna, lacunae
larva, larvae
larynx, larynxes
lens, lenses
locus, loci
madam, mesdames
matrix, matrices
medium, mediums (*or* media)
memorandum, memorandums (*better* memo, memos)
minutia, minutiae
nucleus, nuclei
oasis, oases
octopus, octopuses
opus, opera
parenthesis, parentheses
phylum, phyla
plateau, plateaus
radius, radii
radix, radixes
referendum, referendums
septum, septa
seta, setae
stimulus, stimuli
stratum, strata
stylus, styluses
syllabus, syllabuses (*or* syllabi)
symposium, symposia
synopsis, synopses
terminus, termini
testatrix, testatrices
thesaurus, thesauri
thesis, theses
thorax, thoraxes
vertebra, vertebras (vertebrae, *zoology*)
virtuoso, virtuosos
vortex, vortexes

Note 1: Many of the above forms now have regular plurals (*appendix, appendixes* or *memorandum, memorandums*). However, some editors still prefer the irregular forms (usually based on the word's origin in Latin or another language: *appendices, memoranda*). The longer a word is in English, the stronger the tendency is to make the plural conform to the regular English pattern (adding an *–s* or *–es* to the singular form).

Note 2: *Data* (the plural of *datum*) is now used as both a singular and plural form. In everyday conversation or informal writing, the singular is common:

Data shows that the new . . .

In technical contexts, the plural is usually preferred:

Test data show that a high correlation exists between . . .

See Word Problems.

Note 3: Though many of these words come from other languages, they are now sufficiently English words and do not need underlining or italics. See Italics and Underlining.

2. Add the plural ending (usually *–s* or *–es*) to the most significant word in compound terms:

attorneys at law
bills of fare
brothers-in-law
comptrollers general
daughters-in-law
goings-on
grants in aid
lookers-on
reductions in force
surgeons general
assistant chiefs of staff
assistant surgeons general
assistant attorneys
deputy judges
lieutenant colonels
trade unions
hand-me-downs
higher-ups
pick-me-ups

3. Add an *–es* to form the plural of nouns ending in *–o* with a preceding consonant:

echo, echoes
potato, potatoes
tomato, tomatoes
veto, vetoes

NOTE: This rule has many exceptions, so if in doubt, check a good dictionary. See REFERENCES. Here are some of the common exceptions:

dynamo, dynamos
Eskimo, Eskimos
ghetto, ghettos
halo, halos
indigo, indigos
magneto, magnetos
octavo, octavos
piano, pianos
sirocco, siroccos
two, twos
zero, zeros

4. Add an *–s* or sometimes an apostrophe plus an *–s* to form the coined plurals of abbreviations, titles, figures, letters, and symbols:

ABC's *or* ABCs
COD's *or* CODs
g's
OK's *or* OKs
1 by 4's
SOS's
the three Rs

NOTE: Use the apostrophe only if necessary. Save the apostrophe to show possession, as in *John's hat.* The trend is for the apostrophe to vanish, leaving the simple *–s* to signal that the item is a plural. See APOSTROPHES.

5. Plurals of pronouns, when they exist, are very irregular:

I	we
you	you
he, she, it	they

See PRONOUNS.

Possessives

Possessives are those forms of nouns and pronouns that show ownership or, in some cases, other close relationships:

Ownership

IBM's service booklet
Mr. Vaughan's store
his store
Susan's desk
her desk
Lewis' (*or* Lewis's) report
the engineer's schedule
the Lewises' house

Other relationships

the book's cover
its cover
the corporation's support
his lawyer's consent
the captain's story
a summer's day
a day's absence
a doctor's degree
three years' experience

NOTE: Noun possessive forms routinely require an apostrophe or an apostrophe plus an –*s*. See APOSTROPHES.

Possessives

1. Distinguish between true possessives and descriptive terms.
2. Add an apostrophe and an –*s* to form the possessive of singular nouns not ending in –*s* and for plural nouns not ending in –*s*.
3. Add only an apostrophe to form the possessive for both plural nouns and singular nouns ending in –*s* or an –*s* sound.
4. Use only a single apostrophe plus –*s* when joint possession is intended.
5. Add an apostrophe plus an –*s* to the end of personal and organizational names and abbreviations showing possession.
6. Do not use an apostrophe to form the possessives of personal pronouns and of the relative pronoun *who*.
7. Add apostrophes to form the possessives of indefinite pronouns.
8. Be aware that possessives sometimes occur without a following noun.
9. Use a possessive to modify an –*ing* form of a verb used as a noun.

1. Distinguish between true possessives and descriptive terms:

Possessives (whose . . .)

Exxon's reply
the employee's record
Oregon's laws
the Smiths' house

Descriptive terms (what kind of . . .)

an Exxon reply
the employee record
Oregon laws
the Smith house

NOTE 1: Either form of the above phrases is correct, so decide which form you prefer and then be consistent within the same document. See ADJECTIVES.

NOTE 2: The names of countries, governmental units, and organized groups ending in –*s* usually do not require apostrophes:

the United States plan
a Massachusetts statute
the mineworkers court case
a United Nations publication

2. Add an apostrophe and an –*s* to form the possessive of singular nouns not ending in –*s* and for plural nouns not ending in –*s*:

the cat's paw
Anne's statement
a man's coat
men's coats
an accountant's books
the children's payments

3. Add only an apostrophe to form the possessive for both plural nouns and singular nouns ending in –*s* or an –*s* sound:

General Dynamics' proposal
Sears' 4th Quarter Report
Penneys' reaction
the boss' idea
James' speech
for goodness' sake

NOTE: Some editors and writers prefer to add both an apostrophe and an –*s*, especially if the new word has an extra syllable:

General Dynamics's proposal
the actress's script
the boss's idea
James's speech

Use of this extra –*s* is, however, declining.

4. Use only a single apostrophe plus *–s* when joint possession is intended:

Susan and Harry's proposal

Irene and Sam's book

NOTE: As in the preceding examples, joint possession is usually clear when the two people possess only a single item. But when plural objects or items are involved, the meaning is sometimes ambigious:

Irene and Sam's books

Do Irene and Sam own all books jointly? From this sentence it is hard to tell, so a clearer version would be this one:

Irene's and Sam's books

or

Irene's books and Sam's books

5. Add an apostrophe plus an *–s* to the end of personal and organizational names and abbreviations showing possession:

Sears and Roebuck's policy

Charles F. Shook's decision

the Odd Fellows's initiation (*or* Odd Fellows' initiation)

Dewey, Cheatum, and Howe's corporate policy

HUD's policy

BLM's guidelines

NOTE: Corporate and organizational practices vary, so if possible, check the letterhead or other correspondence for exceptional cases:

American Bankers Association
Steelworkers Union
Investors Profit Sharing

See APOSTROPHES.

6. Do not use an apostrophe to form the possessives of personal pronouns and of the relative pronoun *who*:

My secretary had **mine**.

Your supervisor had **hers**.

The company lost **its** comptroller.

We refused to pay for **ours**.

They lost **theirs** when the market fell.

He mentioned **his**.

Whose report is this?

NOTE 1: These are the possessive forms of personal pronouns: *my/mine, your/yours, his, her/hers, its, our/ours, their/theirs*. In cases with two forms, the first form must come before the noun it possesses and the second form comes after the verb, with its noun implied:

They were her ideas.

The ideas were hers.

NOTE 2: Distinguish between possessive forms without apostrophes and contractions with apostrophes:

The pump had lost its cover.
We decided that it's (*it is*) time to redrill.

He had changed his job recently.
He's changing his job.

Whose idea was this?
Who's going to be at the meeting?

See APOSTROPHES.

7. Add apostrophes to form the possessives of indefinite pronouns:

anyone else's task

one's ideas

the other's notion

the others' schedules

anybody's recommendation

someone else's job

8. Be aware that possessives sometimes occur without a following noun:

My ideas are like Sue's.

His nose is like a bloodhound's.

I'll be at Jim's.

She was at the doctor's.

NOTE: Large, established companies sometimes violate the usual rule (singular form plus an apostrophe and an *–s*):

We conducted a survey at Woolworths.

Harrods is splitting its stock.

9. Use a possessive to modify an *–ing* form of a verb used as a noun:

Bill's speaking to my boss helped.

I objected to his working on the rig overnight.

I admired Sue's planning.

Prepositions

Prepositions are words that connect or relate nouns and pronouns to preceding words and phrases:

The engineer moved **from** his desk.

The plans **for** the new substation have yet to be completed.

The case **against** her became even more convincing.

There is truth **in** what you say.

The firm submitted a summary **of** the specifications.

The simple prepositions are *at, by, in, on, down, from, off, out, through, to, up, for, of,* and *with*. More complex, even phrasal, prepositions also exist: *against, beneath, in front of, on top of, on board, according to, on account of, by means of,* etc.

Although they number less than a hundred, prepositions are essential words in English. Most normal sentences contain one or more prepositions.

Prepositions

1. Do not be overly concerned if you end a sentence with a preposition.
2. Omit certain prepositions if you can do so without changing the meaning.
3. Distinguish between the prepositions *between* and *among*.

1. Do not be overly concerned if you end a sentence with a preposition.

For over a thousand years English has had normal sentences that ended with prepositions:

That's something we can't put up with.

The Universal acquisition is the most difficult deal we've gotten into.

This is the report I've been telling you about.

She's the accountant you spoke to.

This project is the one that you objected to.

Winston Churchill was once corrected for ending a sentence with a preposition, and he supposedly replied, "That is the sort of English up with which I will not put."

As Churchill's witty reply indicates, many English sentences sound awkward if you try to avoid ending them with prepositions:

The Universal acquisition is the most difficult deal into which we've gotten.

This is the report about which I told you.

She is the accountant to whom you spoke.

This project is the one to which you objected.

Make your sentences as simple, smooth, and direct as possible, and don't worry about such misconceptions as not ending a sentence with a preposition. The fact is, a preposition is a fine word to end a sentence with.

2. Omit certain prepositions if you can do so without changing the meaning.

In each of the following sentences, the preposition within parentheses is unnecessary. In the last sentence, *at* is both unnecessary and incorrect.

All (of) the engineers visited the site.

We moved the pipe off (of) the loading dock.

The filing cabinet is too near (to) the door.

We met (at) about 8 p.m.

Where are they (at)?

Generally, native speakers of English can be guided by their innate sense of what constitutes smooth and clear uses of prepositions. See Modifiers.

3. Distinguish between the prepositions *between* and *among*.

Between usually refers to two things, while *among* refers to more than two things. However, *between* can also refer to more than two things if each of the things is compared to all the others as a group:

The judge divided the land between the two parties. (*not* among)

The judge divided the land among a dozen parties. (*not* between)

We had trouble deciding between the two pumps. (*not* among)

We carefully analyzed the differences between the six alternatives. (*or* among)

We had trouble deciding among all the possible small trucks. (*or* between)

See *among/between* in Word Problems.

Presentations

Presentations have much in common with documents. Both should have at least one specific purpose and an identified audience. If you want to accomplish the purpose of your presentation and get your audience to know, do, and feel what you want, you must plan your presentation with care and deliver it with confidence.

Presentations

1. Develop your presentation according to standards of effective design and delivery.
2. Identify clearly the purpose of your presentation.
3. Analyze your audience and plan your presentation to meet their needs.
4. Assess the situation in which you will deliver your presentation.
5. Build the evidence necessary to support your points.
6. Organize your message logically.
7. Design graphics that support your message.
8. Deliver your presentation engagingly and with confidence.
9. If appropriate, invite your audience to participate.

1. Develop your presentation according to standards of effective design and delivery.

Presentations consist of two equally important elements—design (organization) and delivery.

The design and delivery standards that we recommend in rules 2 through 9 apply generally to effective presentations. However, if you are delivering a presentation for a particular professional society or for a technical organization, you may find that they have special requirements you should strive to follow.

2. Identify clearly the purpose of your presentation.

Nothing is more important for an effective presentation than having a clear purpose—knowing specifically what you want your audience *to do, to know,* and *to feel.*

Identifying the purpose is important—not only for yourself in the planning phase of the presentation, but for the aid of your audience as well:

> Today I intend to explain the benefits of the TTCL proposal and recommend that we accept it.
>
> The purpose of my remarks this evening is to encourage you to reject the formation of a new water district.

Unless you, as a presenter, have a clear understanding of the purpose of your presentation, you cannot expect your audience to know what you hope to achieve with it. See ORGANIZATION.

3. Analyze your audience and plan your presentation to meet their needs.

To be truly effective as a presenter, you must know as much about your audience—their needs, desires, biases, etc.—as you can, even before you plan your presentation.

In addition to general information about audiences (e.g., that the attention span of the average adult is somewhere between 20 and 30 minutes), you would be wise to learn as much as you can about your specific audience. Ask yourself and others these questions:

—Who will be there? Why?

—What are their expectations?

—What do they need or want from me?

—What do they already know about the topic? How much or how little?

—What biases do they have?

—What questions will they have?

Many otherwise well-planned presentations have failed simply because the presenter knew too little about the audience, as in the case of the politician who unwisely called for an end to public old-age assistance at a convention of retired people.

4. Assess the situation in which you will deliver your presentation.

By "situation," we mean the time, setting, and occasion of the presentation.

If your purpose is to win audience support, for example, you may find that you could do just the opposite—simply by allowing your presentation to run 15 minutes overtime.

The setting can be a major factor in the overall effectiveness of your presentation. If possible, go beforehand to the place where you will deliver your presentation. That will give you a good idea of the advantages and disadvantages of the setting. Consider any possible changes that may contribute to a better situation (room setup, seating, lighting, your position in relation to the audience, screen visibility, etc.).

The occasion will often dictate certain elements of your presentation, such as the degree of formality, the use of visual aids, and even the type of clothing you choose to wear.

5. Build the evidence necessary to support your points.

Once you have determined the main points of your presentation, make sure that you provide enough (but not too much) evidence to support those points adequately. Among the types of support available, you may select the following:

Definition

Increased control means you'll be able to make your own decision about your investments.

Explanation

Each of you will be able to direct your money into 10 available funds.

Example

I divided my money among three different funds and am very pleased with the returns.

Testimony

Our comptroller, Brenda Lewis, estimates your return will be between 5 and 7 percent.

Statistics

Our former plan returned only half a percentage point last quarter.

See THINKING STRATEGIES.

6. Organize your message logically.

Organization is one of the keys to effective presentations. Unless you can lead your audience logically through your presentation, they may miss the points you are trying to make.

Preview your main points when you present your message. Basically, tell the audience very briefly what you intend to talk about:

These are the three reasons my committee chose Acme Plumbing:

1. Next to lowest price
2. Existing contracts with similar firms
3. Branch office less than 10 minutes away

One organizational technique that works very well is what we call the "Triple-S formula":

State your point.

Support your point.

Summarize your point.

Use and even repeat this organization—to ensure a logical flow, a solid format, and a convincing message. This kind of repetition increases the likelihood that your audience will remember your key points. Repetition improves retention. The Triple-S formula will work for your entire presentation or for each subpoint or subtopic.

7. Design graphics that support your message.

Graphics (graphs, charts, tables, photographs, etc.) can be a valuable means of supporting your message. But avoid the tendency of using them too much. Some suggestions to keep in mind about the use of visuals in presentations:

—Use one primary visual to communicate your main point. This visual might be a comparison of two courses of action, a description of a process, or a graph that reveals a key trend. Whatever your main point is, visualize it in a simple and memorable way for your audience. Return to this primary visual at the end of your presentation to summarize the point.

—Keep the visuals simple, clear, and uncluttered. If a complicated graph or chart is essential to your purpose, distribute a handout of it so audience members can see and understand it.

—Use no more graphics than you need to convey your message. A common mistake presenters make is to "firehose" the audience with dozens of slides.

—Consider taking a low-tech approach instead of using a computer projector. A simple whiteboard or paper flip chart allows you more spontaneity in drawing visuals and capturing information from the audience.

—Be cautious about using multimedia for your presentation. The more complex the stew of video, animations, and sound effects, the more likely you are to have a technical problem. Your message can get lost in the technology. Unless the situation requires multimedia, most business audiences prefer a simple visual presentation.

—Put informative headings on graphics, not just mysterious one- or two-word labels. See HEADINGS.

For additional suggestions regarding the use of graphics, see GRAPHICS FOR PRESENTATIONS, CHARTS, GRAPHS, ILLUSTRATIONS, LISTS, MAPS, PHOTOGRAPHS, and TABLES.

8. Deliver your presentation engagingly and with confidence.

Delivery is just as important as good organization and design.

Without an effective delivery, even an excellent message may suffer.

Excellent delivery results to a great extent from practicing effective delivery techniques. Among the techniques that contribute to effective stand-up presentations are the following:

—Know your material.

—Approach and leave the podium with confidence.

—Begin your presentation with a *relevant* attention-getter. A brief personal story is often the most effective way to start, or you could begin with a startling fact or a provocative question.

—Do not read your presentation (although you may refer to notes for specific information).

—Speak clearly, confidently, and with enthusiasm.

—Vary your voice in pitch, volume, and rate of speed.

—Use appropriate (and natural) gestures, facial expressions, and body movements.

—Maintain good eye contact with members of your audience. Talk to individuals, not to a crowd.

—End with a powerful call to action.

9. If appropriate, invite your audience to participate.

Tell your audience when and if you would welcome comments or questions.

In formal situations, you might decide to reserve questions until after your presentation. This strategy allows you to finish all of your prepared content.

In less formal situations, plan to stop to ask for an exchange of ideas. If you have just finished with an example of cost-saving techniques, you might turn to the audience with a question:

> What are some cost-saving steps you noticed?
>
> *or*
>
> What steps would be necessary to implement the cost-saving techniques I just outlined?

Notice that both of the preceding questions are substantive, requiring examples or facts. Avoid dead-end yes/no questions:

> Do any of you have ideas?
>
> How many of you here thought the same thing?

Project Management

Project management includes several processes and skills. A project manager must know how to choose and assign team members. She must know how to prepare and track useful schedules and accurate budgets. She must know how to monitor project accomplishments and how to give constructive feedback. She must also know how to report to senior management.

Also, an often overlooked aspect of project management is documentation. Documents for a typical manufacturing project would include design or engineering plans, marketing surveys, sales brochures, maintenance manuals, and even day-to-day project team records. See Managing Information.

The project team might believe that they are producing a new drug or the latest in mobile device apps, but these products will not be viable without supporting documents.

Documents, then, are an important product for any project team.

The following rules for project management are generic and universal, not unique to any professional group or academic specialty.

Project Management

1. Begin by developing a vision of the final product and of associated documentation.
2. Choose a broad, well-balanced project team to help with all phases of the project.
3. Identify quality standards for the project.
4. Make task assignments and set up an achievable schedule.
5. Provide for ongoing reviews for the project products—both the physical products and the documents.
6. Provide for open, rich, and safe communication.
7. Develop strategies for improving project management within your department or company.

1. Begin by developing a vision of the final product and of associated documentation.

Skilled athletes know that if they repeatedly visualize a successful final game or the final winning seconds of a race, they are more likely to be successful in the game or race. Their visualizations guide and strengthen them as they train and practice day by day.

An intellectual vision of a product is an essential starting point for a project manager and for a project team. This vision needs to be as comprehensive and as rich as possible, answering these key questions: What does success look like? What is the job to be done for the customer? What features are essential to this product or service? What questions of technology or science remain to be answered? Who will use this product and how will they use it? What quality standards are important?

This vision should also include the dead ends, engineering problems, disputed data, and anything else the project team expects to delay the project.

As early as possible, the team will be making engineering sketches and designs. At some point, the team will even produce a full prototype of the product, often one that works or functions as the final product would. Early development of prototypes enables early identification of issues that might affect the final outcome.

In a similar way, the team should work early with document prototypes (often called mock-ups). Documents to mock up include marketing plans, product descriptions, websites, and sales collateral. These document prototypes mature and develop as the team members answer questions and make decisions. See Writing and Revising.

Both physical prototypes and document prototypes are essential tools for capturing, as early as possible, a vision of the entire project.

2. Choose a broad, well-balanced project team to help with all phases of the project.

Your goal is to include in the team anyone who has a say or a stake in the success of the project. Your method is to get the fullest and richest input possible as early as possible.

There should be two project teams: a core team and a review team. The core team consists of those who actually carry out the project. The review team comes in and out of the project, evaluating early prototypes and helping with changing requirements.

A good review team includes interested people with many different perspectives—customers, managers, sales, marketing, finance, legal professionals, and members of the community. The more diverse the perspectives of your reviewers, the more robust the outcome will be.

If you miss a key contributor, you will wind up redoing phases of the project and, quite likely, wasting valuable time and money.

3. Identify quality standards for the project.

Quality standards establish your expectations as to what the team is to produce and how they will produce it. Often the best practice is to make the team responsible for identifying its own quality standards.

Quality standards apply to three different areas of project management:

—Project-management standards

—Product standards

—Document standards

Project-Management Standards. These standards are essentially the seven rules in this discussion of project management.

Figure 1 presents these project-management standards in a checklist format. Use this checklist to guide you as you manage a project or as you help someone else be a successful project manager.

Product Standards. Product standards are physical features and requirements that apply to the product to be produced. These are the criteria of a successful outcome. For example, one criterion of success for a new drug might be that it can be taken only once a day because government regulations require it. Another criterion of success might be that the new drug must comply with regulatory requirements in Japan, Europe, and the United States simultaneously.

Documents that accompany the drug must also be designed to a standard. For example, the package insert that comes with a trade medication must meet certain industry and regulatory criteria.

As in this example, product standards include document standards, but they are often overlooked.

Document Standards. Document standards begin with the quality of a writer's thoughts. Are major points and conclusions convincingly explored? Is the language (and the thought behind it) credible? Quality documents reflect the intellectual acuity of their writers.

These questions are the basis for the quality standards recorded in figure 2. For each quality standard, figure 2 lists common user (reader) questions and the criteria for measuring a document's response to each question.

Project Management Checklist

Check off each item in this list when you are satisfied that you (and your project team) have completed it. The rules in parentheses refer to rules from this entry on project management.

- ❑ Begin by developing a vision of the final product and its documentation. (Rule 1)
  - —Product prototypes
  - —Document prototypes
- ❑ Choose a broad well-balanced project team to help with all phases of the project. (Rule 2)
  - —Technical and scientific experts
  - —Marketing and financial representatives
  - —Customers (both external and internal)
  - —Others?
- ❑ Identify and record (in writing) quality standards for the entire project. (Rule 3)
  - —Product standards
  - —Document standards
  - —Project-management standards
- ❑ Make clear task assignments and set up an achievable schedule. (Rule 4)
- ❑ Provide for usability tests. If appropriate, repeat these. (Rule 5)
  - —Tests for the product
  - —Tests for the documents
- ❑ Identify and praise each person's contributions to tasks and to project milestones. (Rule 6)
- ❑ Develop strategies for improving project management within your department or company. (Rule 7)

Figure 1. Project Management Checklist. *Use this checklist to monitor your progress, either as a project manager or as a project team member. Remember that project management is complex enough to require your constant efforts to improve and refine its steps.*

A new drug application to the FDA (Food and Drug Administration) is an illustration of the role of document standards. As its primary content, an application must describe and analyze clinical trials (tests) of the drug. These data and their analyses must be credible and consistent, but also, data must be arranged and presented to comply with FDA submission requirements (FDA document standards). Besides test data, the application will also include other documents—for example, the proposed package insert to be included with each bottle of pills.

Each separate document has its own quality standards. Many of these standards are unique, as in this new-drug example, but see the rules in the EMPHASIS entry for a list of document standards that are applicable to most documents.

Quality documents, if prepared during the early prototyping phase, can influence the development of the product itself. Early documentation forces team members to commit to ideas; and as the review team evaluates these ideas, new insights will be discovered.

To return to the new-drug example, if the project team prepares an early prototype of the package insert, the team might discover certain features of the product that need further analysis or attention. The one-a-day requirement may be one such feature. Another might be that the drug should be taken with milk or a dairy product to avoid stomach cramps. If the team explores such issues early, they can investigate options. For instance, is a timed-release format medically effective? Or can the drug be mixed with an antacid so that it need not be taken with milk?

4. Make task assignments and set up an achievable schedule.

The team members should generate their own task assignments and set their own schedules. If a manager or the organization dictates these decisions, the project team will be more likely to fail.

Some project managers find it helpful to document a process flow, showing the relationships among key tasks and subordinate tasks, as well as determining which tasks are dependent on other tasks. Such a flowchart is often called a "critical path" description. The more complex the project, the more helpful such a flowchart might be. Another useful chart is a schedule chart (also known as a Gantt chart) that enables team members to track progress. On such a chart, key phases of the project are known as milestones. See CHARTS.

An achievable schedule is usually based on negotiation. Some people will always overestimate the time needed for a task; others will underestimate. And what is achievable is always uncertain.

As one project manager said about documentation tasks, "Make your best, most reasonable estimate and then double it." The time required will usually match the doubled figure!

5. Provide for ongoing reviews for the project products—both the physical products and the documents.

The bane of a project team is changing requirements, known as "scope creep" or "feature creep." Requirements can and do change with time, market or political realities, or new insights. A project team can become frustrated with changes and fail, or recognize that changes are inevitable and provide for them in the project-management process.

Ongoing reviews are the best mechanism for dealing successfully with changing requirements. The review team should do periodic evaluations early and regularly, on an iterative basis. The core team should expect to hear about important issues, new insights, and changing expectations. Adjustments follow, and the next review will raise more issues.

The review team employs a method generally called "usability testing" to evaluate the project team's products. Many techniques are included in usability testing: benchmarking, testing of prototypes, performing structured walkthroughs, etc. See WRITING AND REVISING for a detailed discussion of ongoing reviews.

Testing the physical product is an obvious step. Typically, a sample group of customers receives a concept description or a version of the product for testing—whether it is a new type of breakfast food or a redesigned electric drill. All the customers have to do is taste the new breakfast food or drill a few test holes. However, parallel testing of the product documentation should also occur.

Figure 3 lists six tests for evaluating the usability of a document. Choose one or more of these tests when usability is an issue.

6. Provide for open, rich, and safe communication.

Open, rich, and safe communication is critical to good project management, not to mention organizational health and growth.

Without good input and feedback, members of a project team are isolated and tasks performed by the team will be fragmented and often contradictory. If individual team members have a limited grasp of the project purpose and features, what they work on will be similarly

Document Standards		
Standard	**User Questions**	**Measurement**
Purpose	• Do readers know when they begin what they are going to read and why?	• Each section of the document begins with a statement of purpose. See ORGANIZATION.
Content	• Are regulatory, industry, and experiential standards met? • Can the reader easily identify the main messages?	• The document conforms to standards and guidelines. • Detail and "big picture" are in balance. • The document defines what the reader should do, know, or feel. See ORGANIZATION.
Logic	• Are the message components logically connected?	• In all appropriate places, the document presents conclusions and assesses their significance. • Information is consistent from one portion of the document to another. See REPETITION. • All conclusions clearly evolve from the data presented.
Organization	• Does the reader find key messages in emphatic positions? • Can the reader find information easily?	• Each part of the document begins with a summary overview. See SUMMARIES. • Text is organized deductively, from general to specific. See ORGANIZATION. • Organizational devices guide the reader and improve access to information. See EMPHASIS. • The document has a consistent internal format and organization. See EMPHASIS.
Context	• Can the reader recognize the context of the information being presented?	• Information is placed within a process or industry context. • The relationship between the information in this document and others is clear. See CITATIONS.
Presentation	• Can the reader easily comprehend the information presented? • Are verbal and visual techniques used effectively?	• Information is presented in the most appropriate visual form (text or figure). See GRAPHICS FOR DOCUMENTS. • The graphics are high-quality and clearly related to the text. See GRAPHICS FOR DOCUMENTS. • Emphasis techniques help the reader access important information. See EMPHASIS.
Language	• Do language mistakes distract the reader?	• The reader does not have to interpret. • The message is conveyed through good writing techniques and concise language. See STYLE and TONE. • Grammar, spelling, and punctuation are correct.

Figure 2. Document Standards. *These standards are generic guidelines for writers and team members to follow. Team members would need to adjust or expand these standards to match specific writing tasks.*

limited. Much reworking of products and documents will be inevitable.

Regular and frequent feedback should provide team members with a rich view of their roles and contributions. They should know if they are "winning"—if the project is on track. Their contributions should be celebrated. The lessons they are learning should be shared. These team members are part of true learning organizations.

A learning organization (or team) gives everyone broad access to information and the tools for sharing this information. Shared online workspace is now common for project-management teams. The newest team member can text or email the most senior member, perhaps even the company chief executive officer. Such communication links help to flatten and to transform organizational hierarchies.

Above all, team members must feel safe in sharing their thoughts and perspectives. Beyond the cost to team relationships of stifling opinions, the final outcome might be less than robust as important issues are suppressed. A good project manager won't allow political or interpersonal conflicts to interfere with open, transparent dialogue on her team. The success of the project depends on it.

Document Usability Tests

These documentation tests enable you to gauge how a document performs in the hands of a user.

Test 1: Reading Time. How long does it take for the user to read the document in real time?

Method: Time the user on various versions of the document. An alternative: Time how long it takes to access relevant or critical information.

Test 2: Prediction. How easily can the user predict the sequence and content of the message simply by scanning foreground materials?

Method: Limit the user's access only to previews, introductory statements, headings, and visuals. Ask the user to predict the main points of the message.

Test 3: Memory/Summary. How easily and thoroughly can the user summarize the message?

Method: Ask the user to summarize in writing or orally the content of the message.

Test 4: Message Mapping. How easily and thoroughly can the user outline a pattern or map of the message?

Method: Ask the user to sketch an outline of the message and compare it to a standard outline.

Test 5: Performance to Procedure. How helpful is the document in carrying out a procedure?

Method: Monitor both experienced and inexperienced users as they follow the document in actually carrying out the procedure described. Note hesitancies, recursions, confusions, errors, random access, and inferences made.

Test 6: Access and Retrieval. How easily can the user access and retrieve the information?

Method: Evaluate protocols for filing and retrieving information. Are file names systematic and consistent? Does keyword search enable quick and accurate access.

Figure 3. Document Usability Tests. *Use one or more of these tests to evaluate the usability of a document.*

7. Develop strategies for improving project management within your department or company.

The preceding six rules of project management reflect common sense. Surprisingly, few organizations consistently apply these rules to their ongoing projects.

These rules will go unheeded unless a project manager makes a conscious effort to adopt them.

Develop your own strategies for working to improve the project-management processes within your organization. By way of an illustration, consider some of the following ways to adapt these strategies to your own situation:

—Experiment with different team memberships. Not everyone works well with everyone else.

—Develop personal strategies for organizing and monitoring team accomplishments.

—Experiment with different project-management applications to gauge their effectiveness for your team.

Pronouns are words that take the place of more specific nouns or noun phrases:

George completed the drawings.
He completed the drawings.

The young engineer spoke up.
She spoke up.

John's survey was efficient.
His survey was efficient.

First State Bank went bankrupt.
It went bankrupt.

John, Sue, Esther, and George left the party early.
They left the party early.

Who completed the drawings?
(Who = *somebody unknown*)

The man hit himself.
We ourselves paid for the damage.

The bid and the interest were issues.
Those were issues.

John, Sue, Esther, George, etc., left the party early.
Everyone left the party early.

As the above examples indicate, pronouns include several types of words: **personal** pronouns (*he, she, his, they,* etc.), **interrogative** and **relative** pronouns (*who, that,* and *which*), **reflexive** and **intensive** pronouns (*himself, ourselves,* etc.), **demonstrative** pronouns (*this, that, these,* and *those*), and **indefinite** pronouns (*everyone, anybody, someone,* etc.).

Personal Pronouns

Personal pronouns are those that commonly replace the names of individuals or objects: *I, we, you, he, she, it,* and *they*:

- *I* and *we* are first-person pronouns—that is, they are used for the person(s) speaking.
- *You* (singular and plural) is the second-person pronoun—that is, it is used for person(s) spoken to.
- *He, she, it,* and *they* are third-person pronouns—that is, they are used for persons or objects spoken about.

Pronouns

1. Use personal pronouns in business and technical documents to establish a personal, human tone.
2. Use subjective-case pronouns for the subject of a verb or when the pronoun follows a form of *be (am, is, are, was, were, be, been)*.
3. Use objective-case pronouns when the pronoun is the object of a verb or the object of a preposition.
4. Use possessive cases correctly.
5. Choose pronouns that agree with their antecedents in number, in gender, and in case.
6. For interrogative and relative *who*, use the subjective case (*who*) for subjects and following a form of *be*; use the objective case (*whom*) for objects of verbs and prepositions.
7. Avoid the unnecessary use of intensive/reflexive forms when an ordinary personal pronoun would suffice.
8. Avoid vague uses of demonstrative pronouns (usually when the pronoun is used without the noun it is describing).
9. Indefinite pronouns used as subjects should agree in number with their verbs.

Most personal pronouns also have different forms for the singular and the plural, for different uses in sentences, and for possessives. Table 1 on the next page summarizes these different forms of the personal pronouns.

While the differences between the singular and plural forms are obvious, the differences between the cases are often confusing. Different cases (pronoun forms) have different uses in sentences:

I surveyed the site. (I *is the subjective case.*)

The committee quizzed me. (Me *is the objective case.*)

My report was too long. (My *is the possessive case—the one that comes before its noun.*)

The report is mine. (Mine *is the possessive case—the one that follows its noun.*)

See PLURALS.

1. Use personal pronouns in business and technical documents to establish a personal, human tone.

Pronouns—especially *I* and *you*—establish the identities of the writer and the reader. Such pronouns bring a human tone to a letter or memo:

this

Based on my review, I recommend that you sell your DataDeck stock and buy IBM stock.

not this

Based on a careful review, the sale of your DataDeck stock is recommended. IBM would be a good replacement stock.

See TONE and ACTIVE/PASSIVE.

Pronouns

2. Use subjective-case pronouns for the subject of a verb or when the pronoun follows a form of *be* (*am, is, are, was, were, be, been*):

Subject of a verb

Jan and I analyzed the blueprints.

We discussed the design options.

You were omitted from the roll.

He called two colleagues.

She and her employee came to the 11 a.m. meeting.

It was a poor choice.

They and Harold met about the legal problem.

NOTE: In cases where a pronoun and a noun are both subjects of the same verb, you can test the pronoun by removing the noun. In the sentence *She and her employee came to the 11 a.m. meeting*, no one would read it as follows: *Her came to the 11 a.m. meeting*. So *she* is the correct pronoun.

Following a form of be

The engineer chosen was she.

The contractor who won is he.

Was it they who called?

It could have been I.

NOTE: Pronouns following forms of *be* often sound strange:

It is I.

This is she.

Normally, objective case pronouns follow verbs, except the verb *be*, as mentioned above. For that reason, people often put objective-case pronouns after the various forms of *be: It is me* or *This is her*. These sentences may be acceptable in informal speech, which is where they ordinarily would appear anyway. In formal speech and in most writing, the subjective-case forms are correct. (Of course, if your sentence sounds stiff or strange, rephrase it to avoid the problem. You could say *I am here* instead of *It is I.*)

Table 1. Personal Pronouns

			Subjective Case	Objective Case	Possessive Case
First Person	Singular		I	me	my/mine
	Plural		we	us	our/ours
Second Person	Singular		you	you	your/yours
	Plural		you	you	your/yours
Third Person	Singular	Masculine	he	him	his
		Feminine	she	her	hers
		Non-gendered	it	it	its
	Plural		they	them	their/theirs

3. Use objective-case pronouns when the pronoun is the object of a verb or the object of a preposition:

Object of a verb

The committee chose her.

Carol did not include Jane and him.

The pollution affected them.

The President contacted Harry and me.

NOTE: When a noun and a pronoun both follow a verb, you can check the pronoun by omitting the noun. So *The President contacted Harry and me* becomes *The President contacted me*. Few people would be comfortable with *I*, which is the wrong pronoun, but many speakers often use *I* when a noun or another pronoun comes before it: *The President contacted Harry and I. The clerk gave it to him and I.* These sentences are incorrect.

Object of a preposition

Frank walked by her.

The team studied with me.

On account of me, the project ended.

The proposed schedule depended on Jane and her.

Acme Inc. worked against Jason and me.

NOTE: In the last two sentences, you can test for the correct pronoun by removing the noun or preceding pronoun that comes between the preposition and the pronoun. (Wrong: *The proposed schedule depended on she. Acme Inc. worked against I.*)

4. Use possessive cases correctly:

That was her report.
The report was hers.

The problems were ours, not yours.
They were our problems, not your problems.

I rejected your arguments.
The arguments were yours, not mine.

Note 1: In cases where two possessive forms exist for a single pronoun, the simple form (without an *–s*) comes before the noun; the other form (with an *–s)* follows a form of *be* (*am, is, are, was, were, be, been*). *His* and *its* have only a single form.

Note 2: Do not confuse the possessive pronouns with simple contractions. Possessive pronouns do not have apostrophes; contractions do have apostrophes:

Possessive Pronoun	Contraction
its	it's (it is)
his	he's (he is)
their/theirs	they're (they are)
your/yours	you're (you are)

See Possessives and Contractions.

5. Choose pronouns that agree with their antecedents in number, in gender, and in case:

Agreement in number

The men worked on the design plan all night, but they were unable to finish. (men *[antecedent] = plural;* they = *plural*)

Cheryl Higgins prepared a revised safety procedure, but she failed to get managerial approval. (Cheryl Higgins *[antecedent] = singular;* she = *singular*)

Agreement in gender

Sidney Brown was eager to take his first actuarial exam. (Sidney *[antecedent] can be either male or female, so the pronoun is a key sign of which is intended. Often the female spelling is* Sydney.)

Note: When the antecedent is an indefinite pronoun, writers often don't know whether to use a singular or plural pronoun, much less a masculine or feminine pronoun:

Everyone should arrange their (his *or* her?) desk before leaving.

Everybody is responsible for their (his *or* her?) own time cards.

The forms with *their* have become more acceptable in speech, even though *everyone* and *everybody* are usually considered singular pronouns. To avoid any sexist language and to eliminate the clumsy *his or her,* the best option would be to rewrite the sentences as plurals:

All employees should arrange their desks before leaving. (*or simply,* Employees . . .)

All clerks and secretaries are responsible for their own time cards.

See Agreement, Bias-Free Language, and the discussion below of indefinite pronouns.

Agreement in case

Betty argued with them—especially Susan and him. (them = *objective case;* him = *objective case*)

Interrogative and Relative Pronouns

Interrogative and relative pronouns are similar in their uses, and some interrogative pronouns are identical to the main relative pronouns: *who, whom, whose,* and *which.* Relative pronouns also include *that, whoever,* and *whomever.* Interrogatives also include *what, why,* and *where.*

Many interrogative and relative pronouns have only one form, but *who* changes its form just as the personal pronouns do:

Subjective case:	who, whoever
Objective case:	whom, whomever
Possessive case:	whose

That and the other relative and interrogative forms do not change to reflect the different cases.

6. For interrogative and relative *who,* use the subjective case (*who*) for subjects and following a form of be; use the objective case (*whom*) for objects of verbs and prepositions:

Subjective case

Who is the project engineer?

Whoever writes the report gets all the credit.

The agent who spoke to us was courteous.

The supervisor determined who would be the representative.

The representative is whoever has the most years of service.

Objective case

Whom did you nominate?

Whom did you wish to speak to?

Or: To whom did you wish to speak?

Whomever you nominate, we'll find someone to balance the ticket.

The man whom you spoke to is our president.

Note 1: The first two examples under the objective case sound almost too formal, even stiff. Most speakers of English are more comfortable if a subjective *who* begins the sentence or question. For this reason, two other versions of these questions are acceptable in informal, spoken English:

Who did you nominate?

Who did you wish to speak to?

Similarly, even with non-questions, subjective *who* sometimes replaces the more correct *whom*:

Whoever you nominate, we'll find someone to balance the ticket.

These informal forms with *who* or *whoever* are incorrect in writing even though acceptable in informal speech.

NOTE 2: Deciding on the correct case is especially difficult when the relative pronoun is part of a complex sentence. The trick is to isolate only the clause containing the relative pronoun:

> who wants to study chemistry
>
> whom the proposal team chose
>
> whoever is the top candidate

These are correct uses of *who* and *whom*, so you can insert them into any sentence, and they'll be correct:

> John, who wants to study chemistry, is our lab technician.
>
> We left with Sandra, whom the proposal team chose.
>
> Whoever is the top candidate will be our speaker.
>
> We planned to interview whoever is the top candidate.

So, to decide on the case of a troublesome use of *who* or *whom*, try to isolate the clause containing the relative pronoun (underlined below):

> We decided to abandon the direction established by the geologist who/whom we first talked to.

The proper form then is *whom*, used as the object of the preposition *to*.

NOTE 3: Adjective clauses introduced by a relative pronoun are often difficult to punctuate. The simplest rule is to enclose such clauses with commas when they are nonessential:

> Jack Craven, who is our project coordinator, has been with the company for 15 years.
>
> We presented the proposal to the team from AirFlo Inc., which is a firm based in Denver.

When the clauses are essential, no commas are required:

> The proposal that we sent to the Department of Transportation has been canceled. *(Without the clause* that we sent to the Department of Transportation, *the sentence would not be clear to most readers, especially if more than one proposal were possible.)*
>
> The Governor proposed banning all autos that did not have a safety inspection. *(The clause* that did not have a safety inspection *is essential to the meaning.)*

See COMMAS and *that/which* in WORD PROBLEMS.

Reflexive and Intensive Pronouns

Reflexive and intensive pronouns are identical in appearance: Both end in *–self* (singular) or *–selves* (plural). Personal pronouns have reflexive and intensive forms: *myself, ourselves, yourself, yourselves, himself, herself, itself,* and *themselves.*

Reflexive pronouns "reflect" back to a noun mentioned earlier in the same sentence:

> George injured himself.
>
> The team quarreled among themselves.
>
> Amy cut herself.
>
> The dog licked itself.

Intensive pronouns usually follow the nouns they intensify:

> George himself was injured.
>
> The team themselves were arguing.
>
> Amy herself ran the errand.
>
> I myself checked on the data.

7. Avoid the unnecessary use of intensive/reflexive forms when an ordinary personal pronoun would suffice:

> With our permission, Jeff became the spokesman for Jim and myself. (*better:* Jim and me)
>
> John and myself are both uncomfortable with the proposal. (*better:* John and I)

Demonstrative Pronouns

Only four demonstrative pronouns exist:

	Singular	Plural
"near"	this	these
"far"	that	those

Demonstrative pronouns sometimes appear before a noun and sometimes they can replace the noun or noun phrase entirely:

> This design plan has problems.
>
> This has problems. (*This sentence assumes that the reader or listener knows what the word* this *refers to.*)
>
> That proposal for the DOE demonstrates effective graphics.
>
> That demonstrates effective graphics.

8. Avoid vague uses of demonstrative pronouns (usually when the pronoun is used without the noun it is describing):

> This is something to consider. (*better:* This shortfall in payments is something to consider.)
>
> These are difficult. (*better:* These exercises are difficult.)

NOTE: Such vague sentences are particularly annoying when, for example, a *this* refers back to an entire sentence or several things in a sentence:

> *not this*
>
> The travel plans were a jumble and even turned out to be beyond our budget. This meant that we could not make connections as planned.
>
> *this*
>
> The travel plans were a jumble and even turned out to be beyond our budget. The jumble (*or* This jumble) meant that we could not make connections as planned.

Indefinite pronouns include a number of words that have unspecified or vague meanings, usually because the writer or speaker does not want to identify or can't identify the person(s) or thing(s) referred to. These sentences illustrate some of the common indefinite pronouns:

> **Everyone** should proofread the report before we send it out.
>
> We cleaned out **everything** before vacating the office.
>
> **Both** were involved in the design tests last summer.
>
> We determined that **somebody** had changed the entry code in the computer.
>
> The procedures required that **anyone** leaving the security area sign and note the date and time.
>
> **Nobody** had a pass to be on the construction site after normal work hours.

A list of indefinite pronouns

everyone	everything
everybody	every
each	all
both	
someone	something
somebody	somewhere
some	another
one	
anyone	anything
anybody	anywhere
either	any
no one	nothing
nobody	nowhere
none	neither
many	few
several	others
more	most

NOTE: Many of these indefinite pronouns can act very much like adjectives:

> Several books were stolen from our library.
>
> Most participants would want to attend the final session.
>
> *or*
>
> Several were stolen from our library.
>
> Most would want to attend the final session.

9. Indefinite pronouns used as subjects should agree in number with their verbs:

> Anyone likes to receive a positive performance review. (Anyone = *singular;* likes = *singular*)
>
> Several were contacted before we chose a final candidate. (Several = *plural;* were = *plural*)
>
> All of the sugar was tainted. (All of the sugar = *singular;* was = *singular*)
>
> All of the employees were notified of the new vacation policy. (All of the employees = *plural;* were = *plural*)

NOTE: Sometimes the indefinite pronoun, even though considered grammatically singular, requires a plural pronoun:

> Everyone should abandon their attempt to discover the error.

Strict editors would argue that *everyone* is singular and *their* is plural, so traditionally the correct form has been this version:

> Everyone should abandon his attempt to discover the error.

Feminists and others have criticized this version as being sexist because females are ignored. One option would be to replace *his* with *his or her*. A more reasonable option is to rewrite the sentence so that it is clearly plural:

> All engineers should abandon their attempts to discover the error.

See AGREEMENT and BIAS-FREE LANGUAGE. Also see the discussion above on personal pronouns.

Punctuation

Punctuation marks are as much a part of language as words. Like words, punctuation is a code that allows writers to communicate with readers who know the code.

Like words, punctuation rules (conventions) are not firm, fixed, and unchanging. The rules are evolutionary, but about 95 percent of the rules are relatively fixed and not debated as, for instance, the rule that requires the comma in *June 15, 2010.*

Punctuation rules are subject to tradition and local usage. The rules differ slightly from industry to industry and from publication to publication. The punctuation style that one publisher prefers might differ slightly from the style of another publisher.

Writers who use every permissible mark of punctuation follow a **mandatory** (formal) style. Writers who omit **optional** punctuation follow an optional (informal) style. Other terms can be and have been used to describe these two styles of punctuation: conservative vs. liberal and closed (close) vs. open.

The following examples illustrate the differences between the mandatory and optional styles.

Example 1: Commas in a Series

Writers following the optional style usually omit the comma before the *and* joining the last two items in a series:

> We requested employment figures for 2004, 2005 and 2006.
>
> The report analyzed the possible market, the projected labor costs and the supply of raw materials.

Writers following the mandatory style always include the comma before the *and* joining the last two items in a series.

Punctuation

1. Use the mandatory style for most technical, scientific, and legal documents.

In some sentences, the comma before the *and* is not necessary, but in some sentences, omitting the comma can create ambiguity:

> The maintenance people replaced the rocker arm bracket, hinge pin and wheel assembly.

Are the hinge pin and the wheel part of the same assembly? A comma after *pin* clearly signals that the three items are separate:

> The maintenance people replaced the rocker arm bracket, hinge pin, and wheel assembly.

Because clarity and precision are so important in technical and scientific writing, we recommend that writers adopt the mandatory style and retain the comma before the *and* in a series.

Example 2: Commas in Dates

Punctuating the day, month, and year is usually simple and straightforward. A comma separates the day and the year:

> June 15, 2010

What happens, however, when the date appears in the middle of a sentence?

> We moved on June 15, 2010 into our new office.

Writers following the optional style would not insert a comma after 2010. Writers following the mandatory style would add a comma after 2010:

> We moved on June 15, 2010, into our new office.

Both versions are "correct." Both are acceptable because the presence or absence of the comma after 2010 does not affect clarity.

1. Use the mandatory style for most technical, scientific, and legal documents.

The mandatory style allows for more precision and can help you avoid ambiguity, so we recommend it.

We also recommend it for any formal or critical documents. In letters and memos to familiar readers, however, you might wish to follow the optional, less formal style.

Mandatory Punctuation

The following rules are those that allow for optional punctuation. Writers following the optional style might omit punctuation in the cases cited below except when clarity would suffer. If omitting punctuation would make the writing less clear, more ambiguous, or at all confusing, then use the appropriate mark of punctuation.

—Use an apostrophe and an *–s* to form the singular possessive of words of more than one syllable ending in *–s:*

> Harris's report
>
> Davis's plan
>
> Optional Style: These possessive forms would be *Harris'* and *Davis'*. Note that the apostrophe is not optional.

—Use a comma before a coordinate conjunction to separate two independent thoughts, even when they are quite short and simple:

The plan was finished, and the budget calculated.

Optional Style: You would omit the comma after *finished.*

—Use a comma before the *and* that joins the last two items in a series:

We ordered two water pumps, a fan, and three replacement belts.

Optional Style: You would omit the comma after *fan.*

—Use a comma to enclose parenthetical expressions, even very short ones such as *thus*:

He was, thus, surprised by the answer.

Optional Style: You would omit both commas around *thus.*

—Use a comma after introductory phrases and clauses, even when they are very short and simple:

After we wrote the report, we submitted it.

Optional Style: You would omit the comma after *report.*

—Use a comma after the year in a date and the state in an address when either appears in the middle of a sentence:

The time sheets for July 6, 2010, show no overtime.

His speech in Joplin, Missouri, was most forgettable.

Optional Style: You would omit the commas after *2010* and *Missouri.*

—Use hyphens to form compound words that modify other words:

We will install a high-tension line above the Bradley overpass.

Optional Style: You would omit the hyphen between *high* and *tension.*

—Use periods after all abbreviations, including the names of government agencies, colleges and universities, and private organizations:

I.B.M.
S.U.N.Y. (State University of New York)
N.A.A.C.P.

Optional Style: You would omit the periods in these and most other abbreviations.

—Use a semicolon along with a coordinate conjunction to join independent clauses or complete thoughts that already contain a comma, even if they are clear without the semicolon:

Although new rain gauges helped us monitor total precipitation, we could not have anticipated the heavy spring runoff; and the resulting floods caused considerable damage to the watershed.

Optional Style: You would use a comma after *runoff,* not a semicolon.

For further information on commas, periods, semicolons, colons, dashes, and other marks of punctuation, see the alphabetical entries elsewhere in this *Style Guide.*

Question Marks

Question marks in English, unfortunately, are placed at the end of sentences. Spanish, to make reading easier, places an upside-down question mark at the beginning and a right-side-up question mark at the end of question sentences.

Question Marks

1. Use question marks to indicate direct questions.
2. Use question marks to indicate questions within a sentence.
3. Use question marks within parentheses to indicate doubt.
4. Place question marks inside quotation marks, parentheses, or brackets only when they are part of the quoted or parenthetical material.

1. Use question marks to indicate direct questions:

Will analysis modeling be required?

NOTE 1: Use a separate question mark for each in a series of incomplete or elliptical questions:

When will the preliminary targeting studies be due? In 30 days? 60? 90?

NOTE 2: Use question marks at the end of statements written as declarations but intended as questions:

These are the final figures?

Regulating the supply pressure was the only realistic solution?

NOTE 3: Do *not* use question marks for indirect questions or for statements written as questions but intended as courteous requests:

During our preliminary design studies, we asked whether the weapon parameters would be an integrated part of the overall system requirements. (*The underlined clause is an indirect question. Its phrasing is not identical to a direct question.*)

Will you please forward five copies of DD Form 1425.

See PERIODS.

2. Use question marks to indicate questions within a sentence:

Your project managers have the authority, don't they, to reallocate resources based on changing needs?

The payment is always late, isn't it?

NOTE 1: When the question follows an introductory statement, capitalize the first word of the question and use a question mark:

The remaining question is, Can shortfalls in determining performance results be quantified with the Phase 2 synthesis procedure?

Today's manufacturer asks, How can Q&A expenses be distributed across project and functional lines?

NOTE 2: If the question precedes a concluding sentence remark, the question mark goes after the question, and the sentence remark begins with a lowercase word:

Which mission would prove most productive? was the remaining question.

3. Use question marks within parentheses to indicate doubt:

CAD/CAM was first investigated at MIT in 1964(?).

Question Marks With Other Punctuation

4. Place question marks inside quotation marks, parentheses, or brackets only when they are part of the quoted or parenthetical material:

During the preproposal conference, an Allied representative asked, "Does the Statement of Work represent a minimal subset of requirements?"

What did the contracting officer mean when she said, "The Statement of Work is the minimal subset of requirements"?

When did she ask, "Who is the project manager?" (*not* manager?"?)

The engineering program includes analysis, design, testing (fabrication?), and integration of the unit into the helicopter.

Doesn't the Program Plan call for Task F completion no later than 75 days ARO (August 17)?

See PARENTHESES, BRACKETS, and QUOTATION MARKS.

Quotation Marks

Quotation marks have multiple uses. The most common use (enclosing direct quotations) gives them their name. In earlier centuries, quotation marks did not exist and the reader could not tell easily the writer's words from the quoted words. Today, the conventions for using quotation marks are well established. See QUOTATIONS.

Quotation Marks

1. Use quotation marks to enclose direct quotations.
2. Use quotation marks to indicate the title of an article, section, volume, and other parts of a longer document.
3. Use quotation marks to indicate that a word is used in a special or an abnormal sense.
4. Always place periods and commas inside closing quotation marks.
5. Always place semicolons and colons outside closing quotation marks.
6. Place dashes, exclamation marks, and question marks inside quotation marks if they are part of the quotation; otherwise, place them outside quotation marks.

1. Use quotation marks to enclose direct quotations:

> The RFP says, "All pages in the proposal must be numbered."

Direct quotations include the actual words and phrases from a document or from a person speaking.

Indirect quotations do not take quotation marks:

> The RFP says that all pages in the proposal have to be numbered.

Indirect quotations do not give every word and phrase in the direct quotation. Often an indirect quotation is only a paraphrase:

> The RFP says that we should number all pages in the proposal.

See QUOTATIONS.

NOTE 1: Long quotations have two equally acceptable conventions: (1) They may be enclosed by quotation marks, or (2) they may be indented from both the left and right margins (in which case they do **not** require quotation marks). If quotation marks are used and the quotation extends for more than one paragraph, quotation marks should appear at the beginning and ending of the entire quotation and at the beginning of each new paragraph within the quotation.

NOTE 2: Single quotation marks are used to indicate a quotation within a quotation:

> According to the NASA report, "National Aerodynamics argued for a 'differential scale of evolution.'"

NOTE 3: In most British publications, single quotes replace the double quotes used in U.S. publications:

> The RFP says, 'All pages in the proposal must be numbered.'

In British publications, double quotations are used to indicate a quotation within a quotation:

> According to the NASA report, 'National Aerodynamics argued for a "differential scale of evaluation.".'

2. Use quotation marks to indicate the title of an article, section, volume, and other parts of a longer document:

> The contracting officer must approve all items specified under "Special Equipment" in the cost proposal.
>
> This volume must include a completed and signed Standard Form 33, "Solicitation, Offer, and Award."

See TITLES.

3. Use quotation marks to indicate that a word is used in a special or an abnormal sense:

> The 2008 study suggested that NASA's definition of "suitability" contradicts the goals of the program.
>
> Only in English do "thin chance" and "fat chance" mean the same.

NOTE: Italics (or underlining) can replace quotation marks when you want to refer to a word as a word:

> In the contract, *boundaries* refers only to those property lines surveyed after July 2010.

See ITALICS and UNDERLINING.

Quotation Marks

Quotation Marks With Other Punctuation

4. Always place periods and commas inside closing quotation marks:

> We have completed the section entitled "Representations, Certifications, and Acknowledgments."

> The logarithmic decrease of the differential threshold is sometimes mistakenly called "Fechner's law."

NOTE: This convention seems illogical, but it is now standard in the United States. The convention developed because printers wanted to put the smaller periods and commas inside of the closing quotation marks for a cleaner appearance on the page.

British usage places commas and periods inside or outside the quotation marks, depending on whether they are or are not part of the quotation:

> The logarithmic decrease of the differential threshold is sometimes mistakenly called 'Fechner's law'.

> A phrase on the cover, 'Recommended by the Royal Academy', caught her eye.

5. Always place semicolons and colons outside closing quotation marks:

> The corporation's experience belongs under "Related Experience"; the project manager's experience belongs under "Resumes of Key Personnel."

> Include the following under "Manhours and Materials":
>
> a. Work statement tasks
> b. Estimated completion schedule
> c. Materials/equipment required

6. Place dashes, exclamation marks, and question marks inside quotation marks if they are part of the quotation; otherwise, place them outside quotation marks:

> The section entitled "Personnel Qualifications"—the only part of the proposal where we can address the team's APL experience—is restricted to five pages.

> He said, "Can we improve the unload utilities without losing the language interface?"

See QUESTION MARKS and SPACING.

Quotations

Quotations are an effective tool when you have to refer persuasively to data and conclusions from another document.

When you quote from another document, be careful not to misrepresent the content or the intent of the original passage. The following rules are suggestions. You must judge for yourself if you are accurately representing the original document.

> **Quotations**
>
> 1. Quote only the key or relevant passages.
> 2. Cite the sources for any quoted material—from words and phrases to whole sentences and paragraphs.
> 3. Distinguish carefully between direct and indirect quotations.
> 4. Use brackets to insert comments or corrections in quoted material.

1. Quote only the key or relevant passages:

Your letter made an excellent case for the "procedural lapse" that caused the double billing to your account.

In the analysis of the data, Jameson (2010) argues for three "equally persuasive hypotheses."

NOTE 1: Sometimes a two- or three-word phrase is sufficient to capture the flavor of the original document. Rarely do you need to quote whole sentences or paragraphs from the original document.

Long quotes distract the reader and often signal that the writer has not done the work necessary to boil the quotation down to its essentials.

NOTE 2: As illustrated in the two examples above, all quoted words, phrases, and sentences should normally be enclosed by quotation marks. See QUOTATION MARKS.

NOTE 3: In cases where words are omitted from the middle of quoted material, an ellipsis signals material has been omitted. See ELLIPSES.

Your letter made an excellent case for the "procedural lapse . . . and the sloppy record keeping" that caused the double billing of your account. (*The original read:* ". . . the procedural lapse when my account was opened and the sloppy record keeping ever since.")

2. Cite the sources for any quoted material—from words and phrases to whole sentences and paragraphs.

Inexperienced writers sometimes are careless in their citations.

No material from another source, especially copyrighted material, should ever appear in another document without full and adequate credit being given to the original author(s). See ETHICS and INTELLECTUAL PROPERTY.

No reader should ever have to guess what is original and what the writer is borrowing. Accurate citations and accurate use of quotation marks and ellipses can remove uncertainties. See CITATIONS and FOOTNOTES.

3. Distinguish carefully between direct and indirect quotations.

Direct quotations contain only the original words and phrases of the document being quoted.

Indirect quotations are a summary of someone else's words. Some minor words may come from the original, but the writer has made significant changes. Even then, key words should appear within quotation marks.

Below is an original passage. Following are examples of a direct quotation and an indirect quotation:

Original passage

Hank Stevens was over 30 minutes late three times during the week of November 3. He called in on one of these mornings, but we received no calls on the other two mornings. Although it is now November 15, he has given no satisfactory excuse or explanation of his lateness.

Direct quotation

According to Hank Stevens' supervisor, Hank has failed to give a "satisfactory excuse or explanation of his lateness" on three mornings during the week of November 3. Hank did call in one of the mornings, but he did not call in the other two mornings.

Indirect quotation

According to Hank Stevens' supervisor, Hank has not explained adequately his three instances of lateness during the week of November 3. Hank was over 30 minutes late in each case, and he only called in one time.

NOTE 1: As this example illustrates, the direct quotation is often embedded in a passage that contains some indirect quotations. As rule 1 above indicates, you should quote only the pertinent words and phrases.

NOTE 2: Often changing from direct to indirect quotations involves changing syntax and wording (especially in the pronouns):

Direct quotation

"We need to receive your response by no later than January 15."

Indirect quotation

MOGO Oil said that they must have our written response by January 15.

Direct quotation

"I have been unable to locate the original data collected in 1999. The files seem not to have been moved when we moved in 2004 from the old building to our present building."

Indirect quotation

Gene Sayers has been unable to find the original 1999 data. He suspects that the files were not moved in 2004 when we moved into our present building.

4. Use brackets to insert comments or corrections in quoted material:

"The measured length [8.35 feet] differs from the length specified in the specification."

"Ralph Stevenson [Stephenson] was the representative elected from Curry County."

See BRACKETS.

Redundant Words

Redundant words or phrases unnecessarily qualify other words or phrases.

For instance, in the expression *basic fundamentals*, the word *basic* is unnecessary because, by definition, all fundamentals are basic.

Past experience is redundant because for experience to be experience, it must have been acquired in the past. *Past history* is redundant for the same reason. You can't have present or future history. All history is past. However, *ancient history* and *recent history* are acceptable because they both refer to specific parts of the past.

Redundant words or phrases are not the same as deliberate repetition. Repetition occurs when writers choose to repeat key words and phrases throughout a document. These key words and phrases usually appear in major headings or in summaries. See REPETITION.

Redundant Words

1. Eliminate redundant words.
2. Use redundant words to emphasize or dramatize a situation or condition.

1. Eliminate redundant words.

A fundamental of good writing style is to eliminate unnecessary words. Do not say *main essentials*. Just say *essentials*. Do not speak of the *final conclusion*. It is simply the *conclusion*. Documents are not *attached together*. They are simply *attached*. See WORDY PHRASES.

2. Use redundant words to emphasize or dramatize a situation or condition:

> Mailing the ramjet study by Friday is absolutely essential.

The only proper use of redundancies in writing is to emphasize. The expression *absolutely essential* is redundant because nothing can be more essential than *essential*.

If you need to heighten the sense of urgency in a situation by exaggerating, then redundant words are acceptable. However, do **not** overuse redundancies for emphasis. The effect diminishes quickly. You can become so exaggerated in your style that readers pay less attention to your ideas. Too much emphasis becomes no emphasis. See EMPHASIS and GOBBLEDYGOOK.

A List of Redundancies

The following list of redundancies will help you identify those you habitually use. The redundant expression appears in the left column; in the right column are possible substitutes:

Absolutely complete	complete
absolutely essential	essential
absolutely nothing	nothing
accidentally stumbled	stumbled
a.c. current	a.c./alternating
current	
adequate enough	adequate/enough
advance forward	advance
advance planning	planning
aluminum metal	aluminum
and etc.	etc.
any and all	any/all
arrive on the scene	arrive
ask the question	ask
assembled together	assembled
attached hereto	attached
attach together	attach
Basic fundamentals	fundamentals
before in the past	before/in the past
betwixt and between	between
blue in color	blue
brief in duration	brief/quick/fast
Check up on	check
circle around	circle
close proximity	proximity
collect together	collect
combine together	combine
completely destroyed	destroyed
completely opposite	opposite
connect together	connect
consensus of opinion	consensus
consequent results	results
consolidate together	consolidate
continue on	continue
continue to remain	remain
contributing factor	factor
cooperate together	cooperate
couple together	couple
Desirable benefits	benefits
diametrically opposite	opposite
disappear from sight	disappear
disregard altogether	disregard
Each and every	each/every
early beginnings	beginnings
empty cavity	cavity
enclosed herewith	enclosed
endorse on the back	endorse
end product	product
end result	result
entirely destroyed	destroyed
equally as good	as good/equally
good	
exactly identical	identical
expired and terminated	expired/terminated
extremely immoderate	immoderate
Fast in action	fast
few in number	few
filled to capacity	filled
final completion	completion
final conclusion	conclusion
finally ended	ended
first beginnings	beginnings
following after	following/after
funeral obsequies	obsequies
fused together	fused
Heat up	heat
hidden pitfall	pitfall
hopeful optimism	hope/optimism
Important essentials	essentials
Joint cooperation	cooperation
join together	join
joint partnership	partnership
just exactly	just/exactly
Large in size	large
large-sized	large
lift up	lift
living incarnation	incarnation

Redundant Words

Redundant	Concise
Main essentials	essentials
melt down	melt
mingle together	mingle
mix together	mix
more preferable	preferable
mutual cooperation	cooperation
Necessary requisite	requisite
new innovation	innovation
One and the same	the same
one definite reason	one reason
one particular example	one example
one specific case	one case
Part and parcel	part
past experience	experience
period of time	period
personal friend	friend
personal opinion	opinion
pervade the whole	pervade
plan ahead	plan
plan for the future	plan
plan in advance	plan
postponed until later	postponed
presently planned	planned
prolong the duration	prolong
Qualified expert	expert
Really and truly	really
reason is because because	reason is that/
recur again	recur
red in color	red
reduce down	reduce
regress back	regress
remand back	remand
repeat again	repeat
resultant effect	effect
Same identical	same
seems apparent	seems/is apparent
separate and distinct	separate/distinct
shuttle back and forth	shuttle
single unit	unit
skirt around	skirt
small in size	small
small-sized	small
specific example	example
still continue	continue
still remains	remains
suddenly collapsed	collapsed
summer months	summer
surprising upset	upset
surrounding circumstances	circumstances
surround on all sides	surround
Ten miles distant from	ten miles from
three hours of time	three hours
throughout the entire	throughout
throughout the whole	throughout
total of ten	ten
to the northward	north/northward
traverse across	traverse
true fact	fact
Ultimate end	end
universal the world over	universal
unsolved problem	problem
Visit with	visit
Ways and means	ways/means
where you are at	where you are
where are you at	where are you

References

Many good reference sources online and in books help writers make proper choices, including dictionaries, thesauruses, and industry-specific style manuals.

This *Style Guide* will answer many of your stylistic questions and will help you to become a more effective writer, particularly in business and technical fields. This entry will help you find more detailed and specialized treatments of some stylistic issues. These references are among the finest available in their special areas. For rules on how to cite a reference, see CITATIONS.

References

Dictionaries

A reputable dictionary is an essential writer's tool. Besides providing correct spellings, a dictionary gives definitions, pronunciations, word origins, synonyms, and guidance on word division and usage.

Most dictionaries are abridged dictionaries written for general readers and college students. The exception is *Webster's Third New International Dictionary of the English Language, Unabridged*; this dictionary was completed in 1961 and has been reprinted and updated many times since. This dictionary is still the best reference for American English. Another exception is the *Oxford Dictionary of English*, also often updated, which is the best reference for British Commonwealth and international English speakers. (This dictionary should not be confused with the *Oxford English Dictionary*, a multivolume work that traces the evolution of English words.)

Some general dictionaries are directed toward specific users. For example, the *Oxford Student's Dictionary of American English* is designed for non-English-speaking students. The entries are brief and very clear, and special attention is given to simple idioms (phrases) that have special meanings.

Online dictionaries vary in quality and usefulness. The standard for Americans is *Merriam-Webster's Online Dictionary*, based on *Webster's Collegiate Dictionary*. This useful reference provides not only complete definitions, usage, and citation rules, but also links to current articles of interest. For example, if you look up the word *engagement*, you might well find a link to an article about employee engagement, an important issue in business.

Other useful websites, such as yourdictionary.com and onelook.com, enable you to search many dictionaries at once and compare entries.

Specialized Dictionaries

In addition to those dictionaries listed above, many professional groups or disciplines have specialized dictionaries that list the technical terms and jargon particular to their professional area. Following are a few of these specialized dictionaries, many of them available online or in digital form and as mobile-device apps. Most of these dictionaries are periodically updated, so look for the latest editions. The title and publisher follow:

Black's Law Dictionary. West Publishing Company.

Dictionary of Geological Terms. Anchor Press.

The Illustrated Petroleum Reference Dictionary. Pennwell Books.

Mathematics into Type. American Mathematical Society.

McGraw-Hill Dictionary of Scientific and Technical Terms. McGraw-Hill Professional.

Webster's Geographical Dictionary. Merriam Webster.

Thesauruses

A thesaurus provides synonyms and often antonyms. A thesaurus in standard writing software will feature a limited number of common synonyms. Use a thesaurus when you need to find optional ways of expressing an idea, when you can't think of a word but know the word exists, or when the only word you can think of is not exactly correct and you need to find a more precise alternative. Do not use a thesaurus to find bigger words than the ones you can think of. In other words, don't try to sound impressive by finding and using big words. See GOBBLEDYGOOK.

The standard thesaurus for business is *Roget's International Thesaurus* (Collins Reference), also available online.

General Style Guides

This *Style Guide* is a general style guide. However, it differs from those listed below in that it provides more focus on effective writing techniques and the writing process for business and technical professionals, and it provides models of effective business documents.

General style guides are, next to an up-to-date dictionary, the best resources for writers. They usually cover punctuation, abbreviations, spelling problems, capitalization, and special signs and symbols. In addition, each of those listed below has special features. *The Chicago Manual of Style*, for instance, has a complete discussion of publishing conventions. The *United States Government Printing Office Style Manual* has guidelines useful for government contractors. Some of these resources are available online as indicated.

The Chicago Manual of Style. University of Chicago Press. Available online.

The Chicago Guide to Communicating Science. University of Chicago Press.

The Economist Style Guide. Profile Books. Available online.

The Complete Guide to Citing Government Information Resources: A Manual for Social Science and Business Research. Congressional Information Service.

The Gregg Reference Manual: A Manual of Style, Grammar, Usage, and Formatting. McGraw Hill/Career Education. Available online.

The McGraw-Hill Style Manual: A Concise Guide for Writers and Editors. McGraw-Hill Book Company.

Administrative Assistant's and Secretary's Handbook. AMACOM.

United States Government Printing Office Style Manual. Claitor's Law Books and Publishing Division. Available online.

Specialized Style Guides

Specialized style guides are intended for a particular professional group. Despite their special audience, they usually also cover basic punctuation and other items of general interest. These titles are standard references for scientists, academics, and journalists.

American Chemical Society. *The ACS Style Guide: A Manual for Authors and Educators.*

American Psychological Association. *Publication Manual.*

Council of Biology Editors. *Scientific Style and Format: The CBE Manual for Authors, Editors, and Publishers.*

Modern Language Association. *MLA Handbook for Writers of Research Papers.*

Associated Press. *Associated Press Stylebook.*

Grammar and Usage Handbooks

Grammar handbooks survey the principles of English grammar and provide many dos and don'ts for writers. You should have such a handbook in addition to the grammar checker on your computer. Often your computer will give you grammar suggestions that might be confusing to you or even mistaken. A grammar handbook will help you understand what the computer is suggesting and why you should or should not accept the suggestion. You can also find some free grammar- and spell-checking websites online that can be quite useful.

References

Usage handbooks provide rules for the proper use of words. Such rules quite often reflect the authors' prejudices, so keep that in mind as you use such handbooks.

Writers of international English should refer to an updated version of Henry Watson Fowler, *Dictionary of Modern English Usage.* (Oxford University Press.)

Writers of American English should refer to an updated version of *Webster's Dictionary of English Usage.* (Merriam-Webster, Inc.)

Casey Miller and Kate Swift, *The Handbook of Nonsexist Writing for Writers, Editors, and Speakers* (iUniverse) can be helpful when you aren't sure how to phrase something to be inclusive and to avoid offending a diverse audience.

For tips on writing effective emails and instant messages, see Dom Sagolla, *140 Characters: A Style Guide for the Short Form* (Wiley).

Books on Writing

The following books cover writing, both as a process and as a final product. Strunk and White's book is perhaps the most famous and the most readable of those listed, but the others are also valuable.

Katz, Michael J. *From Research to Manuscript: A Guide to Scientific Writing.* Springer.

Lamott, Anne. *Bird by Bird: Some Instructions on Writing and Life.* Anchor.

Strunk, W., Jr., and E. B. White. *The Elements of Style.* Longman.

Truss, Lynne. *Eats, Shoots and Leaves.* Gotham.

Williams, Joseph M. *Style: Ten Lessons in Clarity and Grace.* Longman.

Repetition

1. Repeat key words and phrases throughout a document without changing them in any way.
2. Design documents with clear and deliberate repetition in mind.
3. Use the inherent repetition in formal report structure to reinforce major ideas and to strengthen logic and impact.
4. Use repetition when you want to emphasize the logic behind a discussion, especially when details are parallel in their intent.
5. Be careful not to repeat an idea too soon after stating it the first time. The rule applies to more traditional text, where one paragraph flows into another.

Repetition is sometimes considered a trait of poor writing. Actually, repetition can be a valuable emphasis technique. It can enhance the impact and readability of a document.

Effective repetition of a fact or an idea reinforces it in the reader's mind. See Organization, Emphasis, and Key Words.

1. Repeat key words and phrases throughout a document without changing them in any way.

Paraphrasing an idea, no matter how well done, is not as good as exact repetition. In technical and business documents, exact repetition should be the goal.

For example, if you open a document with a list of topics to be discussed, the wording of this list should be identical to the wording of subheads in the document.

In the Introduction on page 1:

Strategies for the Control of Costs

Subhead 3 on page 5:

3.0 Strategies for the Control of Costs

2. Design documents with clear and deliberate repetition in mind.

As you design a document and organize your ideas, build in effective repetition. Consider repeating (1) key requests or recommendations, (2) major conclusions, and (3) most important or most convincing facts.

Most professionals in the world of work have more documents crossing their desks than they have time to read. Research shows that the majority of intended readers skim through the documents they receive. Rarely do they read the documents in depth, and almost never will they read something more than once.

Consequently, you must convey your important ideas quickly and emphatically. Repeating key requests, recommendations, conclusions, and facts helps to ensure that readers who skim will not miss the most important ideas in your documents.

3. Use the inherent repetition in formal report structure to reinforce major ideas and to strengthen logic and impact.

Technical and other formal reports are structured for deliberate repetition. The writer's conclusions, for instance, typically appear in these sections of the report: abstract, executive summary, conclusions, and discussion. The recommendations will appear one way or the other in the abstract, executive summary, and perhaps the discussion. They will also appear in their own section.

Repetition in reports is deliberate. It allows readers to read selectively. Some readers will read the abstract and will not need to read further. Others will read the executive summary and perhaps the conclusions and recommendations sections. Still others will glance at the results section and then read through the discussion.

Because reports are deliberately repetitive, all of these readers will have encountered the major conclusions and recommendations. Those who read the major conclusions and recommendations more than once will have had those ideas reinforced. See Reports.

Repetition

4. Use repetition when you want to emphasize the logic behind a discussion, especially when details are parallel in their intent.

Repetition of a key word or phrase indicates that a sequence of ideas is parallel:

> Skillful writers have a clear sense of purpose. Skillful writers are aware of their readers' wants, needs, and concerns. Skillful writers can choose between many stylistic options. In short, skillful writers focus their task and use the right techniques to create the right effect.

This kind of repetition is effective in emphasizing key points and limiting sentence length. (Imagine how difficult this passage would be to read if it were a single sentence.) Beware of using this kind of repetition too often, however. With much repetition, the effect diminishes and the writing begins to sound unnatural. See PARALLELISM, KEY WORDS, and EMPHASIS.

5. Be careful not to repeat an idea too soon after stating it the first time. The rule applies to more traditional text, where one paragraph flows into another.

Consider the following example:

this

> In my opinion, the Rothskeller algorithm will not alleviate error detection inaccuracies. The mathematical model proposed does not adequately account for errors introduced by migrant electrons. These electrons interact randomly with bits and can change bit patterns in fundamental and disastrous ways. Superior error detection techniques not only must minimize the mushrooming effect of errors compounded by repetition of a bit sequence (which Rothskeller does very effectively) but also must account for random errors that may inadvertently be replicated hundreds of times.
>
> The Rothskeller algorithm for random error detection identifies logical bit pattern inconsistencies and applies logic correction.
>
> However, its identification strategies assume that errors will be of predictable types. Nonpredictable logic errors, caused by random migrant electrons, can go undetected. Therefore, I do not believe that the Rothskeller algorithm will alleviate error detection inaccuracies.

The repeated idea (first and last sentences) is separated by lengthy discussion. Someone reading the preceding passage will be less likely to find the repetition obtrusive.

not this

> In my opinion, the Rothskeller algorithm will not alleviate error detection inaccuracies. The mathematical model proposed does not account for errors introduced by migrant electrons. Therefore, I do not think that the Rothskeller algorithm will alleviate error detection inaccuracies.

This is poor repetition because the repeated idea (first and last sentences) is repeated too quickly.

Reports cover a broad range of business and technical documents, including formal reports, scientific reports (often published), corporate technical reports (following internal guidelines), progress reports, trip reports, laboratory and research reports, accident reports, memorandum reports, and financial reports.

Many of these reports are periodic documents that convey the status of a program, project, task, study, or other organizational effort. Other reports are written in response to specific needs and situations.

Some reports are more informal than others, but readers of reports generally expect to find certain information (summaries, conclusions, recommendations, analyses, supporting facts, etc.), and they expect the report's tone to be businesslike—not officious or bureaucratic, but objective, factual, and honest. See TONE.

Formal reports have traditional components, which are discussed below. For further information relevant to reports, see MEMOS, ORGANIZATION, and GRAPHICS FOR DOCUMENTS.

Reports

Scientific Reports

Scientific reports are tightly controlled by scientific convention and tradition. Many scientific reports are published by professional groups and conferences, and many appear in technical or scientific journals. These professional organizations usually have explicit editorial guidelines that writers must follow.

The tradition of the scientific method dictates the format of most scientific reports:

Abstract
Introduction
Materials and Methods
Results and Discussion
  Fact 1
  Fact 2
  Fact 3
  (therefore)
Conclusions
Recommendations (if any)
Summary (optional)

This format is roughly chronological—moving from the problem to be solved, to the test design (including materials as well as methods), to the test results, to an analysis of the results, and finally to the conclusions.

This logical pattern roughly duplicates the process the scientist used while conducting the study. Ideally, readers should be able to duplicate the process themselves and reach the same conclusions.

In scientific reports, the process of arriving at the conclusions is generally as important as the conclusions themselves. And the pattern of the scientific report tends to reinforce the equal importance of process and conclusions. In doing so, however, it delays the conclusions, which many readers would consider the most important ideas in the report.

The logic behind scientific reports is inductive: *fact, fact, fact (therefore) conclusions.* This pattern is effective, but it is also suspenseful—and that's the major drawback to the scientific format. See ORGANIZATION.

Most business and technical readers are too busy to be held in suspense. They want to know what's important right away. If they want to read a good mystery, they'll read Agatha Christie.

So avoid the scientific format unless you are a scientist writing for other scientists.

In fairness, we should note that reports of all kinds are typically divided into clearly marked sections, and most readers never read the entire report anyway. They read selected sections, depending upon their needs. If readers of scientific reports want to know the conclusions first, they go to the the section entitled "Conclusions."

Reports

The inductive mode of thought behind the scientific format is contrary to the way most readers want to encounter information. For most readers, you should use the far more common format found in standard technical reports.

Technical Reports

Technical and scientific reports often share many features, but technical reports usually differ from scientific reports in several crucial respects:

—They are distributed within an organization and are generally not formally published.

—They are intended for internal use only (many are even proprietary) or have a very limited distribution outside the parent organization.

—Their readers are decision makers and others who need to have all the important information (key findings, conclusions, and recommendations) right away. They may not be at all interested in how the writer arrived at that important information.

Technical report formats usually follow a managerial format, which emphasizes the conclusions and recommendations by placing them at the beginning of the report and subordinates the results and discussion:

Executive Summary
Introduction
Materials and Methods (Optional)
Conclusions
Recommendations (if any)
    (because of)
Results and Discussion
    Fact 1
    Fact 2
    Fact 3
Summary (optional)

The managerial format opens with a summary (often called an executive summary) that presents a distillation of the report's most important ideas. Following the summary, writers often provide a list of conclusions and recommendations so thaat busy managers and supervisors have to read no further to discover the essence of the report.

The managerial format follows an inverted logic: *conclusion (based on) fact, fact, fact.* If readers wish, they can read the facts to determine how the writer arrived at the conclusion. But they don't have to. They can read the conclusion alone and then go on to something else. See ORGANIZATION.

Decision makers are almost always part of the audience for technical reports, and they are usually the primary readers. So most technical reports should follow the managerial format.

Memorandum Reports

Memorandum reports are less formal and usually shorter versions of technical reports, although both types of reports may be very similar in content. Memorandum reports almost always follow the managerial format, but they usually do not include all of the components of a standard technical report. They do not, for instance, have covers, title pages, tables of contents, lists of figures, abstracts, and other formal sections required for either scientific or technical reports. These formal sections are necessary only when reports are widely circulated or published. See MEMOS.

Parts of Reports

Scientific, technical, and some memorandum reports might include the following:

Letter of Transmittal
Cover
Abstract
Title Page
Preface or Foreword
Table of Contents
List of Figures
Body
Bibliography or List of References
Appendix

Letter of Transmittal

A letter of transmittal accompanies and introduces a report. It might explain what the report is about, why it was written, how it relates to previous reports or projects, what problems the writer encountered, why the report includes or excludes particular data, and what certain readers may find of interest.

A letter of transmittal can provide information that would not be appropriate in the report itself, especially sensitive or confidential information. Hence, different letters of transmittal may be written for different readers of the same report.

Letters of transmittal are usually brief. They tend to be less formal than the reports they transmit. However, the more formal the report, the more formal the letter of transmittal is likely to be.

Cover

Covers are appropriate on formal reports and on those intended for widespread or public distribution. The information on the cover is usually similar to the information found on the title page (see below). However, covers are usually well designed and often include artwork.

Abstract

An abstract is a very brief distillation of a report's content. It is intended to describe the report's content and sometimes to provide information about key findings, conclusions, and recommendations.

Some abstracts, especially those for published scientific reports, are primarily useful in data banks and library catalogues. Such abstracts will be printed in catalogues or bibliographies. Prospective readers should be able to determine from reading the abstract whether they would profit from reading the entire report.

Less formal abstracts often function as one-page summaries of corporate technical reports.

Actual summaries may be longer and include more information. In addition, summaries are part of the report and should not be separated from it. Abstracts, on the other hand, are not considered part of the report and should always be understandable in and of themselves.

Abstracts are usually either **descriptive** or **informative.** In either case, they present the key information in a brief paragraph or two (usually no more than about 250 words).

Descriptive Abstracts

Descriptive abstracts describe the content of the report but do not include interpretive statements, conclusions, or recommendations:

> The report analyzes the effects of caffeine on three groups of heart patients: (1) those with diagnosed hypertension and initial signs of heart trouble, (2) those using blood pressure medication but who have not had surgery, and (3) those having had heart surgery. The report discusses the correlation between caffeine and variations in blood pressure for these three groups.

This abstract describes the general scope of the research but does not provide results or conclusions.

Informative Abstracts

Informative abstracts are generally longer and more comprehensive than descriptive abstracts. Typically, they describe the research or project and summarize key results and conclusions:

> Caffeine, in moderate amounts (no more than two cups of coffee per day), has no significant impact on patients with heart problems (ranging from those with diagnosed hypertension to those having had actual heart surgery). Beyond two cups, however, the impacts become increasingly severe. Patients with recent heart surgery showed the most effects, including very high blood pressure and chest pains. Patients on blood pressure medication could cancel the effects of the medication by drinking more than two cups of coffee. Patients with diagnosed hypertension showed elevated blood pressure for up to 3 hours after drinking over two cups of coffee. In conclusion, the effects of caffeine increased substantially with every cup of coffee beyond the two-cup threshold.

In this (fictitious) informative abstract, readers interested in the subject can determine if they would want to read the full report. Informative abstracts provide key results and conclusions. Consequently, if the research techniques are obvious, knowledgeable readers may need no more than the abstract.

If you have a choice, always write an informative abstract. See SUMMARIES.

Title Page

The title page can contain the following information:

—The title of the report

—The name of the person(s) writing the report

—The name of the person(s) for whom the report is prepared

—The date of submission

—The name of the division, group, or department, as well as the name of the organization

—A research number or other documentation aid

—A list of key words for cataloging or for a computer search

—A copyright notice and other special notations (such as *SECRET* or *PROPRIETARY INFORMATION*)

Preface or Foreword

Prefaces or forewords (they are the same) generally appear only in formal or published reports—and often not even there. Informal and memorandum reports rarely include a preface or foreword. If used, a preface or foreword can include the following:

—References to other researchers or reports to which the author is indebted

—Background information regarding the origin of the report—such as who requested it, who funded it, what the goals were, and so on

—Acknowledgment of contributors, including other researchers, managers, technicians, reviewers, editors, proofreaders, and so on

—Financial implications

—Observations regarding unusual conclusions or recommendations

—Miscellaneous personal comments about the contents, including areas for future study

Table of Contents

The table of contents is an outline of the report.

It helps readers understand the structure of the report and locate particular sections. It helps writers organize their thoughts (or check their organization).

A table of contents should contain enough second- and third-level headings to capture the actual content and approach of the chapters. Chapter headings by themselves are often too cryptic:

not this

I. Introduction

II. Preliminary Conditions

III. Governmental Controls

this

I. Introduction

- A. Corporate Policy on Experiments with Animals
- B. Precedents for this Research
- C. Guidelines and Goals of this Research

II. Preliminary Conditions

- A. Physiological Profiles of the Test Animals
- B. Structure of Control and Experimental Groups
- C. Checks and Balances in the Research Procedures

III. Governmental Controls

- A. Documentation Needed for Report to the FDA
- B. External Verification of Results
- C. Legal Penalties for Failure to Report

See TABLES OF CONTENTS.

List of Figures (or Tables or Maps)

A list of figures (or tables or maps) is necessary only if the report is extensive and contains many of these or other visual aids. See GRAPHICS FOR DOCUMENTS.

If used, the list of figures appears on a separate page following the table of contents. (NOTE: The table of contents should include the list of figures and its page number.) See TABLES OF CONTENTS.

Tables and figures are usually listed separately, so if you list tables, do so in a list of tables.

Figures include charts, graphs, maps, photographs, and diagrams. If you have a large number of any particular type of figure, you can list them as separate types of visuals:

List of Maps
List of Charts
List of Photographs

Number tables and figures (and other specific types of visuals) separately, and number them sequentially as they appear in the report. See CAPTIONS.

NOTE: If your report has large separate sections, you can number visuals sequentially within each section. See CAPTIONS.

Body

The body of a report can follow either the scientific format or the managerial format. (See the opening section of this discussion of REPORTS, and also see ORGANIZATION.)

We recommend the managerial organization for most internal reports and many external reports.

Scientific Organization

Abstract
Introduction
Materials and Methods
Results and Discussion
Conclusions
Recommendations (if any)
Summary (optional)

Managerial Organization

Executive Summary
Introduction
Conclusions
Recommendations
Materials and Methods (optional)
Results and Discussion
Summary (optional)

See ORGANIZATION and SUMMARIES.

Summary and Executive Summary

Traditionally, the summary appeared at the end of the report. In the scientific format, the summary still appears at the end (if it appears at all). Its purpose at the end of a report is to "sum up" the major ideas presented, and to remind readers what was important about what they read—the key findings, the conclusions, and any recommendations.

Summaries at the beginning of a report are becoming more common. They are highly desirable in reports directed to managers and supervisors.

Quite often, this opening summary is called an "executive summary." The title indicates clearly how this summary is meant to be read and who is meant to read it. It opens the body of a report written in the managerial format. If well done, the executive summary includes everything a busy manager or supervisor needs to know to make a decision. The detailed results and data often appear only as an appendix. Consequently, a good executive summary in effect makes the rest of the report superfluous—and it should be. If you are writing well, readers should not have to read beyond your summaries unless they have a particular need for the detail that follows.

The information in a summary must be consistent with information appearing throughout the body of the report. Furthermore, you should have nothing in the summary that does not also appear elsewhere in the report.

Summaries are always part of the body of a report. Abstracts, on the other hand, should always be able to stand by themselves. See SUMMARIES.

Introduction

The introduction sets the stage. It normally includes the historical background of the report (and the project or program being reported) and establishes the scope of the report. The introduction may also define special terms and discuss the report's relation to other reports or research efforts.

Introductions also discuss the content and organization of the report. In other words, the introduction tells readers what the report contains and where to find it. In essence, the introduction is a road map.

If the report does not contain a preface, some of the items covered in the preface may also appear in the introduction:

—Person or group authorizing the research

—Contributors, especially other researchers

—Financial implications

—Noteworthy points about the conclusions and recommendations

—Other special items of interest

One major difference between an introduction and a summary is that the introduction does not contain the conclusions and recommendations. Another major difference is that a summary does not provide background information or lay out the structure of the report. See INTRODUCTIONS.

Materials and Methods

This section includes the materials and methods used during the experiment, study, or project. Limit this section to those materials and methods unfamiliar to knowledgeable readers. If the materials and methods are standard, you can mention them briefly in the introduction and then omit this section.

Results and Discussion

This section presents relevant data, discusses the meaning and significance of the data, makes inferences, and states the conclusions. If you have a lot of raw data to present, place it in an appendix and extract only the most important data to present in this section.

In some reports, especially formal scientific reports, the results are separate from the discussion.

Conclusion

This section brings together everything in the report and states your convictions. Every conclusion should grow out of information elsewhere in the report. Without such logical support, readers will justifiably feel that you have failed to accomplish the goals of the research.

Recommendations

Recommendations are suggestions for future actions—either managerial action or future research. Recommendations are almost always present in reports directed to corporate managers and supervisors.

In some cases, conclusions and recommendations are presented in a single section.

Bibliography or List of References

A bibliography or list of references is necessary only in more formal reports or in reports with a number of references.

If you have only two or three references, cover them fully in the text:

> As George Stevens established in "The Life Cycle of the Toad" (*Animal Physiology*, X [March 1983] 234–237), toads have very low metabolic levels.

See CITATIONS and BIBLIOGRAPHIES for full information on the use of parenthetical citations and for different ways to list bibliographic entries.

Appendix

The appendix is for information that is not properly part of the text or is too lengthy to be included in the text (voluminous data, computer programs, lengthy descriptions of methods, etc.).

If information in the appendix is of more than one kind, use two or more appendices, each identified by letter and title:

> Appendix A—Graphs
> Appendix B—Photographs
> Appendix C—Programs

Ensure that you always mention the appendices in the text. Where a reference to an appendix would help readers, identify which appendix is appropriate and what the reader can expect to find there:

> Appendix E presents raw distillation data.

> For further information regarding these formulas, see appendix B.

> The names and addresses of all of those who responded are listed in appendix H.

See APPENDICES.

Resumes

A resume is a document that briefly presents your skills, education, and experience and is intended to get you an employment interview with the right person. Like a product brochure in which you are the product, it must answer the employer's question "What can you do for me?"

The days of the self-centered resume are over. The old-fashioned resume starts with the applicant's self-serving objective—"A rewarding position in a firm that can make me rich and successful."

Decision makers don't have the time or inclination to study your life story and make your dreams come true. They have real problems and opportunities, and they need help with them. Therefore, your resume needs to address those problems and opportunities directly. Your resume needs to show *specifically* and *quantitatively* how life will be better for them if they bring you on board.

Resumes

1. Research the employer and the job before writing your resume.
2. Choose a resume format that emphasizes your skills and abilities. If possible, limit resumes to one page.
3. Prepare a detailed working list of all positions, activities, or other experiences that can validate your skills and abilities.
4. For each job, activity, or experience you decide to include in your resume, list accomplishments or skills using an action verb and specifics.
5. List job information and education as accurately as possible, and include specifics.
6. Omit references from your resume, but prepare a separate list of references for submission when requested.
7. Review your resume a final time before submitting it to verify the completeness and accuracy of the information and to avoid inappropriate information.
8. Attach a cover letter when you submit a resume.

1. Research the employer and the job before writing your resume.

There are two kinds of resumes: **generic** and **targeted (job-specific)**. A **generic resume** contains your contact information, work experience, and educational background. You can post a generic resume on your social-media site and hand it to prospects as you encounter them.

By contrast, a **targeted resume** is aimed at *one* prospective employer and is the product of serious research into that organization. It shows how your capabilities, experience, and education match up with the job requirements as well as the mission and values of the organization. If you really want the job you've researched, you send a targeted resume. See the model generic and targeted resumes in Model Documents.

To draft a targeted resume, study the job description carefully. List the job requirements and sketch in your relevant skills and experiences next to each requirement. Perhaps most important, find out what problem or opportunity has motivated the organization to post the job opening. Few organizations today simply "fill positions"; often, some new need is behind the opening. Talk to insiders, if possible, to discover what that need is.

Do not send a "one size fits all" generic resume to your targeted employer. If you already have a generic resume, rewrite it to target specifically the needs and requirements of the job you're applying for.

2. Choose a resume format that emphasizes your skills and abilities. If possible, limit resumes to one page.

Useful resume templates are available on the Internet and in your word-processing program. Choose a template that gives a clean and open impression. Avoid templates that include flashy graphics, multiple boxes and borders, unusual fonts, or large contrasts in font size or style.

Two resume formats are common: the **chronological (traditional) format** and the **functional format**. We recommend that you use the functional format. See figure 1 on the next page for a contrastive view of these two formats.

A functional format uses your skills and abilities as subheadings so that your skills and abilities are immediately visible to readers.

A chronological format often uses the dates of employment as headings. The dates can demonstrate the continuity in your work history, but they fail to emphasize your skills and abilities. Even if you decide to use a chronological format, add boldface and other emphasis techniques to help make the resume more effective.

Functional Format

111 Any Street
City, State 00000
111-555-1111 (W) 111-555-2222 (H)

Skills and Accomplishments

- Coordinator of shipping department
- Experienced with WordPerfect
- Ability to manage projects

Education

Chronological Format

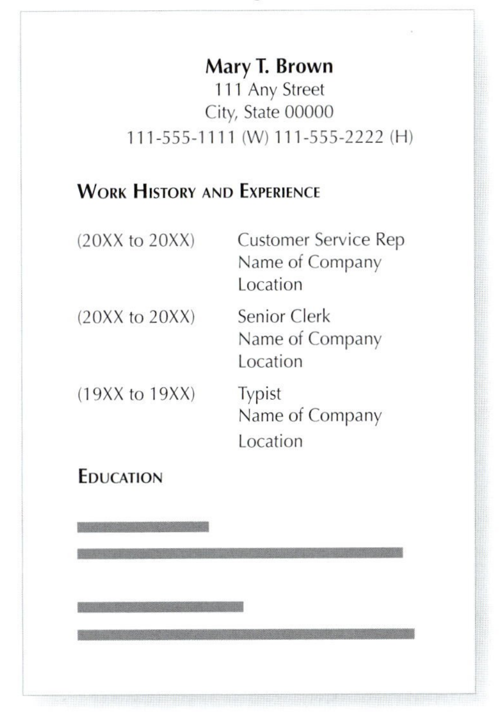

Mary T. Brown
111 Any Street
City, State 00000
111-555-1111 (W) 111-555-2222 (H)

Work History and Experience

(20XX to 20XX) Customer Service Rep
Name of Company
Location

(20XX to 20XX) Senior Clerk
Name of Company
Location

(19XX to 19XX) Typist
Name of Company
Location

Education

Figure 1. *Choose a functional format when you want to emphasize your skills and accomplishments. Choose a chronological format if the sequence of your jobs is more important to the prospective employer.*

3. Prepare a detailed working list of all positions, activities, or other experiences that can validate your skills and abilities.

This list is merely a preliminary source of information. Later, you can choose and refine items to emphasize in your resume.

Positions are only the beginning. Especially if you have a limited job history, you need to consider any non-job activities or experiences:

—Have you been active in a local club—maybe even served as an officer?

—Do you sew and tailor your own clothes?

—Do you make routine repairs on automobiles?

—Can you speak, read, or write a foreign language?

—Are you an amateur photographer?

—With which computer programs are you experienced?

In each of these activities, you will have used or developed skills. These skills will be highlights in your resume.

4. For each job, activity, or experience you decide to include in your resume, list accomplishments or skills using an action verb and specifics.

This write-up of accomplishments or skills is where you really begin to sell yourself. Remember to target your list to the specific job you are applying for.

As an example of how to develop skill descriptions, assume that you were the treasurer for a local sports club. You could initially list this fact as follows:

Treasurer, Loma Linda Sports Club, 2009–2010.

Unfortunately, the title of the position and the name of the club are not really informative. So, to beef up the preceding information, you should add specifics, especially numbers, if they apply:

- Collected and kept account records for over 100 members.
- Prepared the club's 2009-2010 annual budget ($300,000).
- Handled payments and prepared financial reports for monthly meetings.

Note 1: The preceding three phrases begin with an action verb—that is, a verb that suggests actions or accomplishments. Phrases of this sort are both more forceful and shorter than a full-sentence narrative. You can also use these phrases in a displayed list in your resume.

not this

- As club treasurer, I worked to prepare the club's annual budget for 2009–2010, which was roughly $300,000.

nor this

- In 2009–2010 I was responsible for planning and preparing the annual club budget, which provided roughly $300,000.

Both of the preceding are wordy, and they each bury key points in the middle of the sentence.

NOTE 2: Most phrases on resumes will begin with the past-tense verb, but if you are still performing the task or you still hold the job, then use a present-tense verb:

- Collect and record dues for over 1,000 members.
- Prepare the club's annual budget ($300,000).

5. List job information and education as accurately as possible, and include specifics.

Both your list of jobs and your list of educational credentials should be complete and accurate. Usually, you should explain any obvious gaps or other unusual information.

Lying or omitting information would be grounds for dismissal later if your employer discovers the problem. See ETHICS.

- Director, Sales Operations, Argent Consulting, Vancouver, BC 2006–present
- Sales Manager, QuadB Software, Toronto, ON 2005–2006
- Client Partner, Haxton-Sokol Consulting, Chicago, IL 2001–2005

When appropriate (considering the job you are applying for), add specific skills or accomplishments:

B.S. (2005), Business Administration, University of Maryland—(Emphasis: International Financial Markets and Exchange Rates). London Internship, Spring 2006, Barclays Bank.

LL.M. (2011), World Commerce Law, University of Bern (CH). Thesis compared legal requirements for drug approvals in Europe, Japan, and the United States.

Be specific about your accomplishments and skills, including numbers and dates:

- Boosted leads in fiscal 2010 from 30 to 238 and divisional revenues from $6.2 million to $12.4 million.
- Created, in one year, a proactive marketing program that increased positive media coverage 650%.
- Led cross-functional team that cut costs 50% by implementing preconfigured hosting solutions.
- Earned green belt certification in Six Sigma quality-improvement process and applied it to reduce cycle time 20% within 3 months.

6. Omit references from your resume, but prepare a separate list of references for submission when requested.

Most resumes nowadays don't list references. Even mentioning in a resume the availability of references is now unnecessary. Every employer assumes that applicants are prepared to provide references when requested.

Always ask references for permission to use their names. Sometimes, a person would rather not be listed. The person might not know enough about you or might feel that he or she can't be positive about your skills.

Usually, the list of references includes one or more recent job supervisors, as well as perhaps one non-job reference who can talk about your character.

7. Review your resume a final time before submitting it to verify the completeness and accuracy of the information and to avoid inappropriate information.

A final review is an essential step when a single error or unintentional omission in your resume may eliminate you from consideration for a job.

If possible, ask one or more friends to review your resume, both for content and for mere mechanics (spelling, punctuation, and consistency). Someone else's eyes will see your errors more readily than your eyes would.

One priority is to check all company names, personal references, and numbers to verify that you have names correctly spelled, any titles accurately listed, and current contact information. Make sure your contact information is for telephones and Internet connections that will not embarrass you if accessed.

A second priority is to avoid mentioning information that is now illegal. Do not provide any information about your race, age, sex, religion, ethnic background, or family situation. You should also not mention your physical condition unless you have a problem that would limit your ability to perform a certain job.

The best way to handle topics that you should avoid is to remove some specifics from your application. If your age might be a decision factor, you might choose not to list the graduation dates for your academic degrees. If you need to mention a church activity, do so without mentioning the name of the denomination:

this

Social Chair for Church Adult Forum (over 100 active members) 2006–2009. Planned monthly socials and three annual spring weekend retreats.

not this

Social Chair for the Presbyterian Adult Forum (over 100 active members) 2006–2009. Planned monthly socials and three annual spring weekend retreats.

8. Attach a cover letter when you submit a resume.

The cover letter is your best opportunity to tailor and personalize your application.

In your cover letter, you introduce yourself and your resume, tell why you're applying for the job, and ask for an interview. Many employers are more interested in your cover letter than your resume because it tells them why you're applying and gives insight into your personality.

Don't submit an offhand or "one size fits all" cover letter. You must target your cover letter to the specific needs and values of the prospective employer. Study the job description and list the requirements. Study the mission, vision, and values statements of the organization to get a feeling for the culture. Study the organization's financial information, which is readily available in the case of a public company. Talk to their customers. Talk to insiders if possible. Incorporate what you learn into your draft.

Ideally, your cover letter is a one-page proposal to fill an important business need that you've identified in your research about the job. Here is a basic format to follow:

1. Start by summing up the problem or opportunity your prospect faces and how you plan to help solve it. Use numbers that are meaningful to the prospect.
2. Give evidence that you *can* solve the problem and, if applicable, that you have solved it in the past. Evidence includes work and educational accomplishments that would lead your prospect to conclude that you're right for the challenge.
3. Provide contact information and request a meeting.

Let the job requirements govern the form of your letter. For example, if the job posting mentioned attributes such as "attention to detail," "hard working," and "team player," you might use these three phrases in your letter as headings in the body of the letter. See Letters.

If, for example, you are addressing the "team player" requirement, you might have a brief paragraph like this one:

> **Team Player**
>
> I work well with people and can coordinate activities effectively. My job as a retail clothing buyer (at Meyers Clothing) required me to work closely with retail sales clerks, with management, and with the financial planners. My last performance review at Meyers rated my team skills as excellent.

In a cover letter, you might also include personal details that would set your application apart from the dozens of other applications. Here, for example, is a possible closing with a personal comment:

> In our phone conversation, you mentioned that Ajax employees set their own working hours. My current job—at Hoffmann's Condiments—allows us to work flexible hours. I usually get in by 6:45 a.m. so that I can be home by late afternoon for quality time with my family.
>
> Ajax sounds like a great place to work. Please call me at (555) 445-3396 (my cell) if you would like me to come in for an interview.

For an example of an effective cover letter, see Model Documents.

Most organizations now accept soft (electronic) copies of your resume and cover letter when you apply for a job, but it's a good idea to send hard copies as well. Print these documents on high-quality paper—not ordinary copy paper.

Scientific/Technical Style

Scientists and technical specialists write and speak a different language. They do use many of the words and sentence patterns that nontechnical writers use, but their language is sufficiently different to be called a scientific or technical style. See STYLE.

Most obviously, scientists and technical specialists use technical terms: *joule, volt, electron, ion, protozoa, electroencephalogram, uterine tube, ethyl ethers, hypochlorous acid,* etc. Such terms have specific meanings and are generally foreign to lay readers. Often lay readers consider such technical terms to be jargon because they do not readily know and perhaps cannot easily understand them. See Jargon.

Less obviously, scientists and technical specialists sometimes use common words in uncommon ways. Often such uses are more confusing to lay readers than technical words, which lay readers expect to be foreign to their experience. Here are two examples of uncommon uses of common words:

> A continuous function in an *n*-dimensional vector space is considered **smooth** when the function has certain well-defined features.
>
> We were unable to calculate the **work** because the granite outcrop would not budge, even after a standard charge exploded in a hole drilled into its base produced a hairline fracture evident throughout the circumference of the **neck**.

In the first example, mathematicians have borrowed the intuitive notion of smoothness to describe an abstract mathematical concept. In the second example, physicists define *work* to be the effect when a force moves an object a certain distance. Without movement, work (according to the definition) has not taken place.

Scientific/Technical Style

1. Make your use of technical and scientific terms appropriate for your readers.
2. Use concise, direct sentences.
3. Design scientific and technical documents for visual impact and readability.
4. Above all, be as accurate as possible.

Similarly, *neck* describes that portion of an outcropping where the base of the outcropping most clearly resembles the border between the outcropping and the mass of rock from which the outcropping protrudes. The resemblance to a human neck is clear, but lay readers may not make the connection as readily as geologists.a

Both technical terms and common words used in uncommon ways complicate scientific and technical writing. In both cases, lay readers may become lost, even when the writing is clear, concise, and logical. Therefore, if you are writing a scientific or technical document that is intended, at least in part, for lay readers, try to define technical terms (or don't use them), and avoid using common terms in uncommon ways.

If you are writing for scientific or technical readers, you must still try to make the writing as clear, concise, and logical as you can. Technical word usage may not hinder technical readers, but writing that is poorly organized, clumsily phrased, and inaccurately conceived will still be difficult if not impossible to read.

1. Make your use of technical and scientific terms appropriate for your readers.

Avoid unusual or overly technical terms if possible. If you cannot avoid them, and if your readers will not understand them, then define the terms. Do not use unfamiliar abbreviations unless your readers will understand them or unless you carefully explain the abbreviations first. See ABBREVIATIONS and ACRONYMS. Once you have established the terms and abbreviations, be consistent in your usage throughout the document.

Do not assume that your readers (all of your readers) know and understand your terms as you are using them. This is a trap.

First, rarely do all readers of a document have the same kind and level of technical background. You should usually write for the lowest common denominator—that is, for the least technically sophisticated reader.

Second, even well-established technical terms, concepts, and abbreviations may be subject to dispute, so you may have to stipulate definitions so that your readers know exactly what you intend when you use a particular word.

One of the easiest ways to handle difficult or unfamiliar technical terms is to provide an informal definition when the terms first appear:

> A further health problem in many African countries results from filariasis, which means that the blood contains small threadlike worms.
>
> *or*
>
> The presence of small threadlike worms in the blood (called filariasis) is a further health problem in many African countries.

Third, writers sometimes use different words to describe essentially the same thing, and this creates substantial confusion, even among scientific readers. Do not speak of *aspects* of a procedure in one paragraph, and then refer to the same concept as *elements* or *features* later.

2. Use concise, direct sentences.

Sentences are the building blocks of effective writing of any kind. However, they are especially important in writing that is inherently complicated and includes jargon. So strive to make your sentences clear, direct, and concise. See SENTENCES.

The principles of writing direct, concise sentences are summarized below. Elsewhere in this *Style Guide*, you will find each topic discussed in depth.

- **Choose active rather than passive sentences:**

  this

  We determined that the coefficient of friction varied most at extremely cold temperatures.

  not this

  It was determined that the coefficient of friction varied most at extremely cold temperatures.

  See ACTIVE/PASSIVE.

- **Avoid wordiness:**

  this

  Our revised proposal presented further justification for development costs while offering to reduce our profit fee by 20 percent.

  not this

  Further justification for development costs, along with an offer for the 20 percent reduction of our profit fee, was presented in our revised proposal.

  See WORDY PHRASES.

- **Use strong verbs:**

  this

  After lengthy study, we adjusted the effluent guidelines to accommodate projected economic hardships.

  not this

  After lengthy study, we made an adjustment of the effluent guidelines to effect greater accommodation with projected economic hardships.

  See STRONG VERBS.

- **Avoid false subjects:**

  this

  Five-spot pattern recoveries are probably less efficient in sand trap reservoirs that have been depleted to within 15 percent of recoverable reserves.

  not this

  It is probable that five-spot pattern recoveries are less efficient in sand trap reservoirs that have been depleted to within 15 percent of recoverable reserves.

  See FALSE SUBJECTS.

- **Use pronouns to make your writing more personal and direct:**

  —Pronouns are appropriate when a scientific or technical writer is recommending something, drawing conclusions, or conveying deliberate decisions or choices:

  > The evidence suggests that this species is in fact indigenous to the Everglades. Therefore, I recommend broadening the scope of the USDA's habitat study before it writes the Los Puertos EIS.
  >
  > The globule had a specific density of 6.78, more than twice what the OCS recovery team had estimated, so we concluded that zinc, not iron, was its major constituent.
  >
  > Once we had analyzed the data from the pilot tests, we determined how to control temperature variations in succeeding tests.

  —Some technical and scientific writing does not lend itself to first-person pronouns. The historical description of a laboratory procedure, for instance, should not mention the person who performed the procedure unless the focus of the discussion is on that person and not the procedure. If the procedure itself is more important, do not use personal pronouns:

  > The ore sample is washed in a weak solution of hydrochloric acid. Next, the ore is cleansed thoroughly in a water bath and then dried in a heat chamber.

Note: If this passage appeared in a set of instructions, second-person pronouns and imperative sentences would be appropriate:

> Wash the ore sample in a weak solution (no more than 5 percent) of hydrochloric acid. Then you should clean the sample thoroughly in a water bath and dry it in a heat chamber.
>
> *or*
>
> Wash the ore sample in a weak solution (no more than 5 percent) of hydrochloric acid. Then clean the sample thoroughly in a water bath and dry it in a heat chamber.

The use of pronouns in technical writing has become more common in the last several decades, but don't overuse pronouns. Most scientific and technical writing is meant to convey information objectively, not to establish the writer's personality. If you are writing for the general public, however, use more pronouns and try to convey both clarity and warmth. For an example, read anything by Stephen Hawking,

Michio Kaku, and Deborah Cadbury, three successful scientific writers whose readers are primarily nontechnical. See PRONOUNS.

3. Design scientific and technical documents for visual impact and readability.

Visual appearance has become more and more important as a feature of good scientific and technical style. Well-conceived charts, graphs, tables, photographs, and illustrations are essential if a document is to look fully professional.

In essence, the text supports the graphics, not the reverse. So time spent planning and designing graphics is critical for successful scientific and technical documents.

Below are some key principles of using graphic aids. For a much more complete discussion, see GRAPHICS FOR DOCUMENTS, GRAPHICS FOR PRESENTATIONS, CAPTIONS, CHARTS, GRAPHS, ILLUSTRATIONS, TABLES, MAPS, and PHOTOGRAPHS.

- Create your graphics before you write your text.
- Select graphics that are appropriate for your readers.
- Focus your graphics on key points and keep them simple and uncluttered.
- Introduce graphics before they appear in the text.
- Use clear, interpretive captions on your graphics.
- Use emphatic devices (underlining, *italics*, **boldface**, shading, etc.) to emphasize important ideas in your graphics as well as your text.

4. Above all, be as accurate as possible.

Good scientific and technical writing is precise and accurate—in its facts, its references, its procedures, and its analyses.

Check every detail. If your document mentions that the site of the project is southwest of Dry Gulch, then the map should not show the site to be northwest of Dry Gulch. Similarly, if you mention production data in the text, the same figures (rounded to the same significant figures) should appear in accompanying tables. Every detail in the document should be as accurate and consistent as the writer can humanly guarantee.

A single inaccurate figure or fact can destroy the credibility of a document and its writer.

Note: Some writers defend weasel wording or hedging by stating that we can't know anything for sure; therefore, we can make no absolute statements. According to them, we can never state a fact because we can't be sure that anything is irrevocably true. The words typically used to hedge are *generally, usually, likely, possibly, notion, surmise, speculation, conjecture, indicate, suggest, appear, seem, believe,* etc.

If necessary, use such words to indicate honest doubt or uncertainty, but do not overdo them. If you are certain of something, say so. Weasel wording weakens your document and your image:

this

Malodin can be effective as a maintenance therapy for osteoarthritis.

not this

We believe that the evidence favors an interpretation that Malodin is likely to be a potential maintenance therapy for osteoarthritis.

Semicolons have two primary functions—linking thoughts and separating thoughts.

Semicolons

1. Use semicolons to link complete thoughts that could otherwise stand alone as separate sentences.
2. Use semicolons to separate items in series when one or more of the items has a comma.

1. Use semicolons to link complete thoughts that could otherwise stand alone as separate sentences:

> Western Aeronautics has completed more than 15 avionics contracts in the last 5 years; in 11 of those contracts we used CAD/CAM techniques to minimize development costs and improve both reliability and performance.

Typically, the complete thoughts linked by a semicolon are equal in structure and importance. Writers could separate the complete thoughts with a period and create two sentences; however, the semicolon shows a closer relationship between the thoughts than a period does. The semicolon says, "These thoughts are closely related."

NOTE 1: Semicolons used to link complete thoughts do not require conjunctions (transitional or connecting words like *however, consequently, furthermore,* and *thus*):

> Detailed trade studies helped us determine the relative importance of the various technologies in designing a supersonic cruise fighter; similar performance studies helped us evaluate the trade study findings in air-to-air operations in a simulated combat environment.

> High maneuverability was our primary design consideration; our secondary considerations included short take-off and large payloads.

NOTE 2: The following transitional words and phrases can be used with semicolons if the writer needs to indicate or clarify the relationship between the thoughts before and after the semicolon: *accordingly, consequently, for example, for instance, further, furthermore, however, indeed, moreover, nevertheless, nonetheless, on the contrary, on the other hand, therefore, thus.* See CONJUNCTIONS.

> We use aircraft geometry to generate survivability parameters such as radar signature; furthermore, we add performance, weapons, and avionics characteristics from the developed aircraft design to evaluate military effectiveness.

> The SDC is a multiterminal facility dedicated to a variety of projects; however, its size and flexibility, coupled with strict management controls, guard against fragmentation and crossover. *(Note the comma following* furthermore *and* however.)

NOTE 3: Do **not** use the shorter conjunctions (*and, but, or, for, nor, so,* and *yet*) with semicolons; use these simple conjunctions only when linking two complete thoughts with a comma.

However, if the complete thoughts are lengthy and already contain commas, then use a semicolon with these shorter conjunctions to join the two complete thoughts:

> The committee began its scrutiny of the preliminary design study, which General Avionics submitted upon just 2 weeks' notice; and they found that the transformational analyses, although obviously hurried, were still superior to those of other designers who had had much more time.

See COMMAS, CONJUNCTIONS, and TRANSITIONS.

2. Use semicolons to separate items in series when one or more of the items has a comma:

> Our cost breakdown demonstrates our cost consciousness; our commitment to low overhead, especially through our direct contract-costing procedures; and our desire to minimize risk by using proven resources.

Without the semicolons to separate each major item in the series, readers might not understand how many items the series contains. The need for clarification varies from sentence to sentence. Sometimes it is not crucial; sometimes it is:

> Our avionics designs incorporate state-of-the-art subcomponents, including the Leon2 CPU core; Marquette frequency stabilizers, the most advanced anti-ECM devices currently available; and Barnett Industries' redesigned RFX regulators.

If the three items in this series had been separated with commas, the purpose of the phrase *the most advanced anti-ECM devices currently available* would not be clear. It would seem to be a fourth item in the series, although its true purpose is to describe the Marquette frequency stabilizers. The semicolons clarify its purpose in the sentence.

NOTE: With a longer and complex series of items, the more desirable option would be a displayed list:

> Our avionics designs incorporate state-of-the-art subcomponents, including:
>
> - Leon2 CPU core
> - Marquette frequency stabilizers, the most advanced anti-ECM devices currently available
> - Barnett Industries' redesigned RFX regulators

See LISTS.

Sentences

Sentences are the building blocks of thought. Without sentences and the context in which they appear, communication would be impossible.

Sentences are fundamental to language, yet they are hard to define. *Webster's New Collegiate Dictionary* says that a sentence is a "grammatically self-contained speech unit." Many teachers would call it "a complete thought" and perhaps add that it usually contains, at a minimum, a subject and a verb. That definition would seem to rule out the following, all of which are sentences:

Yes.
No.
Maybe.
Stop.
Hello.
Me?
When?
Where?
Why?
How?
OK.
Oh!

All written sentences begin with a capital letter and end with a mark of punctuation, usually a period but sometimes a question mark or an exclamation mark. This convention applies to all sentences, even those with only one word and those in which words that are understood have been left out.

Most sentences are longer than one word and do include both the subject and the verb:

The invoice was late.

Because the invoice was late, we could not include it in accounts payable for November.

When did the invoice arrive?

It arrived late.

The invoice, which was late, missed the deadline for November accounts payable.

The invoice was late, so we could not include it in November accounts payable.

Next time, send us the invoice promptly.

What happened with the late invoice was that we could not include it in November accounts payable.

In some cases, parts of the sentence are understood but not stated:

Too late. (The invoice was too late.)

Late again! (The invoice was late again!)

These examples reveal several things about sentences. First, they are self-contained, although they might rely heavily on something said earlier (or later) for readers to fully comprehend them. Second, they consist of a meaningful word or group of words. A word constituting a sentence does not have to be a particular kind of word, nor does it have to be meaningful in normal contexts. *OK* and *Yes* are clearly self-contained expressions, but *String* could also be a sentence:

What did you use to secure the box?

String. (I used string. *The* I used *is understood.*)

Single verbs can also function as sentences:

What do you suggest I do during lunch?

Run. (I suggest that you run.)

So, must sentences express complete thoughts to be sentences? Yes, although the completeness of the thought usually depends on the context in which the sentences appear. For an utterance to be a sentence, it must either state or imply a complete thought, given its context.

Must sentences contain a subject and a verb to be sentences? Yes and no. Sentences do have a grammatical structure, including a subject and a verb, but either or both can be understood:

When should I leave?

Now. (You should leave now. *In this sentence, the subject* [you] *and the verb* [should leave] *are both understood.*)

Sentences

1. Use declarative sentences to make statements of fact and opinion. Usually such sentences follow the subject-verb word order, and they end with a period.
2. Use interrogative sentences to ask questions. Interrogative sentences usually begin with a question word *(who, which, where, when, why,* and *how)* or with a verb.
3. Use exclamatory sentences to make strong assertions or surprising observations. Exclamatory sentences usually end with an exclamation mark.
4. Use imperative sentences to give directions or commands. Imperative sentences usually begin with a verb and end with a period (although an exclamation point is also occasionally possible).
5. Limit average sentence length to about 20 words for typical business and technical writing.
6. Use a variety of sentence types and sentence lengths.
7. Strive to make all sentences direct.

Sometimes the subject and/or verb are complicated and don't convey the primary meaning or central thought in the sentence:

> What happened with the late invoice was that we could not include it in November accounts payable.

The subject is *What happened with the late invoice*. The verb is *was*. The subject is a complicated noun clause, and the main meaning is not in the main verb, but in the final clause: *that we could not include it in November accounts payable.*

Still, in most business and technical writing, sentences do have a subject and verb, even if these grammatical slots are filled with many words and have complex grammatical relationships.

Perhaps because sentences are difficult to define, most grammar handbooks settle for two fairly simple yet practical systems for cataloguing sentences:

- Purpose or Intent
  - —Declarative Sentences
  - —Interrogative Sentences
  - —Exclamatory Sentences
  - —Imperative Sentences
- Grammatical Structures
  - —Simple Sentences
  - —Compound Sentences
  - —Complex Sentences
  - —Compound-Complex Sentences

Purpose or Intent of Sentences

1. Use declarative sentences to make statements of fact and opinion. Usually such sentences follow the subject-verb word order, and they end with a period:

> We reviewed the report.
>
> Because of the detailed analyses involved, our review of the report is likely to take several days.
>
> The report is only five pages long.
>
> The report, which is only five pages long, will still take several days to review because the analyses are lengthy.

See PERIODS.

2. Use interrogative sentences to ask questions. Interrogative sentences usually begin with a question word (*who, which, where, when, why, what,* and *how*) or with a verb:

> Who is the engineer in charge?
>
> Which plan is likely to be approved?
>
> When will the construction project end?
>
> How often do they propose to inspect the site?
>
> Have you filled in all the necessary forms?
>
> Were the construction specifications adequate?

See QUESTION MARKS.

3. Use exclamatory sentences to make strong assertions or surprising observations. Exclamatory sentences usually end with an exclamation mark:

> What a surprising conclusion!
>
> That's wrong!
>
> What a field day for the lawyers that will be!
>
> Oh, I doubt that!

NOTE: Exclamatory sentences often have grammatical structures very different from normal declarative sentences. In fact, they may be only a word or two long:

> No!
>
> How surprising!
>
> A shame!

See EXCLAMATION MARKS.

4. Use imperative sentences to give directions or commands. Imperative sentences usually begin with a verb and end with a period (although an exclamation mark is also occasionally possible):

> Check the Internet for additional information.
>
> Adjust the flange on the steam connection to prevent leakage.
>
> Do not submit the pink copy of this form!
>
> Stop. (*or* Stop!)

Grammatical Structures of Sentences

Sentences have four grammatical structures—simple, compound, complex, and compound-complex. These four structures, however, can be formed into an infinite number of unique sentences.

Simple Sentences

Simple sentences are sentences that express one complete thought. Essentially, they contain a single subject and a single verb, although both subject and verb may be compound:

> The pump failed. *(single subject and verb)*
>
> The new steam pump failed after only 3 weeks of service. *(single subject and verb)*
>
> We analyzed the blueprints. *(single subject and verb)*

Sentences

James Hawkins and I analyzed the blueprints for the new maintenance facility. *(compound subject, single verb)*

James Hawkins and I analyzed and revised the blueprints for the new maintenance facility. *(compound subject and compound verb)*

NOTE 1: As indicated above, simple sentences can contain compound subjects, such as *James Hawkins and I*, and compound verbs, such as *analyzed* and *revised*. Although compound, such subjects and verbs form a single unit, at least for the purpose of the sentence in question, so the sentence is still considered simple.

NOTE 2: A quick test for a simple sentence is that you cannot logically break the sentence at any point and come up with two other simple sentences. For example, the following simple sentence, even with compound subject and compound verb, cannot be broken into two simple sentences, so the sentence is a single, simple unit:

My supervisor and I joked about the assignment and then worked on ways to accomplish it.

Compound Sentences

Compound sentences are a union of two or more simple sentences. These simple sentences are usually linked by one of the simple coordinating conjunctions: *and, but, or, nor, for, yet,* and *so*.

The project was expensive, and management still hadn't decided to proceed with it.

Our supervisor wanted to increase office productivity, but the turnover in personnel made such an increase unlikely.

See COMMAS and CONJUNCTIONS.

Sometimes the simple sentences are linked by a semicolon or by a semicolon and a conjunctive adverb:

The data confirmed our initial assumptions about the problems in prototype production; with these problems, the project will almost certainly exceed the budget.

The surveying was to have been completed by October 15; however, construction must start on or before November 15.

See SEMICOLONS and CONJUNCTIONS.

NOTE 1: A quick test for a compound sentence is to see if you can divide the sentence into two or more simple sentences:

The surveying was to have been completed by October 15. Construction must start on or before November 15.

NOTE 2: Sometimes three or more simple sentences can combine into a compound sentence:

The site was ready, the construction crew was ready, and the materials were ready, but the weather was not cooperative.

Complex Sentences

A complex sentence is a simple sentence with a dependent (subordinate) clause attached to it. The dependent clause can appear before the main clause (the otherwise simple sentence), in the middle of the main clause, or following the main clause.

Here are some dependent clauses:

Although our bid was the lowest . . .

. . . who was the most expensive candidate . . .

. . . because the tailings pile is virtually inert.

Adding a simple sentence to each of these dependent clauses forms a complex sentence:

Although our bid was the lowest, another contractor had more experience.

Cameron Blake, who was the most expensive candidate, did have the most impressive credentials.

Reclamation of the mine site will be difficult because the tailings pile is virtually inert.

NOTE 1: A test for a complex sentence is to separate the dependent clause and the main clause (simple sentence). This test will work for the sentences above, but not for this sentence:

What Jack wanted to discuss with us became clear once the meeting got under way.

What Jack wanted to discuss with us is a noun clause that functions as the subject of the sentence. Trying to separate it from the rest of the sentence would result in two sentence fragments, neither of which can stand alone. Although this sentence fails the test, it is still a complex sentence.

NOTE 2: Dependent clauses are usually introduced by these words:

Subordinate conjunctions

Because, since, although, even though, after, before, so that, while, when, etc.

See CONJUNCTIONS.

Relative pronouns

Who, whom, whose which, that, whoever, whomever, why, when, where, etc.

See PRONOUNS.

Compound-Complex Sentences

Compound-complex sentences are a combination of the two previous sentence types. They are both compound and complex. Therefore, a compound-complex sentence has two attached independent clauses—a compound sentence (two simple sentences attached to each other)—and at least one dependent clause:

> Because the firm's manufacturing capacity could not be increased rapidly enough, they were unable to fill their orders; consequently, competitors gained a significant foothold on the market.

NOTE: Because this sentence is compound, the semicolon separates two independent clauses, each of which could stand alone as a complete thought. The introductory dependent clause beginning with *because* makes the sentence complex.

Sentence Length and Readability

All readability formulas include average sentence length as one measure of the readability of a piece of writing. Some authorities argue that average sentence length for any level of reader should be kept below some maximum (15 to 20 words). For younger or less sophisticated readers, the average sentence length should be even less (8 to 12 words, depending on how young the readers are).

Sentence length is one important factor in readability. Readers have to read, comprehend, remember, and interpret information sentence by sentence. The longer a sentence, the more they have to hold in their minds to comprehend, remember, and interpret the thought being expressed. Long sentences place an unnecessary burden on readers.

Another factor in readability is sentence syntax (structure). Long sentences can still be easily readable if they are constructed so that the sentence is easy to follow and easy to understand:

> We accepted the bid from the Cranston Construction Company—although it was the highest bidder—because (1) it has the manpower and equipment to start the project immediately, (2) its personnel are experienced in this type of construction, (3) it has the necessary permits in hand, (4) it is a local company using local resources, and (5) its management was willing to post a substantial performance bond.

This 67-word complex sentence is long but readable because it states its central point immediately: *We accepted the bid from Cranston.* The dependent clause enclosed by dashes clearly states a fact contrary to our expectations: Cranston was the highest bidder. We don't expect the contract to go to the highest bidder, and we wonder why Cranston received the award. So the numbered list provides the rationale for the decision. The numbering of the list helps clarify the syntax.

However, long sentences must be written this clearly to be readable. If your average sentence becomes too long and you do not have the skill to structure each sentence with great clarity, then your writing will be difficult, if not impossible, to read. See STYLE.

5. Limit average sentence length to about 20 words for typical business and technical writing.

The 20-word average is based on readability experiments and studies that reinforce the discussion of sentence length presented above. Obviously, in arriving at a 20-word average, you will have some longer sentences and some shorter sentences.

Longer sentences were acceptable many years ago. Writers in the 19th century produced some mammoth sentences. Today the trend is toward short, concise sentences intended for busy readers who don't have the time for or the interest in long-winded prose. The short sentence is the better sentence.

6. Use a variety of sentence types and sentence lengths.

Writing that is uniform is tedious. Make some sentences long and some short. Make most of them moderate in length. Remember that long sentences are useful for presenting involved concepts and for elaborating on a point that requires some thought. Short sentences are useful for stating clear, crisp thoughts.

Create variety, too, in your choice of sentence types. Use mostly simple sentences, but do not avoid the compound, complex, and compound-complex sentences. Compound and complex sentences are very useful for expressing related ideas. If you use only simple sentences, you will be producing Dick-and-Jane writing.

7. Strive to make all sentences direct.

Keep the syntax (structure) of the sentence as uncomplicated as possible. You can do this by keeping the subject as close as possible to the verb; by keeping modifiers as close as possible to the words they modify; and by using conjunctions and transitions to show progress, sequence, connection, and contrast. See TRANSITIONS, PRONOUNS, MODIFIERS, ADJECTIVES, ADVERBS, NOUNS, VERBS, and CONJUNCTIONS.

Signs and Symbols

Signs and symbols are increasingly important in today's visually oriented world. Television stations use visual images (and accompanying theme music) as identifying signs for a particular video production. Corporations spend great sums of money designing logos that symbolize the corporation's mission or image.

> **Signs and Symbols**
>
> 1. Choose standard signs and symbols and ensure that readers understand them.
> 2. Use the written-out version of signs and symbols in text unless the nature of the text is such that readers would expect to see signs and symbols.
> 3. Repeat a symbol when you have a series of numerical values and the symbol would appear without a space between it and the numerical values.

Internationally, signs and symbols have crossed national boundaries. For example, the dashboards of automobiles, wherever produced, now use international symbols for the horn, the lights, and so on. A less obvious example are the internationally recognized symbols of distress when the crew of a downed aircraft wish to communicate with a plane overhead. If the crew need food and water, the symbol is a block capital F, which can be recorded in snow, in vegetation, or with rocks. As in this example, many international standards ultimately derive from English, which has become the international language of trade and business.

Within sciences and technology, many symbol systems are now international—for example, the symbols used for an electronic schematic, the astronomical symbols for the planets, or the chemical symbols for the elements.

For a valuable source of information on visual signs and symbols, as well as the names for visual objects or features, see "The Diagram Group," *The Macmillan Visual Desk Reference* (New York: Macmillan Publishing Company, 1993).

The following rules are general guidelines about the use of signs and symbols. If you need specific information about signs and symbols for a single technical area, refer to a publication that covers that area. If you merely need to identify a sign or symbol, check a recently published collegiate dictionary, such as *Merriam-Webster's Collegiate Dictionary*, (Springfield, Mass.: Merriam-Webster, Inc.). See References.

1. Choose standard signs and symbols and ensure that readers understand them.

Some symbols are so well known that you don't need to explain them, regardless of where they appear: =, +, –, $, %, x, ±, and ÷.

Remember that some symbols may be obvious to you—for example, the K for Potassium—but to a layperson, the connection between K and Potassium will be far from obvious.

So if you doubt that **all** of your readers (or viewers) will be familiar with a sign or symbol, explain it either in parentheses within the text or in a footnote. Consider including a glossary of signs and symbols in your document if you think readers would find such a glossary helpful.

2. Use the written-out version of signs and symbols in text unless the nature of the text is such that readers would expect to see signs and symbols:

Poisoning from lead or other heavy metals was common among the ancient Romans and Greeks.

not

Poisoning from Pb or other heavy metals was common among the ancient Romans and Greeks.

Most commercial textbooks are routinely copyrighted.

not

Most commercial textbooks are routinely ©.

The examples above come from nontechnical contexts and from ordinary text, not from graphics or equations. Therefore, symbols are not appropriate.

The symbol Pb would be appropriate in a chemical equation or in a graphic. But even within the text of a chemical report, the symbol Pb would not usually appear.

Similarly, the © would be appropriate in the copyright information on the back of a title page or as in the copyright notice at the bottom of this page, but not within ordinary text.

NOTE: Some symbols, such as the percent sign (%), do appear in ordinary text within such documents as economic surveys, accounting reports, or internal banking reports:

> First State Bank offers to finance the proposed mall at 9.2%, which is 0.8% less than the rate currently offered by Home Savings Bank.

These uses of the percent sign probably reflect a desire to avoid having to spell out the full word a dozen times per page. As is often the case, the purpose and readability of a document are the best indicators of whether or not to use signs and symbols in ordinary text.

3. Repeat a symbol when you have a series of numerical values and the symbol would appear without a space between it and the numerical values:

> $10 to $15 per visit (*or* $10–$15 per visit)
>
> from 6′ to 8′ sections (*not* 6–8′ sections)
>
> 18% to 20%
>
> *but*
>
> 18 to 20 percent (for *nonfinancial* text)
>
> 38 to 45°C

NOTE: Editors sometimes differ in how to handle such repeated symbols. For instance, some editors would prefer to use this version of the temperature example:

> 38° to 45°C

In cases where usage seem to be mixed, check with some of the common references. See REFERENCES. Then decide what pattern you wish to use, and be consistent throughout your document. See EDITING AND PROOFREADING.

Slashes

Slashes (/) have several meanings:

And
Or
Both
To
End of a line in poetry
Mathematical division
Per (mathematical division again)

Be careful when you use a slash. Your readers might not understand the meaning of this apparently simple little symbol.

Note 1: The slash sometimes is called a solidus and sometimes a virgule.

Note 2: A backslash (\) is now common in computer languages. Its special uses there do not carry over into ordinary text.

Slashes

1. Use a slash when a season or a time period extends beyond a single year.
2. Use a slash to replace *per*.
3. Use a slash in mathematical division when an equation appears on a single line.
4. Use a slash to separate lines of quoted poetry presented in prose form.
5. Avoid using *and/or* in most contexts.

1. Use a slash when a season or a time period extends beyond a single year:

winter 2009/2010

fiscal year 2009/10

425/424 B.C.E.

Note: An en dash sometimes replaces the slash:

Winter 2009–2010

425–424 B.C.E.

See DASHES.

2. Use a slash to replace per:

yards/mile

feet/second *or* ft/sec

3. Use a slash in mathematical division when an equation appears on a single line:

$x/a - y/c = 1$ for $\frac{x}{a} - \frac{y}{c} = 1$

$(C/D)/(C+2D)3$ for $\frac{\frac{C}{D}}{(C+2D)^3}$

See MATHEMATICAL NOTATIONS.

4. Use a slash to separate lines of quoted poetry presented in prose form:

As Shakespeare so aptly noted: "There is a tide in the affairs of men, / Which, taken at the flood, leads on to fortune."

See QUOTATION MARKS.

5. Avoid using *and/or* in most contexts:

We asked for a rebate and/or an explanation.

Writers and editors object to all uses of *and/or*, arguing that the expression is ambiguous. In the above example, the meaning includes *a rebate and an explanation* as well as *rebate or an explanation*. Many editors would rewrite the sentence:

We asked for a rebate or an explanation or both.

See *and/or* in WORD PROBLEMS.

Spacing decisions range from the arrangement of text and graphics on the page to the space left after each paragraph and the number of spaces following the final period in a sentence. Proper, even creative, spacing can help guarantee that a final document looks professional and is easy to read.

Spacing

1. Leave ample white space on your pages, especially around important ideas or data.

Page Formats

Decide early in the writing process how you want your pages to look. Options, especially with word processing, include variable margins, a variety of typefaces, and variable line spacing. For important writing projects, you might consult professional designers before writing the text. Sometimes design considerations affect how much and what you write. See PAGE LAYOUT, EMPHASIS, and WORD PROCESSING.

1. Leave ample white space on your pages, especially around important ideas or data.

Writers often cram too much writing onto a page, trying to stay within page limits or adhering to custom, an arbitrary format, or just plain myth about how pages ought to look. Some writers ignore the appearance of the document because they have never had to think about it.

Without careful and early attention to the desired page layout, the text might not fit within prescribed limits, and the tables, figures, and other graphics might not complement the text.

Although every page layout is different, here are some principles to follow as you work with available space:

- Adjust the margins in letters and memos so that the top and bottom margins are roughly equal and the left and right margins are also equal. Your goal is to center the letter or memo on the page. See LETTERS and MEMOS.
- In single-spaced text, double-space between paragraphs. Even triple-spacing might be desirable for extra space on a page or for highlighting key paragraphs.
- Avoid excessively long paragraphs or a series of short, choppy paragraphs. See PARAGRAPHS.
- Add lists, tables, figures, or other graphics to break up long stretches of text. See LISTS and GRAPHICS FOR DOCUMENTS.
- Design a system of headings that allows you to divide the text frequently and to highlight key ideas. See HEADINGS, EMPHASIS, and WORD PROCESSING.

NOTE: A rough prototype of your document is an excellent planning aid. A prototype typically consists of a series of blank pages, one for each page in the proposed document, with titles, headings, lists, graphics, and perhaps paragraphs sketched in. The prototype suggests where and how long each of the major textual units will be. See WRITING AND REVISING.

You should produce the prototype well before writing the actual text.

A variation of the prototype is to prepare a style sheet prior to writing the text. A style sheet would contain guidance on the margins, the number of columns, the typefaces, and the spacing of all levels of headings. Such style sheets are especially valuable when teams of writers are working on a document. See WRITING AND REVISING, PAGE LAYOUT, and WORD PROCESSING.

Spacing and Punctuation

Standard and consistent spacing are signs of quality in any well-produced document. The following conventions reflect current desktop publishing/word-processing programs, which routinely use proportional spacing. These conventions differ from those that many writers learned who gained their typing skills on typewriters, not personal computers.

- Leave one space after any mark of punctuation that ends a sentence.

NOTE: Documents printed with proportionally spaced typefaces, as most are today, should not have two spaces following ending punctuation. For consistency, therefore, we recommend that organizations use a single space after ending punctuation regardless of how the document will be printed. Following this recommendation means that someone will not have to go through your draft of a document removing one of the two spaces after each mark of ending punctuation.

- Place colons, semicolons, and dashes outside quotation marks. Place question marks and exclamation marks inside or outside quotation marks, depending on whether they are or are not part of the quotation.

Spacing

- Always place commas and periods inside quotation marks.

NOTE: British usage places commas and periods inside or outside quotation marks, depending on whether they are or are not part of the quotation. See QUOTATION MARKS and BRITISH ENGLISH.

- Leave no space before a semicolon and one space after it.
- Leave no space before or after dashes. Most word-processing programs have a code to allow for a solid dash, not two hyphens typed together. Use this code whenever possible so that your dashes are solid and do not have a space before or after them:

  The plan—a method for extracting iron ore—is cost effective.

  See DASHES.

- Leave no space before a colon and one space after it within a sentence.
- Leave one space before an opening or left parenthesis within a sentence. Leave one space when the opening parenthesis follows another sentence; if it is a complete sentence, the parenthetical material opens with a capital and the final punctuation comes before the final parenthesis. See PARENTHESES.
- Leave no space before a closing or right parenthesis and one space after it within a sentence. When an entire sentence is enclosed within parentheses, the final parenthesis goes outside of the closing punctuation and one space follows the right parenthesis.
- Leave one space before and after each of the three periods in an ellipsis. If an ellipsis concludes a sentence, use three spaced periods followed by the end punctuation mark for a sentence. See ELLIPSES.

NOTE: Each of the punctuation marks discussed in this section has its own entry in this *Style Guide*. Refer to those entries if you have questions or wish to see additional examples.

Spelling every word correctly is the final step in writing any professional document—from formal report to everyday letter or memo. Fortunately, your computer will flag many misspellings for you and suggest corrections. Still, automatic spell check is far from foolproof and does not relieve you from the need to check your own spelling.

Spelling is important for the sake of both clarity and credibility. Most misspellings do not cause readers to misinterpret the sentence in which the misspelling occurs. But the misspelled word draws attention to itself, which slows down readers and diverts their attention away from the ideas being expressed.

Language is a medium. When the medium draws attention to itself, it detracts from the message.

Misspelled words may also cause readers to question the writer's competence, intelligence, and credibility. How much confidence would you have in this writer's engineering abilities?

> Raw seewater has been considured as a posible alternet sorce for the consentrater principle water supply. However, bench scale tests indacate that the high consentration of disolved salts in seewater interfeer in the eficenct recovary of minaral from the ore.

Misspellings in a document make the writer and the writer's organization look incompetent, sloppy, careless, and potentially untrustworthy.

However, spelling in English is far from simple. Roughly 90 percent of the words in English are regular, but the other 10 percent are demons. Which words in the following pairs are correct?

Spelling

1. Challenge the spelling of every word in your document, especially those words you have difficulty with.
2. Keep a list of the words you have trouble spelling.
3. Use those spelling rules that you find helpful.
4. Form plurals carefully. Many irregular forms exist.

accomodate/accommodate
committment/commitment
concientious/conscientious
changable/changeable
imperceptable/imperceptible
indispensible/indispensable
inevitible/inevitable
irresistable/irresistible
occurence/occurrence
offerred/offered
preceed/precede
prefered/preferred
prevalant/prevalent
privlege/privilege
seperate/separate
similiar/similar
transfered/transferred
truely/truly

If you are like most people, you had to pause on at least two or three of the above pairs. Perhaps you still aren't sure. Did you look up any of them in a dictionary?

English is a hybrid language. It evolved over centuries of influence from the languages of the armies that invaded England: the Romans, the Saxons, the Normans, the Vikings, and so on. English is "impure" in this regard and consequently has many inconsistent words. That's why English pronunciation and spelling are inconsistent.

1. Challenge the spelling of every word in your document, especially those words you have difficulty with.

The best proofreaders and editors challenge every word, especially those that are known to be difficult (such as those listed above).

If the word is common enough, you can trust yourself to recognize correct spelling. If you are unsure, however, check a dictionary or spelling dictionary. See REFERENCES.

2. Keep a list of the words you have trouble spelling.

Remembering spelling rules and their exceptions is difficult. A simpler and nearly foolproof method for improving your spelling is to keep a list of the words you commonly misspell. Look up the correct spellings and list the words alphabetically.

When you see that you have misspelled a word, add it to your list. Then refer to the list when you need to use one of those words. Over time, your mind will come to recognize the look of the word with its correct spelling, and you will no longer need the list.

Until you no longer need it, keep the list in a convenient place: tucked inside your dictionary, on the wall in front of your desk or writing area, under the glass on top of your desk, or taped inside your notebook. Keep it where you can see it easily as you write.

3. Use those spelling rules that you find helpful.

Spelling rules are difficult to remember, and most have many exceptions. If you take the time to memorize the rules, you should probably also memorize the exceptions. At some point, the exercise becomes tedious, and the rewards are questionable.

However, you should use those rules that you have found helpful and that you remember well enough to apply.

Probably the most well-known and most useful rule is "*i* before *e* except after *c*." Here are some of the exceptions:

counterfeit
foreign
freight
height
neighbor
sleigh
weigh
weight

Some of the other common rules are briefly summarized below:

- Change a final *y* to *i* before adding a suffix to a word, but keep the *y* before *–ing*:

activity	activities
deny	denies, denying
happy	happily, happier, happiest, happiness
likely	likelihood
study	studies, studied, studying

- Drop a silent final *e* before suffixes beginning with a vowel but not before suffixes beginning with a consonant:

age	aging
desire	desirable
mobile	mobility
notice	noticing
scarce	scarcity
care	careful
manage	management
safe	safety
wife	wifely

exceptions

acreage
argument
changeable
courageous
judgment
lineage
mileage
ninth
truly
wholly

- Double a final consonant before a suffix beginning with a vowel (1) if the consonant ends a stressed syllable (or a single-syllable word) and (2) if the consonant is preceded by a single vowel:

bag	bagged
brag	bragged
gun	gunned
shop	shopped, shopper
stop	stopped
begin	beginning
occur	occurred
prefer	preferred
regret	regretting, regretted

4. Form plurals carefully. Many irregular forms exist:

man	men
ox	oxen
analysis	analyses
matrix	matrices
potato	potatoes
piano	pianos

See PLURALS for a discussion of these irregular forms as well as a list of the most common irregular plurals.

A List of Common Spelling Demons

Below is a list of some of the most common spelling demons. These words, interestingly enough, are not technical ones because most of us learn to spell technical words as we learn our technical subjects.

Common words are the problem because their irregularities are often difficult to predict and almost impossible to remember. Also, we often see common words misspelled, so we remember the look of the misspelling, not the look of the correct spelling.

absorb
acceptable
accessible
accommodate
accompanied
accuracy
accustomed
acetic *(acid)*
achievement
acoustic
acquire
acreage
adapter
adsorb
aegis
affect *(usually a verb)*
affected
aggression
aging
aid *(help)*
aide *(assistant, helper)*
aisles
all ready *(all prepared)*
all right
all together *(all those in group)*
all ways *(by every means)*
a lot of
already *(previously)*
altogether *(entirely)*
aluminum
always *(all the time)*
amateur
analogous
announcement
anonymous
antibiotics
any one *(any specific person or object)*
anyone *(any person)*
appall, appalled
apparent
appearance
appraise *(estimate value)*
apprise *(inform)*
appropriate
aquatic
archaeology
artisan
ascetic *(austere)*
aspirin
athletics
attendance
authentic
a while *(noun)*
awhile *(adverb)*

bargain
basically
beneficial
benefited
beside *(next to)*
besides *(in addition)*
beveled
biased
breath *(noun)*
breathe *(verb)*
bulletin
bureaucracy
business

caffeine
calendar
caliber
caliper
calk
calorie
canceled, canceling
cancellation
candor
canvas *(cloth)*
canvass *(solicit)*
capital *(city)*
capitol *(building)*
carat *(gem weight)*
caret *(arrow mark)*
category
cemetery
census
challenge
changeable
channel
characteristic
chisel, chiseled
choose *(present tense)*
chose *(past tense)*
coarsely
commitment
committee
competent
competition
complement *(complete)*
compliment *(praise)*
conceited
conceive
condemn
confidant *(person)*
confident *(sure)*
conscience
conscientious
consensus
consistent
continuous
controlled
controversial
councilor *(of council)*
counselor *(advisor)*
courteous
criticism
criticize
curiosity
curious

deceive
decision
definitely
descend
descendant
description
desirable
despair
desperate
despicable
device *(noun)*
devise *(verb)*
dietitian
disappoint
disapprove
disastrous
discipline
discreet *(prudent)*
discrete *(distinct* or *separate)*
disease
distill, distilled
distinct
doctor
dyeing *(coloring)*
dying *(death)*

easily
ecstasy
effect *(usually a noun)*
efficient
eighth
elaborately
elicit *(to draw)*
embarrass
emigrate *(go from)*
employee
enroll, enrolled
ensure *(guarantee)*
entirely
envelop *(verb)*
envelope *(noun)*
environment
equipment
equipped
especially
every day *(each day)*
everyday *(ordinary)*
evidently
exaggerate
except
exhaust
existence
experiment
explanation
eying *or* eyeing

familiar
farther *(distance)*
fascinate
favorite
February
fiber
finally
financially
flammable (not *inflammable*)
fluorescent
fluorine
foreign
foresee
foretell
forgo *(relinquish)*
forego *(precede)*
forty
forward *(ahead)*
foreword *(preface)*
fulfill, fulfilled
further *(degree)*
fuselage

gauge
generally
glamour
government
governor
grammar
guaranteed
guerrilla

happened
harass
heard
height
heroes
hindrance
hoping
humane
humorous
hurriedly
hypocrisy
hypocrite

ideally
idiosyncrasy
ignorant
illicit *(illegal)*
illogical
imaginary
imagine
imitate
immediately
immensely
immigrate *(go into)*
incalculable
incidentally
incredible
indispensable
inequity
influential
initiative
innocuous
insurance
insure *(guarantee financially)*
integrate
intelligent
interference
interrupt
irrelevant
irresistible
irritated

jealousy
jewelry
judgment

kilogram
knowledge

laboratory
laid
lath *(wood)*
lathe *(machine)*
led
leisure
length
lenient
leukemia
liable
library
license
lightning

Spelling

likelihood
liquefy
liveliest
logistics
loose *(adjective)*
lose *(verb)*
luxury
lying

magazine
magnificent
maintenance
manageable
management
maneuver
mantel *(shelf)*
mantle *(cloak)*
margarine
marijuana
marriage
material *(goods)*
materiel *(military goods)*
mathematics
meant
medicine
meteorology
mileage
miniature
minor
mirror
mischievous
missile
morale
mortgage
mucous *(adjective)*
mucus *(noun)*
muscle
mysterious

naturally
necessary
nevertheless
nickel
niece
nineteen
ninety
ninth
noticeable
nowadays
nuclear
nuisance
numerous

occasion
occasionally
occurred
occurrence
occurring
off
offense
official
omission
omitted
omitting
oneself
opponent
opportunity
opposite
oppression
optimism
ordinance *(law)*
ordnance *(military)*
ordinarily
originally

pamphlet
parallel
paralleled
parole
particle
particularly
pastime
peaceable
peculiar
penetrate
perceive
performance
perhaps
permanent
perquisite *(privilege)*
personal *(individual)*
personnel *(employees)*
perspective *(viewpoint)*
persuade
pertain
phosphorous *(adjective)*
phosphorus *(noun)*
physical
picnicking
pigeon
poison
politician
pollute
possession
possibly
practical
practically
precede
precedence *(priority)*
precedents *(prior instances)*
predominant
preferred
prejudice
prerequisite *(requirement)*
prevail
prevalent
preventive *(not preventative)*
principal *(chief* or *main)*
principle *(theory* or *idea)*
prisoner
privilege
probably
procedure
proceed
processes
professor
programmed
programmer
programming
prominent
pronounce
pronunciation
propaganda
prophecy *(noun)*
prophesy *(verb)*
prospective *(expected)*
psychology
publicly
pursue
pursuing
pursuit

quandary
quarreled
quarreling
questionnaire
quiet
quite
quizzes

rarefy
rarity
rebel
receipt
receive
recession
recipe
recommend
reconnaissance
reconnoiter
recyclable
referring
regular
regulate
rehearsal
reinforce
relief *(noun)*
relieve *(verb)*
religious
remembrance
reminisce
repellant *(noun)*
repellent *(adjective)*
repetition
resemblance
resistance
restaurant
rhythm
ridiculous

sacrifice
safety
salvage *(save)*
satellite
scarcity
scenery
schedule
secede
secretary
seismology
seize
selvage *(edging)*
separate
sergeant
sheriff
shining
shrubbery
signaled
signaling
significant
similar

sincerely
sizable
some time *(some time ago)*
sometime *(formerly)*
sometimes *(at times)*
souvenir
spacious *(space)*
specious *(deceptive)*
sponsor
stationary *(fixed)*
stationery *(paper)*
statistics
stepped
stopped
straight
strategy
strength
strenuous
stretch
studies
studying
subpoena
subpoenaed
subtlety
suburban
succeed
succession
suicide
sulfur
superintendent
supersede
suppress
surely
surreptitious
surround
surveillance
susceptible
suspicious
synonymous

technical
technique
temperature
temporary
their *(pronoun)*
themselves
there *(adverb)*
therefore
thorough
though
through
tie, tied, tying
tobacco
too
totaled, totaling
tragedy
transferable
transferred
traveled
traveling
tremendous
truly
twelfth
typical
tyranny

unanimous
unconscious
undoubtedly
until
usage
usually

vacuum
various
vengeance
villain
violence
visible
vitamin

warrant
warring
weather
Wednesday
where
wherever
whether
whichever
whiskey
wholly
woman *(singular)*
women *(plural)*
writing
written

yacht
yield
yogurt *or* yoghurt

zenith
zephyr
zodiac

Strong Verbs

Strong verbs shorten sentences and usually convey direct, memorable messages.

A common stylistic problem is using weak, rather than strong, verbs. Weak verbs are those simple verbs that occur so frequently in our language that they have little impact: *is, are, was, were, can, could, has, had, have, do, did, done, make, use, come.* See VERBS.

Obviously, these verbs are essential to English. Using them, either as primary or auxiliary sentence verbs, is inescapable. However, writers often use them unnecessarily to create wordy, weak sentences:

> The system has wide applicability for a variety of industrial cogeneration situations. *(12 words)*

In this sentence, the writer has transformed *apply*, a much stronger verb than *has*, into an awkward, bureaucratic noun, *applicability*, and used the weaker verb as the sentence verb. Using *apply* as the sentence verb creates a shorter, stronger sentence:

> The system applies to a variety of industrial cogeneration situations. *(10 words)*

See SCIENTIFIC/TECHNICAL STYLE.

Strong Verbs

1. Use strong verbs.

1. Use strong verbs.

Strong verbs are less common; therefore, readers tend to pay more attention to them. They have more impact in a sentence, and they help writers avoid long, bureaucratic nouns:

this

Since 2005, our Basic Development Department has studied membranes for the separation and enrichment of gas mixtures. (17 words)

not this

Membranes for the separation and enrichment of gas mixtures have been under study by our Basic Development Department since 2005. (20 words)

---

this

We have vastly improved our reaction mechanisms. (7 words)

not this

We have made vast improvements in our reaction mechanisms. (9 words)

---

this

We would like to investigate further the pilot program before proposing a factory location. (14 words)

not this

We would like to conduct further investigations into the pilot program before giving a proposed factory location. (17 words)

---

this

We will emphasize evaluating plating techniques for depositing amorphous or glassy metal coatings. (13 words)

not this

We will give special emphasis to the evaluation of plating techniques for the deposition of amorphous or glassy metal coatings. (20 words)

---

this

Before developing the catalyst, we would assess the technical and economic advantages of the source materials. (16 words)

not this

Before proceeding with catalyst development, we would make technical and economic assessments of the advantages of the source materials available. (20 words)

Style is the sum of the choices, both conscious and unconscious, that writers make while planning, designing, writing, and editing documents.

These choices include the type of document, the words chosen, the structure and length of sentences, the length and type of paragraphs, the document's organization, the use of emphatic devices (headings, lists, white space), the use and kind of graphics, the typeface and type size, the paper, and so on. See PAGE LAYOUT.

Such choices give each document a unique style, tone, and feeling. Each fax, field note, business letter, quick memo, formal report—they are all unique because they differ in tone, attitude, perspective, and style from every other fax, note, letter, memo, and report.

Style

Style and Tone

Style and *tone* are often confused. The terms are similar, and some speakers use them interchangeably. However, style is the cause and tone is the effect:

- **Style** refers to those choices writers make that create the tone conveyed to readers.
- **Tone** refers to the feeling or impression a document conveys to its readers. See TONE.

Style is often categorized as being either **formal** or **informal**. These distinctions typically depend on the type of document being written, the intended readers, the writer's relationship to the readers, the document's purpose, and the message being conveyed.

Formal documents, such as legal agreements, technical reports, and many letters, are written for readers with whom the writer has no personal relationship. Therefore, familiarity or humor is generally unacceptable. It would seem inappropriate to the circumstance and would interfere with the message. Formal documents are usually written to convey information objectively to readers who may not know (or care to know) the author.

The tone of formal documents tends to be impersonal, objective, restrained, deliberate, and factual. Formal documents that are intended to convince readers to do or accept something are often also forceful, dynamic, and perhaps intensive. Most documents that convey negative or unpleasant information are formal.

Informal documents, such as personal letters, newsletters, trip reports, and most memos, are generally written for people the writer knows or feels comfortable with, perhaps only through employment in the same organization. Familiarity, humor, and wit are often acceptable. The informality creates a more relaxed atmosphere, which makes the message warmer and more easily accepted. Informal documents are often informative, friendly, and subjective.

The tone of informal documents is generally relaxed, informative, helpful, casual, personal, positive, nonthreatening, and perhaps cheerful. Informal documents may be persuasive, but they are rarely forceful or aggressive. When writers need to be aggressive and when they need to convey negative or unpleasant information, they generally become more formal in their approach. See TONE.

Style also refers to styles associated with particular disciplines: geologic style, geophysical style, legal style, medical style, engineering style, auditing style, academic style, scientific style, social science style, bureaucratic style, and so on. See SCIENTIFIC/TECHNICAL STYLE.

In each of these styles, the writer uses jargon particular to the discipline and writes according to a long tradition of document preparation and appearance. We are perhaps most familiar with the legal and bureaucratic styles, but all disciplines have a set of standards and traditions that affect the way people communicate in writing. See JARGON.

Word Choice and Style

Each word you choose helps establish the style of your writing. Words are one of the most visible traits of style.

Style

You may, for instance, wish to discuss the effects of a decision. The word *effects* is a fairly neutral choice. Instead of *effects*, you might speak of *impacts, consequences*, or *results. Impacts* suggests some negative connotations, as does *consequences*, which is more formal sounding (perhaps because of its length) than *impacts. Results*, on the other hand, has a positive feeling to it: *We're going to get results.*

You have other less common choices: *aftermath, corollary, end product, eventuality, outcome, sequel, upshot. Aftermath* has definite negative implications, besides being almost dramatic in its tone. *Corollary* has limited usefulness, if for no other reason than its mathematical echoes (and hence almost too educated tone). *End product* seems plain, yet still wordy when compared to *effects* or *results. Eventuality* implies some final or ultimate result; again, its length makes it sound more formal. *Outcome* is about as neutral as any of the words in this list, but it may have negative connotations, as in the outcome of a medical test. *Sequel* implies a second follow-up event, not a real effect. And *upshot* suggests surprise, even chaos. Even more words are possible: *development, fruit, outgrowth, ramification, repercussion, conclusion.*

Part of the richness of English is its large vocabulary, which offers multiple possibilities for expressing any idea. Yet no two words mean exactly the same thing, so when you choose one word rather than another, you change the style of the document, if only slightly. Within a few lines, then, you will make dozens of choices, all of which combine to establish the style (and resulting tone) of the document.

Obviously, some word choices make a bigger difference than others. Selecting *effects* rather than *results* changes the style very little. But if you use *impacts, consequences*, or *aftermath*, the document may shift in tone and effect:

> What is the effect of altering course in midflight?
>
> What are the consequences of altering course in midflight?

> What were the effects of the pipeline installation on the salmon in Little Middle River?
>
> What was the aftermath of the pipeline installation on the salmon in Little Middle River?

As you can see, some word choices are important stylistic signals. Others are inconsequential:

> What is the effect of altering course in midflight?
>
> What is the result of altering course in midflight?

Writers in various technical professions have a body of technical words that affect the style:

Legal

tort, legatee, real property, contract, conveyance, *amicus curiae*, party, sue, brief, witness, jurisdiction, plaintiff, etc.

Medical

curette, mammillary, uvulae, amoebic, gastric hernia, leucoplast, dermatitis, proboscis, etc.

Computer

batch processing, cursor, default, field, file, logon, real time, sign off, etc.

Construction

sill, head, transom, fascia, neoprene spaces, support mullion, jamb, seat board, glazing gasket, butt glazed, hopper sash, soffit, rowlock, etc.

Such specialized scientific and technical terms are unavoidable given today's complex technologies. See JARGON. The presence of this jargon is the most visible sign to readers that a document reflects a particular style.

Sentences and Style

Writers have virtually an infinite number of ways to express ideas. Even in an ordinary 15- to 30-word sentence, the possibilities run into the millions, both in terms of word choice and sentence structure. The sheer range of possibility means that every sentence except the shortest and most trivial is potentially unique (never having been written or said before).

Sentence options—**length** and **structure**—are the most important features of a writer's style. However, these options are often less obvious to readers than the choice of a particular word or technical term. Readers notice the style of sentences only when something goes wrong, as in an awkward sentence or one where something is deliberately unusual:

> That is something up with which I shall not put.
>
> *or*
>
> Turning into the wrong driveway, a tree was hit by me which I don't have.

Sentence Length

Sentence length by itself normally will not establish a definite style, but it will contribute to style. Most sentences average anywhere from 12 to 25 words. Readers are accustomed to those lengths, so they are likely to notice only those sentences that are either extremely short or extremely long:

> We refuse.
>
> *or*
>
> Science is nothing but trained and organized common sense, differing from the latter only as a veteran may differ from a raw recruit: and its methods differ from those of common sense only as far as the guardsman's cut and thrust differ from the manner in which a savage wields his club. (From Thomas Huxley's *Collected Essays*)

An individual sentence, even if potentially noteworthy, won't be noteworthy unless it stands out from the sentences surrounding it. So **average sentence length** is probably more of a direct indication of style than the length of a single sentence. Readability formulas always include average sentence length as a measure of readability because length is a good indicator of the difficulty of a document. See SENTENCES.

Long sentences can reflect different styles (and tones), depending on other features within them. Long, well-structured sentences with a sophisticated vocabulary usually convey an educated or thoughtful quality. But if the sentence seems longer than necessary for the ideas being expressed, and if the vocabulary is more sophisticated than necessary, then the sentence may seem stuffy, extravagant, or pompous:

> If biological populations or habitats that may require additional protection are identified by the DCMOFO (Deputy Conservation Manager, Offshore Field Operations) in the leasing area, the DCMOFO will require the lessee to conduct environmental surveys or studies, including sampling, as approved by the DCMOFO, to determine existing environmental conditions, the extent and composition of biological populations or habitats, and the effects of proposed or existing operations on the populations or habitats that might require additional protective measures.

This sentence's length—77 words—is surely excessive, but other features contribute to its bureaucratic, stiff, faintly legal style:

- The use of the unfamiliar acronym *DCMOFO* gives the sentence a bureaucratic touch, especially with its repetitions of the acronym.
- The repetition within the sentence, especially of the phrase *populations or habitats*, reinforces the bureaucratic, even stuffy, tone.
- The delay of the main subject and verb *(the DCMOFO will require)* until after the long introductory *if* clause forces readers to absorb, comprehend, and remember too much information at once. Consequently, the sentence is more difficult to read than it should be.
- Some of the phrasing is clumsy and ill-placed. The phrase *in the leasing area* comes so late in the opening clause that its meaning is fuzzy. Is the DCMOFO in the leasing area? Are the populations and habitats in the leasing area? Have any or all of these been identified in the leasing area?

For an example of a well-structured long sentence, see SENTENCES.

Sentence Structure

Sentence structure—including grammatical structure, the sequence of ideas, and the various repeated word patterns—all contribute to the style of a sentence or passage. The following versions of the same basic sentence say much the same thing, but their different structures create different tones:

1. We considered how best to present the conflicting data and our interpretations of these conflicts.
2. How to present the conflicting data, as well as our interpretations of these conflicts, was under consideration.
3. Because of conflicting data and differing interpretations, we were considering different presentation strategies.
4. We were considering different strategies for presenting the conflicting data and our interpretations of the data.
5. It was difficult to decide on strategies for presenting the conflicting data and the differing interpretations of the data.

Sentences 1 and 4 are the most direct (and they happen to be the ones that most closely follow normal English word order). Sentence 2 is formal, even stuffy, because its opening clause is so long that the verb *was* is almost lost. Sentence 3 is fairly ordinary, even though it opens with the conditional *because* clause. Sentence 5 is perhaps the most stuffy; it opens with a false subject (see FALSE SUBJECTS), and it avoids all pronouns. See PRONOUNS.

The structural patterns for a single sentence present a broad range of possibilities. Putting sentences together increases the possibilities exponentially.

A string of formal, oddly structured sentences not only slows down readers but also conveys a tone of formality or stuffiness. A string of short, direct sentences can sound clean and efficient (or abrupt and efficient, depending on the context).

Other Stylistic Choices

Many other features in a document besides words and sentences can convey a particular style.

The basic format of a document is usually significant. A document with narrow margins, single-spaced text, and long paragraphs conveys a dense, information-packed but potentially dull image. Readers may consider the language heavy and ponderous. A document with generous margins and lots of open space makes readers feel that the writing is open and inviting, easier to read. See PAGE LAYOUT.

Besides format choices, many other features influence a reader's perception of a document:

- The typeface used for the text
- The type of paper—both weight and texture

Style

- The number and quality of the graphics
- The care with which the editing and proofreading has been done
- The professionalism of the binding and the quality of the printing
- The presence or absence of color

Individual Style

We all have styles of speaking and writing that are unique to us, regardless of the circumstances in which we write. This fact reflects the basic and pervasive nature of style.

In speech, an individual's style is easy to identify. Most of us can recognize a close friend, not only from the sound of the friend's voice, but also from the structure and content of the speech, from the words chosen, and from the sentence patterns used. We know our friends as talkative, quiet, abrupt, cheerful, depressed, thoughtful, humorous, or tactful. In writing, an individual's style may be harder to identify, and yet it exists in each choice the individual has made to produce a document.

Style and Ineffective Writing

Ultimately, style **is** the writer. We can't describe a universally preferable style because the decisions writers make depend on the context: on the subject or content, the purpose of the document, the readers, previous or related documents, and the situation or climate in which the document is produced.

Nevertheless, good writing is distinguishable from bad writing, and you should never confuse bad writing with style.

Good writing is clear, emphatic, well organized, and concise. Bad writing is often vague or confusing, unemphatic, chaotic, and wordy.

Good writers obey the principles of effective writing—regardless of subject matter, purpose, readers, context, style, or tone.

For a review of those principles, see this *Style Guide*, particularly the sections on SENTENCES, PARAGRAPHS, ORGANIZATION, EMPHASIS, ACTIVE/PASSIVE, STRONG VERBS, FALSE SUBJECTS, and KEY WORDS.

For information on writing specific types of documents, see LETTERS, MEMOS, REPORTS, and SUMMARIES.

Summaries

Summaries are abridgments or compendiums of the important points in a document. Summaries are essential for readers who don't have time to read the entire document, are not interested in reading the entire document, or need to review the important points without reading the entire document.

Traditional summaries appear at the end of documents, especially those organized scientifically. See ORGANIZATION. These summaries present the main points from the preceding discussion.

Executive summaries appear at the beginning of documents, especially in documents organized according to the managerial format. See ORGANIZATION. These summaries preview the main points that will appear in the discussion that follows.

Summaries

Traditional Summaries

Traditional summaries briefly repeat the major ideas, especially conclusions and recommendations. These summaries include little, if any, background information and no supporting data or detail of any kind. They are typically shorter than executive summaries, which may include some background and supporting information.

Traditional summaries should not be separated from the rest of the document. They are a final summation, so they depend on information presented earlier in the document. Everything in them must already have been stated earlier.

The writer of a traditional summary often assumes that readers will use the summary after having read the rest of the document.

Executive Summaries

Executive summaries—also known simply as summaries—appear at the beginning of documents. They usually include the following:

- **Background/introduction to the document.** Although very brief, this section gives readers enough information so they will understand the reason for the document, the key problems addressed, and any special conditions or situations that the reader should be aware of. See ORGANIZATION and INTRODUCTIONS.
- **Main conclusions.** These may be a little longer than in traditional summaries because they are not repeated from conclusions presented earlier. Therefore, in addition to the conclusion itself, you may need a little explanation or elaboration. Just keep them short.
- **Recommendations, if any.** Again, you are not repeating recommendations presented earlier, so each recommendation might require some explanation. If appropriate, tie each recommendation to the specific conclusion or conclusions that prompted it.
- **A review of data (optional).** This section is limited to pertinent items, not all the data. The complete data usually appear in an appendix. However, you may need to present key data in an executive summary so that readers are aware of the key supporting information that your document presents. Remember that readers may read only the executive summary, not the entire document. Give them enough detail to substantiate your conclusions, but not so much that the rest of the document becomes unnecessary.

In some instances, business and technical documents may consist only of an executive summary, with supporting or explanatory material located in appendices or attachments. Such documents reflect the attempt of many businesses to limit documentation to the essentials.

Summaries vs. Abstracts

Scientific and technical writers often include abstracts with their reports. Abstracts are condensations of a document, usually written so that readers can preview the content of the document to determine whether they are interested in reading the entire document. In a sense, abstracts are extended titles.

Summaries

Abstracts may contain much of the same information found in summaries, but abstracts are meant to be detached from their documents. Frequently, abstracts are published separately, in an online catalog or list of abstracts. Sometimes, journals request that potential authors submit abstracts of their articles. The editors read the abstracts to determine which articles they want to read in full.

Summaries, on the other hand, are an integral part of the documents they belong to, and they are not normally detached.

By convention, abstracts are rarely longer than one paragraph (about 250 words). Summaries have no conventional length restrictions. They should be concise, of course, but summaries may range from 50 to 5,000 words, depending on the length of the parent document. An extensive report or multivolume proposal could very well have an executive summary of 20 or 30 pages. See REPORTS and ORGANIZATION.

How to Write Summaries

You may wish to produce a preliminary summary of a long or complicated document before you have written the whole document. A preliminary summary will help you organize your thoughts and determine your most significant points. However, the final summary (the one appearing in your document) should be written last. Here is a procedure for writing good summaries (especially executive summaries):

1. Read through the entire document. Ensure that you have a firm grasp of the document's purpose, scope, point of view, and major ideas.
2. Identify the major ideas and data. You can underline or circle major ideas, or put stars beside them, or highlight them with a highlighting pen. Identify all of the major statements: observations, conclusions, recommendations, key supporting data, key facts, etc.
3. Pull the major ideas together and note how they are developed sequentially throughout the document. If necessary, refer to details in the text to clarify any points not absolutely clear.
4. Condense by combining sentences, generalizing, eliminating unnecessary supporting information, and eliminating unnecessary words and phrases. Use simple, direct sentences, and eliminate all unnecessary jargon and big words. Keep your writing simple and straightforward.
5. Use transitional words and phrases to link ideas and provide a smooth flow of thought from one sentence to the next.
6. Test the result by challenging every word, phrase, sentence, and idea. If something isn't pulling its weight, get rid of it.
7. Challenge the overall summary. Does it accurately reflect the content of the whole document? Have you left out major ideas? Does the summary distort any facts, relationships, conclusions, or recommendations? Does the summary provide ample information for readers to comprehend the ideas without reading the whole document? Can the summary stand by itself? If readers read nothing but the summary, will they be adequately informed?

Remember that the purpose of the summary is not to convey everything. Summaries convey the essentials. If readers want additional support or proof, they should read the rest of the document. But make sure that the summary provides readers with a firm grasp of the major ideas.

Tables are information displays organized by rows and columns. Three types of information displays are common:

- Formal data tables
- Matrix (text) tables (see rule 15)
- Run-in tables (see rule 16)

Formal data tables, as traditionally used, present precise data: *3.1415, 9.8690, 31.0035,* etc. Such data cannot be presented as precisely in any other type of graphic aid. See GRAPHICS FOR DOCUMENTS. Tables allow for quick and accurate comparisons and can also depict trends and relationships, although not as well as charts and graphs. See CHARTS and GRAPHS.

Use formal data tables when you must present a large amount of information, when you need to give readers the exact figures, or when you want readers to be able to compare figures or other information presented in different rows and columns.

Spreadsheet software is so widely available and easy to use that virtually anyone can create good-looking data tables. However, there are three important pitfalls to avoid in using spreadsheet software. First, remember the adage "garbage in, garbage out." The results are only as good as the data you input, so you must proofread important tables to ensure that you have entered the right information. It's easy to get an answer from a data table, but it might not be the right answer.

Second, flaws in the software itself might produce inaccurate results; so as you proofread, watch for anomalies that are not due to mistakes in data input. See EDITING AND PROOFREADING.

Tables

1. Keep tables as simple as possible.
2. Place important tables in the body of the document; place unimportant tables in appendices or attachments.
3. Use table numbers and captions to identify tables and to help readers understand the information presented.
4. Clearly label tables that extend beyond one page, and use continuation headings on continued pages.
5. Use white space, boxes, or lines to separate tables from surrounding text.
6. Logically organize rows and columns so that they reflect the purpose of the table and make the table easy to read.
7. Use row headings (in the stub) and column headings (in the boxhead) to identify the information listed in each row and column.
8. Use rules to separate the boxhead from the field and, as necessary, use rules to separate groups of data columns or rows from each other.
9. Where space is limited, use abbreviations.
10. Base comparable numerical amounts on the same unit of measurement, and convert all fractions to decimals.
11. Align decimals vertically within columns.
12. Use zeros, dashes, ellipses, or *NA* to indicate that information is missing or not applicable.
13. Use footnotes, comments, or hyperlinks to explain or clarify table headings and entries.
14. Identify all data sources.
15. Use a matrix (text) table to present complex textual information.
16. Use informal or run-in tables to replace ordinary text, especially lengthy paragraphs.

Third, you can't retrace what you've done as you enter, delete, change, and manipulate data in an automated spreadsheet. Particularly in scientific and technical contexts, you should be able to produce "working papers" containing variations on your treatment of data. Therefore, consider saving evolving versions of a spreadsheet so you can explain your line of reasoning.

For a much more elaborate discussion of formal tables, including printing considerations, see the *United States Government Printing Office Style Manual.* See REFERENCES.

Matrix (text) tables. A useful variation of the formal table is a matrix that presents textual information, not precise data. This

Table Number and Caption

Boxhead

Stub

Field

Table 119. Frequencies of Selected Circulatory Diseases: United States, 2008

Selected characteristic	All persons 18 years of age and over	All types	Coronary	Hypertension	Stroke
Numbers in thousands					
Total	**225,227**	**26,628**	**14,428**	**56,159**	**6,460**
Sex					
Male	108,755	13,170	8,439	26,031	2,965
Female	116,472	13,457	5,989	30,128	3,495
Age					
18–44 years	110,615	5,067	1,258	9,643	714
45–64 years	77,335	9,469	5,205	25,126	2,260
65–74 years	19,869	5,287	3,337	10,774	1,248
75 years and over	17,409	6,805	4,628	10,616	2,238
Race					
One race	222,430	26,233	14,210	55,482	6,360
White	182,651	23,082	12,511	45,054	5,286
Black or African American	26,765	2,436	1,307	*7,900	832
American Indian or Alaska Native	2,178	230	*119	478	*80
Asian	10,501	472	261	1,985	153

Footnotes

*Estimates preceded by an asterisk have a relative standard error of greater than 30% and less than or equal to 50%.

Source Line

Source: U.S. National Center for Health Statistics, Summary Health Statistics for U.S. Adults: National Health Interview Survey, 2008.

Figure 1. Sample Table Layout. *A clear, correct table layout assists readers to understand and interpret the data.*

type of table often looks like a hybrid between formal tables and more informal run-in tables. A matrix table may have column and row labels just like a formal table. The matrix will also have data in some of its squares. Other squares, however, may have words, phrases, or even short paragraphs.

A matrix table can often replace several pages of text. See rule 15 for an example of a matrix table.

Run-in tables are useful ways to break up dense text and to emphasize a point. The simplest run-in table has no figure number, no caption, and may be only two or more columns inserted into a passage of ordinary text. The sentences preceding such a table introduce the information. See rule 16 for an example of a run-in table.

Parts of Tables

The standard parts of formal tables are the following: the **table number** and **caption**, the **boxhead** (containing column headings), the **stub** (containing row headings), **the field** (**body**), **rules**, **footnotes**, and the **source line**. See figure 1 for a sample table layout.

Table Number and Caption

Number tables sequentially as they appear in the document, and number them separately from figures.

Unless you are presenting uninterpreted data, make the table caption both a title and an informative caption (see tables 1 through 5). The table caption should clearly identify the table and tell readers how to read or interpret the table:

> **Table 1.** *The industrial gas shipment decline since 2006.*
>
> **Table 2. Wood Panel Products (2002–11).** *Production more than doubled in 10 years.*
>
> **Table 3.** *Increasing arsenic concentrations in groundwater, Sonoma County, California, 2008.*

These captions indicate both what the tables are about and how readers should interpret them.

If you are presenting a large amount of uninterpreted data and your purpose is to present, not interpret, the data, then use shorter, title-like captions:

> **Table 1. Industrial Gas Shipments (2006–11)**
>
> **Table 2. Wood Panel Products/Production (2002–11)**
>
> **Table 3. Arsenic Concentrations in Groundwater, Sonoma County, California, 2008**

Most table titles and their associated captions should appear above tables, especially when, as in the preceding examples, the titles/captions are little more than a brief phrase. In some instances, writers place short titles (and brief captions) above a table and then include a detailed explanation under the table. *Scientific American* uses this pattern occasionally, both for tables and figures. Their short title above a table or a graphic becomes a headline; detailed interpretations then appear under the table or figure.

Whichever pattern you choose for placing the titles/captions, be consistent throughout a document. See CAPTIONS.

If a unit of measurement applies throughout the table, you may state the unit of measurement in the caption or within parentheses beneath the caption. If the unit of measurement appears below the caption, you can subordinate it by printing it in a smaller type size. See CAPTIONS.

Boxhead

The boxhead contains the stub heading and the column headings (see figure 1). Place a rule (line) above and beneath the boxhead to separate it from the table number and caption and from the field. Make column and stub headings as concise as possible. Use more than one line, if necessary, to state a column heading, but try not to exceed three lines for any heading. Orient the headings horizontally.

If appropriate, include units of measurement in the headings or enclose the units of measurement within parentheses below the headings. If the column headings require more than one line and you place units of measurement below the headings, separate the headings and units with a thin rule (see tables 2 and 3). Use thicker rules around the boxhead itself.

Stub

The stub is the left column. Use it to label the rows. Make the row headings as concise as possible. As necessary, put row units of measurement either within parentheses following the row heading or in another column beside the stub (see table 4).

If the rows consist of major and subordinate items, place the major items flush left and indent the subordinate items (see table 1).

Table 1. U.S. Crude Oil Supply and Disposition: July–December 2009. *Although domestic production levels remained stable, exports more than doubled between July and December.*

Download Series History   Definitions, Sources, and Notes

Process	Jul-09	Aug-09	Sep-09	Oct-09	Nov-09	Dec-09	View History
Supply							
Field Production (Commercial)	162,213	163,857	163,327	168,069	163,974	169,246	1920–2009
Alaskan	17,088	17,733	19,562	20,401	19,743	20,296	1981–2009
Lower 48 States	145,125	146,125	143,765	147,668	144,231	148,950	1993–2009
Imports	286,024	275,371	276,701	265,536	261,269	252,128	1920–2009
Commercial	286,024	274,872	275,728	265,536	260,217	251,323	2001–2009
Strategic Petroleum Reserve (SPR)	NA	499	973	NA	1,052	805	1977–2009
Imports into SPR by Others	NA	499	973	NA	1,052	805	2005–2009
Adjustments (Commercial)	4,828	3,507	167	2,941	-3,331	2,685	1981–2009
Disposition							
Stock Change	-1,515	-9,951	-1,594	-1,349	4,707	-11,308	1981–2009
Commercial	-1,511	-9,949	-2,588	-1,344	3,659	-11,795	1993–2009
Strategic Petroleum Reserve (SPR)	-4	-2	994	-5	1,048	487	1993–2009
Refinery Net Inputs	453,609	451,595	440,524	435,654	415,840	433,344	1981–2009
Exports	971	1,091	1,265	2,241	1,365	2,022	1920–2009
Products Supplied	0	0	0	0	0	0	1981–2009
Ending Stocks							
Ending Stocks	1,071,247	1,061,296	1,059,702	1,058,353	1,063,060	1,051,752	1920–2009
Commercial	347,153	337,204	334,616	333,272	336,931	325,136	1920–2009
Cushing, OK	33,916	30,273	25,505	25,718	32,978	35,645	2004–2009
Strategic Petroleum Reserve (SPR)	724,094	724,092	725,086	725,081	726,129	726,616	1977–2009

Field

The field (body) consists of the data rows and columns below the boxhead and to the right of the stub. If necessary for clarity, use leaders (rows of periods) between rows to facilitate reading. Align the data presented in each column by placing words flush left within columns, integers (e.g., *40*) right flush by digit, and real numbers (e.g., *40.0* or *40.068*) vertically by decimal point.

If your table appears online, you can hyperlink items in the field to helpful information located on other sites (see hyperlinked items in tables 1 and 3).

Rules

Also known as borders, rules are horizontal or vertical lines that separate parts of the table. Always place rules between the boxhead and the field. Also, place a rule above the boxhead to separate the boxhead from the table number and caption. If you use footnotes or a source line, place a rule between them and the bottom of the field. See PAGE LAYOUT.

Use rules only when necessary to help readers distinguish among parts of the table. A simple table rarely needs rules, which can make a table look dense and boxy.

If your table is large, you might need to place rules between groups of rows. Typically, the rules appear between groups of five. These rules help readers follow information down large tables.

The rules surrounding the box head should be thicker than those appearing within the field.

NOTE 1: Instead of placing rules between every fifth row, you may leave an extra line to separate groups of rows.

NOTE 2: Some authorities also use vertical rules to separate the stub from the field and groups of columns from each other. The trend today is to eliminate all vertical rules.

Footnotes

Use footnotes to clarify the headings, to identify unfamiliar abbreviations and units of measurement, and to explain the data appearing within the field.

Use superscripted footnote numbers or letters ([1],[2],[3] or [a],[b],[c]) or symbols (*, **, ***) to link the footnoted information in the table and its explanation (see tables 2 through 4).

If the footnote applies to the entire table, you do not need a footnote symbol (see the note beneath table 1).

The footnotes should appear below the table, beginning flush left. If a footnote extends across the table and you have many footnotes, break the footnote references into two columns and print them in a smaller type size, if possible (see table 4).

Source Line

The source line identifies the source of the information presented in the table. Source lines always appear below footnotes and should be aligned with the footnote references (see tables 2 through 4). If your table appears online, you can provide

Table 2. World Average of Crude Oil Prices, 1945–2010. *Major leaps in price occurred during the energy crisis of the 1970s and Middle Eastern wars of the 2000s.*

	(in $/bbl. in 5-year increments)	
Year	Nominal	*Inflation Adjusted
1945	$1.63	$17.73
1950	$2.77	$24.84
1955	$2.93	$23.56
1960	$2.91	$21.24
1965	$3.01	$20.59
1970	$3.39	$18.84
1975	$12.21	$48.91
1980	$37.42	†$98.07
1985	$26.92	$53.98
1990	$23.19	$38.17
1995	$16.75	$23.71
2000	$27.39	$34.29
2005	$50.04	$55.21
2010	$83.40	$83.10

*Prices are adjusted for inflation to November 2009 prices using the Consumer Price Index (CPI-U) as reported by the U.S. Bureau of Labor Statistics.

†Represents the high point of price inflation during the energy crisis.

Source: Independent Statistics and Analysis, U.S. Energy Information Administration. http://www.eia.doe.gov/oil_gas/petroleum/historical_average.html [accessed March 1, 2011].

Deposits, assets, and losses in $000

Name	FDIC #	Date	Deposits	Assets	Losses (est.)	Percent Losses
Abberline Bank	23124	2/6	95,106	11,0361	18,000	18.5
America West Bank	35461	5/1	286,040	281,564	119,400	42.4
Bank of Wyoming	22754	7/10	665,98	70,188	27,000	38.5
Colonial Bank	9609	8/14	20,020,047	25,455,112	2,800,000	11.0
Guaranty Bank	32618	8/21	12,000,000	13,000,000	3,000,000	23.1
Integrity Bank	35469	8/29	962,456	110,7514	210,779	19.0
Mutual Bank	18659	7/31	1,546,525	1,595,657	696,000	43.6
North River Bank[a]	00450	2/13	422,708	77,689	201,500	38.5
Silver State Bank	34194	9/5	173,3091	1,957,120	553,095	28.3

[a]Actually failed 8/31/08, but included because assets were impounded until 2/13/09.

Source: U.S. Federal Deposit Insurance Corporation.

Table 3. U.S. Bank Failures 2009. *Size had little to do with bank failures in 2009: the financial crisis affected both small community banks and major national banks.*

hyperlinks to sources (see table 1). See INTELLECTUAL PROPERTY.

Rules for Using Tables

1. Keep tables as simple as possible.

Tables can become complicated quickly, so simplify them as much as possible. If the information you are trying to present becomes too complicated, break it up into two or more tables.

2. Place important tables in the body of the document; place unimportant tables in appendices or attachments.

Tables conveying critical information must appear in the text. However, if you are presenting tables of uninterpreted data, consider placing them in an appendix or attachment. Do not force readers to ponder uninterpreted data unless they want to. See APPENDICES and EMPHASIS.

3. Use table numbers and captions to identify tables and to help readers understand the information presented.

Table numbers help readers track tables through a document and help them find important tabular information.

Captions label the information found in a table and can tell readers how to read the table. The caption is your opportunity to influence the reader's perception of the information you present. You should take advantage of that opportunity and write an informative caption. See CAPTIONS.

4. Clearly label tables that extend beyond one page, and use continuation headings on continued pages.

If you must continue a table from one page to another, write *continued* on the bottom right corner of the table on each page to be continued. Then on the top of each new page where the table has been continued, write, for instance, *Table 4, continued.*

Finally, repeat the boxhead on each continued page.

5. Use white space, boxes, or lines to separate tables from surrounding text.

Tables are graphics and should be placed as thoughtfully as any other graphics. Separate them from surrounding text by leaving three lines above and below tables (more white space on the page), by placing the tables within boxes (or frames), or by using thin rules above and below the tables to separate them from the text. See EMPHASIS and PAGE LAYOUT.

6. Logically organize rows and columns so that they reflect the purpose of the table and make the table easy to read.

The arrangement of rows and columns depends on the information being presented and on the purpose of the table.

Use a chronological order for information presented by time, date, or sequence. Use a whole-to-parts pattern for major and subordinate items. Use logical grouping for items that should appear together because of type, size, or relationship. See ORGANIZATION.

Place information that you want readers to compare in adjacent rows or columns. As much as possible, avoid forcing readers to compare or contrast information separated by other rows or columns.

Place information that must be numerically tallied or compared in columns. Mathematical operations across rows are harder to grasp than the same operations down columns.

When time is one of the variables, you can arrange it along rows or columns. However, time is traditionally displayed on the horizontal or x axis of charts, so display time across columns (see table 1). (Note, however, that table 2 is also easy to read. Time is displayed in rows in this table because displaying time in columns would create a wider, rather than a longer, table.)

7. Use row headings (in the stub) and column headings (in the boxhead) to identify the information listed in each row and column.

Tables without adequate headings are often incomprehensible. Ensure that your headings are concise and yet descriptive. If necessary, use footnotes to explain or fully describe the headings.

8. Use rules to separate the boxhead from the field and, as necessary, use rules to separate groups of data columns or rows from each other.

Complex tables are much easier to read and to follow when the information is separated by rules. Use thicker rules to separate the boxhead and the stub from the field and thinner rules to separate groups of rows and columns.

9. Where space is limited, use abbreviations.

Abbreviations are appropriate in tables (as well as other graphics). Use them where space is limited. If the abbreviation is uncommon, explain it in a footnote. See ABBREVIATIONS.

10. Base comparable numerical amounts on the same unit of measurement, and convert all fractions to decimals.

Numbers that readers will want or need to compare must be presented in the same unit of measurement. Do not present some numbers in meters and similar numbers in centimeters, and some data in feet and comparable data in yards.

Further, convert all fractions to decimals and use real numbers (numbers with decimals) as necessary for accuracy. However, do not make numbers more precise than accuracy allows. In other words, do not give more than the significant digits in a real number. If you do, you will convey a false sense of accuracy. See DECIMALS and FRACTIONS.

Table 4. Minimum Viable Population Levels for Management Indicator Species for the Targhee National Forest.

(Unless marked, population numbers reflect the total number of individual birds or animals.)

Species	Habitat Unit	Total Acres Habitat on Forest	Potential[1] Population	Minuimum[2] Viable Population
Grizzly Bear	Variable	340,000[3]	Undetermined	Undetermined
Bald Eagle	500 ac/nest site	25,000	50 pairs	Undetermined
Williamson's Sapsucker	20/ac/territory	467,000	23,350 pairs	9,340 pairs
Pika	1 ac/territory	53,000	53,000 pairs	21,200 pairs
Brewers Sparrow	5 ac/territory	121,000	24,200	9,680 pairs
Goshawk	15 sq mi/territory	870,000	90 pairs	36 pairs
Aquatic Invertebrates	per/sample[4]	25,000	Undetermined	Undetermined
Beaver	2 colonies/sq mi	46,000	144 colonies	58 colonies
Trout	lb/ac	25,000	Undetermined	18 lb/ac
Antelope	1/25 ac	77,000	4,000	1,000
Elk	6,000 ac	1,329,000	11,075	5,000[5]
Bighorn Sheep	Undetermined	77,000	600	150[6]
Mountain Goat	Undetermined	77,000	400	50[6]

[1]Within multiple use constraints and using currently available acres.

[2]40 percent of the potential is considered minimum viable population level on nongame species. Game species is the population at which a general hunt can be sustained.

[3]Situation 1 habitat and the occupied portion of Situation 2.

[4]Species diversity will be the measure of habitat quality.

[5]This level with about 40 bulls/100 cows preseason and 20 bulls/100 cows postseason.

[6]Idaho Fish and Game will begin to hunt a population at this level in any given herd area.

Source: *Final Environmental Impact Statement for the Land Management Plan for the Targhee National Forest.*

11. Align decimals vertically within columns.

Your spreadsheet software will generally do this for you; still, except where you are mixing types of numbers, ensure that decimals are aligned vertically. The exception occurs in columns containing different types of real numbers, such as decimal numbers and percentages.

12. Use zeros, dashes, ellipses, or *NA* to indicate that information is missing or not applicable.

Do not leave entries in rows or columns blank where the information is unavailable or not applicable (NA). Readers will not know whether the blank entry was intentional.

Instead, indicate missing or inapplicable information by entering a dash, a zero, an ellipsis mark, the abbreviation *NA*, or a comment like *undetermined* (see table 4).

NOTE: Use the abbreviation *do* (for *ditto*) where a data entry in a column is the same as the data entry directly above it.

13. Use footnotes, comments, or hyperlinks to explain or clarify table headings and entries.

Tabular information is governed as much by space limitations as it is by content necessity, and you may not be able to explain fully a table heading or entry in the space available. So use footnotes, comments, or links to other sites where necessary for clarification. See GRAPHICS FOR DOCUMENTS.

Insert explanatory comments for readers to click or mouseover in an online table. Your computer will position footnotes for you. Footnotes belong below the table and are usually flush left. Footnotes usually have the following format: superscripted reference number or letter indented, second or additional lines flush with the left margin, etc. See tables 2, 3, and 4.

Table 5. Three Potential Sites. *Each site has its strengths, so a decision will likely depend on possible tax incentive packages for each site.*

	A—Omaha, Nebraska	B—Joplin, Missouri	C—Ft. Worth, Texas
Total Construction Cost, Including Site Preparation	$4.385 M. Full site prep is necessary.	$3.9 M. Site is already prepared; costs are minimal.	$4.450 M. Full site prep is necessary.
Accessable to Major Transportation (Rail) Lines	Omaha is a major rail hub with connections both north/south and east/west.	Joplin has good north/south connections, but east/west connections are not as well developed.	Ft. Worth has good east/west connections. Southern location makes connections to northern states less convenient.
Airline Connections	Direct connections to major cities, but not a hub airport.	Flights must connect through St. Louis or Tulsa—a clear weakness.	Major hub airport with excellent national and intl. connections.
Quality of Life	Moderate-sized urban area with good schools and many recreational opportunities, esp. Missouri River.	Small-town setting with access to rural areas and associated opportunities. Many streams/reservoirs nearby. Branson is about 40 miles to the southeast.	Large urban center with all its advantages. Dallas/Ft. Worth area is highly desirable place to live.
Cost of Living Base = $100 Market *Basket Survey*	$102.80	$96.30 (This figure parallels a depressed real estate market.)	$105.20 (This figure reflects higher transportation costs.)

If you are referencing numerical entries located within the field, use footnote letters or symbols rather than numbers, which some readers may confuse with the number being referenced. However, use a consistent system of reference. If you establish a system of footnote letters, then use that system throughout all of your tables. See the example tables within this section. See also FOOTNOTES.

14. Identify all data sources.

For the sake of clarity, as well as substantiation, identify all data sources. The source line may follow the standard bibliographic format (see table 2) or may identify the source by name alone (tables 3 and 4). See BIBLIOGRAPHIES and CITATIONS.

Whether you provide full source information depends on the nature of the source documents, the extent to which you have summarized from the original, the purpose of your document, and the purpose of the table. If readers will want to examine the original sources, then provide full source information. However, if you have gathered data from a variety of sources or documents, you might not be able to identify specific documents.

If you have examined a number of EPA documents, for instance, and collated data from many of those documents, your source line should not list every single source. Instead, you should simply indicate:

Source: Environmental Protection Agency.

Begin source lines with the word *Source* or *Sources*, followed by a colon. End the source line with a period.

15. Use a matrix (text) table to present complex textual information.

A matrix is a visual tool for reproducing information that needs to be matched or tracked. For example, a manager wanting to consider three sites for a new plant might decide to construct a summary matrix. Table 5 is an example of what such a summary matrix might look like.

Matrices similar to table 5 are especially common in environmental reports, where the writers want to compare or assess the feasibility of various alternatives. These environmental matrices usually have alternatives listed along one axis and environmental factors (or criteria) listed along the other axis.

16. Use informal or run-in tables to replace ordinary text, especially lengthy paragraphs.

An informal or run-in table is just one such emphasis technique. A run-in table is unnumbered and appears without all the trappings of formal tables. Run-in tables are useful for comparing and contrasting information. Use them to make clearly visible the differences among points of view or courses of action.

The sentences immediately preceding a run-in table introduce the table and, as in the following example, tell readers what they are supposed to see in the table.

People looking for a job have one of two paradigms or mind-sets. If you have a Job Seeker paradigm, you see yourself as a prepackaged product that you're trying to push on a buyer. Your resume is your product brochure. On the other hand, if you have a Contributor paradigm, you see yourself as a solution to a genuine problem, concern, or opportunity. You have a proposal that will make a difference.

Job Seeker Paradigm	Contributor Paradigm
"I'm a product."	"I'm a solution."
"Because I need a job, will you please hire me?"	"Because you have a significant need, I want to offer you strengths that match that need."
"Here is my resume."	"Here is my proposal to help you."

Tables of contents help readers in two ways: (1) they outline the structure of the document and thus provide insight into the document's organization, and (2) they provide the page numbers for all sections and subsections, thus helping readers to locate parts of the document.

Editors often choose to shorten *Table of Contents* to merely *Contents*. We retained the full phrase as the title of this entry, but we recommend simply *Contents* when you are labeling a page in a document.

Creating a preliminary table of contents is a useful writing technique because writers have to think carefully about the document's organization. Gaps, illogical order, and misplaced emphasis become more apparent when the writer is forced to clarify and complete a preliminary table of contents.

The final table of contents cannot be prepared until the document is finished and the pages numbered.

Your computer will create a table of contents for you. The following rules will help you decide what goes into your table of contents.

Tables of Contents

1. Use a table of contents for any report longer than 10 pages.
2. Include major divisions (often chapter headings) and the next level of subdivisions in the table of contents.
3. Use letters and numbers with division and subdivision titles in the table of contents only if you also use them in the text.
4. Use blank lines, indentation, and leader dots to lay out the table of contents and help readers find page numbers.
5. Include preliminary material and appendices or attachments in the table of contents.
6. If appropriate, supplement your table of contents with a matrix or checksheet showing how and where you have dealt with key requirements or issues.

1. Use a table of contents for any report longer than 10 pages.

The 10-page figure is arbitrary, but remember that a table of contents helps readers to see the overall organization of the document as well as to find key sections. Documents under 10 pages are generally short and uncomplicated enough for readers to determine the structure by skimming through the document before reading. However, skimming through longer documents may not provide an adequate sense of structure because the reader's mind is being asked to comprehend too much information spread over too great a distance.

2. Include major divisions (often chapter headings) and the next level of subdivisions in the table of contents.

Major divisions and the next level subdivisions are essential if readers are to comprehend the document's structure. Make the division titles as specific as possible:

this

2. Testing for Flammability

    2.1 Temperatures

    2.2 Duration

    2.3 Flash Points

not this

2. Testing

NOTE: Ensure that all of the divisions and subdivisions that appear in the table of contents also appear in the body of the document. See HEADINGS.

3. Use letters and numbers with division and subdivision titles in the table of contents only if you also use them in the text.

A table of contents can resemble an outline with page numbers, but such an outline structure (*I, A, 1, a,* etc.) is not necessary unless the chapter or section titles and the headings in the document reflect the same numbering system. Tables of contents often have a decimal numbering system (*1.1.1, 1.1.2, 1.1.3,* etc.) in place of the standard outline system. Either system is acceptable. See OUTLINES and NUMBERING SYSTEMS.

4. Use blank lines, indentation, and leader dots to lay out the table of contents and help readers find page numbers.

Leave **blank lines** between major entries, and ensure that the spacing reflects the logical structure of the document. For instance, you may want to leave two lines between major divisions (first-level headings) and one line between major subdivisions (second-level headings). Readers should be able to tell where major divisions occur simply by noting the number of lines between entries.

Use **indentation** to show levels of subordination. Major divisions (first-level headings) should be flush left. Major subdivisions (second-level headings) should be indented five spaces. Minor subdivisions (third-level headings) should be indented 10 spaces, and so on.

Finally, consider using **leader dots** (a row of unspaced periods) to connect entries with their page numbers. Leader dots allow readers' eyes to trace the connection across the page.

To emphasize major divisions or subdivisions, you might also use all capital letters, boldface type, and other emphatic techniques. See EMPHASIS.

The table of contents example in figure 1 shows one method of displaying the organizational structure of a document and of providing page numbers to help readers locate the document's parts. For another example, see the table of contents to this *Style Guide*.

CONTENTS

iii

Figure 1. Sample Table of Contents. *This table of contents includes second-level headings so that readers can easily find the relevant sections.*

5. Include preliminary material and appendices or attachments in the table of contents.

Some writers ignore the preliminary material such as the preface or foreword and attachments when they construct their tables of contents. However, a table of contents should reflect the structure of the entire document, so include all of the document's parts.

6. If appropriate, supplement your table of contents with a matrix or checksheet showing how and where you have dealt with key requirements or issues.

Such matrices or checksheets do not replace the standard table of contents or an index, but they can help guide readers to see how you have organized your document. They are especially valuable when you need to show readers that you have met all legal or procedural requirements. See TABLES.

These matrices or checksheets have various formats, but as figures 2 and 3 on the next page show, most requirements or issues are usually listed vertically down the left with the document sections or chapters horizontally across the top. Page numbers (or subsection numbers) appear in the matrix.

Place a matrix where readers can easily use it to track content questions. Often the most obvious placement is for a matrix to follow the contents page(s). As an option, consider printing the matrix on a foldout sheet positioned at the end of the document. Readers can open the foldout and use it to track points as they read.

Statement of Work Requirements	Proposal Section Numbers
3. Survivability	4.4, 8.2, and 9
3.1 Stress Resistance	4.4.1 and 8.2.2
3.2 Temperature Constraints	4.4.2 and 8.2.3
3.3 Humidity	4.4.3 and 8.2.4
3.4 Electromagnetic Pulse	4.5 and 9

Figure 2. A Matrix from a Proposal. *A two-column matrix may be sufficient to track the way the document responds to each requirement listed in a client's statement of work. An option would be to list all section titles and numbers across the top and then fill in the matrix with page numbers.*

Chapters (page numbers)

Issues	Summary	I. Purpose and Need	II. Alternatives	III. Affected Environment	IV. Environmental Consequences	Appendix A Public Involvement	Appendix B
1. Soil Stability	2	4–5	8, 10–11	18–21	49–52	A–3, A–5	B–6, B–11
2. Water Quality	2–3	5	8, 11–12	21–24	53–57, 68–69	A–3, A–5, A–14, and A–20	B–5, B–8, and B–20
3. Wildlife	3	5–6	8, 13	27–30	60–64, 69–70	A–5, A–14 A–20, and A–24	B–1, B–2, B–7, and B–15

Figure 3. A Matrix From an Environmental Document. *The issues are crucial to the organization (and legality) of the document. By showing that each issue appears in each chapter, the writer is able to show readers that all issues have been consistently addressed. In a sense, a matrix like this could replace the index.* *See* INDEXES *and* TABLES.

Thinking Strategies

Writing and thinking are to a very great extent the same process. A piece of writing readily reveals the quality of thinking behind it, and writers who cannot think clearly—no matter how cosmetically attractive their documents and grammatically flawless their language—cannot produce good writing.

Thinking Strategies

1. Choose one or more thinking strategies to help you address a problem or resolve an issue.
2. Brainstorm schematically or visually your answers to each chosen question.
3. Use your visualizations to guide your writing.

Note how the rules of critical thinking are phrased in the form of writing challenges:

- Frame a problem clearly so that it can be analyzed.
- Reason logically from one point to another.
- Make clear information out of raw facts and data.
- Evaluate alternative viewpoints, including underlying assumptions and implications.
- Arrive at reasoned conclusions based on firm logic and evidence.

The flaws that weaken writing are the same failings that afflict unsound thinking: unclear framing of the issue, breaks in logic, undigested data, unexamined biases, and wobbly conclusions. By the same token, substantive writing is evidence of high-quality thinking.

The following discussion shows how common thinking strategies can help you to develop a cogent and convincing message. For example, the simple strategy of listing pros and cons might help you organize your thoughts to create an effective document.

1. Choose one or more thinking strategies to help you address a problem or resolve an issue.

What are the possible strategies? In ancient Greece, rhetoricians first described methods or strategies of thinking that enabled them to communicate logically and persuasively. The questions listed in figure 1 are merely an updated version of the strategies known and practiced by the ancient Greeks.

When you are dealing with an issue or a problem, begin by surveying the questions in figure 1. Which ones might be helpful? Often you will find that several of them have promise.

As an example of how to use the strategy questions, assume that you need to decide whether to close a branch office. In this instance, you decide that questions 2, 5, and 11 might be helpful.

The most influential writing often arises from thinking through question 14, and the worst writing—no matter how well reasoned and cosmetically pleasing—often arises from the failure to challenge assumptions.

For example, in 19th-century science, many carefully reasoned and "beautifully written" books used the Newtonian paradigm to try to solve certain problems in physics. But because the authors failed to challenge Newton's assumptions, much of their work collapsed when Einstein's paradigm of relativity superseded Newton's thinking.

2. Brainstorm schematically or visually your answers to each chosen question.

Working alone or in teams, brainstorm answers to a given question.

Create a schematic or visualization of your thinking, using pad and pencil, a whiteboard, multiple flip-chart pages, or a computer projection. In a teleconference, take advantage of online programs that can help everyone see the shared thinking.

This visualization is not merely a list of words. As shown in figure 2, make your thinking visible by sketching tables, diagrams, or charts. Also, don't try to write connected text, which can come later. Instead, your goal is to generate and capture your best thinking.

Figure 2 shows several common visual thinking strategies that you might use as you develop your ideas. Pick one and flesh it out until the flow of ideas dries up. Then, if you still need more information, try a second strategy.

Retain (or record) these visualizations for later use. Show them to interested people—customers, bosses, members of your online community—to get input on your ideas. Such a visualization

Strategy Questions. Choose one or more of these strategy questions when you need help thinking about a problem or an issue. Often you will discover that you need to work through several strategies before you find one that will help you generate productive ideas.

1. How would you define the issue or frame the problem? (definition)
2. What caused this problem, and what are the effects? (cause and effect)
3. What category would the problem or issue fit into? (classification)
4. What are the constituent parts of the issue? (division)
5. How important are the issues in comparison to each other? (ranking or prioritization)
6. What steps or options would permit an event to occur? (process)
7. How feasible—with pros and cons—is one or more of the options? (comparative advantage)
8. What situations are similar to the problem? (analogy)
9. How does this problem or issue compare (or contrast) with others? (comparison/contrast)
10. What objections can be made to our position, and how valid are they? (argument)
11. What specific instances help with understanding the issue? (examples)
12. What is the story of this issue? (narration)
13. What related ideas, solutions, or questions can we come up with quickly about this issue? (brainstorming)
14. What assumptions need to be challenged in relation to this issue? (paradigm shifting)
15. What evidence and reasoning can we marshal to support our position? (deduction)
16. What position should we take based on this evidence and reasoning? (induction)

Figure 1. Thinking Strategies. *Each of these questions drives a different strategy for thinking through an issue and also suggests methods for writing about the issue.*

might serve as an initial prototype for an important document. See Writing and Revising.

Whatever way you decide to retain or record information, you should add the following facts, as appropriate:

- Who teamed with you to generate the ideas
- When and where you generated and recorded the ideas
- What comes next and who is responsible for this next step
- How the information should be filed or cross-referenced for retrieval and later use (or expansion)

Recording and retrieval are crucial steps, especially given the increasing amount of information available. See Managing Information.

3. Use your visualizations to guide your writing.

Visualizations should drive every step in the thinking process. Early in the process, visualization helps you discover ideas. Visualization should also guide you as you communicate your message. See Writing and Revising.

The early visualization often becomes a primary graphic for what you intend to write. See Graphics for Documents and Graphics for Presentations.

For example, if you have spent time analyzing the effects of a proposal, you might decide to use as a primary graphic a cause-and-effect flow chart (example 5 in figure 2).

Or to consider another example, if you are analyzing the root causes for a problem, such as the delivery of the wrong food in a hospital, you might choose to list these causes as reasons arranged around a circle with the problem stated in the middle (example 2 in figure 2).

Even if you don't choose to use an actual graphic in your document, the form of the early visualization should help you design and organize your document. If you determined that the solution to a problem required three key actions, you would want to design a document with three major subdivisions (example 4 in figure 2).

As in these examples, use the initial visualization as the basis of your final documentation or presentation. Good documents must be well-designed visual constructs if they are to communicate effectively and memorably. See Emphasis, Graphics for Documents, and Organization.

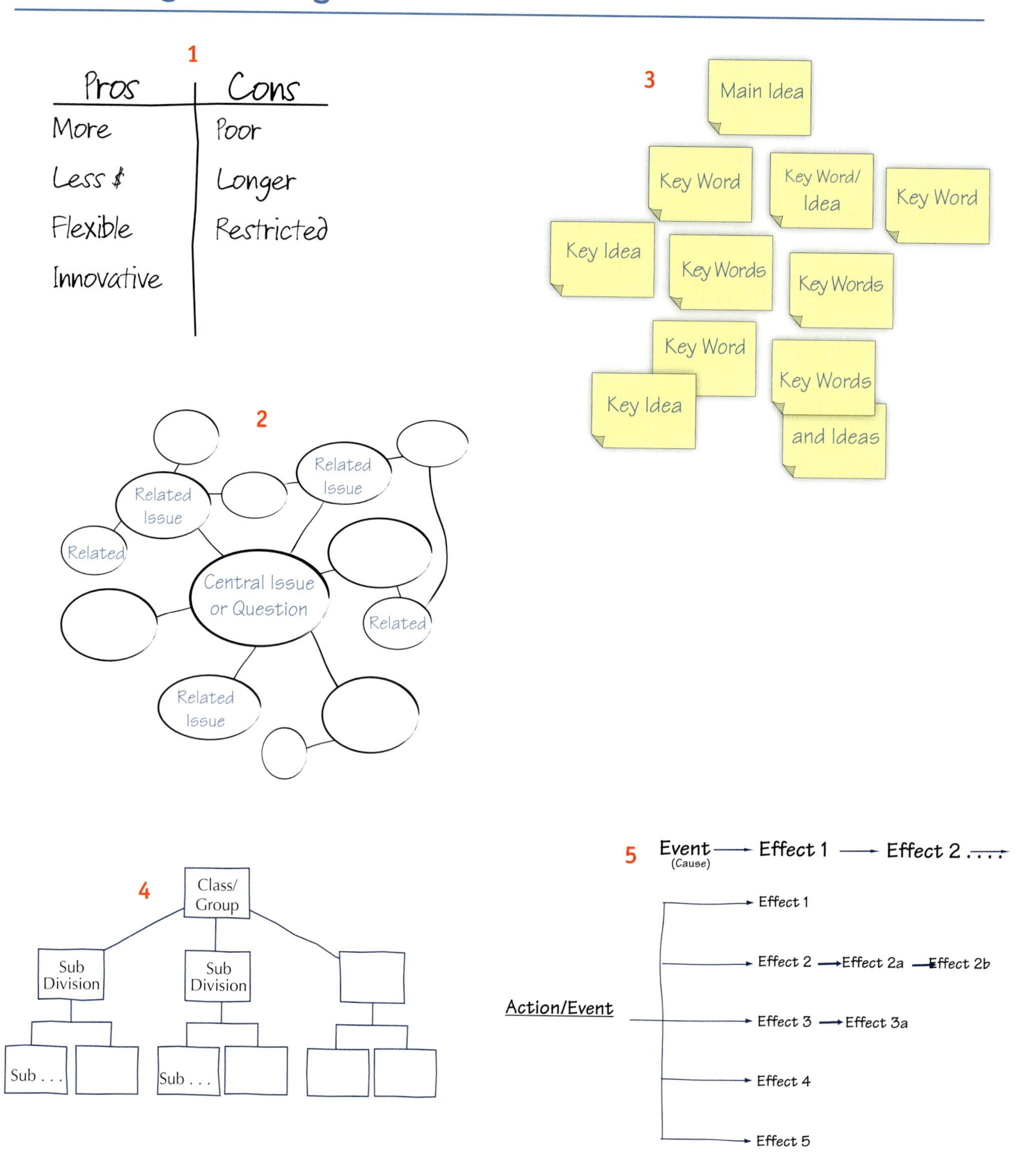

Figure 2. Visualization Patterns. *As you record your answers and thoughts relating to the strategy questions, use one or more of these visualization patterns. Or create your own visualization to fit your specific data.*

Titles of people, organizations, governments, and publications often require special capitalization, punctuation, and other format conventions. See Nouns and Capitals.

1. Capitalize the first letter of titles when they immediately precede personal names, but do not capitalize the first letter when they follow personal names:

Mrs. Robert T. Evans

Mr. Edward Johnson

Miss Sylvia Smead

Ms. Josephine Kukor

President Amy Kaufmann

Assistant Professor Ned Davies

Mayor-elect Boon Hollenbeck

General Laswell Hopkins

Lieutenant Cynthia Wagner

the Reverend John Tyler

Rabbi Tochterman

Amy Kaufmann, the president of Union College, spoke to the press.

Ned Davies is an assistant professor at Columbia University.

We voted for Boon Hollenbeck, who is now mayor-elect.

Laswell Hopkins was our general for only 2 months.

Our lieutenant was Cynthia Wagner. *(Here,* was *separates the title from the name.)*

Note 1: The titles of high-ranking international, national, and state officials often retain their capitalization, even when the name of the individual is either absent or does not follow the title:

The President spoke before the Congress.

We wrote the Vice President.

The Pope toured South America.

The Governor still had 2 years to serve.

The Prime Minister of Ghana was invited to the White House.

Titles

1. Capitalize the first letter of titles when they immediately precede personal names, but do not capitalize the first letter when they follow personal names.
2. Capitalize the first letter of names of companies, schools, organizations, and religious bodies.
3. Capitalize the first letter of names of government bodies.
4. Capitalize and italicize (or underline) the titles of books, magazines, newspapers, plays, movies, television series, blogs, and other separately published works.
5. Capitalize and use quotation marks for chapters of books, articles in magazines, news stories or editorials, acts within a play, episodes of a television series, or other sections of something separately produced or published.

Note 2: Titles of company or corporation executives as well as titles of lesser federal and state officials are sometimes capitalized. Such capitalization is unnecessary, but you should follow company or agency practice:

The Mayor announced an end to the New York transit strike. (*or* The mayor)

The Vice President for Finance is resigning Monday, September 18.

The Superintendent refused to approve our budget request.

Note 3: Titles used in a general sense are not capitalized:

a U.S. representative

a king

a prime minister

an ambassador

2. Capitalize the first letter of names of companies, schools, organizations, and religious bodies:

the Johnson Wax Company

the University of Glasgow

the World Trade Organization

the Urban League of Detroit

the Liberal Party

St. John's Lutheran Church

Note: The words capitalized are those normally capitalized in any title. See Capitals. In contrast with book titles, an initial *the* in most such titles is not capitalized unless the company, school, organization, or religious body has established *the* as part of its legal name: *The Johns Hopkins University, The Travelers Insurance Company.* See Capitals.

3. Capitalize the first letter of names of government bodies:

the United Nations

the Cabinet

the Medicines Control Agency

the California Legislature

the Manitoba Department of Education

the Devon County Council

the House (*for* House of Representatives)

the Department (*for* Department of Agriculture)

the Court (*for* the U.S. Supreme Court)

Titles

NOTE 1: Except for international and national bodies, shortened forms of these government bodies or common terms are not capitalized.

the police department

the county council

the board of education

NOTE 2: As explained in rule 2, initial uses of *the* do not require a capital *T*.

4. Capitalize and italicize (or underline) the titles of books, magazines, newspapers, plays, movies, television series, blogs, and other separately published works:

Oliver Twist (book or movie)

Newsweek

The Guardian

West Side Story

Superman

NOVA

Lifehacker

See CAPITALS, ITALICS, and UNDERLINING.

5. Capitalize and use quotation marks for chapters of books, articles in magazines, news stories or editorials, acts within a play, episodes of a television series, or other sections of something separately produced or published:

The last chapter was called "The Final Irony."

"The Colombian Connection" was the lead article in last week's *Time*.

We supported his editorial, "A Streamlined Election System."

We watched "The Fatal Circle" last night on *BBC Presents*.

See QUOTATION MARKS, UNDERLINING, and ITALICS.

Tone reflects your attitude toward your subject and your readers. Your writing may strike your readers as personal or impersonal, friendly or distant. You may sound warm and engaging or cold and abrupt.

Tone and *style* are often confused. Some people use the terms interchangeably, but one is the cause and the other the effect:

- **Tone** refers to the feeling or impression a document conveys to its readers. It is one of the products of the writer's style.
- **Style** refers to the writer's choices that create the tone readers perceive. Style is the writer's "manner of speaking," the way the writer uses language to express ideas.

See STYLE and SCIENTIFIC/TECHNICAL STYLE.

Tone, then, is the impression readers receive from your writing and the attitude conveyed in your treatment of the subject. We usually describe the tone of a piece of writing with these words:

abrasive	formal
aggressive	forthright
assertive	friendly
authoritative	impersonal
blunt	informal
bureaucratic	informative
casual	objective
cold	officious
condescending	personal
courteous	polite
demanding	sincere
discourteous	stiff
distant	subjective
earnest	threatening
engaging	warm

Tone

1. Use pronouns to establish a personal, human tone in letters and memos.
2. Make your letters and memos sound very much like the language you would choose if you actually talked with the readers.
3. Choose sentence structures that reflect a friendly, conversational tone.
4. Include personal information and personal references.
5. Choose your paper, font, and format to reflect a personal, friendly tone.

Desirable Business Tone

Most of the time, your business documents should be:

courteous	informative
forthright	personal
friendly	polite
helpful	sincere
informal	warm

The extent to which your documents are personal will depend on your relationship with the reader. But never fail to be courteous, polite, informative, sincere, and helpful—especially when you don't know the reader.

If you need to convey negative information, as in a poor performance appraisal or reprimand, or in a document threatening legal action, your document may need to be:

assertive	impersonal
formal	objective

However, you should never write documents that are:

abrasive	blunt
discourteous	cold
bureaucratic	officious
condescending	

Good business documents—no matter how tough or adversarial they are—should **never** be discourteous.

Below are the stylistic choices that will help you write business letters and memos that have an effective tone. Not coincidentally, these are the same rules that make writing clear, concise, and easy to read.

1. Use pronouns to establish a personal, human tone in letters and memos.

Probably no single language choice is as effective in making business documents sound human and personal as well-chosen pronouns. Of the pronouns possible, *you* is the most important. You should always be aware of your readers and address them directly:

this

During the discussion of your April bill, you mentioned that you had called your local service representative at least three times during the month. Do you remember the representative's name and the dates when you called?

not this

Concerning the April bill, the local service representative may have been called, but these calls cannot be verified unless the representative's name and the dates when the representative was called are provided to this office.

Tone

The ineffective version has no personal pronouns; consequently, the reader is ignored. Omitting personal pronouns makes the letter cold and informal. The passive verbs contribute to the impersonal tone and make the letter sound unfriendly at best. See ACTIVE/PASSIVE.

Next to *you*, the pronouns *I* and *we* are essential for effective letters and memos. Some people argue that the writer should not be mentioned in documents. They argue that documents should not reflect personal opinions or the personality of the author. This argument fails to distinguish between personal opinions and personal responsibility for one's actions. Contrast these two examples:

this

Based on the data, I (*or* we) conclude that MOGO should plug and abandon the Gilbert Ray Well 3.

not this

Based on the data, it is concluded that MOGO should plug and abandon the Gilbert Ray Well 3.

The second version is mechanical, almost robotic. No person seems to have acted. The conclusion simply occurred, like something out of the twilight zone. The result is a faceless, anonymous tone, one calculated to avoid responsibility and perhaps to confuse readers or keep them deliberately in the dark.

2. Make your letters and memos sound very much like the language you would choose if you actually talked with the readers.

Read your document aloud. Would you be comfortable saying those words to someone in person? If you delivered the message to readers orally, would you express yourself this way?

The tone of a good business document is a natural one. It isn't full of slang or homey conversational expressions (*Well shucks, I reckon we ought to drill anyways*), but it should sound natural, not forced or contrived. If the document does not sound natural, if it is stiff and complex, if it is formal and faceless, you should rethink your tone. See LETTERS AND MEMOS.

A business document is not a transcription of your actual words, pauses, corrections, and other verbal lapses (*Well, uh, I think that, uh, if we, uh . . .*). It should, however, be similar to the way you talk. Even contractions are useful in letters and memos but not in more formal reports. See APOSTROPHES and CONTRACTIONS.

The words, phrases, and sentences you use should be simple and direct, even though you have edited and revised them:

this

I recommend that we immediately replace the roof on the Bradley building.

not this

It is recommended that the roof on the Bradley building be replaced forthwith.

---

this

Before leaving the room, turn out the lights.

not this

Prior to evacuating the premises, ensure that the illumination has been terminated.

The simplest remedy for overly stiff, bureaucratic writing is to write like a human being. Don't write like an official, faceless bureaucrat. Just be yourself. Imagine that you're talking to other people in person. Try to sound human, not mechanical.

Using personal pronouns will help considerably. Here are some other suggestions:

- **Keep your sentences short and direct.** Challenge any sentence that is longer than 30 words, and try to limit average sentence length to about 20 words. You can often break a long sentence into two shorter sentences. Ensure that each sentence, whatever its length, is as clear and direct as possible. See SENTENCES.
- **Avoid long, unnecessarily complex words and unnecessary technical terms.** See GOBBLEDYGOOK and JARGON. Never use words because you think those words are impressive. The writer who struggles to sound intelligent and educated (often by using a thesaurus) winds up sounding silly:

this

We think your water pipes have corroded so much that only a trickle of water can flow through them.

not this

Our hypothesis is that your water supply system has undergone severe corrosion and reached the debilitating point where water normally available is unavailable in the quantity and at the pressure provided for in the original specifications for your domicile.

Leaving aside the laughable words (**debilitating** and **domicile**), the second version still suffers from terminal wordiness. Isn't *pipes* better than *water supply system*? And *corroded* so much better than *undergone severe corrosion*? What do the inexact references to quantity and pressure accomplish that the word *trickle* doesn't do more vividly? The idea being expressed does not require technical terms, especially if the primary reader is a homeowner, not an engineer.

Legal documents often suffer from the same kind of wordiness and unnecessary complexity:

this

According to the procedure outlined above, please sign all three copies of the conveyance and have your signature notarized. Then return the completed forms to this office by Friday, May 23.

not this

In accordance with the provisions of the aforementioned procedure, the attached conveyance should be executed by you in triplicate, with the signature duly witnessed and attested to by a Notary Public, and the executed set of conveyance forms should then be returned to this office on or before, and no later than, Friday, May 23.

3. Choose sentence structures that reflect a friendly, conversational tone.

- **Avoid passive sentences:**

this

Our review of your claim indicates that you should receive a refund of $72. This refund would apply to December charges.

not this

Your claim has been reviewed, and it has been determined that $72 should be refunded to you for the period January to December.

---

this

We analyzed the data entries for the sources of the discrepancies. Over 90 percent of the discrepancies turned out to be simple errors in daily recording.

not this

The data entries have been analyzed to determine the source of the discrepancies. It is concluded that over 90 percent of the discrepancies were caused by simple errors in daily recording.

See Active/Passive.

- **Avoid false subjects:**

this

Unless we address these issues during this quarter, they will distort our financial report for the entire year.

not this

There are certain issues that we should address during this quarter that will distort our financial report for the entire year.

---

this

Under Pharmtech's policies, we will not acquire additional laboratory equipment until the beginning of the new fiscal year.

not this

It is likely that, given Pharmtech's policies, it will be impossible to acquire additional laboratory equipment until the beginning of the new fiscal year.

See False Subjects.

4. Include personal information and personal references.

Readers like to know that you have addressed their needs. So, if appropriate, include information from previous letters, memos, or discussions in your document. Or include information they have either requested or will need:

I recommend that you file a complaint with the Federal Trade Commission. Your review of the relevant correspondence with the company persuaded me that you have a case.

A mechanical yet easy way to make a letter or memo personal is to include the reader's name in the body of the letter:

So, Beth, if you have any more suggestions, please call me at ext. 3578.

or

If we proceed, Cal, you should ask Product Research for a copy of their assessment form. I think you'd find it helpful.

Note: As in the preceding example, an occasional contraction (*you'd*) makes the tone conversational and informal. See Apostrophes and Contractions.

5. Choose your paper, font, and format to reflect a personal, friendly tone.

Even physical concerns can affect the tone of a business letter or memo.

Choose quality paper (usually 20-pound rag paper) and a pleasing font. With computers, the font options are growing. Know your options. Your goal is to capture the personal, friendly quality of the content of your letter (consider Helvetica or Optima). Some fonts are too rigid and stark (American Typewriter). Others are too strange (Script, which attempts to look like cursive writing). See Page Layout.

Next, design your document so that it has a lean and open look, one conveying a personal, friendly tone. This design usually means using generous margins, short paragraphs (on the average), headings, lists, and lots of white space. See Emphasis.

Transitions

Transitions are words or phrases that connect ideas and show how they are related. Occurring between two sentences or paragraphs, a transition shows how the sentences or paragraphs are connected, thus making the writing smoother and more logical. A transition creates a point of reference for readers, allowing them to see how the writing is organized and where it is heading.

Following is a list of transitions and their functions.

Transitions

1. Use commas to separate transitions from the main body of a sentence.
2. If the transition interrupts the flow of a sentence, place commas on both sides of the transitional word or phrase.
3. If a transitional word occurs at the beginning of a sentence and is essential to the meaning of the sentence, do *not* separate it with a comma.

A List of Transitions

Addition

additionally, again, also, besides, further, furthermore, in addition, likewise, moreover, next, too, what is more

Comparison or contrast

by contrast, by the same token, conversely, however, in contrast, in spite of, instead, in such a manner, likewise, nevertheless, otherwise, on the contrary, on the one hand, on the other hand, rather, similarly, still, yet

Concession

anyway, at any rate, be that as it may, even so, however, in any case, in any event, nevertheless, of course, still

Consequence

accordingly, as a result, consequently, hence, otherwise, so, then, therefore, thus

Diversion

by the way, incidentally

Generalization

as a rule, as usual, for the most part, generally, in general, ordinarily, usually

Illustration

for example, for instance

Place

close, here, near, nearby, there

Restatement

in essence, in other words, namely, that is

Summary

after all, all in all, briefly, by and large, finally, in any case, in any event, in brief, in conclusion, in short, in summary, on balance, on the whole, ultimately

Time and sequence

after a while, afterward, at first, at last, at the same time, currently, finally, first (second, third, etc.), first of all, for now, for the time being, immediately, instantly, in conclusion, in the first place, in the meantime, in time, in turn, later, meanwhile, next, presently, previously, simultaneously, soon, subsequently, then, to begin with

See Conjunctions, Organization, and Paragraphs.

1. Use commas to separate transitions from the main body of a sentence:

> However, uncontrolled R&D efforts that do not design quality into the product, process, or service may be of dubious value.
>
> Consequently, areas of future focus will include software design, packaging enhancements, and marketing efforts in the Pacific Rim.
>
> The regulator poppet is, however, normally held open by the regulating spring.

Transitions interrupt and are generally not part of the main thought of the sentence; therefore, you should separate them from the rest of the sentence. When a transitional word occurs where two complete thoughts are joined, as in the following sentence, use a semicolon where the two complete thoughts join (usually in front of the transition) and a comma after the transition:

> The scope of this study will not permit a review of all technologies; however, we believe our experience in advanced ship design will allow us to maximize the study of those areas with greater operational potential.

See Semicolons and Commas.

2. If the transition interrupts the flow of a sentence, place commas on both sides of the transitional word or phrase:

> Upon examination, for instance, the clinical monitor considered 10 patients to have significant findings.
>
> The sensor probe, on the other hand, contains thermistor sensing elements.

3. If a transitional word occurs at the beginning of a sentence and is essential to the meaning of the sentence, do *not* separate it with a comma:

> However warm air enters the cabin, the conditioned temperature will not rise above the nominal range.

Note: Such sentences can be confusing, so you should consider rephrasing the sentence:

> Regardless of how warm air enters the cabin, the conditioned temperature will not rise above the nominal range.

Underlining can replace italics as a tool for highlighting certain unusual words and phrases. See ITALICS. You can also underline words, phrases, and sentences to emphasize them.

Underlining is becoming less common because word-processing programs provide italics.

So, in rules 1 to 4 below, italics would be preferred if it is available. In rule 5, underlining, not italics, would still be preferred. See ITALICS and WORD PROCESSING.

Editors differ as to whether to underline words and the spaces between them or only the words. Given the influence of word processing, which allows writers to highlight a block of text and then underline it, the pattern of continuous underlining is more common:

> The Making of a President
>
> *not*
>
> The Making of a President

Underlining

1. Underline words used as examples of words.
2. Underline foreign words and phrases that have not been absorbed into English.
3. Underline the titles of separate publications and productions.
4. Underline the names of aircraft, vessels, and spacecraft.
5. Underline words and phrases for emphasis:

1. Underline words used as examples of words:

> The words affect and effect are often confused.
>
> The contract stipulated that monitoring must be continuous, but during negotiation they stated that periodic monitoring would suffice. Should the contract read continual or periodic rather than continuous?

NOTE: Single letters, words, and even phrases should be underlined to separate them from the ordinary words within a sentence:

> The phrase come hell or high water has a long and interesting history.

Some editors prefer to use quotation marks for words and phrases used unusually within a sentence. However, quotation marks clutter up a sentence if several words are being talked about:

> The forms "am," "is," "are," "was," and "were" don't even resemble "be," which is the principal form of the verb.

2. Underline foreign words and phrases that have not been absorbed into English:

> After a coup d'oeil, the detective was ready to question the suspect. (Coup d'oeil means "a quick survey.")

NOTE: Contrast *coup d'oeil*, which is clearly not part of English, with *ad hoc*, which is now so familiar that no underlining is necessary. Many modern dictionaries fail to specify whether a word should be considered foreign or not, so you may have to use your own judgment.

3. Underline the titles of separate publications and productions:

> I bought a copy of The Wall Street Journal.
>
> Have you read the novel All Quiet on the Western Front?
>
> The Sunshine Patriot is a pamphlet being circulated by the Republican Party.
>
> We attended a preview of The Story of the Bell System.
>
> We have tickets for the opening night of Aida.

NOTE: Underline separately published or produced items, but use quotation marks for book chapters, articles in magazines, and sections of the separate works:

> The latest issue of the Oil and Gas Journal contained an article entitled "Shearing Problems with Sucker Rods."
>
> The London Times had an editorial entitled "A Bold Proposal."

See QUOTATION MARKS and TITLES.

4. Underline the names of aircraft, vessels, and spacecraft:

> U.S.S. Constitution
>
> H.M.S. Bounty
>
> Gemini 4

5. Underline words and phrases for emphasis:

> Please send two copies to us by Friday, October 13, at the latest.
>
> Unscrew the fitting by turning it clockwise, not counterclockwise.

NOTE: Use underlining for emphasis sparingly. Underlined text is difficult to read, and the effect diminishes quickly if you overuse it. See EMPHASIS.

Units of Measurement

Units of measurement include either English units (also called the U.S. customary system) or metric units. Many U.S. firms still favor English units, but metric units are now widely used by scientists and engineers. See METRICS.

Units of Measurement

1. Use common English units to measure length and area.
2. Use common English units for volume or capacity and weight.

1. Use the following common English units to measure length and area.

NOTE: The abbreviations for these units of measurement usually appear only in tables, charts, graphs, and other visuals. See ABBREVIATIONS.

LENGTH

English Unit	U.S. Equivalents	Metric Equivalents
inch	0.083 foot	2.54 centimeters
foot	1/3 yard, 12 inches	30.48 centimeters
yard	3 feet, 36 inches	0.914 meter
rod	5 1/2 yards, 16 1/2 feet	5.029 meters
mile (statute, land)	1,760 yards, 5,280 feet	1.609 kilometers
mile (nautical, international)	1.151 statute miles	1.852 kilometers

AREA

English Unit	U.S. Equivalents	Metric Equivalents
square inch	0.007 square foot	6.452 square centimeters
square foot	144 square inches	929.030 square centimeters
square yard	1,296 square inches 9 square feet	0.836 square meter
acre	43,560 square feet 4,840 square yards	4,047 square meters
square mile	640 acres	2.590 square kilometers

2. Use the following common English units for volume or capacity and weight:

VOLUME OR CAPACITY

English Unit	U.S. Equivalents	Metric Equivalents
cubic inch	0.00058 cubic foot	16.387 cubic centimeters
cubic foot	1,728 cubic inches	0.028 cubic meter
cubic yard	27 cubic feet	0.765 cubic meter

English Liquid Measure	U.S. Equivalents	Metric Equivalents
fluid ounce	8 fluid drams, 1,804 cubic inches	29.573 milliliters
pint	16 fluid ounces 28.875 cubic inches	0.473 liter
quart	2 pints, 57.75 cubic inches	0.946 liter
gallon	4 quarts, 231 cubic inches	3.785 liters
barrel	varies from 31 to 42 gallons established by law or usage	

VOLUME OR CAPACITY (continued)

English Dry Measure	U.S. Equivalents	Metric Equivalents
pint	1/2 quart 33.6 cubic inches	0.551 liter
quart	2 pints 67.2 cubic inches	1.101 liters
peck	8 quarts, 537.605 cubic inches	8.810 liters
bushel	4 pecks, 2,150.420 cubic inches	35.239 liters

WEIGHT

English Avoirdupois Unit	U.S. Equivalents	Metric Equivalents
grain	0.036 dram, 0.002285 ounce	64.798 milligrams
dram	27.344 grains, 0.0625 ounce	1.772 grams
ounce	16 drams, 437.5 grains	28.350 grams
pound	16 ounces, 7,000 grains	453.592 grams
ton (short)	2,000 pounds	0.907 metric ton (1,000 kilograms)
ton (long)	1.12 short tons, 2,240 pounds	1.016 metric tons

Apothecary Weight Unit	U.S. Equivalents	Metric Equivalents
scruple	20 grains	1.295 grams
dram	60 grains	3.888 grams
ounce	480 grains, 1.097 avoirdupois ounces	31.103 grams
pound	5,760 grains, 0.823 avoirdupois pound	373.242 grams

NOTE: Although the British Commonwealth and the United States both use the same names for units and their abbreviations, the two systems do differ, so be cautious in interpreting publications using these units. Here, for example, are the equivalents between the British Imperial units and the U.S. English or customary units:

British Imperial Liquid and Dry Measure	U.S. English Equivalents	Metric Equivalents
fluid ounce	0.961 U.S. fluid ounce 1.734 cubic inches	28.413 milliliters
pint	1.032 U.S. dry pints 1.201 U.S. liquid pints 34.678 cubic inches	568.245 milliliters
quart	1.032 U.S. dry quarts 1.201 U.S. liquid quarts 69.354 cubic inches	1.136 liters
gallon	1.201 U.S. gallons 277.420 cubic inches	4.546 liters

Verbs are the key action words in most sentences. They tell what the subject has done, is doing, or will be doing, and they indicate the subject's relationship to the object or complement. Because verbs also signal time through their different tenses (forms), they are potentially the most important words in a sentence. For instance, varying only the verb in a sentence produces major shifts in the meaning:

She shows us her report.
She showed us her report.
She will show us her report.

She has shown us her report.
She had shown us her report.
She will have shown us her report.

She is showing us her report.
She was showing us her report.
She will be showing us her report.

These nine sentences only begin to illustrate all the possible verb forms. If we include more complex verb phrases, the possibilities multiply:

She is going to be showing us her report.
She must have been showing us her report.

Verbs

1. Check a recent dictionary to determine the correct forms for any verb you are unsure of.
2. Vary your verb tenses to reflect the varying timing of events in your writing.
3. Ensure that your verbs agree with your subjects.
4. Use a subjunctive verb in *if* clauses to state a situation that is untrue, impossible, or highly unlikely.
5. Use subjunctive verbs in sentences making strong recommendations or demands, or indicating necessity.

Principal Verb Forms

Verbs commonly have several standard forms (called principal parts) from which all the other verb forms are built:

base form	call eat cut
–s form (third person singular present—he, she, it)	calls eats cuts
past form	called ate cut
–ed participle (past participle)	called eaten cut
–ing participle (present participle)	calling eating cutting

Regular verbs, like *call*, routinely require only an *–s*, *–ed*, or *–ing* to change the base form. If a dictionary does not supply any forms except the base form, the verb is regular, like *call*.

Irregular verbs, like *eat* and *cut*, are unpredictable, so writers have to know the different forms, not just follow the regular pattern. Dictionaries include these irregular forms in their entries for these verbs.

NOTE: Unfortunately, not all verbs are clearly regular or irregular. The following verbs, for example, have two different forms of the past participle, one regular, the other irregular:

mow, mowed, mown (*or* mowed)
show, showed, showed (*or* shown)
swell, swelled, swollen (*or* swelled)

1. Check a recent dictionary to determine the correct forms for any verb you are unsure of.

Here, for instance, are the main forms for some of the common irregular verbs:

buy, bought, bought
cost, cost, cost
drink, drank, drunk
freeze, froze, frozen
keep, kept, kept
lead, led, led
lie, lay, lain
light, lighted/lit, lighted/lit
rise, rose, risen
sell, sold, sold
sit, sat, sat
speak, spoke, spoken
spoil, spoilt/spoiled, spoilt/spoiled
take, took, taken
tear, tore, torn
think, thought, thought
wet, wet/wetted, wet/wetted
write, wrote, written

NOTE: Where two forms exist, the regular forms (with *–ed*) are becoming more common. Over time, many irregular verbs have changed and are changing into regular verbs.

Verb Tenses

Verbs have the following basic tenses or times:

Basic Verb Tenses

Present:	They study.
Past:	They studied.
Future:	They will study.
Present Perfect:	They have studied.
Past Perfect:	They had studied.
Future Perfect:	They will have studied.

In addition, a parallel set of progressive forms exists, which indicates that the action is continuing:

Progressive Verb Tenses

Present:	They are studying.
Past:	They were studying.
Future:	They will be studying.
Present Perfect:	They have been studying.
Past Perfect:	They had been studying.
Future Perfect:	They will have been studying.

Finally, a parallel set of passive verb tenses also exists:

Passive Verb Tenses

Present:	The report is studied.
Past:	The report was studied.
Future:	The report will be studied.
Present Perfect:	The report has been studied.
Past Perfect:	The report had been studied.
Future Perfect:	The report will have been studied.

NOTE: As in the above sentences, passive verb sentences highlight the object or thing receiving the action, not the person or thing performing the action. The passive sentences above do not identify the person who is, was, or will be studying the report. In most cases, you should favor the active verb and avoid the passive. See ACTIVE/PASSIVE.

2. Vary your verb tenses to reflect the varying timing of events in your writing.

This rule contradicts what you may have learned in high school: "Don't mix your tenses." Actually, you can and should vary your tenses to reflect the often diverse time relationships of your subject:

> Yesterday, we analyzed *(past)* the samples for any traces of zinc ore. We found *(past)* none. Today, however, we were reexamining *(past progressive)* the sample when we found *(past)* two promising pieces of rock. They have *(present)* veins like zinc ore, although their color is *(present)* not quite right. Our report will therefore show *(future)* the potential presence of zinc.

Most writers choose their tenses unconsciously, but several basic conventions exist for selecting tenses in technical writing:

—Record in the past tense experiments and tests performed in the past:

> The second run produced flawed data because the heating unit failed. We failed to detect the failure until the run was almost over.

—Use the present tense for scientific facts and truths:

> Water freezes at 32°F, unless a chemical in the water changes its freezing point.
>
> Newton discovered that every action has an equal and opposite reaction.

—Use the present tense to discuss data within a published report:

> The slope of the temperature curve decreases sharply at 20 minutes. The figures in table 3–14 document this decrease.

—Shift from present to past tense as necessary to refer to research studies and prior papers. When you are discussing an author and his or her research, use the past tense:

> Jones (1976) studied a limited dose of the drug. He concluded that no harmful side effects occurred.

—When you are discussing different current theories, use the present tense:

> Jones (1976) argues that limited doses of the drug produce no harmful side effects. His data, however, are flawed because he failed to distinguish between the natural and synthetic versions of the drug.

Auxiliary Verbs

Auxiliary verbs are the most common verbs in English: *is, are, was, were, be, been, can, could, do, did, has, have, had, may, might, shall, should, will, would, must, ought to,* and *used to*. See STRONG VERBS.

Auxiliaries are crucial to many of the tenses presented above, but auxiliaries also can function by themselves as main sentence verbs:

> The tests are complete.
> He did the primary drawings.
> They have no budget.

See STRONG VERBS.

Verbs and Agreement

The verb should agree in number with its subject. So a plural subject requires a plural verb, and a singular subject requires a singular verb:

> The geologist has completed the tests. *(singular)*
> The geologists have completed the tests. *(plural)*
>
> A test was completed last week. *(singular)*
> Several tests were completed last week. *(plural)*
>
> The investigator has completed the medical tests. *(singular)*
> The investiagtors have completed the medical tests. *(plural)*

See AGREEMENT.

3. Ensure that your verbs agree with your subjects.

The only circumstance in which verbs change their forms to adjust to different numbers is in the third person forms of the present tense:

She works every day. *(singular)*
They work every day. *(plural)*

She is the candidate. *(singular)*
They are the candidates. *(plural)*

He has the answer. *(singular)*
They have the answer. *(plural)*

It is broken. *(singular)*
They are broken. *(plural)*

NOTE 1: Third person singular verbs have an *–s* ending, as in *works* above. The third person plural verbs have no *–s*, as in *work*. So the rule for verbs is the opposite of nouns: The forms with *–s* endings are the singular forms.

NOTE 2: The verb *be* is exceptional because it changes in the present tense to agree with different pronouns:

I am studying.
You are studying. *(singular)*
He, she, it is studying.

We are studying.
You are studying. *(plural)*
They are studying.

Subjunctive Verbs

Subjunctive verbs are special verb forms that signal recommendations or conditions contrary to fact. Centuries ago, subjunctives were very common verb forms, but today they are limited to the instances covered in the following two rules.

4. Use a subjunctive verb in *if* clauses to state a situation that is untrue, impossible, or highly unlikely:

If I were (*not* was) the candidate, I would not agree to a debate.

If it were (*not* was) raining, we couldn't conduct the experiment.

If I were (*not* was) you, I would change banks.

NOTE 1: The above sentences require *were* rather than normal *was*, which would appear to agree with the subjects. This use amounts to a historical survival, so it doesn't fit our modern expectations. (Actually, other verbs in *if* clauses are subjunctive, but only the *were/was* pattern looks or sounds exceptional.)

NOTE 2: Subjunctive verbs are especially useful when a writer has to make recommendations. The *if* clause presents the hypothetical condition, and then the main clause indicates what would, could, or might happen:

If the Accounting Department were reorganized, overall efficiency could increase by perhaps 50 percent.

If rainfall were 40 inches a year or more, most dirt roads would be impassable.

See *would* in WORD PROBLEMS.

NOTE 3: If the *if* clause states something that is possible or likely, then do not use a subjunctive:

If he leaves this job, he'll get $500 in severance pay.

If it was an error, and I suspect it was, then we'll have to pay you damages.

5. Use subjunctive verbs in sentences making strong recommendations or demands, or indicating necessity:

I recommend that the case be settled by Tuesday.

He demands that the money be refunded.

The court has resolved that the witness be found in contempt of court.

It is essential that he leave by noon.

They urge that she return the money.

They resolved that Dan write the termination letter.

NOTE 1: As the first three sentences show, if the verb in the *that* clause would normally be *am*, *is*, or *are*, then its subjunctive form is *be*.

NOTE 2: As the last three sentences show, if the verb in the *that* clause is normally a third person singular verb, then its subjunctive form does not take the usual *–s* ending.

Word Problems

Words are symbols representing persons, places, things, actions, qualities, characteristics, states of being, and abstract ideas. Words are the substance of language.

A major writing problem can occur if you use the wrong word for the context. Which of the words within parentheses in the following sentences are correct?

(You're, Your) supposed to call the manager before leaving.

The (principal, principle) problem with this alternative is its reliance on a second wash cycle between chemical baths.

The plan will (affect, effect) residents living south of the city.

In all three cases, the first word is correct. Using the wrong word might not prevent readers from understanding the sentences. But many readers might begin to wonder if you are literate. And if you seem illiterate, then even the facts in a document might appear questionable. So using correct words is important.

You should be aware, however, that using incorrect or imprecise words in legal documents (or in any document that can potentially be subject to interpretation in a courtroom) is very dangerous. Courts tend to support the word on the page, not what the author claims to have intended.

Sometimes, using the incorrect word causes serious confusion and potential misinterpretation:

Please ensure that the third production cycle is stopped continuously for quality-assurance testing.

Continuously means "uninterrupted or constant." *Continually* means "recurring often." If the production cycle is stopped continuously, then it never runs. If the cycle is stopped continually, it is stopped frequently, but not constantly. Some readers might not make this distinction. For the sake of clarity, the writer would be better off issuing a more specific order:

Please ensure that the third production cycle is stopped every half hour for quality-assurance testing.

Some words are clearly correct or incorrect. You can't write *it's* when you mean *its*, and vice versa. However, some distinctions are more difficult to make. Even the dictionaries do not offer clear-cut distinctions in some cases:

(Since, Because) we had not received payment, we decided to close the account.

They proposed to publish a (bimonthly, semimonthly) newsletter for all employees.

The proposal (that, which) you prepared has turned out to be very profitable.

In the first example sentence, both choices do include the notion of causality, but *since* primarily means "before now" or "from some time in the past until now." Only secondarily does *since* suggest causality. Thus, strict editors consider *because* less ambiguous.

In the second sentence, *bimonthly* is unclear because the ambiguous *bi–* can mean either twice within a time period (twice a month) or every other time period (every 2 months). *Semimonthly*, which has always meant twice a month, has suffered because of the confusion over *bimonthly*. Given the confusion between the two terms, you should avoid both terms and just say *twice a month* or *every 2 months*.

In the third sentence, both words are acceptable to many writers and editors. According to strict editors, however, *that* is more correct because the clause it introduces is essential to the sentence and cannot be removed; if the clause were nonessential, *which* would be the more correct word and the nonessential clause would be separated from the main clause by commas. Many educated writers and speakers violate this distinction, some using *that* and *which* almost interchangeably. So, insisting that *that* is the only correct choice for the third sentence is difficult.

Language and usage rules cannot be and never have been independent from the living language they describe. The spoken and written language that people use is always the final arbiter in disputes over word correctness. And a living language is always changing, so today's correct choice might be tomorrow's error. Good writers must remain alert to changes in a word's meaning and the acceptability of those changes.

The following list of problem word choices reflects some of the main choices facing today's writers. You should supplement this list and update it as necessary to remain current.

Word Problems

Guidance on individual word problems (often called usage problems) appears alphabetically under each word discussed. Thus, guidance on using *affect* versus *effect* appears initially under *affect/effect*. Readers looking up *effect* will find a cross-reference back to the discussion under *affect/effect*.

For information about words not in the following list, or if you want more detailed information, see *Webster's Dictionary of English Usage* (Springfield Massachusetts: Merriam-Webster, Inc. 1989).

Word Problems

accent/ascent/assent. *Accent* is both a verb meaning "to emphasize or stress" and a noun meaning "emphasis." *Ascent* (noun) means "a going up or rising movement." *Assent* means "to agree" or "an agreement":

> He accented the word *demotion.*
>
> His accent on the word *demotion* betrayed his true feelings.
>
> His ascent up the corporate ladder has been rapid.
>
> The trustee's assent is necessary before we sign the agreement.

accept/except. *Accept* is a verb meaning "to receive." *Except* is a preposition meaning "to the exclusion of":

> Did she accept your explanation for the overpayment?
>
> We completed everything except the two proposals for Acme, Inc.

A.D. and B.C. *A.D.* stands for the Latin *anno Domini* (in the year of the Lord). Using *A.D.* is simple—place it before a year and after a century:

> Pope Julius II, whose fertile partnership with Michelangelo produced many fine works of art, reigned from A.D. 1503 to 1513.
>
> Arabians borrowed coffee from the Abyssinians about the twelfth century A.D.

B.C. stands for "before Christ." Using *B.C.* is also simple—place it after the year and after the century:

> King Priam's Troy fell near the end of the Bronze Age, around 1200 B.C.
>
> The high point of ancient Greek civilization, the Periclean Age, was during the fifth century B.C.

Some writers, however, have qualms about using *A.D.* and *B.C.*, which are connected to the birth of Jesus Christ. They use other systems, such as *B.C.E.* (before the common era) for B.C. or C.E. (common era) for A.D. and place them after the date.

adapt/adept/adopt. *Adapt* means "to adjust to a situation." *Adept* means "skillful." *Adopt* means "to put into practice or to borrow":

> Within a week she adapted to the new billing procedure.
>
> She won the promotion because she was so adept at her job.
>
> Just last year we adopted a new method for maintaining inventory.

adjacent/contiguous/conterminous. *Adjacent* is the most general word, usually meaning "close to and nearby" and only sometimes "sharing the same boundary":

> Burger King is adjacent to the Cottonwood Mall.
>
> The adjacent lots were both owned by the same construction company.

Contiguous usually means "sharing the same boundary" even though it includes the notion of "adjacent" in most of its uses:

> The two mining claims turned out to be contiguous once the survey was completed; the owners had originally believed that a strip of state land separated the claims.

Conterminous (also *coterminous*) is the most specific of the terms and also the rarest. Its most distinctive meaning is "contained within one boundary" even though it also includes the senses of "sharing the same boundary" and quite rarely of being "adjacent." Its most distinctive use, however, is as follows:

> The conterminous United States includes only 48 of the 50 states.

These three words are a problem because they share a common meaning: "close to or nearby each other." At the same time, they each have more specific meanings, as illustrated. As with any confused words, writers should choose other phrasing if they wish to be as precise as possible:

> *better*
>
> Our two lots shared a common boundary on the north.
>
> *or with a different meaning*
>
> Our two lots fell entirely within the city boundary.
>
> *not*
>
> Our two lots were adjacent. (*neither* contiguous, *nor* conterminous)

adverse/averse. *Adverse* is an adjective meaning "unfavorable." *Averse* is an adjective meaning "having a dislike or a distaste for something." The two also contrast in how they are used in sentences. *Averse* appears only after the verb *be* or occasionally *feel*:

> We studied the adverse data before making our decision to plug and abandon the well.
>
> An adverse comment destroyed the negotiations.
>
> The President was averse to cutting the Defense budget.
>
> We felt averse to signing for such a large loan given the adverse economic forecasts.

advice/advise. *Advice* is a noun meaning "recommendations." *Advise* is a verb meaning "to make a recommendation":

> My advice was to meet with the client about the service problem.
>
> Did someone advise you to hire a lawyer?

affect/effect. *Affect* is usually a verb meaning "to change or influence." *Effect* is usually a noun meaning "a result or consequence":

> Temperature variations will affect the test results.
>
> The technician analyzed the effects of the new sample on the data.

Note: *Affect* can also be a noun meaning "the subjective impression of feeling or emotion," and *effect* can also be a verb meaning "to bring about or cause":

His strange affect (*noun*) caused the psychiatrist to sign the committal order.

The general manager's directive effected (*verb*) an immediate restructuring of all senior staff operations.

aid/aide. *Aid* is both a verb meaning "to help" or "assist" and a noun meaning "the act of helping someone." *Aid*, as a noun, also means "the person or group providing the assistance":

He aided us by carrying furniture out of the burning building. *(verb)*

U.S. aid included three emergency evacuation flights. *(noun)*

A nursing aid helped to care for my hospitalized mother. *(noun)*

Aide is a noun meaning "a person who acts as an assistant," especially in a military or governmental situation:

The admiral's aide was in charge of all of the admiral's appointments.

Aid and *aide* are frequently confused because of their similar meanings. Remember to limit *aide* to military or governmental contexts.

all right/alright. *All right* is the standard spelling; alright is an informal or nonstandard spelling and is not considered correct. Never use *alright*.

allusion/illusion/delusion. *Allusion* means "a reference to something." *Illusion* means "a mistaken impression." *Delusion* means "a false belief":

His allusion to Japanese management techniques was not well received, but he made his point.

Like the old magician's illusions of the floating lady, the bank manager created an illusion of solvency that fooled even seasoned investors.

The patient had delusions about being watched by the FBI.

alternate/alternative. Confusion comes from competing adjective uses. *Alternate* as an adjective means "occurring in turns" or "every other one." *Alternative* as an adjective means "allowing for a choice between two or more options":

Winners were chosen from alternate lines rather than from a single line.

The alternative candidate was a clear compromise between the two parties.

Sometimes these two adjective meanings almost merge, especially when an alternative plan is viewed as a plan that replaces another:

An alternate/alternative plan provided for supplementary bank financing. *(Either is correct.)*

Note: Strict editors attempt to restrict *alternative* (in its noun and adjective uses) to only two options:

Life is the alternative to death.

Actual (correct) usage, however, has broadened *alternative* to include any number of options:

The planning commission analyzed five alternative sites. (*or* The planning commission analyzed five alternatives.)

altogether/all together. *Altogether* means "completely or entirely." *All together* means "in a group":

We had altogether too much trouble getting a simple answer to our question.

The spare parts lists are all together now and can be combined.

a.m./p.m./m. The abbreviations *a.m.*, *p.m.*, and *m.* sometimes appear in printed text with small caps: A.M., P.M., and M. In most word processing text, lowercase versions are preferred.

Use *a.m.* for times after *midnight* and before *noon*:

12:01 a.m.	(1 minute after midnight)
6:00 a.m.	(early morning)
11:59 a.m.	(1 minute before noon)

Use *p.m.* for times after *noon* and until *midnight*:

12:01 p.m.	(1 minute after noon)
6:00 p.m.	(early evening)
12:00 p.m.	(midnight)

Noon remains a problem. Some guides continue to list *12 m.* as *noon*. However, using *m.* (Latin for *meridies*) for *noon* with readers who do not know this abbreviation is unwise. They might well read *m.* as *midnight*. *Noon* and *midnight* will never be misunderstood; use them in place of *12 m*, *12:00 a.m.*, and *12:00 p.m.*:

The conference will be at noon on July 25.

among/between. *Among* refers to more than two choices. *Between* usually refers to two choices only, but it can refer to more than two:

We had difficulty deciding among the many options—over 200 colors.

The contracting officer has eliminated three bidders, so the Source Selection Authority must choose between us and Universal Data.

Strict editors do try to restrict *between* to two choices only. But occasionally, you can use *between* instead of *among*, especially where *among* would not sound right:

The research group analyzed the differences between the five alternatives.

See Prepositions.

and/or, or, and. Avoid using *and/or*. This term is usually difficult and sometimes impossible to read with surety. See Slashes.

In one court case, three judges ruled three different ways as to the meaning of *and/or*. Courts in different cases have ruled that *and* means *or*, and that *or* means *and*. Be careful when using these three troublesome words. See Conjunctions.

Try to avoid using *and/or*. If you do use it, make sure that the situation you describe has at least three possibilities:

> The road will be made of sand and/or gravel.

This road could be (1) an all-sand road (2) an all-gravel road, or (3) a road made from both sand and gravel.

You can make reading even more difficult by adding other possibilities:

> The road will be made from asphalt, concrete, sand, *and/or* gravel.

The road now has far more construction combinations. Rather than using the shorthand *and/or*, change the sentence to explain to your readers your exact meaning:

> The road will be made from asphalt, concrete, sand, and gravel. *(Implies a combination of all four—with the percent composition variable.)*

> The road will have an asphalt and concrete surface, with fill being a mixture of sand and gravel. *(More specific and less ambiguous.)*

Or has two meanings: inclusive and exclusive. For the most precise use of this word, use only the **exclusive** meaning:

> Use the blue pen or the black pen.

> Use either the blue pen or the black pen.

Do not use *or* in its **inclusive** sense:

> Limestone or calcium carbonate is used to neutralize the effects of acid rain.

Readers who do not know that limestone is calcium carbonate might think that two substances can be used. They cannot tell whether *or* is being used in its inclusive or its exclusive sense. See CONJUNCTIONS.

Readers who know that limestone is calcium carbonate know that *or* is being used in its inclusive sense, but these knowledgeable readers do not need the reminder that limestone is calcium carbonate.

Write the sentence, eliminating *or* and using parentheses:

> Limestone (calcium carbonate) is used to neutralize the effects of acid rain.

See PARENTHESES.

ante–/anti–. *Ante–* is a prefix meaning "comes before" as in *antebellum* ("coming before a war") or *antedate* ("coming before a specified date"). *Anti–* is also a prefix, but with the meaning of "in opposition or against" as in *antiacademic* or *antismoking.* Occasionally *anti* appears as a word, not as a prefix:

> He turned out to be anti to our proposal.

anyone/any one. *Anyone* means "any person." *Any one* means "a specific person or object":

> Anyone who wants a copy of the Camdus report should receive one.

> We were supposed to eliminate any one of the potential mine sites.

appraise/apprise. *Appraise* is a verb meaning "to give the value or worth of something." *Apprise* is a verb meaning "to tell or notify":

> Before the auction, the auto was appraised for over $100,000.

> The defendant's lawyer apprised the court that the defendant had been declared bankrupt a year earlier.

as regards. See in regard to/as regards/in regards to.

assure/insure/ensure. All three words mean "to make certain or to guarantee." *Assure* is limited to references with people:

> The doctor assured him that the growth was nonmalignant.

Insure is used in discussing financial guarantees:

> His life was insured for $150,000.

> The company failed to insure the leased automobile.

Much of the time, *insure* and *ensure* are confusingly interchangeable. For example, in one Federal document in the late 1970s, *insure* and *ensure* both appeared numerous times with the same meaning; chance seems to have guided which spelling appeared in which sentence. To try to end such confusion, the *U.S. Government Printing Office Style Manual* (March 1984) defines insure as "protect" and *ensure* as "guarantee":

> Your life insurance will insure (protect) your family from financial ruin.

> To ensure (guarantee) that the drill bit does not overheat as it penetrates through the rock layer, keep the drilling fluid flowing at a maximum rate.

Reality suggests that despite such tidy distinctions, *insure* and *ensure* will continue to be confused. Try, however, to be consistent within a single document by choosing *insure* or *ensure* and then sticking with your choice.

bad/badly. *Bad* is primarily used as an adjective, but in informal English it is also an adverb. *Badly* is an adverb. Now, however, *badly* has begun to function in a few sentences the same way *bad* has. This overlap is a problem when you use the verbs of the senses (*feel, look, smell,* etc.):

> ***Originally and currently acceptable***

> Harold felt bad *(adjective)* all day from the blow on his head.

> The machine worked badly *(adverb)* despite the overhaul.

> *or*

> The machine worked bad *(adverb)* despite the overhaul.

> ***Currently acceptable***

> Harold felt badly all day from the blow on his head.

Word Problems

Note: Only the verb *feel* allows for either *bad* or *badly*, as in the preceding sentence. Other confusions between the adjective *(bad)* and the adverb *(badly)* are not acceptable, especially in written English:

Not acceptable

He looked badly after the bachelor party. (*correct form:* bad)

The lab smelled badly after the drainage samples arrived. (*correct form:* bad)

B.C. See A.D. and B.C.

between. See among/between.

biannually/biennially. *Biannually* means "two times a year." *Biennially* means "every 2 years":

Because we meet biannually, we will have 10 meetings over the next 5 years.

Our long-range planning committee meets biennially—on even-numbered years.

bimonthly/semimonthly. *Bimonthly* can mean either "every 2 months" or "twice a month." *Semimonthly* means "twice a month." Because of the potential confusion surrounding *bimonthly*, you should avoid the word and write *every 2 months* or *twice a month. Semimonthly* has only one meaning and should not be confusing. Still, it has suffered from the ambiguity of *bimonthly*:

Our bimonthly newsletter appears in January, March, May, July, September, and November. (*better:* Our newsletter appears every 2 months, beginning in January.)

We proposed semimonthly meetings of the legislative committee. (*better:* We proposed meetings twice a month of the legislative committee.)

can/may. *Can* means (1) "ability," (2) "permission," and (3) "theoretical possibility." *May* means (1) "permission" and (2) "possibility":

Ability

She can speak German, but she can't write it very well.

Permission

Can I help you with your project?

May I help you with your project?

Note: *May* sounds more formal than *can*, so if you wish to sound formal, use *may*.

Possibility

George can make mistakes if he's rushed. (*or* may)

The project can be stopped if necessary. (*or* may)

The trail may be blocked, but we won't know until later.

Your objection may be reasonable, but we still don't agree.

capital/capitol. *Capital* means "the central city or site of government," "invested money," and "an uppercase letter." *Capitol* means "the main government building":

Paris is the capital of France.

The necessary capital for such an elegant restaurant is $1.5 million, but I doubt that investors will put up that much.

THIS SENTENCE IS WRITTEN IN CAPITALS.

The legislature authorized a complete renovation of the capitol dome.

carat/caret/karat. *Carat* means "the weight of a gem." *Caret* means "a mark showing an insertion." *Karat* means "a unit for the purity of gold":

The ring had a 2.2-carat diamond.

I've used a caret to indicate where to insert the new sentence.

The ring is made of 18-karat gold.

cite/sight/site. *Cite* means "to quote." *Sight* means "vision." *Site* means "a location":

During the trial, our attorney cited earlier testimony.

Most of the tunnel was out of sight, so we could not estimate the extent of the damage.

The contractor prepared the site by bulldozing all the brush off to the side.

complement/compliment. *Complement* means "completing or supplementing something." *Compliment* means "an expression of praise":

The report's recommendations complement those made by the executive committee last year.

The manager passed on a compliment from the vice president, who was impressed with the proposal team's efforts.

comprise/compose. Strict editors carefully distinguish between these two words—that is, *comprise* means "to include or contain" and *compose* means "to make up from many parts":

The U.S. Congress comprises the House of Representatives and the Senate.

The House of Representatives and the Senate compose the U.S. Congress.

Such sentences, especially those with *comprise*, are beginning to sound stiff and overly formal, and passive alternatives are more and more common, although not accepted by all editors:

The U.S. Congress is comprised of the House of Representatives and the Senate.

As with other disputed word uses, choose an alternate version whenever possible:

The House of Representatives and the Senate constitute the U.S. Congress.

concerning/worrying. Avoid using *concerning* as a synonym for *worrying* or *alarming*: "We find it concerning that the Web site has misled clients." Use *concerning* as a synonym for *about*:

We contacted you in July concerning your overdue payment.

We find it worrying that the website has misled clients.

contiguous. See adjacent/contiguous/conterminous.

continual/continuous. *Continual* means "intermittent, but frequently repeated." *Continuous* means "without interruption":

Because the pipes are so old, continual leaks appear despite our repair efforts.

Because of a short in the wiring, the horn sounded continuously for 10 minutes.

contractions. Do not use contractions (*don't, couldn't*) in formal documents. However, when you want to create a personal tone, do use contractions in informal documents such as letters and memos:

John, don't forget that you must finish this work by Tuesday; otherwise, Bill can't get your figures into his report, which must be done for Vice President Stern by Thursday.

See Style and Contractions.

council/counsel/consul. *Council* means "a group of people." *Counsel* means "to advise" (verb), "advice" (noun) or "an attorney." *Consul* means "a foreign representative":

The safety council passed a motion to ban smoking in shaft elevators.

The consultant counseled us in ways to improve our management of ID team efforts.

His counsel was to rewrite the proposal.

MOGO's counsel made the opening statement in the hearing.

The French consul helped us obtain an import license.

councilor/counselor. *Councilor* means "a member of council." *Counselor* is "an advisor or lawyer":

The councilors decided to table the motion until the next meeting.

Our staff medical counselor has a PhD in clinical psychology.

could. See would/probably would/could/might/should.

credible/creditable/credulous. *Credible* means "believable." *Creditable* means "praiseworthy." *Credulous* means "gullible":

His revised report was more credible, chiefly because the manpower estimates were scaled to match the price.

In spite of some short cuts, the proposal team wrote a creditable proposal; in fact, they won the contract.

He was so credulous that anyone could fool him.

data. *Data* (the plural of *datum*) is now often used as both the singular and plural forms of the word. In some technical and scientific writing, however, *data* is still traditionally plural only. If the convention in your discipline or organization is to use *data* as a plural, then be sure that your sentences reflect correct agreement of subject and verb:

Our production data are being examined by the EPA because a citizen complained about excessive emissions from our plant.

The data have been difficult to analyze, chiefly because of sloppy recordkeeping.

These sentences may sound strange or awkward to many readers. Consequently, some technical writers avoid phrasing that calls attention to the plural meaning of *data*. The two sentences above, for instance, could read as follows:

EPA is examining our production data because a citizen complained about excessive emissions from our plant.

We found the data difficult to analyze, chiefly because of the sloppy recordkeeping.

delusion. See allusion/illusion/delusion.

different from/different than. These two forms can and should be used interchangeably. Strict editors, however, may insist that *different from* is somehow better than *different than*. Actually, well-educated writers and many editors have used both forms for well over 300 years. The argument over these two forms is an example of a preference being mistaken for a rule. In fact, no clear distinction between the two forms has ever existed:

The results were far different from those we expected. (*or* different than)

The study is different than we had been led to expect. (*or* different from the one we had been led to expect)

disburse/disperse. *Disburse* is the verb meaning "to pay out." *Disperse* is the verb meaning "to scatter":

The payroll clerk disburses the petty cash funds as needed.

The reserved top soil was dispersed over the site after the project was completed.

discreet/discrete. *Discreet* means "tactful or prudent." *Discrete* means "separate or individual":

The counselor was so discreet that no one learned we had been meeting with him.

The testing included three discrete samples of the ore body.

disinterested/uninterested. Originally, *disinterested* meant "neutral or unbiased," and *uninterested* meant "without interest." Careful writers and editors still maintain this distinction:

The judge was appointed because he was clearly disinterested in the dispute.

The President was so uninterested in the problem that he failed to act.

effect. See affect/effect.

e.g./i.e. Both *e.g.* and *i.e.* are Latin abbreviations. Most of the time avoid them and use the equivalent English phrases.

E.g. is the abbreviation for *exempli gratia*, meaning "free examples." *I.e.* is the abbreviation for *id est*, meaning "that is." The two have similar, but still slightly different uses:

Many appliances (e.g., clothes washers, dryers, and dishwashers) are so common that they have been standard in all new homes. *(The list consists of several free examples.)*

Most first-time house buyers should be aware of this one principle, i.e., let the buyer beware. (I.e. *introduces a restatement of an idea, not a list of examples.)*

In the preceding examples, *for example* could replace *e.g.* in the first sentence. A simple *that is* could replace *i.e.* in the second sentence. Use the English versions unless you are writing for a publication that prefers *e.g.* and *i.e.*

Both abbreviations use periods in the U.S. style: *e.g.* and *i.e.* In British English, the periods usually are not used, with no space between the letters: *eg* and *ie*. Notice also that both abbreviations are enclosed in commas or by another mark of punctuation. See British English.

elapse/lapse. *Elapse* is a verb meaning "to pass by or to slip"; it usually refers to time. *Lapse* is a verb with many meanings, most derived from the sense of "to drift, to discontinue, or to terminate":

> Two weeks elapsed before we heard from the Internal Revenue Service.
>
> The speaker lapsed into silence after the embarrassing question.
>
> Our contract lapsed before we could negotiate its renewal.

emigrate/immigrate and **emigration/immigration.** *Emigrate* (verb) means "to leave one's home or residence—to go out." *Immigrate* (verb) is the opposite: "to come into a town or country."

> Peter and his family emigrated from Poland over 10 years ago.
>
> When he first immigrated into the United Sates, he had difficulty finding work as a doctor because of his poor English skills.

The similar distinction holds for the two parallel nouns: *emigration/ immigration*. The only difference is in the point of view:

> Over 10,000 emigrants left Haiti last year for the United States.
>
> The U.S. policy on Haitian immigrants is becoming increasingly restrictive.

eminent/imminent. *Eminent* means "outstanding or prestigious." *Imminent* means "impending or soon to happen":

> Only eminent researchers will win Nobel Prizes.
>
> The dam's collapse was imminent, so we evacuated downstream communities.

ensure. See assure/insure/ensure.

envelop/envelope. *Envelop* is the verb meaning "to enclose or to encase." *Envelope* is the noun meaning "something that contains or encloses":

> They proposed to envelop the storage tank with the fire-retardant foam.
>
> We placed in the envelope both the final report and the backup surveys.

etc. Avoid using *etc.* in most of your writing. *Etc.* is the abbreviation for *et cetera*, which is Latin for "other items or persons of the same sort":

> She wrote letters to many governmental figures: the President, Secretary of Defense, Army Chief of Staff, her senator, etc.

As in the preceding sentence, we can only guess about how many people received a letter from her. The use of *etc.* is usually an evasion of the responsibility for listing all the exact facts or details. So avoid using *etc.* in most business or technical writing.

except. See accept/except.

extemporaneous/impromptu. These two words are widely confused, and even some dictionaries have definitions of them that overlap. *Extemporaneous*, especially among speech teachers, means "a speech that is carefully planned and practiced, but not memorized." So its delivery is an extemporaneous action. The common meaning, however, is closer to *impromptu*, as defined below.

Impromptu means that "something that arises without notice or preparation." So an impromptu speech is made without practice or preparation.

As with many confused words, either avoid the two words or use them in contexts where their meanings will be clear:

> During his tour of the plant, the President unexpectedly asked one of the workers to answer impromptu questions from reporters.
>
> President Clinton is an excellent extemporaneous speaker, as illustrated when his teleprompter broke and he didn't have a printed copy of his speech. He still delivered a good speech.

farther/further. As far back as Shakespeare these words have been confused, and in some contexts they are clearly interchangeable—e.g., *farther/further from the truth.*

Strict editors still maintain that *farther* should be restricted to senses involving distance, while *further* includes other senses:

> The assembly site was farther from the testing area than we wished.
>
> A further consideration was the inflation during those years.

fewer/less. *Fewer*, the comparative form of *few*, usually refers to things that can be counted. *Less*, one comparative form of *little*, refers to mass items, such as sugar or salt, which cannot be counted, and to abstractions:

> We analyzed fewer well sites than the government wanted us to analyze.
>
> Less sodium chloride in the water meant that we had fewer problems with corrosion.
>
> Fewer teachers, less education.

See Nouns and Adjectives.

flammable/inflammable. Avoid using *inflammable* because it has two contradictory meanings: "able to burn" and "not able to burn." Dictionaries usually list the first of these two meanings as the proper one, but the other meaning is too common to ignore. These two meanings arise from two different interpretations of the prefix *in–*; one means "not" while the other signals a "presence or possible" (as in "a flame is present"). So never use *inflammable*, but use *not flammable* or *nonflammable* or other phrases:

> The pajamas are nonflammable at temperatures up to 240 °F. (*not* "are inflammable")

> These treated roofing shingles will not catch fire even if glowing coals land on them from a nearby fire. (*or* nonflammable)

Flammable means "able to burn," or "burns easily," so using it is no problem:

> Even the fumes from a gasoline can be flammable.

forward/foreword. *Forward* is an adjective and an adverb, both meaning "at or near the front." *Foreword* is the noun meaning "the introduction to a book":

> The hopper moves forward when the drying phase is nearly finished.

> The foreword to the book was two pages long.

he/she, s/he, (s)he. Avoid these created new words. If possible, recast your sentence so that it is plural:

> *this*

> All engineers must bring their reports to the meeting. *(plural)*

> All new employees should complete the health forms before beginning work. If they have not done so, their insurance will not be in effect.

> *not this*

> Every engineer must bring his/her report to the meeting. *(singular)*

> Each new employee must complete the health forms before beginning work. If (s)he has not done so, his/her insurance will not be in effect.

These newly coined words, which have not been widely accepted, are attempts to create a singular pronoun other than the word it because it in many sentences is too odd and too impersonal to refer to people. See BIAS-FREE LANGUAGE and AGREEMENT.

i.e. See e.g./i.e.

illusion. See allusion/illusion/delusion.

immigrate. See emigrate/immigrate.

imminent. See eminent/imminent.

imply/infer. *Imply* means "to suggest or hint." *Infer* means "to draw a conclusion or to deduce":

> The report implies that the break-even point may be difficult to reach, but it fails to give supporting data.

> Based on our comments, she inferred that we would not give our wholehearted support to her project.

impromptu. See extemporaneous/impromptu.

infinitives. See split infinitives.

inflammable. See flammable/inflammable.

in regard to/as regards/in regards to. The first two forms are acceptable. By convention, the third form (*in regards to*) is unacceptable. Do not use *in regards to*.

> In regard to your report, our firm is still busy analyzing it.

> As regards your second question, we have taken all of the steps necessary to acquire mineral rights on the Bruneau lease.

A better choice in most sentences is to use the shorter and simpler *regarding*:

> Regarding your second question, we have taken all of the steps necessary to acquire mineral rights on the Bruneau lease.

insure. See assure/insure/ensure.

irregardless/regardless. *Irregardless* is an unacceptable version of *regardless*. Do not use *irregardless* in either speech or writing:

> We decided to fund the project regardless of the cash flow problems we were having.

its/it's. *Its* is the possessive pronoun. *It's* is the contraction for *it is*:

> Its last section was unclear and probably inaccurate.

> It's time for the annual turnaround maintenance check.

See PRONOUNS.

karat. See carat/caret/karat.

lapse. See elapse/lapse.

later/latter. *Later*, the comparative form of *late*, means "coming after something else." *Latter* is an adjective meaning "the second of two objects or persons":

> Later in the evening a fire broke out.

> In our analysis of the Lankford and Nipon sites, we finally decided that the latter site was preferable.

NOTE: *Latter* (and its parallel *former*) are sometimes confusing, so rewrite to avoid them:

> We f inally decided that the Nipon site was preferable to the Lankford site.

lay/laid/laid. These three words are the principal parts of the verb *lay*. The verb itself means "to put or to place." It must have an object:

> The contractor promised to lay the sod before the fall rains began. (*object:* sod)

> The manager laid his plan before his colleagues. (*object:* plan)

> Our recent talks with the Russians have laid the groundwork for control of nuclear weapons in space. (*object:* groundwork)

less. See fewer/less.

lie/lay/lain. These three forms are the principal parts of the verb *lie*. *Lie* means "to rest or recline." In contrast with *lay*, *lie* cannot have an object:

The main plant entrance lies south of the personnel building.

The new access road lay on the bench above the floodplain.

That supply has lain there for over a decade.

loose/lose. *Loose* is an adjective meaning "unrestrained" or "insecurely fastened." *Lose* is a verb meaning "to suffer loss":

The mechanical feeder arm rattles because of loose stabilizer rings.

We can't afford to lose the Pristin deal if we're going to meet this year's revenue goal.

marital/martial. These two are confused because they look similar and can, on occasion, be mispronounced.

Marital refers to "marriage or its conditions." *Martial* refers to "war or warriors." So their uses are distinctly separate:

The marital agreement established a not-to-exceed alimony amount.

The martial sound of drums and bagpipes encouraged the Scottish soldiers to fight and, supposedly, frightened their enemies.

may. See can/may.

maybe/may be. *Maybe* is the adverb meaning "perhaps." *May be* is a verb form meaning "possibility":

Maybe we should analyze the impacts before going ahead with the project.

Whatever happens may be beyond our control, especially if inflation is unchecked.

might. See would/probably would/could/might/should.

or. See and/or, or, and.

p.m. See a.m./p.m./m.

practical/practicable. These two words mean much the same thing, and dictionaries disagree on their distinctions. Given this confusion, writers should stay with the common form *practical* and avoid *practicable.*

Strict editors do maintain that the difference between *practical* and *practicable* is similar to the difference between *useful* and *possible.* Practical means "not theoretical; useful, proven through practice." *Practicable* means "capable of being practiced or put into action; feasible":

Despite the uniqueness of the problem, the contractor developed a practical method for shoring up the foundation.

Although technically practicable, the solution was not practical because it would have put us over budget.

The last sentence says that we were technically capable of achieving the solution but that this solution was not feasible because of financial constraints.

Similarly, building a house on top of Mt. St. Helens is probably possible (it is practicable), but doing so is not feasible (it is not practical) for obvious reasons.

precede/proceed. *Precede* means "to go ahead of or in front of," so it implies an order or sequence of some kind. *Proceed* simply means "start" or "go on." The two are probably confused because they have similar spellings.

A dentist's deadening the nerve should precede any drilling or other painful processes.

The guide proceeded carefully along the bank because the wet soil seemed unstable.

Occasionally *proceeding* appears in a sentence where *preceding* is the correct form:

The preceding speaker ran over her allotted time, so we got off schedule. (*not* the proceeding speaker)

precedence/precedents. *Precedence* is the noun meaning "an established priority." *Precedents* is the plural form of the noun meaning "an example or instance, as in a legal case":

The Robbins account should take precedence over the Jackson account; after all, Robbins gives us over 50 percent of our business.

The Brown decision was the precedent for many later decisions involving racial issues and education.

principal/principle. *Principal* is a noun or adjective meaning "main or chief." *Principle* is a noun meaning "belief, moral standard, or law governing the operation of something":

The principal technical problems we faced were simply beyond current technologies.

The principles of electricity explain the voltage drop in lines.

If we act according to our principles, we will not allow the transaction to proceed.

NOTE: Some writers are confused by *principal* because it also means "the head of a school" and "the money borrowed from a bank." These are noun forms of a word that used to be only an adjective. The noun forms of *principal* come from noun phrases: *the principal teacher and the principal amount.* Over time, the nouns *teacher* and *amount* were dropped, and *principal* assumed the full meaning of the original phrases. Now, *principal* is a noun as well as an adjective.

probably would. See would/probably would/could/might/should.

raise/raised/raised. These three forms are the principal parts of the verb *raise. Raise* means "to move (something) upward." *Raise* always requires an object:

They will raise the funds by January 1, 2010. (*object:* funds)

We raised the water level some 20 feet to accommodate the changing use patterns. (*object:* water level)

They had raised the amount to cover the travel costs. (*object:* amount)

rational/rationale. *Rational* is an adjective meaning "logical or well thought out." *Rationale* is a noun meaning "the supporting facts or proof." Confusion arises because the two words are so similar in their spelling.

> The business team followed a rational seven-step process for evaluating each potential site for the new plant.
>
> His rationale for selling the franchise rested more with his time constraints than with financial considerations.

regardless. See irregardless/regardless.

respectfully/respectively. *Respectfully* means "with deference and courtesy." *Respectively* means "in the sequence named":

> Our representatives were not treated very respectfully.
>
> According to production data for March, May, July, and September, the number of cases were, respectively, 868, 799, 589, and 803.

NOTE: *Respectively* often makes sentences difficult to interpret, so avoid *respectively* whenever you can.

rise/rose/risen. These three forms are the principal parts of the verb *rise. Rise* means "to stand up or move upward." *Rise* does not take an object:

> The balloon rises (*or* will rise) once the air heats up.
>
> Because the water rose, we had to evacuate the ground floor.
>
> The moisture level in the gas has risen substantially over the last week.

said. The word *said* often becomes a shorthand term for a document or item previously mentioned:

> We have examined said plans and can find no provisions for the clay soils on the site.

Such uses of *said* are not appropriate in normal business and technical writing. Only in legal writing (and maybe not even there) should writers ever use *said* in this way. The above sentence could be rewritten as follows:

> We have examined the plans and can find no provisions for the clay soils on the site. *(Readers will usually know from the context what plans the writer is referring to.)*

semimonthly. See bimonthly/semimonthly.

set/set/set. These three forms are the principal parts of the verb *set. Set* means "to put or to place (something)." *Set* must have an object:

> They set the surveying equipment in the back of the truck. (*object:* equipment)
>
> Yesterday we set up the derrick so that drilling could start at the beginning of today's shift. (*object:* derrick)
>
> After we had set the flow, we began to monitor fluctuations from changes in pressure. (*object:* flow)

shall/will. Use *will* for the simple future with all of the personal pronouns:

> I/We will leave.
>
> You will leave.
>
> He/She/It/They will leave.

Shall is rarely used for simple future, at least in American English, but it does retain some sense of extra obligation or force, as in legal contexts:

> The vendor shall provide 24-hour security at the site.

Shall (or *should*) is also used for some questions:

> Shall/Should I stop by your office tomorrow?
>
> Shall/Should I sign the document now?

Neither *will* nor *would* can replace *shall/should* in these questions.

Some grammarians, beginning in the seventeenth century, formulated a supposed rule for *shall* and *will*: For simple future, use *shall* with *I* and *we*; *will* with *he, she, it,* and *they*. For obligation and permission, reverse the choices: *will* with *I* and *we; shall* with *he, she, it,* and *they*.

This rule was not accurate in the seventeenth century, and it has never been true of actual English sentences. British English does use *shall* somewhat more frequently for the simple future than American English, but even in England, the seventeenth-century rule is not consistently followed.

she. See he/she, s/he, (s)he.

should. See would/probably would/could/might/should.

sic. *Sic*, from the Latin, means "thus." Use *sic* when you are quoting something and want to show that, yes, I have copied this ungrammatical or odd language as it was originally written or spoken:

> The computer experts in their memorandum stated: "We took for granite [sic] that everyone knew about their [sic] weekend changes to the system."
>
> The stationary [sic] bus was coming from the opposite direction.

Do not use *sic* to embarrass someone by highlighting harmless mistakes, which we all make.

sight/site. See cite/sight/site.

sit/sat/sat. These three forms are the principal parts of the verb *sit. Sit* means "to rest or to recline." *Sit* does not take an object:

> The well sits at the foot of a steep cliff.
>
> The committee sat through the long session with very few complaints.
>
> The oil drums must have sat on the loading dock all weekend.

split infinitives. Many people who often can't even recognize a split infinitive still believe a split infinitive is a grammatical crime. Perhaps because of its memorable name, a split infinitive has become part of the folklore about what good writers should avoid.

Actually, split infinitives have been acceptable in English for hundreds of years. Only in fairly recent times have editors even worried whether sentences like the following should be accepted because they contain split infinitives:

> The company's goal was **to rapidly retire** its investment debt before moving into new markets.
>
> **To totally avoid** splitting infinitives, the writer decided to eliminate all infinitives from the document.

In the first of these examples, rephrasing to move *rapidly* before or after the infinitive *to retire* is easy: *rapidly to retire* or *to retire rapidly*. Strict editors would choose one of these rephrasings and thus eliminate the split infinitive, but the sentence is actually correct with or without the split infinitive.

In the second example, however, moving *totally* behind the infinitive to *avoid* changes the sentence:

> To avoid totally splitting infinitives, the writer decided to eliminate all infinitives from the document. *(Is the splitting being done totally or does* totally *continue to modify* avoid?)

In this second example, a writer would have to either retain the split infinitive or recast the entire sentence.

To conclude, split infinitives are not worth worrying about. Leave them in your document when the context seems to require a split infinitive.

stationary/stationery. *Stationary* is an adjective meaning "fixed in one spot, unmoving." *Stationery* is a noun meaning "paper for writing on":

> Because the boiler was bolted to the floor, it remained stationary despite the vibration.
>
> The new letterhead on our stationery made our company seem more up to date.

than/then. *Than* is used in comparisons. *Then* is an adverb meaning "at that time":

> George's report was shorter than Mary's.
>
> We then decided to analyze the trace minerals in the water samples.

that/which. After centuries of competition (and confusion), the two words continue to be often interchangeable:

> The connecting rod that failed delayed us for two days.
>
> *or*
>
> The connecting rod which failed delayed us for two days.

In both of these correct sentences, the clauses *that failed* and *which failed* identify which specific rod—that is, the rod that failed, not the other rods. So these clauses are both identifying (restricting) the meaning. As in these two sentences, the restrictive clause is not set off by commas. When *that* and *which* introduce restrictive clauses, the choice between them is merely stylistic—that is, choose the one that sounds the best.

Which is the proper choice, however, for nonrestrictive clauses:

> The Evans report, which took us several months to finish, is beginning to attract attention.
>
> The U.S. Senate, which many consider the most exclusive club in the world, does follow some quaint rules of decorum.

In both of these examples, the *which* clauses provide additional but unnecessary information about the Evans report and the Senate. Both *which* clauses are thus nonrestrictive because neither helps to identify the report or the Senate. An informal test is if the *which* clauses were deleted, the basic meaning of each sentence would not change. Note, also, that such nonrestrictive clauses are enclosed with commas. See Commas.

Strict editors argue that because *which* is clearly the choice to introduce nonrestrictive clauses, *that* should be used for all restrictive clauses. The tidy distinction is not true of spoken English, and many careful writers continue to use *that* and *which* interchangeably for restrictive clauses.

This is another case where a supposed rule (actually part of a rule) is best ignored. To summarize, use either *that* or *which* to introduce restrictive clauses (with no enclosing commas). Use *which* for nonrestrictive clauses (with enclosing commas).

their/there/they're. *Their* is a possessive pronoun. *There* is an adverb meaning "at that place." *They're* is the contraction for *they are*:

> The engineers turned in their reports for printing.
>
> The well site was there along the base of the plateau.
>
> They're likely to object if we try to include those extra expenses in the invoice.

to/too/two. *To* is the preposition or infinitive marker. *Too* is both an adverb meaning "excessively" and a conjunctive adverb meaning "also." *Two* is the numeral:

> The proposal went to the Department of the Interior for approval.
>
> She wants to revise the proposal.
>
> The design was too costly considering our budget. (too = *excessively*)
>
> The issue, too, was that technology is only now beginning to cope with these low-temperature problems. (too = *also*)

toward/towards. *Toward* and *towards* are merely different forms of the same word. *Toward* is the preferred form in American English. In British English, *towards* is more common than *toward*.

uninterested. See disinterested/uninterested.

which. See that/which.

while. *While* is best used as only a time word to show simultaneity—"at the same time as another event":

> While the wash water flows over the screening plates, measure the water's temperature.
>
> Check the level of the car's transmission fluid while the engine is running.

In other instances, *while* can mean "though," "although," "even though," "but," or "and." These meanings are particularly common in spoken English. In written English, however, replace *while* with its equivalent word:

> *this*
>
> Although windmills are economical, they are too often destroyed by severe storms, and in calm weather, they produce no electrical power.
>
> At some places the coal layer is 4 feet wide, but at other places it narrows to 10 inches.
>
> *not this*
>
> While windmills are economical, they are too often destroyed by severe storms, while in calm weather, they produce no electrical power.
>
> At some places the coal layer is 4 feet wide, while at other places it narrows to 10 inches.

who/whom. See the discussion of interrogative and relative pronouns in Pronouns.

who's/whose. *Who's* is the contraction for *who is. Whose* is the possessive form of the pronoun *who*:

> Who's the contractor for the site preparation work?
>
> She was the supervisor whose workers had all that overtime.

will. See shall/will.

would/probably would/could/might/should. Today these words are usually used as one form of the subjunctive mood, stating future probability. Long ago, *would* and *could* were used as past tense verbs.

These words, except for *would*, never convey an exact probability. The context in which these words are used gives the reader a sense of the probability:

- Would = certain, 100 percent

  If we were to drill the well, we would get 100 barrels of oil per day.

- Probably would = very high, 80 percent

  If we were to drill the well, we probably would get 100 barrels of oil per day.

- Could = reasonably high, 50 percent

  If we were to drill the well, we could get 100 barrels of oil per day.

- Might = moderate, 30 percent

  If we were to drill the well, we might get 100 barrels of oil per day.

Do not use *should* for "high probability." Use *should* in its principal meaning—"an ethical or moral obligation":

> You should not allow your graph to have an appearance of precision greater than your data allow.
>
> This final part of the president's speech should not be quoted out of context.

See the discussion of subjunctives in Verbs.

your/you're. *Your* is the possessive form of *you. You're* is the contraction for *you* are:

> Your letter arrived too late for us to adjust the original invoice.
>
> If you're interested, we can survey the production history of that sand over the last decade or so.

See Pronouns.

Word Processing

Word processing is the writing, editing, sharing, and printing of documents using computers, networks, and printers. Word processing allows the writer or editor to experiment freely with different formats and to revise and edit the text easily.

Word Processing

1. Save your work on a backup disk.
2. Save your work using intelligible file names.
3. Decide on a format for your document before you begin writing.
4. Plan your document before you begin to write (input).
5. Brainstorm your content.
6. Organize your content.
7. Use word-processing techniques to simplify drafting your document.
8. Revise (edit) efficiently by setting priorities and sticking to them.
9. Use word-processing techniques to help you proofread your document.

Word-processing software includes many helpful features that users often fail to benefit from.

Templates. Predefined formats for many purposes, from legal contracts to engineering reports to memorandums and resumes. You can also turn documents into PDF (portable document format) files for sharing on the Internet.

Styles. Predefined styles for headings, margins, fonts, column width, tabs, boxes, italics, boldface, headers and footers, and graphics. You can select styles to apply to each section and subsection of your document and then apply them automatically.

Graphics. Predefined charts, graphs, tables, clip art, and shapes for virtually any purpose, including custom drawing tools.

Version Tracking. Options for tracking, accepting, and rejecting changes as you go through various versions of a document.

Media Connections. Options for file sharing and emailing; embedding links, videos, graphics, and spreadsheets from "cloud" (i.e., Internet) sources.

Word processing has become a nearly universal skill in the 21st-century workplace. The options and features available to writers are increasing daily, expanding exponentially their power to communicate via email, blogs, instant messaging, ebooks, and streaming content. Additionally, with the rise of open-source word-processing applications, some of which are free of charge, writers have many more choices than ever before.

However, the flexibility and power of word processing can be a trap. Clients or supervisors begin to believe that everything can be created, changed, and made better in no time. They believe the tools have made writing easy, and they certainly have made it easier; but it is still time-consuming and often frustrating to the person doing the writing when others have unreasonable expectations. In the workplace, efficient word processing means knowing when the document is okay as it is—when it will do the job and enough changes are enough!

Although word processing makes writing much faster and easier than ever before, writing is still a mystery. Even the computer cannot reveal just how someone comes up with a particular organization or a certain train of logical thought. Some writers still need to sit down with a notepad and a favorite pen in order to work.

The following rules are suggestions about how to write using word processing. You need to ask yourself what works for you and what doesn't. You might still need, for example, to plan on paper before turning to the keyboard. Or you might decide that you can only edit or proofread effectively if you have a hard (printed) copy in front of you.

As you write using word processing, test the following suggestions. Be alert for shortcuts. See also WRITING AND REVISING.

1. Save your work on a backup disk.

Remember that the computer's memory is temporary and may vanish when (not if) the hardware or software fails. Therefore, the cardinal rule of word processing is to save to a backup disk. An associated rule is to keep your security program up to date. Nothing is more distressing than to watch your hard work on a document vanish or become corrupted because of a computer virus.

2. Save your work using intelligible file names.

Name your files so that material saved will be accessible to you or a colleague days, weeks, even months later.

Electronic word-processing files are similar to the files in a filing cabinet. Well-organized, clean files are a pleasure to work with; messy, chaotic files are a disaster.

Each word-processing file should have its own name (code), including, as appropriate, alphabetical or numerical indications of its subject, its writer, and the version saved. For example, you might choose to label a file on travel expenses with the abbreviation *travex*. If different writers work on the same subject, you could add the writer's initials: *travexbe*.

Your software program will indicate the date and time when the file was last opened, but this information might not always help you, especially if you have multiple versions of the same document with the same title. After all, you may have opened an early version of a document while you were trying to find the current version; this early version will automatically carry the most recent date. Consider, therefore, some system for identifying the version or draft—A, B, C, or 1, 2, 3, etc.

File Names for a Team Document. File names are crucial if a number of writers and editors are working as a team on a document. Teams usually share electronic drafts for review and revision. The more sharing, the more a recent revision is likely to be lost or replaced with an earlier version.

Also, agree as a team on how to manage access to the files. Consider privacy issues and the problems of handling client and proprietary information. Decide beforehand who will know passwords and have access privileges.

3. Decide on a format for your document before you begin writing.

The simplest format is usually the one that is already set for you by your software program. This basic format will have certain standards (default settings)—for example, default margins, typeface, and tabs. Use this format for routine documents or change the default settings to reflect your own or your company's standards and preferences. These changes will become the default settings.

Still, the needs of your audience or your purpose might require a different format—such as wider margins, two-column format, different font, etc. Word-processing software allows you to program such style variations so they become automatic. See Page Layout.

You can also develop templates of your own with special styles to use whenever you prepare a standard document such as a proposal, newsletter, marketing flyer, etc. Preparing a template for each type of document will allow you to begin writing immediately without having to worry about the styles you used last time.

Consider also creating a primary or boilerplate file with merge codes for regularly used documents instead of generating a new document each time. A primary or boilerplate file is a standardized document or preset form (e.g., memo, sales letter, proposal, customer service letter, invoice, manual, procedure, report) that you change or update frequently. To update such a file, you create a secondary or merge file attaching the same merge codes used in your primary to the new information. These codes permit you to jump immediately to your preselected points of insertion (names, addresses, dates, particular responses or new text, personalized closings).

Format Decisions for a Team. If a team is working on a document, the earlier the team members can agree on a format, the more efficient will be the writing.

A simple team technique is to develop a single template for everyone to use. A template enables you to decide together on margins, spacing, fonts, etc., and special features (boxes for visuals, shaded quotations, etc.). See Page Layout for guidance on selecting styles.

4. Plan your document before you begin to write (input).

Planning is important to avoid false starts or dead ends. Try to identify a few of the givens about the document before you begin to write.

Do you have your task or assignment clearly in mind?

- What is your main purpose?
- What do you want your readers to do, to know, and to feel?
- Who will read your document and what are their goals and priorities?
- What sort of document (length, format, content) do you expect to write?

These and similar questions will surely be in the back of your mind as you are writing. Do you need to write them down before beginning—perhaps on a notepad or on a whiteboard?

Will writing down some of these givens help you focus your efforts? Do what you must to make your writing as efficient as possible. See Writing and Revising.

5. Brainstorm your content.

Your computer can be a valuable brainstorming tool. Don't be afraid of writing down something you feel is stupid or something that you think won't fit. After all, you can move ideas around at will, and you can later delete anything you don't need with the touch of a key.

Word Processing

Experiment with a brainstorming phase before you actually begin to write the text—that is, **before** you try to write complete sentences and fleshed-out paragraphs.

Such a brainstorming phase might be a rapid inputting of key words and phrases without any order or sequence. Let your mind race as fast as it will, and don't block even the dumb ideas. Don't worry about spelling or punctuation. See Thinking Strategies.

Brainstorming is most applicable to documents that are not regular and routine. For routine documents, you already know much of the content, and you may even have a format (template) to start with. See rule 3 above.

After you finish brainstorming, you have to choose what you want to do next. You might, for example, print a copy of your brainstorming and use it as a guide for writing your draft. Your software may even display your brainstorming notes in a window on the screen while you are writing your text.

6. Organize your content.

The more complex the document, the more you will need to organize your brainstorming notes.

This process might be merely a sequencing or tidying up of your brainstorming notes. Or you might decide to work up a fairly complete outline. Using the "sort" function, you can give instant structure to your notes.

Many word-processing programs have an outlining function. This function usually gives you a standard outline format to work with, and once you finish your outline, you can usually view the outline in a window on your screen while you are writing your text. See Outlines and Organization.

If you have always liked to outline before writing, you will probably want to continue using an outline as part of word processing. If you never liked writing from an outline, you might want to move directly from your brainstorming notes to the writing of your draft.

7. Use word-processing techniques to simplify drafting your document.

Writing rough drafts used to be a more daunting task before word-processing software. Now you can write, tweak, move text around, insert, and delete ideas at any point in the writing process.

The computer can help you overcome "writer's block." If you write quickly, not trying to be perfect the first time, the computer can cut your writing time. You can also avoid anxiety about having something with errors, misspellings, and other "warts."

Nevertheless, ease of editing on the computer means you will be tempted to rework sections or whole drafts before you finish writing the initial version.

Try not to allow such revision work to interrupt your thinking or cause you to forget ideas you should record.

Keep these things in mind while writing on a computer:

- Remember, you can always change text later, so avoid editing grammar, spelling, and punctuation before you have all your ideas on the screen.
- Make on-screen notes of ideas that occur to you but may not be relevant to the passage you're working on. Some word-processing programs permit you to make "text notes" (i.e., text that is invisible to the printer but appears on-screen).
- Code or mark points in the text where missing data needs to be inserted or more work needs to be done. For example, you can insert a non-word symbol like < > or ~. Then with your Search feature, you can tell the computer to search for places where this symbol appears.

Team Writing of a Draft. Recognize that rough drafting is not a group activity. Instead, assign each team member a section to write, and remind them of the style they should follow. See rule 3 above. When team members finish with their drafts, assign a single person to integrate the drafts so that the team can review the whole document.

8. Revise (edit) efficiently by setting priorities and sticking to them.

Take care of the big issues such as purpose, content, and emphasis before you begin to polish sentences and mold your paragraphs. See Writing and Revising.

Proofreading comes last.

Setting priorities makes sense because it forces you to address the more global issues rather than spending too much time on trivial details when the document has failed to achieve its main purpose.

Revision (editing) begins when you reread what you've written. Decide, first, if you need to see the document in hard copy. If so, print a copy, preferably with double spacing and wide margins.

Next, review the whole document with these major questions in mind:

- Is my main purpose clear and up-front?
- Have I emphasized my key ideas?

- What else should I tell my readers?
- Is my tone and approach appropriate and effective?

After you have asked these hard questions (and fixed any problems), begin working on your paragraphs and sentences.

Here is where the computer is a boon. You can rewrite, move, or delete text so easily that you can experiment with different ways to say the same thing. For example, you might make a second copy of a paragraph and rewrite the copy. After rewriting, you can return to the original to see if you have improved it. Finally, you can delete the version you decide not to use.

To help with rewriting, consider using the thesaurus feature if you are stumped for an appropriate word or phrase.

Also, during rewriting, you might want to code words or phrases that you wish to appear in your index or in a table of contents. Then, when you finish editing, you can command your software to sort the words and phrases and to prepare the table of contents or the index. See INDEXES and TABLES OF CONTENTS.

Even with the ease of editing on the computer, you might still wish to edit and proofread a hard (printed) copy of a document instead of on a monitor. The monitor screen can prevent you from perceiving the overall organization and structure of a document—the "feel" of the document.

A Final Caution About Revision. No text is ever finished or perfect, at least in the minds of most writers. Word processing often lures writers into spending excessive time revising and polishing their drafts or including flashy graphics. The secret is to know when to quit revising. Decide when the draft is adequate, even good, but not necessarily perfect, and then stop.

9. Use word-processing techniques to help you proofread your document.

Although the computer makes writing and especially editing much more flexible and easier than ever before, it also adds new possibilities for errors. The ease of changing things makes consistency of content, data, grammar, and spelling a major problem. See EDITING AND PROOFREADING.

Proofreading is essential after a writer has tinkered with a text for hours and hours because for every change the writer has made, probably 10 others should have been made for consistency with the single change.

Follow these general suggestions for proofreading on a computer:

- **Proofread on the Screen.** An advantage of the computer is that you don't have to print a hard copy to proofread or to review your text. But even experienced editors have trouble proofreading using only the screen. The text is often too small, and it is sometimes awkward to check the document for overall consistency when you can see only a dozen or so lines of text at a time.

  If you do proofread on the screen, consider changing the text to larger, more readable type and adjusting the line spacing to double spaces. Turn on paragraph marks and hidden formatting symbols. These changes will help you identify proofreading problems.

- **Proofread a Hard Copy.** Some people find it difficult to proofread on a screen because screens do not always show entire pages, and the visual resolution of characters on a screen is not always clear and sharp. Actually, reviewing a hard copy will give you a second chance to catch things you've already missed on the screen.

- **Check for Consistency.** When you change your text at some point, odd problems may appear elsewhere in the text. Some subjects no longer agree with verbs, spellings become inconsistent, and pronouns no longer agree with their antecedents. Unnecessary spaces appear, punctuation is doubled or dropped, and heading levels become garbled.

  Be particularly careful about global search-and-replace commands and text insertion. If you order all the *threes* in your document to become *fives*, you had better be sure that you mean **all** the *threes*. When you copy and paste, insert new information, or update the boilerplate, be careful that the old text is consistent in every way with the new. Fonts, line spacing, lists, and margins can become incompatible.

- **Run a Spelling Check.** Your computer checks your spelling by matching every word in your text with words in its internal dictionary, flagging those it doesn't recognize. Then the machine gives you the option of changing the questionable word or leaving it as is.

  Spellcheckers are wonderful for finding typographical errors and plain old misspellings—but be careful. The spellchecker may overlook words that are spelled right but are wrong for the context (e.g., *there* instead of *their*), so proofreading is still necessary. You can add words from your professional vocabulary to the computerized dictionary; make sure you spell them correctly when you do.

- **Run a Grammar Check.** Grammar checkers are designed to catch grammar and punctuation errors by comparing your writing with lists of common phrases that often present problems.

  For example, most grammar checkers can identify whether you have used a passive or an active verb in a sentence. The checker cannot, however, tell you whether the passive is the correct choice or whether you should change it to an active verb. See ACTIVE/PASSIVE.

  Grammar programs are far from reliable in their ability to point out errors in grammar, punctuation, or word usage. They may flag only a small percentage of the errors in your copy, and what they flag may not even be errors. If you're uncertain, check a reliable grammar-and-usage reference site or book. See REFERENCES.

  Some grammar checkers (many are available online) also purport to analyze style. Such style checkers measure sentence length and word length to determine a "reading level" for your text. These levels may be expressed in terms of an index of readership or grade level in school terms. Often style checkers can help you gauge whether your sentences are short enough and your word choice simple enough to make your point effectively to your intended audience. See SENTENCES and GOBBLEDYGOOK. Style checkers will not really identify basic problems of inconsistent or incoherent sentence structure.

Special Word-Processing Resources for Writers

The following resources can help business and technical professionals create high-quality documents. Many are available online.

Bibliographic Databases. Scientific and technical researchers who must work with or provide reference lists in reports may find bibliography programs helpful. These programs permit you to enter information about references and then to format the information according to any bibliographical style you choose (e.g., for formal reports, catalogs, journals, conference proceedings, or dissertations). Professionals who use online databases can transfer reference information directly into the bibliographic database. See BIBLIOGRAPHIES, CITATIONS, and FOOTNOTES. Remember to proofread reference lists just the same as you proofread text.

These databases are also useful for researching books and publications on virtually any topic. For example, ISBNdb.com lists nearly all books ever published or available in the United States.

Mathematical Expression Checkers. Aimed at technical professionals who present mathematical proofs in their writing, these programs enable them to pull mathematical symbols into their copy; to highlight equations, making editing easier; and to work within templates that provide preset mathematical structures for many common equations. See MATHEMATICAL NOTATION.

Graphics Programs. Many sophisticated, easy-to-use programs exist for generating graphics for business and technical documents. You can scan or create your own graphs, diagrams, and charts with a painting or drawing program, and then paste them into your copy. Or you can use one of the many sources of predesigned graphics, photos, or illustrations and just plug them into your copy (be careful here—the quality of these images ranges widely). See GRAPHICS FOR DOCUMENTS and GRAPHICS FOR PRESENTATIONS.

Design and Layout Programs. If you need a professional look with highly developed, eye-catching graphics, you may want to step from word processing to design-and-layout software, which provides more sophisticated page layout and graphics opportunities. See PAGE LAYOUT.

Online Publishing Resources. If you need to publish documents online, a wide range and variety of resources are available to you. Most of these applications are low cost or even free of charge. Hosted blog sites enable you to disseminate virtually any kind of content in many different formats. Sites vary greatly in the features they provide; for example, some offer templates, archiving capabilities, or automatic RSS feeds to notify your audience of a new posting.

More ambitious documents, such as ebooks, can be helpful in publicizing your business or providing guidance to your clients; for example, technical manuals are generally available online as ebooks for download or just for reference. Publishing an ebook can be as easy as converting an existing document into a PDF and making it available for download. More sophisticated programs permit converting the book for a proprietary reader and formatting the book into pages instead of flowing text.

Wordy Phrases

Wordy phrases are phrases that use too many words to express an idea. Many of them incorporate or are similar to redundancies, and many wordy phrases have been overused so much they've become cliches:

all of a sudden
at a later date
beyond a shadow of a doubt
in light of the fact that
in the environment of
in the neighborhood of
kept under surveillance
on two different occasions
reported to the effect
the fullest possible extent

None of these phrases is necessary, but many writers use them habitually, so they seem "natural" somehow. In informal speech and writing, these phrases may in fact be preferable, depending on your audience. In technical and business writing, however, you should avoid them. See REDUNDANT WORDS, CLICHES, GOBBLEDYGOOK, and SCIENTIFIC/TECHNICAL STYLE.

Wordy Phrases

1. Avoid wordy phrases.

1. Avoid wordy phrases.

A fundamental of good writing style is to avoid unnecessary words. Do not say *all of a sudden*. Just say *suddenly*. Instead of *at a later date*, say *later*. *In light of the fact that* simply means *because of*.

The shorter, simpler words or expressions make your writing more concise and, consequently, make it look and sound more professional.

A List of Wordy Phrases

The following list of wordy phrases will help you identify those you habitually use. The wordy phrase appears in the left column; in the right column are possible substitutes.

According to the law	legally
add an additional	add
add the point that	add that
afford an opportunity	permit/allow
a great deal of	much
a greater number of	more
a great number of times	often/frequently
a large number of	many
a little less than	almost
all of a sudden	suddenly
along the lines of	like
a majority of	most
a number of	several/many/some
any one of the two	either
a period of several weeks	several weeks
as a general rule	usually/generally
as a matter of fact	in fact
a small number of	a few
as of now	now
as of this date	today
as regards	about
as related to	for/about
assuming that	if
as to	about
a sufficient number	enough
at a later date	later
at all times	always
at an early date	soon
at hand	here
at present	now
at regular intervals of time	regularly
at that time	then
at the conclusion of	after
at the present time	now
at the rear of	behind
at the same instant	simultaneously
at this time	now
at which time	then
Based on the fact that	due to/because
beyond a shadow of a doubt	doubtless
brought to a sudden halt	halted
by means of	by
by the time that	when
by the use of	by
by way of illustration	for example
Called attention to the fact	reminded/noted
came to a stop	stopped
cannot be possible	impossible
come to an end	end
cost the sum of	cost
Despite the fact that	although
detailed information	details
draw to a close	end
due to the fact that	because
during the course of	during
during the time that	when/while
during which time	while
Estimated at about	estimated at
estimated roughly at	estimated at
exactly alike	identical
except in a small number of cases	usually
exhibit a tendency to	tend to
expose to elevated temperature	heat
Few in number	few
for a short space of time	for a short time
for the purpose of	for/to
for the reason that	because/since
for this reason	so
from the point of view of	for
from time to time	occasionally
Having reference to this	for/about
If and when	if/when
if at all possible	if possible
if that were the case	if so
in accordance with	by/under
in addition (to)	also/besides
in a number of cases	many/some
in a position to	can/may
in a satisfactory manner	satisfactorily
inasmuch as	because/since/as
in back of	behind
in case of	if
in close proximity	near/close
in conjunction with	with
in consideration of the fact	because
in excess of	more than
in favor of	for
in few cases	seldom
in few instances	seldom
in lieu of	instead of/in place of
in light of the fact that	because

Wordy Phrases

Wordy phrase	Replacement
in many cases	often
in most cases	usually
in order to	to
in other words	or/that is
in rare cases	rarely
in (with) reference to	about/concerning
in regard to	about/concerning
in relation to	with
in respect to	about/concerning
in short supply	scarce
in terms of	according to
in the absence of	without
in the amount of	of/for
in the case of	for/by/in/if
in the course of	during
in the environment of	around/near
in the event of	if
in the event that	should/if
in the first place	first/primarily
in the instance of	for
in the majority of cases	usually
in the matter of	about
in the nature of	like
in the near future	soon
in the neighborhood of	about/near
in the proximity of	near/nearly/about/close to
in the vicinity of	around/near/close to
introduced a new	introduced
in view of the fact that	considering
involve the necessity of	requires
is/are in the possession of	has/have
is defined as	is
is in the process of making	is making
is of the opinion that	believes
is representative of	typifies
it is apparent that	apparently
it is clear that	clearly
it is evident that	evidently
it is obvious that	obviously
it is often the case that	often
it is plain that	plainly
it is unquestionable that	unquestionably
it would appear that	it seems/apparently
Kept an eye on	watched
kept under surveillance	watched
Last of all	last
leaving out of consideration	disregarding
Made an investigation of	investigated
major portion of	most of
make application to	apply
make a purchase	buy
make contact with	meet/contact
more and more	increasingly
Notwithstanding the fact that	although
Of considerable magnitude	big/large/great
off of	off
of no mean ability	capable
of small diameter	fine/thin
of very minor importance	unimportant
on account of	because
on a few occasions	occasionally
on an ongoing basis	continually
on a stretch of road	on a road
on behalf of	for
once in a great while	seldom/rarely
one after another	alternately
one by one	singly
one part in a hundred	one percent
on the basis of	by
on the grounds that	because
on the part of	by/for
on two different occasions	twice
ought to	should
outside of	except/outside
owing to the fact that	since/because
Period of time	interval/period
pertaining to	about
possibly might	might
prior to	before
probed into	probed
proceed to investigate, control, study investigate	analyze/study/
provide a continuous indication of	show continuously
pursuant to	following
Range all the way from	range from
reduced to basic essentials	simplified
refer back	refer
relative to	about
repeat again	repeat
reported to the effect	reported
revise downward	lower/decrease
Seldom if ever	rarely
separate into two equal parts	halve/divide
since the time when	since
started off with	started with
subsequent to	after/following
Take appropriate measures	act
taking this factor into consideration	therefore
that is to say	that is
the foregoing	the/this/that/these/those
the fullest possible extent	mostly/fully completely
the only difference being that	except that
there is no doubt that	doubtless/no doubt
there is no question that	unquestionably
through the use of	by
to be cognizant of	to know
to summarize the above	in summary
total operating costs	operating costs
turn up	turn
two by two	paired/in pairs
Until and unless	unless
until such time as	until
up to now	formerly
Went on to say	added/continued
when and if	if
with a view to	intending to
with full approval	approved
within the realm of possibility	possibly/possible
without variation	constant/stable
with reference to	about
with regard to	about/regarding
with respect to	about/respecting
with the exception of	except
with the object of	to
with the result that	so that

Writing and Revising

Writing has always been a challenging personal intellectual skill; in the workplace it is also an organizational activity. Organizations now rely on teams of experts and writers to communicate important messages to customers internal and external. In the 21st-century Knowledge Age, writing is often the primary work of both individuals and teams as they generate "intellectual products" such as reports, proposals, memorandums, articles, white papers, press releases, marketing collateral, legal documents, training materials, and technical manuals. Thus, when we talk about writing in the workplace, we cover a vast spectrum from emails dashed off in a few seconds to voluminous team projects requiring months of work.

As a personal intellectual skill, writing proficiencies vary from person to person. Many people find writing difficult and intimidating. They procrastinate often and they apologize frequently for what they've written. Others write more willingly because they find satisfaction in the process and the results of their work.

Each person approaches writing differently because each person's intellectual skills are unique.

Some people prefer to plan, to discuss, and to think out their entire content before writing a word of text. With computers, many others begin writing text while ideas are growing and changing. Computers make it easy for them to move language around and to edit as their thoughts flow. See Word Processing.

No matter how you personally approach the task of writing, you likely will find it cyclic, even chaotic. For most writers, writing is an unruly, unpredictable intellectual skill. Unpredictable ideas and implications surface as you struggle to form ideas into words.

Writing and Revising

Frontloading

1. Determine the givens for a document as early as possible.

Prototyping

2. Develop an evolving prototype of your potential document and use this prototype to monitor your progress on the document.
3. Build prototypes collaboratively and with brainstorming techniques.
4. Use the prototype to define content issues that need to be resolved.
5. Integrate ongoing reviews into each prototyping session.

Drafting

6. Develop your own personal strategies for making the writing of a draft a comfortable, repeatable task.

Reviewing

7. Provide for early and frequent reviews from a broad sample of reviewers (including managers).
8. Meet face to face for most reviews, and record the results of such a meeting in writing.
9. Develop strategies for who will work on document defects and who will research content issues.

Revising

10. Set revision priorities both for yourself and for others who can help you revise and edit.
11. Use up-to-date references to guide your revision decisions.
12. Know when to stop revising.

Using your own approach is okay, assuming that you are writing efficiently. Still, individual writers will find helpful guidance for improving the quality and efficiency of their work in the following suggestions.

As an organizational activity, writing is a critical but little understood skill. Writers within organizations often work inefficiently and ineffectively because of the complexity of the task and the changing environment around them. Most technical and business professionals have little writing training to begin with. Many constituencies only complicate things further. The marketing, sales, production, and executive teams might all have different views of the issues. Reviews necessitate costly rework. Mistakes and inconsistencies plague the documentation. Baffled customers try angrily to make sense of the result.

Thus, a crucial writing project—for example, the production of the technical specifications for a new drug or a multivolume proposal to get the contract for a billion-dollar aircraft—can become an overwhelming team challenge.

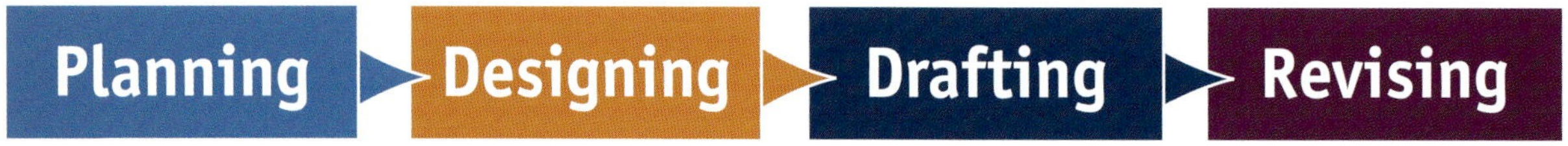

Planning **Designing** **Drafting** **Revising**

Figure 1. A Simple, Linear Writing Process. *This process works well for short, routine documents. For example, a routine memo or letter would require only the four steps listed above. Writing such a document might take 20 minutes or less. Usually, one person plans, designs, drafts, and revises the memo.*

On the other end of the spectrum, most writing done in the workplace is probably short, quick messages in the form of emails. Although emails are often unclear and intrusive, they typically don't require much thought. On the other hand, many emails are quite significant and do need more planning and revision than they usually get.

This section offers general suggestions for all workplace writers, although the focus is primarily on consequential writing projects involving more than a quick email or short memo. For more guidance on how to write and manage email, see Electronic Mail and Managing Information.

Writing—Linear Process or Separate Activities?

Like any process, writing can be improved. Researchers into the writing process have tried to describe and generalize an effective sequence of steps, such as the procedure illustrated in figure 1. They theorize that if people learn these steps, they'll write efficiently.

This view of writing as a linear sequence of steps will work for short, routine documents like letters and memos. Writers of short documents often move step by step, with few if any iterations to account for new inputs.

A simple, linear process does not work if the written document is complex or nonroutine.

Instead of staying with a strict linear model, the following guidelines focus on several skill sets or techniques:

- Frontloading (Planning)
- Prototyping (Designing)
- Drafting
- Reviewing
- Revising

These techniques, drawn from the world of quality project management, enable writers to work quickly and ensure a better outcome, thus avoiding delays and costly rework. As shown in figure 2, successful writers and writing teams usually will draw on each skill set several times during their work on

Figure 2. Writing Process for Nonroutine, Complex Documents. *Nonroutine, complex documents are usually long and data-rich. They usually require multiple contributors and reviewers. The process for such documents requires overlapping, concurrent steps, and often steps will be repeated when new givens (constraints) appear.*

a document. The exact nature and sequence of the steps, however, will vary. Even a single writer might cycle through these techniques several times before the project is complete. See Project Management.

Frontloading

Frontloading means providing as much input to the process as possible from as many concerned people as possible as early as possible.

Quality experts know that frontloading is the key to a robust final product. The more and the richer the input early in the process, the more likely is the outcome to stand up to scrutiny and to meet the needs of all users.

The science of quality focuses on doing something right the first time. If properly designed, for example, the fuel pump in an automobile can be installed only the right way, not upside down or backward.

In a similar fashion, if you frontload a projected document, then what you write should need no major late-stage rethinking of strategy, content, and format.

1. Determine the givens for a document as early as possible.

This step is more inclusive than the traditional notion of document planning. Yes, in this step you'll ask questions and set assignments, but you should also begin to visualize your document even before you've written a word.

This initial visualization is a provisional target, one that will evolve and change as you work on the document. With an early visualization in your mind's eye, your text will grow from and yet be constrained by the format and appearance of your final document. If you are preparing a two-column, four-sided newsletter, the format will limit the sorts of articles and graphics you can include.

As in this newsletter example, format and appearance affect and limit the content. This approach reverses the traditional model, where content comes first and is then formatted and polished later on.

To begin your frontloading of the document, ask the following questions:

—What is your main point? (What do you want readers to do? to know? to feel?)

—Who will read your document and what are their priorities?

—What sort of document (length, format, graphics, and content) do you intend to write?

—What organizational goals (or constraints) exist?

—Who inside or outside the organization can help you meet these goals?

Always answer each of the preceding questions. The more complex the writing task, the more desirable you'll find the habit of answering such questions in writing, either in a formal checklist or an informal list of reminders.

Don't assume that if you are working alone you needn't address these frontloading questions.

If your writing is an organizational activity, you, your collaborators, and your manager(s) need to agree on answers to the frontloading questions. The longer you delay involving these key participants, the more likely you'll face costly rewrites. As early as possible, convene a team with as many contributors as feasible. If possible, include both internal and external customers. See Word Processing.

Be sure to record the answers to these questions in writing and circulate these answers to all the contributors. What you will find is that you have already begun the writing because material in your answers will likely flow directly into the document you are planning.

Prototyping

Prototypes are full-scale visualizations of a document. Often document prototypes are called mock-ups, wireframes, or storyboards. Regardless of the terminology used, a document prototype has these features:

—A full page-to-page correspondence between the prototype and the final document

—A page layout (or style sheet) to guide how text is written (See Page Layout.)

—Headings and subheadings that show the organization and scope of the text to be written

—Graphics and captions to highlight key points

A document prototype differs from an engineering prototype, which is a full-scale working model. A document prototype is not full or complete, so it is not a "working" model in the engineering sense. The document prototype is, however, a full visualization of the final document.

2. Develop an evolving prototype of your potential document and use this prototype to monitor your progress on the document.

A document prototype evolves gradually, through iterations. A document prototype should be a team activity, although a single writer sometimes prepares a prototype for discussion or team review.

A document prototype will save time and money if it guides all later stages of document preparation, including the writing of text. A document prototype is an evolving and maturing tool as a team's vision of the final document evolves and matures.

The **initial prototype** is very skeletal—perhaps little more than blank pages or screens with a few key headings and several key graphics sketched in. The initial prototype is a visualization of the document to come. This visualization functions as a working hypothesis. See figure 3 for a few pages from an initial prototype.

Several iterations later, an **interim prototype** will contain most headings and subheadings, as well as sketches of most graphics. But this prototype still has little text. See figure 4 for an example of an interim prototype.

A **mature prototype**—which may be the product of weeks of teamwork—begins to look like a rough draft. A mature prototype has full headings and subheadings, versions of all graphics, and even some text. Some pages are still blank, however. See figure 5 for an example of a mature prototype.

The mature prototype becomes a rough draft when the text is complete, but not final. The design and scope of the document should be final by the time the rough draft is ready for review.

As in most rough drafts, mature prototypes contain misspelled words and grammar and usage problems. These don't matter. Late-stage, careful revision will correct the errors.

3. Build prototypes collaboratively and with brainstorming techniques.

A prototype should include rich input from all concerned and interested people. Invite input from engineers, scientists, managers, legal staff, fabrication specialists, suppliers, marketing, finance, sales representatives, government regulators, and customers (internal and external).

Early prototyping should be a brainstorming activity—that is, almost a game to be played. Remind yourself and your collaborators not to be too judgmental or too negative.

4. Use the prototype to define content issues that need to be resolved.

Even an initial prototype will allow you to define content issues that need further work.

—What job does the customer want the product to do? Which scientific studies are yet to be done?

—Do government regulations require a particular test or sampling approach?

—What financial constraints will govern production decisions?

—What would our followers in the social media have to say about this approach?

Your work with the early prototypes raises significant strategic issues, not superficial late-stage questions about format or language.

5. Integrate ongoing reviews into each prototyping session.

Make reviewing part of the prototyping process, not something done after a version of the prototype is finished. (See the reviewing guidelines in rules 7 through 11.)

Ongoing reviews are just one version of team collaboration. You might, for instance, open a follow-on prototyping session by walking through the prototype from an earlier session. Or you might end a prototyping session by asking a participant to turn pages and review remaining content issues.

Such reviews are different from a late-stage managerial review.

If possible, include your manager(s) in the early prototyping sessions so they have a stake in the evolving prototype as it is produced. If they don't see the prototype until it is mature, they will often require the team to repeat many of the earlier steps—thus wasting much time and money.

Drafting

Writing the draft begins when you make your first frontloading notes. It continues as you work to flesh out your prototype. It does not end until you make your final revisions.

A key tenet of time management is that you manage a complex task by breaking it up into shorter, less unwieldy tasks. Use the same approach in your writing.

6. Develop your own personal strategies for making the writing of a draft a comfortable, repeatable task.

Your strategies will be necessarily unique, but some possible ones to consider are the following:

—Write the easy stuff first. This will get your creative thoughts flowing.

—Don't worry about writing your draft in one sitting. Instead, work on your draft section by section, sentence by sentence.

—Turn off the internal voice that judges each word you write. Instead, view each word as an accomplishment, not something to immediately delete and rewrite.

—Visualize yourself talking to your readers. Your thoughts and language should be just as natural and as free-flowing as your talk would be.

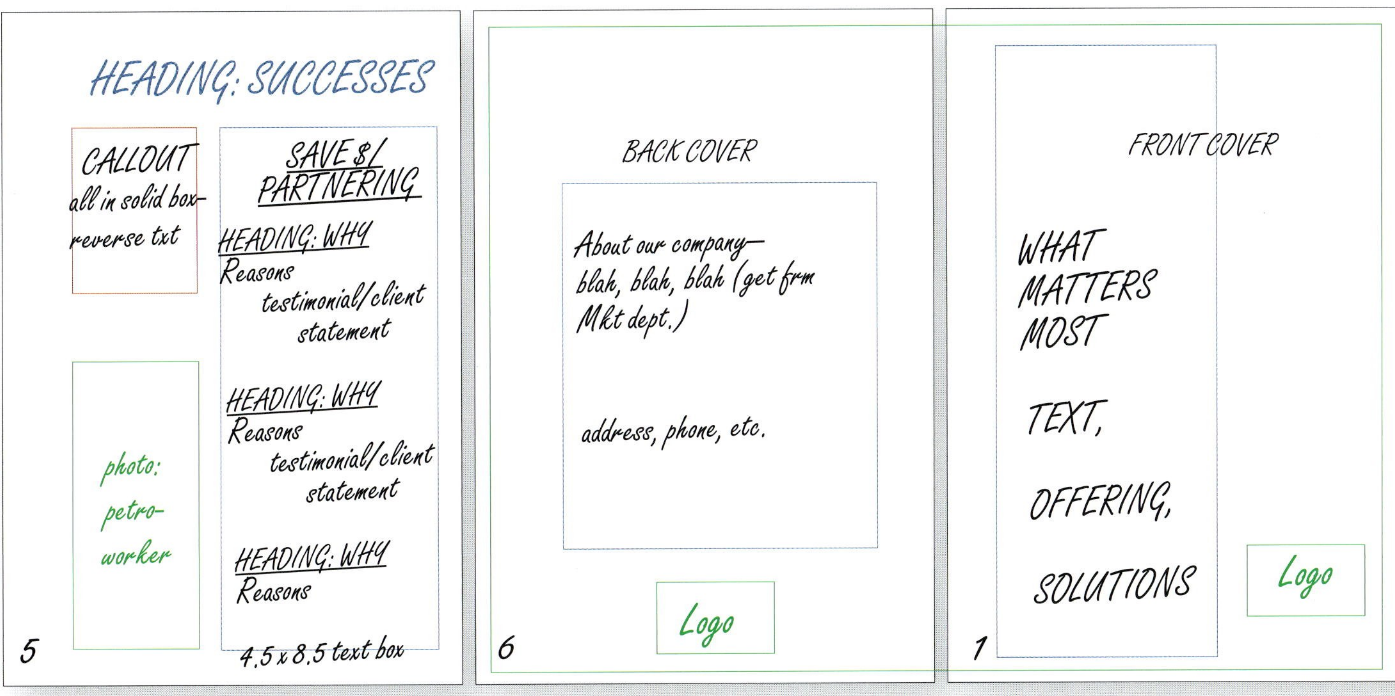

Figure 3. Initial Prototype. *Team members prepare an initial prototype using spaces on a whiteboard, blank pages, or computer screens. For a brochure such as this, they might tape together three 8.5 x 11 sheets to capture the sequence of pages and to help them visualize the final product. As in this example, text in an initial prototype is often sketchy or missing, and graphics remain to be designed or located.*

Outside brochure

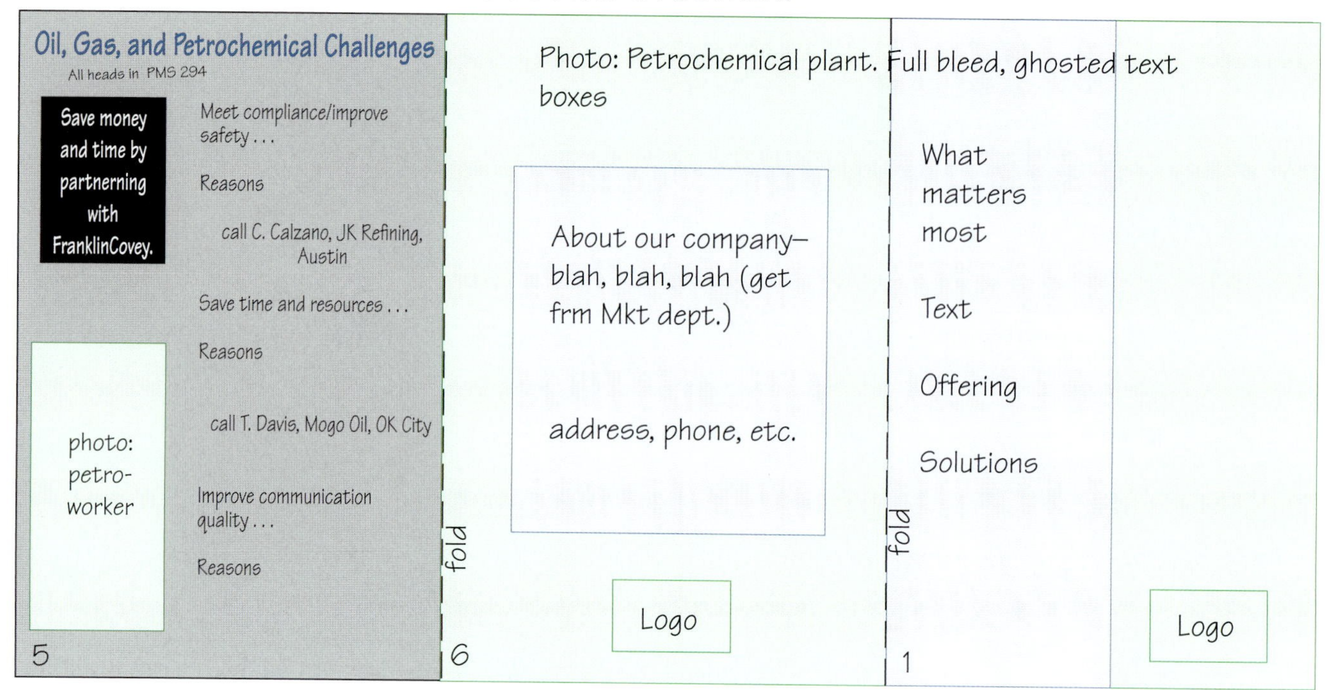

Open dimensions: 25.5 x 11 in.; folded dimensions: 8.5 x 11 in.
100lb. coated stock, full color, varnish

Inside brochure

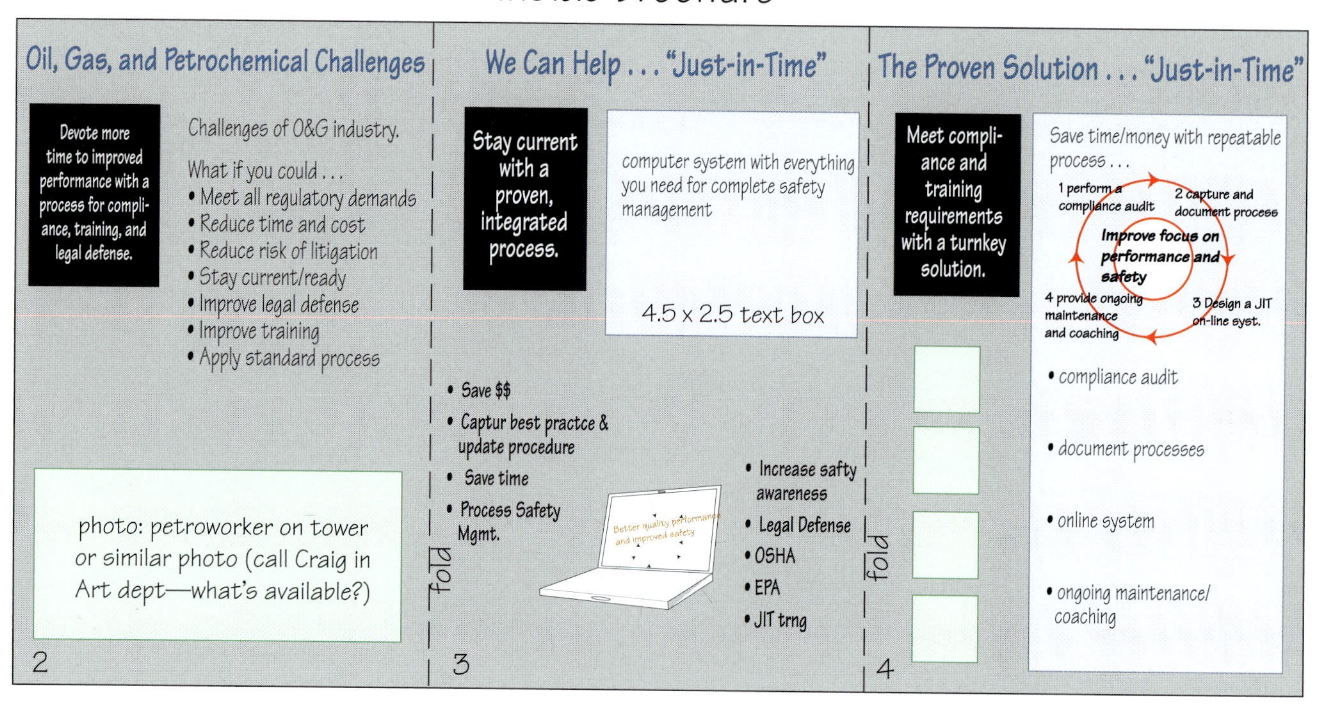

Figure 4. Interim Prototype. *Days or weeks later, the prototype moves over into an established format, as in this example. Text is still sketchy, or missing. Graphics are sketched, but not finished. An interim prototype should be complete enough to allow for review by managers or customers.*

Outside brochure

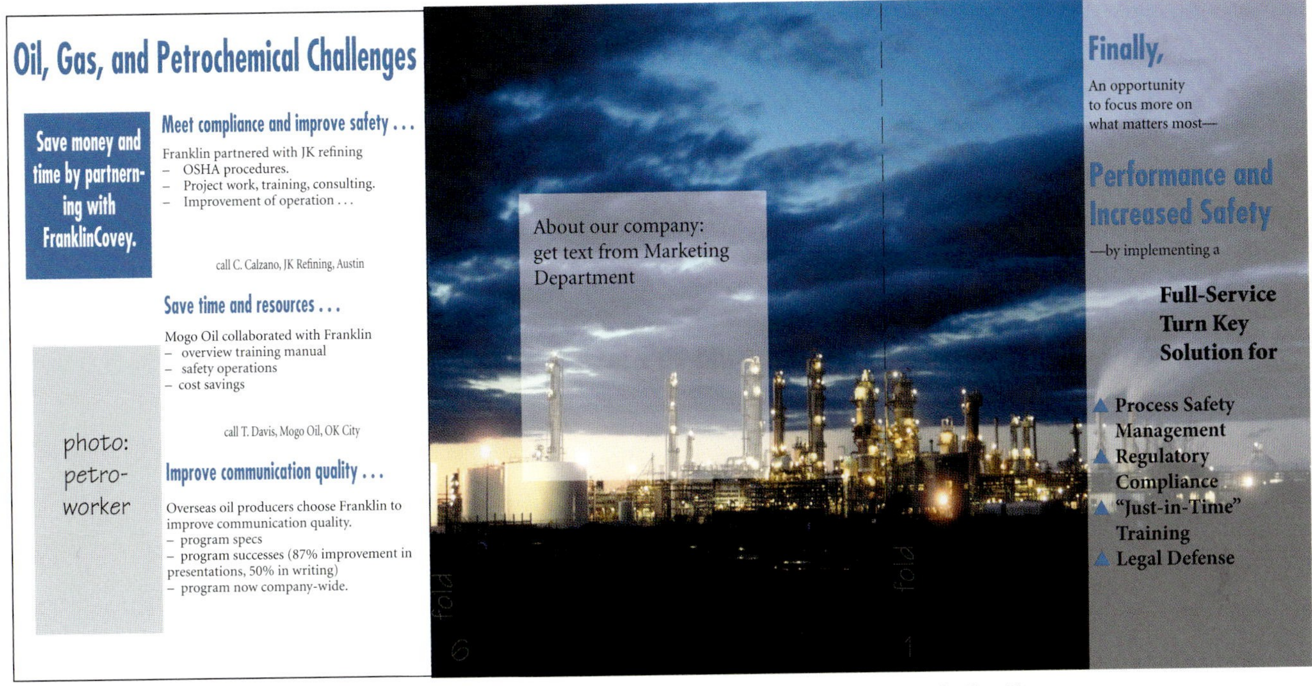

Open dimensions: 25.5 x 11 in.; folded dimensions: 8.5 x 11 in.
100lb. coated stock, full color, varnish

Inside brochure

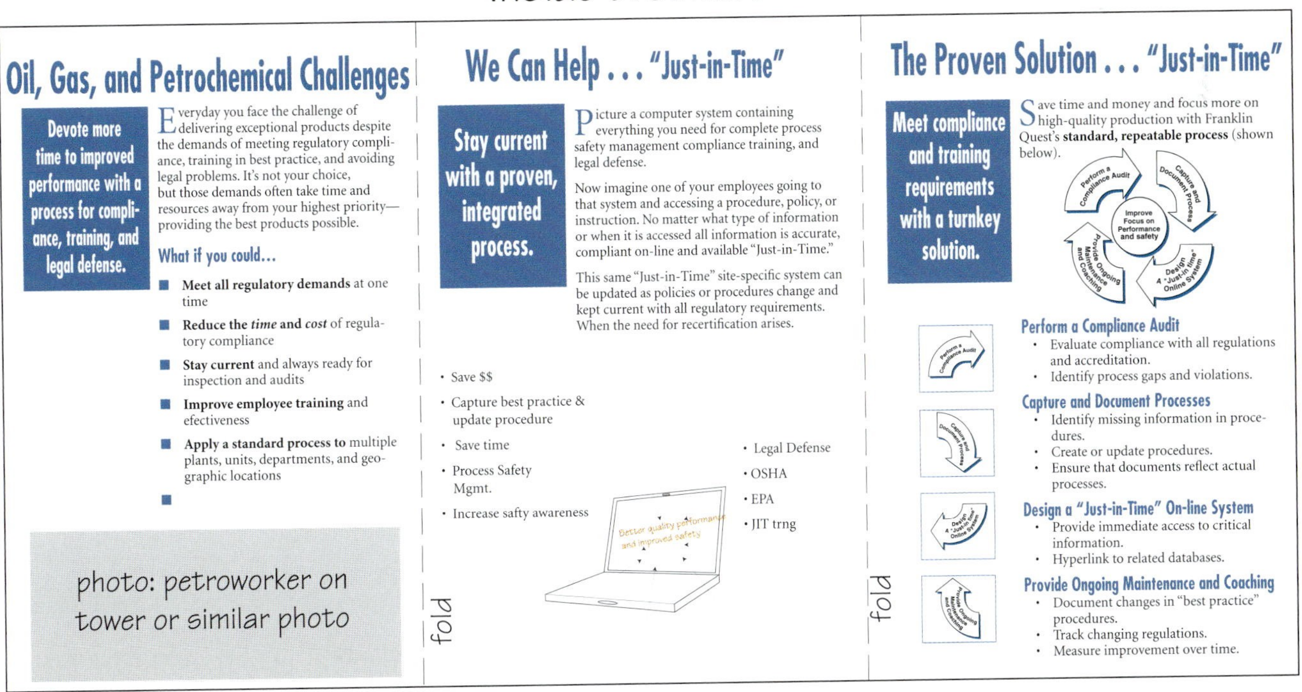

Figure 5. Mature Prototype. *A mature prototype is often the same as a rough draft. Text and graphics are included, even though still rough or unfinished. Color might or might not be used. Revision and editing will change some of the text and, perhaps, even some of the graphics.*

—When you complete a section, take time to congratulate yourself.

Reviewing

Reviewing provides valuable feedback for organizations to use in guaranteeing the quality of intellectual products. Reviewing is, however, an inefficient and costly process unless it is well managed.

Three types of review are often desirable: **customer, peer,** and **managerial.** The numbered rules 7 through 9 apply to all three types of reviews.

Customer reviews can be formal, as with focus groups, or informal. In either case, your aim is to raise issues that actual users of the document will face. The ideal reviewers of a marketing piece or a user manual, for example, are those to whom it's addressed—your customers, whether external or internal. What questions do they have? Do they understand the message? Can they follow the instructions? Are they put off by anything in the document?

For customer reviews, you might want to take advantage of the various document usability tests outlined in PROJECT MANAGEMENT. These tests will tell you if the document is meeting the needs of its intended audience.

Peer review occurs when professional and technical colleagues review a document (either a prototype or rough draft) for its content or its presentation of the content. Peers, by definition, are colleagues who have no managerial authority over the writer(s). Their suggestions to the writer(s) are not binding.

Managerial review occurs when managers within an organization exercise their authority to review and approve or disapprove a document. Their responses to a document are binding unless they choose to allow for negotiations and adjustments.

7. Provide for early and frequent reviews from a broad sample of reviewers (including managers).

Schedule reviews during the development of the initial prototype. Continue to schedule reviews until the final draft is ready to be printed and shipped.

Customer reviews, peer reviews, and managerial reviews should occur early if they are to be effective. Early reviews give reviewers a stake in how an evolving document looks and what it says.

Mid-project or late reviews are wasteful and costly when reviewers change the whole scope and direction of the document. Such mid-project or late reviews are frequently the only time a manager sees the document. Managerial review comments at these later stages are costly to the organization and frustrating to the writer(s).

Include managers, if possible, especially in the early review sessions. Managerial input is essential if you want to head off last-minute adjustments based on managerial preferences that you failed to identify in the early stages of your work on a key document.

The broader your reviewers' backgrounds, the more likely they are to catch all problems or inconsistencies. Include reviewers who are not familiar with the project as well as those who are familiar with it. In customer reviews, consider including non-customers as well as current ones.

8. Meet face to face for most reviews, and record the results of such a meeting in writing.

A face-to-face meeting—or at least a teleconference—among reviewers and the writer(s) is more efficient and productive, if properly conducted, than written reviews.

Written comments on draft materials are time-consuming to write, not to mention being difficult to read and collate, especially if several reviewers disagree about points.

Written comments are often unclear and difficult to interpret. An efficient oral review has the features shown in the checklist on the following page.

9. Develop strategies for who will work on document defects and who will research content issues.

Either during or after the oral review session, prepare a list of content issues, the names of people responsible for resolving the issues, and due dates.

Circulate this list to all participants, including reviewers, and schedule a follow-up session.

Subsequent review sessions should always address these content issues. Some issues, however, will require several follow-up discussions as new information becomes available.

Revising

Traditional advice was to leave revising until you had finished the full draft. Creating the draft supposedly required a different mental approach than revising.

With computers, this advice is less and less true. The computer encourages writers to revise and then revise again as they are working on headings, section introductions, and pieces of the draft.

Oral Review Checklist

- ❑ The reviewers have had ample time to review a copy of the prototype or draft before the oral session. This copy is marked PROTOTYPE or DRAFT.
- ❑ The writer or writers conduct the review session, setting priorities and monitoring the time. Early reviews deal with broad issues; others properly focus on syntax and writing. Tell reviewers what to expect and what you expect!
- ❑ Time devoted to the review is strictly limited, by agreement with the participants.
- ❑ Reviewers give their comments and observations briefly, with no explanations of details or their rationale unless the writer asks for clarification.
- ❑ Comments are as constructive and as positive as possible.
- ❑ Comments need not be repeated unless reviewers disagree about a point. Discussions of the disagreement are not allowed unless the writer asks for clarification.
- ❑ One person records comments. If appropriate, this person prepares and circulates a summary of comments or items to be researched.

Revising on the computer is a problem only if a writer bogs down and wastes time fixing text that is likely to change significantly. Too much revision is the enemy of efficiency. See WORD PROCESSING.

10. Set revision priorities both for yourself and for others who can help you revise and edit.

Revision priorities mean that neither reviewers nor revisers attempt to do everything at once. Most people cannot read and revise everything in a single reading. Instead, plan to use multiple readings, each reading directed toward a different reviewing/revision priority.

The three broad levels of potential revision are as follows:

- **Macro revision.** Rework the overall content, design, and organization of the document. Revise or expand conclusions and recommendations. Tinker with headings and subheadings to make them more informative. Rework or add graphics.
- **Medium revision.** Rewrite data and information for consistency and logical effectiveness. Control the flow of old and new information from sentence to sentence. Your goal is to make sure the language flows logically and smoothly.
- **Micro revision.** Edit and proofread individual sentences and phrases for correctness, including word choice, spelling, punctuation, and grammar.

11. Use up-to-date references to guide your revision decisions.

No one can or should remember all the rules and guidelines of document preparation and language correctness. Maintain online or near your computer a set of basic and up-to-date references. At the least, obtain and use the following:

—A recent dictionary from a major publisher

—One or more general style guides

This *Style Guide* is an illustration of the sort of guide to acquire. For additional suggestions, see REFERENCES.

—Specialized professional guides or glossaries

Engineers and scientists, such as geologists, mathematicians, chemists—these and every professional group have their own guides. Be sure you know what these are and acquire the latest editions.

12. Know when to stop revising.

Know when enough is enough.

Many writers are tempted to revise more than necessary, especially given the editing tools in word-processing software. Hours spent revising and reworking low-priority items are costly. Many writers today waste much of their time with needless revising. They move text around, rewrite passages, and then repeatedly print hard copies for review.

Efficient writing means limiting revisions to the essentials.

Model Documents

STYLE GUIDE
FIFTH EDITION

Using Model Documents

The following models illustrate both effective document design and the best in current business and technical English.

As you survey these models, you will notice that they all use headings, informative subject lines, lists, short paragraphs, color and, whenever appropriate, graphics. The goal of these emphasis techniques is to make the information in these documents 100 percent unambiguous to any and all readers.

This goal is, of course, not achievable. Not all readers read accurately or consistently. Still, writers of documents must do everything they can to assist readers both to discover accurately the information in a document and then to remember this information for later use.

Well-designed documents are the single best tool for making sure readers receive unambiguous information. See EMPHASIS.

The following suggestions should help you use these models to improve your own documents.

- Use these models as a source for document design ideas, not for exact words and phrasing. Most of the time, you will find that your own topic or content will not match the words and phrases in a given model.
- Choose words and phrases that sound like you, not like someone else. The best business or technical style is usually the one that is natural, even conversational. See STYLE, TONE, SCIENTIFIC/TECHNICAL STYLE, and LETTERS.
- Refer frequently to the "Reference Glossary" section in this *Style Guide*. As in the preceding suggestion, all models cross-reference entries in the "Reference Glossary."
- Begin your own collection of models, ones that reflect your own professional content and organizational conventions. Then refine or polish these models each time you use them, referring, as necessary, to the principles illustrated in the *Style Guide* entries and models.

The setup is short but essential. It immediately presents the purpose of the letter. See ORGANIZATION.

The list clearly states what Mrs. Lantham is to do. See ORGANIZATION.

Listing the requirements emphasizes them and provides a ready checklist for Mrs. Lantham to accomplish the necessary prerequisites. See EMPHASIS and LISTS.

The closing summarizes the to do. See ORGANIZATION.

February 16, 2010

Mrs. Louise Lantham
12039 Plaza Drive
Tallahassee, FL 32303

Subject: How to Add Your Husband as a Dependent Under Your Group Insurance Plan

In response to your query, please send me the following:

1. Enclosed enrollment card
2. Enclosed information packet
3. Up-to-date and complete medical report from your physician, including the following information:
   - Current illnesses or injuries, including your physician's prognosis
   - Types and dosages of medications
   - Other information your physician believes might help us to evaluate your husband's physical condition

You may also fill out these forms online at benefits.bcc.chem.com/enroll.

As soon as we receive the enrollment card, the medical information sheet, and the medical report, we will process your application. If you have further questions, please call me at (415) 555-7582.

Sincerely,

Walt Cavanaugh

Walt Cavanaugh
Assistant Manager
Human Resources Department
wcavanaugh@bcc.chem.com

Enclosures

139 Sequoia Park Way, Suite 303, Oakland, California 90022 (415) 555-7500 Fax (415) 555-7150

The bulleted list may not be essential if the attachments or enclosures are clear. See ORGANIZATION and LISTS.

The writer provides all necessary information to ease cooperation in the response—for example, a personal phone number (closing paragraph) and an email address.

Response Letter

To a Concerned Customer

When you don't know the addressee's gender, use no courtesy title, such as *Mr.*, *Ms.*, and *Mrs.*, in the inside address or salutation. See LETTERS.

The first paragraph establishes the context of the response and attempts to develop rapport with the customer.

Acronyms are good shorthand devices, but be cautious about overloading the text with unfamiliar acronyms that hinder reading. When you introduce a new or unfamiliar acronym, use the acronym and then, in parentheses, spell out the name or expression. See ACRONYMS.

Graphics are the best devices for emphasizing information. Because this graphic is highly visual, it provides emphasis and is more memorable than the written text around it. The advantages of the cheese substitute are immediately identified without reading any supplemental text. See GRAPHICS.

January 10, 2010

Sidney Lambert
1667 Willow Avenue
Seattle, WA 96508

Dear Sidney Lambert:

Why do we use a mozzarella cheese substitute in Farmland Frozen Family Foods lasagna?

I appreciate your recent inquiry about the use of a mozzarella cheese substitute in our FFFF (Farmland Frozen Family Foods) lasagna. I, too, am worried about the increasing presence of artificial products and additives in our food and sympathize with your concern.

Let me assure you, however, that the mozzarella cheese substitute and all other ingredients in our lasagna are as wholesome and safe as we can make them. We decided to use the mozzarella cheese substitute only after reviewing the results of extensive testing showing generous benefits to our customers.

Lowered Cholesterol Levels. The mozzarella cheese substitute has no cholesterol but does have all the vitamins, minerals, and protein found in natural mozzarella. Natural cheese has animal fat, which contributes the cholesterol, while the substitute has soybean oil.

Same Delicious Taste. The taste, smell, and appearance of the mozzarella cheese substitute and natural mozzarella are the same.

303 Blossom Avenue, Des Moines, Iowa 50321 (515) 521-4911

Letters responding to customer complaints should be as personal and direct as possible. The writer must address all questions or grievances in the reader's earlier letter and must do so in a friendly, sympathetic, and courteous manner.

This letter illustrates the modified block style, which is not commonly used in business today. The date, complimentary closing, and signature block appear right of center; everything else is flush left. The paragraphs are not indented.

Sidney Lambert
January 10, 2010
Page 2

Quality Ingredients. The ingredients in natural mozzarella and the substitute are almost identical: water, milk protein, and fat. Both products have similar minor additives to enhance the flavor, prevent spoilage, and guarantee consistent quality.

Lowered Product Cost. Using the mozzarella cheese substitute helps us lower the cost of our lasagna, especially given the recent increase in the price of natural cheese. This economic advantage allows us to put more mozzarella flavor into our lasagna for lower cost—something many of our customers want.

I hope this letter has answered your questions about our use of a cheese substitute and will make you feel comfortable about purchasing our lasagna. You mentioned that you enjoy a variety of FFFF products, so I am sending you several complimentary coupons. Thank you for your concern and for your interest in Farmland Frozen Family Foods.

Sincerely,

Martha Frampton

Martha Frampton
Public Relations

MF:sj
Enclosures

The run-in headings match exactly the key benefits shown in the graphic. The reader can glance at the headings to determine where to read more closely and where to skim. See HEADINGS.

The concluding paragraph restates the main objective of the letter and assumes a positive customer reaction.

In the modified block style, the signature block is right of center. See LETTERS.

Although this letter is a response to a concerned customer, it also presents an opportunity to sell a product. The writer designs a message that speaks clearly, effectively, and persuasively to the customer. See PERSUASION.

The personal pronoun *I* is appropriate, even though the writer is speaking for FFFF. She shifts to the collective *we* when she speaks for the company as a whole. Such shifts in pronouns are natural and acceptable. See TONE, LETTERS, and PRONOUNS.

Response Letter

To a Complaint

The inside address follows the same format as the envelope address, allowing both to be printed from the same computer text. This is an optional convention for business letters. The U.S. Postal Service prefers capitals and no punctuation, especially for the last two lines. See LETTERS.

The subject line opens the letter by clearly stating the major point of apology. The boldface type emphasizes this key message. If Mr. Elkins read only the subject line, he could predict what ideas might follow. See LETTERS and EMPHASIS.

The displayed list emphasizes the precautions taken by Farmland Frozen Family Foods. See EMPHASIS, LISTS, and PARALLELISM.

The closing (and the coupon) builds rapport with the customer. The last sentence repeats the opening message. See REPETITION.

June 25, 2010

Mr. Wiley G. Elkins
1456 JACKSON AVENUE
JORDAN MO 64833

Dear Mr. Elkins:

We apologize for the small chunk of burned potato in your Potato Ripples.

Thank you for your recent letter about Potato Ripples. The blackened material you found was a small chunk of burned potato and is not harmful, although I appreciate your dissatisfaction at finding it.

You deserve quality in our products. We do make every effort to prevent chunks of burned potato from being packaged with our Potato Ripples. These blackened chunks occur when particles break loose from potatoes and collect in our deep-fat frying system. To prevent these chunks from being packaged, we perform these steps:

- Continuously filter the oil in our fryers
- Routinely clean the fryers
- Visually inspect all Potato Ripples before packaging

I am enclosing a coupon for a free package of Potato Ripples, Potato Crisps, Zucchini Fritters, or Eggplant Gems. We hope you will use the coupon to try Potato Ripples again or to try one of our other products. Again, thank you for your letter and for your interest in Farmland Frozen Family Foods.

Sincerely yours,

Jackson Blaine

Jackson Blaine
Quality Control Manager

JB/mm
Enclosure

303 Blossom Avenue, Des Moines, Iowa 50321 (515) 521-4911

Tact and a personal touch are crucial in letters to disgruntled customers. Personal pronouns help establish a tactful tone, and the details include specific references to Mr. Elkins's letter without repeating the negative events verbatim. See PRONOUNS and TONE.

The writer avoids technical complexity and jargon (he might have called the chunk of potato a "carbonized carbohydrate particle"). See JARGON, GOBBLEDYGOOK, and SCIENTIFIC/TECHNICAL STYLE.

The lead sentence sets up the problem, and the second sentence asks for a solution. Mr. Van Prooven could have asked for the solution as soon as possible. Instead, "within the next two weeks" softly defines a deadline and his expectations. See LETTERS and ORGANIZATION.

The table quickly presents the supporting information and the approved specifications. See EMPHASIS.

The closing paragraph establishes how serious the writer considers the problem to be and applies some pressure; yet the tone is not blunt or negative. See TONE.

August 8, 2010

Patricia Goodway
Customer Support Manager
The Pacific Baking Company
9924 West Pacific Way
San Francisco, CA 97521

Dear Patricia Goodway:

Request for Adherence to Breader Specifications

Recent samples of your breader contain more fines than our specifications allow. Please investigate and let me know within the next 2 weeks what you will do to solve the problem.

On August 5, our Medford plant inspectors noticed excessive carbon specks on breaded products, so they checked the breader lots on hand (codes 341X, 344X, and 345X). The following table shows the results.

Fines Through a U.S. Standard Sieve No. 80

Breader Lots	Percent Must Fall Within 1% ± 1% or Maximum of 2%	Exceeds Farmland Frozen Family Foods Specifications
341X	3.3%	1.3%
344X	3.6%	1.6%
345X	4.1%	2.1%

As you can see, the fines in these lots exceed our specifications. These levels are not yet serious, but they do cause excessive carbon specks in the frying oil and on the finished products.

We have enjoyed a long relationship with Pacific Baking and hope to continue doing business with you. We are therefore anxious to see this problem resolved and will appreciate your prompt action.

Sincerely,

James Van Prooven

James Van Prooven
Plant Manager

JVP/kls

303 Blossom Avenue, Des Moines, Iowa 50321 (515) 521-4911

Use tables to give readers exact figures or to permit readers to compare figures or other information. Tabular information is usually more understandable than the same information presented in text. See TABLES.

As in this letter, complaint letters usually start with a concise statement of the problem followed by a request for action or resolution. The problem statement should be as brief as possible but long enough to make the request understandable. See the discussion of setups in ORGANIZATION.

Complaint Letter

With a Tactful Request for Aid

A reader can easily predict the message of the letter from these subject lines. This list previews or maps all key content, listing points in order of importance.

The opening paragraph establishes the problem and its seriousness. This paragraph states clearly what Mr. Whitaker wants the director to know and do. See LETTERS and ORGANIZATION.

This description of a delay is concise yet specific. Note the paragraph structure—each sentence leads into the next, and the paragraph ends with a strong statement. See PARAGRAPHS.

The run-in headings match the bulleted subject lines. The director can scan the letter and easily identify content where detailed reading is desired. Each point is supported with additional information. See HEADINGS and ORGANIZATION.

Midland Oil & Gas Operations, Inc.
7000 Jalepeno Boulevard Dallas, Texas 75234
(214) 735-9600

November 15, 2011

District Director
U.S. Immigration and Naturalization Service
2645 S ANNE STREET
HOUSTON TX 77004

Requests:
- **Eliminate Unloading Delays.**
- **Clarify Unloading Policies and Procedures.**
- **Provide Contact Number for Officials.**

I am requesting the three actions listed above because of a recent history of expensive delays. During the past year, MOGO (Midland Oil and Gas Operations) ships delivering crude oil to our Houston refinery complex have been unnecessarily delayed by Immigration officials who have not arrived for boarding as scheduled. These delays cost MOGO (and ultimately consumers) an average of $2,745/hr. We would appreciate whatever you can do to eliminate or at least reduce these delays. Please contact Jack Severenson or me at (214) 555-2233 to discuss possible solutions.

Eliminate Unloading Delays. Occasional delays are inevitable, but we cannot continue to sustain financial losses due to such delays and remain competitive unless we increase the price we charge consumers for our oil products.

The most recent delay occurred on September 28, when the MV *Seaworthy* docked in Gulfview. Boarding was scheduled for 0400 that morning, but Immigration officials did not arrive until 0930. When the officials finally did arrive, they apologized for being late by saying, "We forgot." Our understanding is that dock workers cannot unload a foreign vessel until Immigration officials have cleared it, so unloading the September 28 shipment was delayed over 5 hours. That delay cost MOGO over $15,000.

Clarify Unloading Policies and Procedures. Please clarify for us the regulations regarding the unloading of a vessel prior to boarding by Immigration officials. In particular, can dock personnel unload a vessel that has cleared Customs but has yet to clear Immigration?

Provide Contact Number for Officials. We would appreciate having an after-hours telephone number to call when Immigration officials do not arrive on schedule.

Complaint letters have to be tactful yet firm. They are usually based on one or more incidents, which must be established before you can request a remedy. The opening sentences of the first paragraph provide background information to establish the nature and seriousness of the problem. The request that follows is therefore understandable and reasonable. See ORGANIZATION and LETTERS.

Tone in complaint letters is important. Such letters should not sound shrill, angry, or unreasonable. In this letter, the author writes calmly but directly. He presents a specific case and then makes several understandable requests. The letter ends with a plea for cooperation. See TONE.

The closing paragraph repeats the purpose of the letter. The telephone and Internet address strengthen the pleasant yet direct closing. See LETTERS, ORGANIZATION, and REPETITION.

Reference initials are always flush with the left margin. The writer's initials, *FWW*, can be separated from the secretary's initials, *dfe*, by a colon, as shown here, or a slash (/). See LETTERS.

U.S. Immigration and Naturalization Service
November 15, 2011
Page 2

Please call Jack Severenson or me at (214) 555-2233 to discuss solutions to the unloading delays. I can also be contacted at fwwhitak@somogo.com. We want to do everything we can to cooperate with the U.S. Immigration and Naturalization Service.

Frank W. Whitaker

Frank W. Whitaker
Shipping Operations Coordinator
fwwhitak@somogo.com

FWW:dfe

This letter illustrates the simplified format. The simplified format omits the salutation and the complimentary closing. This format is appropriate in cases where you don't have a name for the addressee, or the addressee is a title or a company department.

Employment Reference Letter

The subject line and opening paragraph announce that this will be a positive recommendation. Don't delay this information. See ORGANIZATION.

Paragraph 1 previews the main reasons for the recommendation. These reasons become the major headings in the letter.

Examples should be specific and detailed enough to be believable to the reader. Vague statements (such as "a good worker" or "courteous manner") mean little without supporting proof. A good test is that this letter could apply only to Regina Sproutly, not anyone else.

June 4, 2011

Josephine E. Longstreet
Human Resources Director
South Shore Clothing, Inc.
3349 Brentwood Street
Houston, TX 99756

Dear Josephine E. Longstreet:

Recommendation for Regina Sproutly

I am pleased to be able to recommend Regina Sproutly for a position as a merchandising manager. Regina has worked in merchandising here at Berkshire for over 3 years. About a year ago, we promoted her to assistant merchandising manager, based on her enthusiastic, take-charge manner and her excellent customer-service ethic.

Her Enthusiastic, Take-Charge Manner

Regina approaches every task—no matter how time-consuming—as a personal opportunity for her to achieve something. For example, a merchandising clerk in another department suddenly resigned, leaving his inventory in bad shape. Records were not just incomplete, but wrong, and many pieces of clothing turned out to be mismarked. We asked Regina to fill in and to clean up that department. She eagerly accepted the assignment, and within less than a month, she had everything up to our standards. She appeared to thrive on the hard work required.

Regina is enthusiastic and positive, no matter the task. Too often, employees just seem to be putting in their time. Regina's manner shows she is not just putting in time.

Her Excellent Customer-Service Ethic

At Berkshire, we take pride in our customer service. To achieve this goal, we give our employees the responsibility and authority to do what it takes to please every customer. Regina has shown that she understands this policy and uses it every time she works with customers.

On one occasion, Regina got a call from a frantic customer saying that he had bought the wrong-sized coat for his wife's birthday celebration that night. Regina immediately volunteered to deliver the correct-sized coat to the man that night.

On a second occasion, we got a letter from one of our longtime customers praising Regina. This customer is well known to Berkshire owners because she usually writes with complaints about our products or services. We were, therefore, surprised when this customer took the time and trouble to praise Regina.

1021 Missoula Ave., Dallas, Texas 59701

Reference letters are a vanishing custom in the business world. Because of possible legal questions, many companies now have policies that prevent any manager from writing a reference letter. At the most, a manager or human resources specialist will write a letter verifying a person's employment status, but making no statements about the person's skills or abilities. (See the verification letter following this reference letter.)

Also, be aware that if you write a reference letter, the person discussed in the letter may have a legal right to see the letter. So do not consider your letter confidential or privileged.

The letter ends with a summary section reinforcing the recommendation. See REPETITION.

BERKSHIRE
DEPARTMENT STORES

Josephine E. Longstreet
June 4, 2011
Page 2

Our Loss, Your Gain

You'll probably have gathered that we are sorry to see Regina leave Berkshire. Unfortunately, we have to close two of our five stores in the Dallas area and can only retain workers with a good deal of seniority. So our loss will be your gain.

Please call me if you would like further information about Regina's abilities. I can be reached at (214) 883-9991 (the number at our central Dallas warehouse).

Sincerely,

Lee S. Baskett

Lee S. Baskett
Merchandising Manager
leesbaskett@sosclothing.com
LSB:cv

1021 Missoula Ave., Dallas, Texas 59701

If you need to write a reference letter, be honest and be specific. You should honestly cite specific examples of the person's abilities and accomplishments. If you do not have positive examples, we recommend that you tell the applicant to find another person to write the recommendation letter. Also, if you praise someone falsely, you open yourself and your company to legal liability if that person is hired based on your letter and then has problems in the new job. See ETHICS.

This letter illustrates the full block format, with all parts of the letter flush left. See LETTERS.

Employment Verification Letter

May 31, 2011

Harold O. Seythe
Loan Officer
Westside State Bank
1413 LaSalle Street
Detroit, MI 34825

Dear Mr. Seythe:

Subject: Letter of Employment Verification for Jason Stoutwell

Jason Stoutwell has been employed as a shipping clerk for Bountiful Chemical since August 18, 2004.

List the key job responsibilities. These should be nonjudgmental descriptions of essential job tasks.

Jason's responsibilities include:

- Packaging items
- Preparing shipping labels
- Entering shipping information on a computerized tracking system
- Assisting team members with back orders

As a courtesy, provide your contact information.

If you need further information about Jason's responsibilities, please contact me.

Sincerely,

Annette Zimmer

Annette Zimmer
Human Resource Director
azimmer@westsidestatebank.com
(801) 555-3343

139 Sequoia Park Way, Suite 303, Oakland, California 90022 (415) 555-7500 Fax (415) 555-7150

Use an employment verification letter when employees need you to verify their position(s) and their dates of employment. Banks often request such letters. Also, insurance companies often request such information when they are verifying insurance status (especially when an employee has double coverage or the employee's spouse has overlapping coverage).

A verification letter is not a reference or a recommendation letter in the traditional sense. See the preceding model in this *Style Guide* for an example of a reference letter.

Bid Solicitation

The first paragraph concisely states the letter's purpose and indicates that the letter has enclosures. See ORGANIZATION.

The table quickly presents the support information, with essential details available at a glance. See EMPHASIS and TABLES.

This paragraph is an effective yet not entirely subtle appeal to the reader to keep the bid price low.

The closing paragraph repeats the deadline date given in the table. See ORGANIZATION and REPETITION.

822 Ocean View Drive,
Long Beach, California 90802
(714) 332-3978

April 24, 2011

Mr. James Quirk
Wizard Machine Tools, Inc.
Flatbush Boulevard
Montrose Island, IL 44572

Subject: Invitation to Bid for Delta Q Indicators

Sky Aviation is soliciting bids for Delta Q indicators for our Foxx 176 business jets. Please examine the attached drawing and specifications and submit your quote to Ernestine Gonzales in our Purchasing Department by May 16. Bid requirements are shown in the following table.

Bid for Delta Q Indicators

Date Due	**Friday, May 16, 2011**
Production Number	**1,980 Delta Q Indicators**
Delivery Rate	**55 per month for 36 months**
Date Delivery Begins	**November 4, 2011**

I understand that the indicators you already provide for our Windstream 88 fixed-wing aircraft are similar to the proposed Delta Q indicators. This similarity should make design and fabrication of the new indicators relatively easy and should therefore reduce development and manufacturing costs.

Please submit your quote so it arrives no later than Friday, May 16. If you have questions, please email or call me or Jim Booth at (714) 332-3984.

Arnold Madsen

Arnold Madsen
Manager, Engineering
amadsen@skyavia.com

AM/mm
Enclosures
1. Drawing DQI-514-B
2. Attachment A (Specifications)

This letter features a classic letter design: a concise opening that clearly indicates the letter's purpose (to solicit bids); a crisp middle with details appearing in table form and a short, well-designed paragraph; and a courteous closing that repeats the deadline and ends with an offer of assistance. See LETTERS and ORGANIZATION.

Some company letterheads display Internet or email addresses as part of the address line. But most frequently, the address is placed in the text of the letter or just below the signature block, as shown above.

This letter illustrates the simplified format. See LETTERS.

Sales Letter

With a Soft Sell

An attention line is used when the inside address does not contain either the name of an individual or the name of a department. The attention line allows the writer to identify a person or department to review the letter as soon as it arrives. See Letters.

The opening is positive yet low-key, and the writer establishes the purpose of the letter in the first paragraph. See Organization.

The bulleted list and run-in headings highlight key points. See Headings, Lists, and Letters.

RFP (Request for Proposal) is an abbreviation that will be familiar to the readers. Therefore, the writer does not have to spell it out. See Abbreviations.

The displayed list emphasizes Sky Aviation's production strengths. See Lists and Emphasis.

822 Ocean View Drive,
Long Beach, California 90802
(714) 332-3978

June 18, 2011

Attention Mr. Boon Hollenbeck
International Aeronautics
PO BOX 1149
GALVESTON TX 41504

Proposal to Assist in Designing the Cabin Pressurization System for the K-38

We are delighted to hear that IA (International Aeronautics) won the contract to develop the K-38 Heavy Cargo Helicopter. As you initiate design studies, we hope you will consider basing cabin pressurization on our patented flowback valve, the VA-321-E, and allowing us to assist in cabin-pressurization design.

Choosing to use the VA-321-E would bring the following benefits to IA's K-38 development program:

- VA-321-E meets or exceeds IA's design requirements.
- VA-321-E increases operational efficiency.
- Sky Aviation's industry track record shows we deliver on time and within budget.

VA-321-E Meets or Exceeds IA's Design Requirements. According to the RFP, the K-38 will require an 8-lb/min valve capable of maintaining a delta P of 0.3 to 0.5 psi. Valve VA-321-E meets and exceeds these requirements. The attached drawing (DRA-321-E) shows the standard SA configuration for the VA-321-E valve, including the outflow, check, and solenoid subsystems. The control modes within this configuration are listed on Attachment A.

VA-321-E Increases Operational Efficiency. The VA-321-E valve is more efficient than any other valve currently used in aircraft-pressurization systems. No other valve regulates differential pressure as accurately as the VA-321-E.

Sky Aviation's Industry Track Record Shows We Deliver On Time and Within Budget. Each of the following production steps is carefully monitored by SA's Production Team.

✓ Accurate sizing of the outflow valve-return spring to match your design specifications. This design-and-test phase would take no more than 4 weeks—less than half the time required by other valve manufacturers.

✓ Customized detailing to your design specifications to assure integration with other design components and to eliminate costly rework.

✓ Careful attention to quality assurance in production to decrease waste, reduce time, and increase cost efficiency.

The attention line has traditionally been placed two spaces below the inside address. Instead of this traditional placement on envelopes, the U.S. Postal Service prefers that the attention line be placed above the company name, the street address, and the city and state. To be consistent in the letter, the attention line is placed as the first line of the inside address. See Letters.

This letter illustrates the simplified format. See Letters.

With a Soft Sell

822 Ocean View Drive,
Long Beach, California 90802
(714) 332-3978

Mr. Boon Hollenbeck
June 18, 2011
Page 2

This paragraph presents relevant past-performance information—justification for selecting Sky Aviation.

Sky Aviation has done much of the pioneering work in cabin pressurization. Our pressurization systems are the state of the art, largely because of our patented flowback valve, the VA-321-E. If you decide to use our valve, you will have all our technical resources at your command, including the engineers who designed the pressurization systems for over 8,000 aircraft flying today.

Please review the enclosed materials. I will call Thursday, July 2, to discuss how we can assist you in the cabin-pressurization design for the K-38. Please call Fred Huber or me at (714) 555-3973 if you have questions or need further information.

Howard C. Patterson

Howard C. Patterson
Vice President
hpatrson@skyavia.com

The list of enclosures helps the reader check to see what should be included. See LETTERS.

HCP:cv
Attachments
1. Drawing DRA-321-E
2. Attachment A, "Control Modes for VA-321-E"

A sales letter is a blend of fact and sales pitch. Presenting information positively is essential, but sincerity and realism are also necessary ingredients of successful sales letters. See PERSUASION. This letter opens and closes with a soft sales pitch. The technical information in the middle is emphasized by boldface headings. The writer's goal is to make the technical information seem substantial and convincing. See EMPHASIS, SCIENTIFIC/TECHNICAL STYLE, and TONE.

The header easily identifies the attachment, and the boxes provide the last revision date and the individual responsible for approval.

Cross-reference numbers allow the reader to connect data from the attachment to the drawing.

822 Ocean View Drive,
Long Beach, California 90802
(714) 332-3978

Attachment A
International Aeronautics
June 18, 2011
Page 3

Engineering Data Sheet
Revision of November 4, 2010

Drawing Reference DRA-321-E	Subject Control Model for VA-321-E
Approval Jason Wells *JWells*	Original Date January 7, 2010

The control modes within the standard Sky Aviation configuration are as follows:

- **Minimum Differential Pressure Mode.** In a 2-psi vacuum with the solenoid valve (3) energized, the outflow valve (2) will be fully open. With an 8-lb/min throughflow, this will result in maximum differential pressure of 0.75 inches of water.
- **Positive Differential Pressure Control.** With the solenoid valve (3) deenergized, the outflow valve (2) will move toward the closed position and regulate the differential pressure to 0.6 psi. The regulation point is achieved by the correct sizing of the main poppet return spring.
- **Negative Differential Pressure Control.** As presently configured, the proposed system incorporates negative differential pressure control—i.e., if the cabin pressure becomes less than the ambient air pressure, the outflow valve (2) opens and admits ambient air into the cabin at approximately 0.4 psi. NOTE: This feature can be eliminated. Follow instructions QR-18 noted on the specification drawings.

Drawing DRA-321-E and Attachment A come from the engineering design team. The drawing and attachment provide necessary information to the sale. To determine if an attachment is justified, ask the following question: "Would a knowledgeable reader need the information in the attachment(s) to interpret the conclusions and recommendations?" If the answer to that question is yes, the attachment is justified. See Appendices.

Note: Drawing DRA-321-E is omitted.

Template/Mail Merge

The subject line is informative. See LETTERS.

The contractions *You're* (first sentence) and *I'll* (second sentence) both signal an informal, friendly tone. See LETTERS and CONTRACTIONS.

The three major headings (benefits) are informative enough that readers can easily scan the letter. See HEADINGS, EMPHASIS, and LISTS.

The next-to-last paragraph repeats the date when the writer plans to call to set up a meeting. See REPETITION.

The lack of a complimentary closing is a feature of the simplified letter format. The short last paragraph functions as an informal closing. See LETTERS.

FranklinCovey

*^date^*

*^name & address^*

Invitation to Enjoy the Productivity Benefits of FranklinCovey's Renowned *FOCUS* Seminar Workshop

Dear *^first name^* :

You're invited to be our guest at our workshop, *FOCUS–Achieving Your Highest Priorities*. I'll call by *^day^* to find a convenient time to talk with you about how you can achieve the following benefits:

Enjoy significant, measurable increases in corporate productivity.

As your employees gain more responsibilities, you're naturally concerned about keeping them focused on corporate objectives. FranklinCovey programs will help employees keep organizational priorities straight and will increase productivity by 10 to 29 percent.

Join the blue-ribbon corporations who have boosted productivity with FranklinCovey training.

You'll benefit from the same productivity increases that these and hundreds of other top companies and government agencies now enjoy as a result of FranklinCovey training:

- Merrill Lynch
- Westinghouse Electric Corporation
- Merck & Co.
- Dow Chemical
- Federal Aviation Administration

More than *^x.x^* million people, mostly professionals in the corporate world, have been trained and can vouch for the productivity-enhancing power of the FranklinCovey programs.

Attend the FranklinCovey workshop as our guest.

We'd enjoy having you at one of our upcoming workshops as our guest (see enclosed flier for dates). This way you can evaluate our program firsthand and decide how it could help your organization—at no cost to you. When you attend, you will receive a complimentary FranklinCovey Planning System.

Again, I'll call you by *^day^* to talk about setting up a meeting to explore how we at FranklinCovey can help you with long-term gains in productivity and effectiveness.

Thanks for your time.

^Name of Account Executive^
FranklinCovey Account Executive
^email address^
1 Enclosure: National workshop schedule

Washington Building
Second Floor
2200 West Parkway Boulevard
Salt Lake City, Utah 84119
(801) 956-1300
Fax (801) 956-1301

A template is a document model or shell that writers can tailor to their own specific needs. Most software programs allow you to copy a template and then rewrite or revise it to fit your needs. Other software programs, similar to the pattern illustrated above, have blanks for you to provide information. This information, once typed in, merges into the letter, so you don't have to do any revising or rewriting.

Customer Service Letter

The subject line is lengthy but informative. See LETTERS.

The focus on *you* in the opening paragraph sets a good customer-service tone for the letter. See LETTERS and TONE.

Paragraph 2 previews the two content subjects to be covered in the letter. See ORGANIZATION.

The displayed list (with bullets) needs no final periods because the listed items are not sentences. See LISTS.

The subheadings are questions, so they are parallel. See PARALLELISM and HEADINGS.

OSAGE GAS & ELECTRICAL COMPANY INC.

February 1, 2011

Miller Door & Gate Company
2000 E. Lake Street
Davenport, IA 54882-2345

Gain Significant Electric Power Cost Savings by Applying for the Interruptible Supply Rate

You can reduce your electric power costs by up to 10 percent through our new Interruptible Supply Program. The reduced rate applies if you permit us to interrupt part of your power supply in rare cases when emergencies or maintenance needs arise. Please send us a letter of application by March 1 if you'd like to participate.

Let me explain how to apply for the reduced rate and how the program affects you.

How to Apply for the Interruptible Supply Rate

To apply, simply send a letter to me by March 1 with the following information:

- A formal request to apply for the Interruptible Supply Rate
- The load to be interrupted (presumably a low-priority area for you)

How the Interruptible Supply Program Affects You

How Much Will the Program Save You? You should save from $18,000 to $21,500 a year, depending on the size of the interruptible load and your past electricity costs.

How Much Power Will Be Subject to Interruption? You should be able to do without 50 kilowatts and up to 2 megawatts temporarily, in the area you designate.

How Long Do Interruptions Last? At no time would your power be interrupted for more than 8 hours a day, or for more than 100 hours a year. We will help you define in advance where and how much power you can afford to have interrupted. (The attached Interruptible Supply Rate tariff sheet contains further information.)

4239 E. Jennings Avenue, Cedar Rapids, Iowa 54336

This customer letter is a combination of service information and an indirect sales request for companies receiving it to apply for the reduced rate. As such, it is more factual than persuasive, although some sentences highlight sales benefits. See PERSUASION and TONE.

The MPSC abbreviation in parentheses follows the full name of the commission. Other editors would reverse the order, using the initials before the full name in parentheses. See ABBREVIATIONS.

Parentheses contain additional information. In this example, they are optional; the text would be fine without them. See PARENTHESES.

The closing paragraph reemphasizes the need for timely action.

OSAGE GAS & ELECTRICAL COMPANY INC.

Miller Door & Gate Company
February 1, 2011
Page 2

Why the Interruptible Supply Program? Uninterruptible power supplies are very expensive to maintain. We don't want you to pay high prices for emergency power at peak use times, and occasionally we must cut power to maintain our system. Therefore, the Michigan Public Service Commission (MPSC) permits us to reduce your power costs if you permit us to interrupt a predetermined portion of your supply when these needs arise. (To help with your planning, you should know that the MPSC will authorize 5 more megawatts of interruptible power each year.)

Apply Early for Reduced Rates

To take advantage of these reduced rates, send me a request letter with the information above by March 1. Apply early because we anticipate that requests will far exceed the limited capacity authorized by the MPSC. I have enclosed my business card if you wish to call for more information.

Elizabeth Johnston

Elizabeth Johnston
Manager
Corporate Power Management
ejohnston@osagege.com

EJ:tm

Attachment: Tariffs Sheet for Interruptible Supply Rates

4239 E. Jennings Avenue, Cedar Rapids, Iowa 54336

The letter follows the four-box letter organization. Box 1 (paragraph l) opens with a request or a statement of purpose. Box 2 (paragraph 2) summarizes or lists the points to be covered. Box 3 (the remaining paragraphs, except for the final one) expands on the points listed in box 2. Box 4 (the final paragraph) closes with a review of the content and a final request for action. See ORGANIZATION.

The letter follows a simplified format. See LETTERS.

Procedure Memo

The subject line and the opening sentence clearly establish the purpose of the memo. See LETTERS and ORGANIZATION.

The side comments, or callouts, provide supplemental information. See PAGE LAYOUT.

Items in the bulleted list are parallel in structure. See LISTS and PARALLELISM.

Use of graphics in procedures aids the reader's understanding of the information being presented. Graphics can emphasize important data and ideas in ways that text cannot. See GRAPHICS.

The split-screen or double-column format makes the text more readable. A 0.5-point divider line separates the main procedure text from the side comments. See PAGE LAYOUT.

822 Ocean View Drive,
Long Beach, California 90802
(714) 332-3978

To: All Department Heads
From: Susan Hall
Date: December 28, 2010
Subject: **Procedures for Storing Records at Trolley Street Warehouse**

Please comply with the following procedures when you ship records for storage at the Trolley Street Warehouse:

1. Obtain a standard records-transmittal box and a records-transmittal slip at skavia.corpserv/rts.

   Call 287-9009 to order the transmittal boxes. Each empty box contains a blank records-transmittal slip.

   Copies of the R-23 policy are available from sskavia.corpserv/rts.

2. Include the following information on each records-transmittal slip:
   - Date of records sent to Trolley Street
   - Description of the contents
   - Name of the responsible supervisor
   - Department number (and extension)
   - Destruction date according to policy R-23
   - Records-transmittal box number

   Box numbers are selected consecutively from the location log at the Trolley Street receiving desk. The numbers indicate the box-storage location at Trolley Street Warehouse.

   Central Records

   Date ____________

   Contents ____________

   Supervisor ____________

   Dept No. ____________
   Destruction date ____________
   Box No. ____________

3. Save and print the records-transmittal slip.
4. Attach the records-transmittal slip to the outside spine of the records-transmittal box within the marked space.

The imperative (command) verbs in the numbered list are essential in procedures. They highlight actions the reader should take, and they allow writers to condense their directions. For the sake of parallelism, state all directions or steps using imperative verbs. See SENTENCES and VERBS.

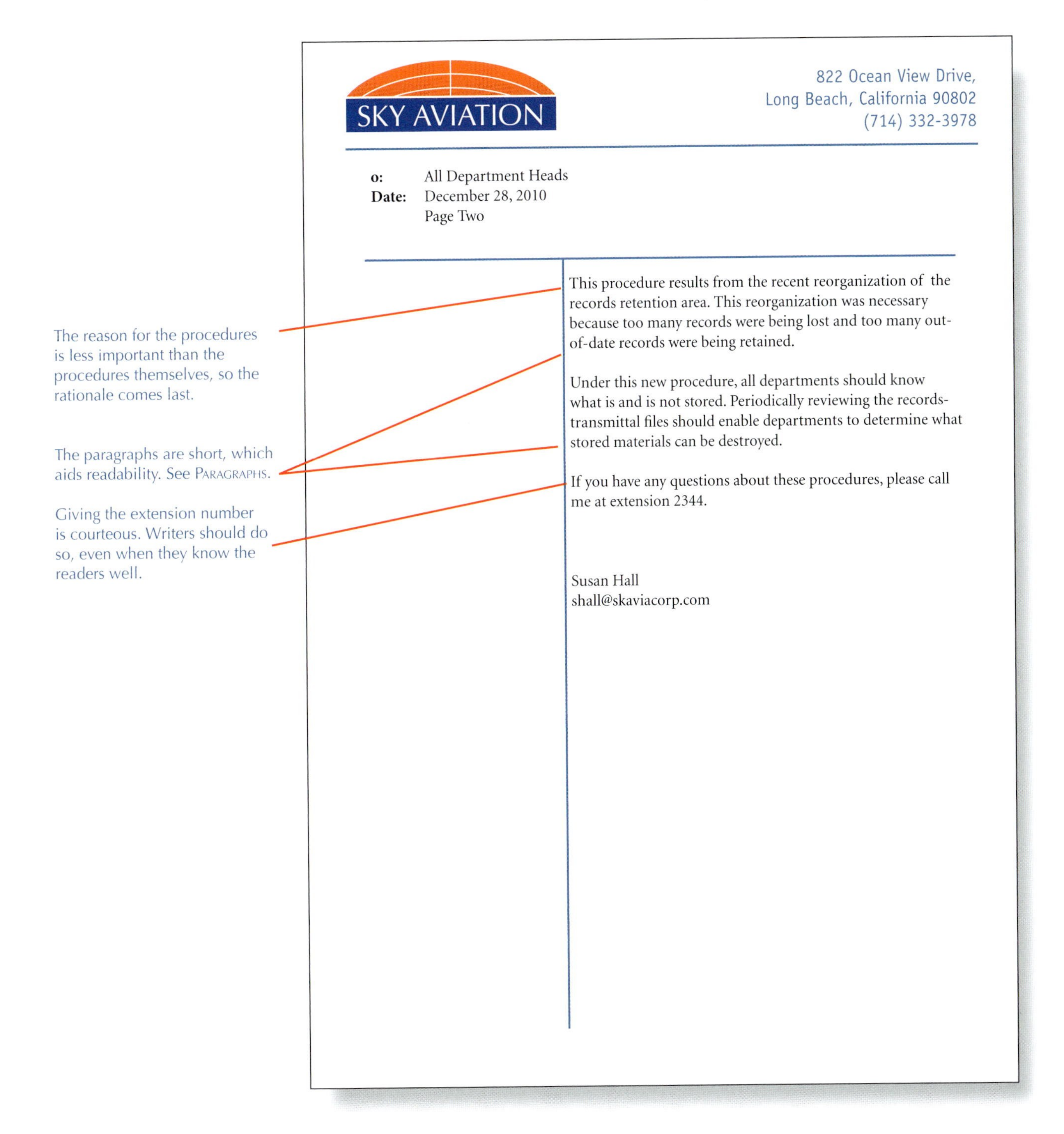

SKY AVIATION
822 Ocean View Drive,
Long Beach, California 90802
(714) 332-3978
o: All Department Heads
Date: December 28, 2010
Page Two
This procedure results from the recent reorganization of the records retention area. This reorganization was necessary because too many records were being lost and too many out-of-date records were being retained.
Under this new procedure, all departments should know what is and is not stored. Periodically reviewing the records-transmittal files should enable departments to determine what stored materials can be destroyed.
If you have any questions about these procedures, please call me at extension 2344.
Susan Hall
shall@skaviacorp.com
The reason for the procedures is less important than the procedures themselves, so the rationale comes last.
The paragraphs are short, which aids readability. See PARAGRAPHS.
Giving the extension number is courteous. Writers should do so, even when they know the readers well.

A clear format is essential in procedures. In this memo, the numbered and bulleted lists are crucial if the procedure is to be clear and readable. Each logical step must be complete and documented. Information gaps leave the reader unsure of what action to take.

See EMPHASIS and LISTS; also see the more formal procedures illustrated elsewhere in these model documents.

Request Memo

For Clarification of a Problem

DATE November 25, 2009

TO Howard Deedy

FROM Charlotte Smart

SUBJECT Request to Investigate Improper Repairs on the Barrett-Woodward Rotary Car Dumper and Positioner

The specific subject line simplifies cross-referencing and filing. See HEADINGS.

The recent repairs on the BW (Barrett-Woodward) rotary car dumper and positioner didn't follow the recommended procedure and didn't fix the problem. Please investigate why the repairs were not made according to BW's recommendations and who was responsible.

The opening paragraph conveys the purpose of the memo and immediately presents the action requested. See ORGANIZATION.

Please contact me by next Tuesday, December 1, with a report of your findings. Here are the facts as I understand them.

The second paragraph establishes a definite deadline and leads into the details. See ORGANIZATION.

Problems Discovered

Early in November, John Sturgees reported many problems with the BW rotary car dumper and positioner as a result of a routine safety inspection. He called the BW Customer Service Engineering Group for help.

Repairs Recommended

Displayed lists are naturally emphatic, so main points are highlighted. Note that listed items are grammatically parallel. See EMPHASIS, LISTS, and PARALLELISM.

The BW consulting engineer analyzed the problems and recommended the following:

1. Replacing the 1-cm steel rails with 2-cm rails
2. Installing the 2-cm rails by grouting under them and continuously welding the AR (abrasion resistant) plate to the rails
3. Installing reinforcing cross beams at 75-cm intervals

Compound modifiers, such as *1-inch*, must be hyphenated. See HYPHENS, MODIFIERS, and NUMBERS.

Repairs Improperly Completed

Management authorized the repairs on November 9. Before the repairs were completed, the BW consulting engineer discovered that the AR plate had been stitch welded, not continuously welded. He claims that he reported this to Howard Beale. However, I learned of the improper welding only after John Sturgees conducted another safety inspection and discovered it more or less by accident.

139 Sequoia Park Way, Suite 303, Oakland, California 90022 (415) 555-7500 Fax (415) 555-7150

While basically a request for information, this memo is potentially critical of the reader or people working for the reader. In such politically sensitive matters, tone is important. The writer must be forceful without being aggressive, direct but not blunt, businesslike but not inhuman. See TONE.

The headings basically provide a chronological list of events. Readers can easily glance at the headings and determine where they need to read more closely and where they can skim. See HEADINGS.

BW's Advice Apparently Ignored

From what I can determine, the improper welding was not replaced. Instead, the repairs were covered with grout. BW believes that the AR plate will eventually curl and become loose because of the improper welding.

Please let me know why the repairs were not made according to BW's recommendations and who was responsible. Please contact me as soon as possible (at least by Tuesday, December 1) to shed whatever light you can on this issue.

139 Sequoia Park Way, Suite 303, Oakland, California 90022 (415) 555-7500 Fax (415) 555-7150

The closing is direct but not harsh. "Please" makes the closing more courteous, and "shed whatever light," which is a cliche, makes it more colloquial and therefore personal. See TONE and CLICHES.

The deadline date is repeated in the closing paragraph. Restating a deadline or key point reinforces or emphasizes that key point in the reader's mind. As you design your document and organize your ideas, build in effective repetition of your most important ideas. See EMPHASIS, KEY WORDS, REPETITION, and ORGANIZATION.

Summary Memo

For an Executive Audience

To: Steward Pollack
Cc:
Subject: Recommendation to Contract with White River Maintenance, Inc.
▶ Attachments: Bid Documents
Arial | 11 | B I U

Based on the attached bids and related analysis, I recommend contracting with White River Maintenance, Inc., to perform this work. The reasons for my recommendations are as follows:

- White River is the lowest bidder of equally qualified contractors.
- White River can meet the scope of work.
- White River can implement now.

I recommend that the work begin on Wednesday, June 28. If you disagree with my recommendation, contact me at ext. 851 by noon, June 15.

White River is the lowest bidder of equally qualified contractors
All bidders, except J & L Incorporated, appear to be equally qualified, but White River is the lowest bidder by an estimated $50,000.

The following mechanical overhaul contractors submitted time-and-material bids to perform the compressor engine and cylinder overhauls at the OG&E plant:

- White River Maintenance, Inc.
- Martin Energy Services
- J & L Incorporated
- Efficiency Production Services

Bids included only time and material, so we analyzed the bids by developing a hypothetical crew size and calculating a composite worker-hour rate. From these calculations, we determined the total labor costs for each contractor. We also calculated probable transportation costs as an add-on amount to labor costs. Finally, we built in an 85 percent contingency amount based on the uncertainty of the parts needed to complete the overhaul. For further details, see the attached bid calculations.

A lengthy setup is unnecessary. The recommendation is presented as a direct statement in the opening sentence (followed by background explanations). See Letters and Organization.

Headings signal key divisions in the executive summary. They highlight the document's organization and allow readers to locate information easily. See Headings and Summaries.

Attachments should always be referred to in the body of the document. See Appendices.

Many memos are actually summaries of attached materials. Such summary memos are similar to executive summaries, which open long, more involved reports. Readers of summaries are often managers and supervisors who already know something about the subject but who do not need extensive background information before making a decision. See Summaries and Reports.

White River can meet the scope of work
The scope of the work includes the overhaul of 5 Hall-Burke HB-76 gas engines and 12 compressor cylinders. White River will unload all equipment and will handle all procurement through their office. They will also pay all transportation costs and for the complete installation of necessary replacement parts. Finally, they will be responsible for aligning and grouting the final crankshaft.

One-sentence paragraphs emphasize key points. See Paragraphs and Emphasis.

A copy of the specifications is available from Sven Nordstrom of Engineering Services.

White River can implement now
White River can implement immediately. Either J. K. Barnes or I will submit a weekly progress report and budget update. J. K. Barnes will be onsite at the White River shop to guarantee the quality of the work.

Please contact me at ext. 851 by noon, June 15, if you disagree in any way with my recommendation to contract with White River Maintenance, Inc. to perform this work (Contract MR-789-651). If I have not heard from you by that date, I will have my support staff prepare the contract for White River Maintenance, Inc., with a starting date of June 28.

The closing repeats the main point of the memo and emphasizes and documents the deadline. See Emphasis, Organization, and Repetition.

Brenda Hamilton

This memo (as in most executive summaries) does not need to contain extensive background data, calculations, supporting documents, and references. Such information should remain as attachments. See Appendices.

Proposal Memo

To an Antagonistic Audience

Because the readers are antagonistic, the memo proposes a meeting, not a solution, and the language is deliberately conciliatory. Note the inclusive *we*. See TONE.

The writer prepares the layout to facilitate note taking, and suggests a dual use of the memo—as an agenda. See PAGE LAYOUT.

The headings act as a map to additional information presented in the preview of key content—the bulleted list. See HEADINGS and LISTS.

OG&E (Osage Gas and Electric Company) is an acronym the readers will be familiar with, so it does not need to be spelled out. See ACRONYMS.

Using the readers' names and the pronoun *we* help make the tone friendly and personal. Consequently, this memo will be easier for the readers to accept. See PRONOUNS and TONE.

To: Jack Ladda, Harvey Dent
Cc:
Subject: Invitation to Discuss Upgrading Substation Testing and Maintenance
Attachments: *none*
Arial 11 B I U

As we found in Tuesday's meeting, we can agree on a number of current problems regarding upgrading substation testing and maintenance. Let's collaborate to discover ways to solve these problems. Could you meet at 2 p.m. on Wednesday, September 6, in my office to discuss ideas and solutions?

During Tuesday's meeting, we identified and agreed upon several specific test and maintenance areas needing improvement. Here are my notes taken at the meeting, and as a starting point, I've included a few preliminary suggestions. Please reply to this email with your suggestions below. I'd like to use it as the agenda for discussion.

- Overall commitment to maintenance as a priority responsibility
- Tests between maintenance intervals
- Maintenance intervals
- Adherence to maintenance procedures

Overall Commitment to Maintenance as a Priority Responsibility
Maintenance is not a high priority in many substations (although it should be, given the complex and expensive equipment in the OG&E electrical system).

MY SUGGESTIONS	YOUR IDEAS
1. Jack could issue email messages from management to increase awareness and stimulate adherence to maintenance procedures.	
2. Harvey, test requests could be issued on a *Priority 1* routine schedule.	

Tests Between Maintenance Intervals
Testing should be performed on a routine schedule, not just when a problem is noted. Currently, company regulations state that testing must be performed once a quarter, but no department is named as a responsible party—Operations has just assumed the task.

MY SUGGESTIONS	YOUR IDEAS
1. Jack, as manager, could be accountable for and issue a 1-year schedule with specific dates assigned for testing and for routine maintenance each quarter.	
2. Harvey, as supervisor of Engineering, could be responsible for and track the testing following Jack's schedule.	

Because of the antagonistic audience, the writer deliberately makes his memo more tentative than it would ordinarily need to be. The opening paragraph is conciliatory, and the writer's ideas are called preliminary suggestions, not proposals. See TONE.

Maintenance Intervals
Maintenance should be performed on a routine schedule. However, maintenance personnel who leave are not being replaced, so many substations have inadequate staff. Consequently, some of the routine procedures have not been completed regularly.

This double-column format allows separation of the writer's and readers' ideas, yet facilitates and implies collaboration. See PAGE LAYOUT.

MY SUGGESTIONS	YOUR IDEAS
1. Jack could be accountable for and issue a 1-year schedule with specific dates assigned for testing and for routine maintenance each quarter (same as above).	
2. We could make a comprehensive workforce and equipment forecast based on present maintenance needs and use this forecast as a basis to hire additional maintenance personnel.	

Adherence to Maintenance Procedures
Many personnel are not properly trained, and maintenance procedures are not posted. Some maintenance crews spend too much time on nonessential tasks.

MY SUGGESTIONS	YOUR IDEAS
1. We could train existing personnel to meet the above forecasts.	
2. We could develop work aids containing basic procedures.	
3. We could use contract or construction personnel, not maintenance personnel, for all construction work.	

Repeating the meeting time reinforces the central message of the memo and solicits the readers' cooperation. See EMPHASIS, ORGANIZATION, and REPETITION.

Please reply and let me know if you can meet at 2 p.m. on Wednesday, September 6, in my office to discuss these and other suggestions. Let's do what we must to reach constructive solutions to our test and maintenance problems.

The first-name-only signature adds continuity to the friendly tone established previously. See TONE.

Carla

Make your closing as simple and direct as your opening lines. You could close with a restatement of your main point, as shown in this memo, or you could refer to questions or problems, such as "Please call or write Neil Thornbush at (415) 555-01234 if you have questions about the proposed manufacturing schedule."

Request Memo

With Informal Instructions

To: Managers and Supervisors List
Cc:
Subject: Request to Prepare for Air-Conditioning Systems Review and Balance April 15–April 26
▶ Attachments: Moniter request; Moniter duties
Arial 11 **B** *I* U

We have contracted with Constant Air, Inc., an air-conditioning specialist, to review and balance the air-conditioning systems in Buildings 3 and 4 between April 15 and April 26. The objective is to ensure that all departments can independently maintain the temperatures appropriate for their areas.

Please complete the following three assignments **before 8 a.m., April 10:**

1. Appoint someone from your department as a monitor to assist Constant Air, Inc.
2. Reply to the attachment "Monitor to Assist Constant Air, Inc., April 15–26.
3. Instruct all employees to remove materials from the ventilation grills.

1. Appoint someone from your department as a monitor to assist Constant Air, Inc.

- Carefully select this individual. (A monitor must have time available to promptly complete all tasks as listed on the attachment.)
- Notify the selected individual. (Confirm their availability to meet with the Constant Air representative at 1 p.m., April 10, and to be working onsite April 15–26.)
- Fill in all information requested on the attachment, "Monitor to Assist Constant Air, Inc., April 15–26."
- Forward to the monitor the attachment. "Duties of the Air-Conditioning Monitor."

2. Reply to the attachment "Monitor to Assist Constant Air, Inc., April 15–26."

Paragraph 1 sets up the request in paragraph 2. Setups are optional, but in this case the managers and supervisors might not understand why the request is being made, so the setup is retained. See Letters and Organization.

The major points in paragraph 2 contain the exact wording of the subheadings in the document. This exact repetition is a good strategy in any document, but especially in longer memos and reports. See Repetition and Organization.

Bullets introduce the displayed list because numbering would be unnecessary, even distracting. See Lists and Numbering Systems.

Bullet 3 references the attachment, as does subheading 2. Using an attachment keeps this cover memo short and to the point. See Appendices.

This memo and its attachments are typical of internal communications within a company or division. In this memo, the writer chose to use the attachments to provide additional information to the appointed monitors. Many long memos (and letters) would be easier to write if they were similarly designed: a cover memo/letter outlining the basics, with details provided in one or more attachments. See Letters and Appendices.

The subheading "NOTE" and the boldface in the following sentence signal a key reminder. As an option, this reminder could come earlier (toward the beginning of page 1). See EMPHASIS and ORGANIZATION.

3. Instruct all employees to remove materials from the ventilation grills.

Many employees have tried to control the temperature in their offices by blocking the vents. Please ask all employees to remove cardboard, tape, and other material from the ventilation grills. Constant Air will not be able to draw valid conclusions about the current system unless the vents are clear.

Note: Only Constant Air, not Building Services, should adjust the air-conditioning systems between April 15 and April 26.

Your monitor will be meeting with the Constant Air representative in the Murri conference center at 1 p.m. on April 10. Please call me (ext. 4556) or Ned Grumer (ext. 4559) before the meeting if you or your monitor have any questions about these assignments.

Jacqueline Burrows

The memo follows the four-box organization. Box 1 (paragraphs 1 and 2) contains the request and some background information. Box 2 (the list in paragraph 2) presents the requested steps. Box 3 (the listed subheadings and text) expands on the steps. Box 4 (the final paragraph) closes the memo and reminds the managers of the deadline (the meeting date and time). See ORGANIZATION.

Request Memo

With Informal Instructions—Attachments

This block identifies the general subject and the date of the main memo. This information on attachments helps identify attachments if they become separated from the main memo.

The second attachment is essentially a numbered list of procedures or duties. The appointed monitor would receive this list and, optionally, the cover memo, which does provide some background information. See LISTS.

Items in the list are parallel in their structure. See PARALLELISM.

Attachment 1
Ai
A

Attachment 2
Air-Conditioning Memo
April 7, 2011

DUTIES OF THE AIR-CONDITIONING MONITOR

Questions? Contact Constant Air at 555-3421.

1. Meet with the Constant Air representative in the Murri conference center at 1 p.m. on April 10.
2. Inspect your department before your scheduled tour to locate all ventilation grills and ensure that all cardboard, tape, or other material blocking ventilation grills has been removed. NOTE: A tour schedule will be given to you when you meet with the Constant Air representative on April 10.
3. Tour your department with the Constant Air representative. Assist in identifying all ventilation grills.
4. Keep an hourly log of temperatures at selected points in your department from April 15 through 26. The Constant Air representative will identify these points during the tour of your department. NOTE: The Constant Air representative will provide log forms at the initial meeting on April 10.
5. Survey employees (in your department) at least once a day from April 15 through 26 to determine their satisfaction with room temperatures. Log their responses.
6. Contact the Constant Air representative each day with any problems so the system can be balanced.
7. Prepare a weekly report summarizing gathered data. NOTE: The Constant Air representative will provide forms for these reports at the initial meeting on April 10.
8. Email all reports to Ned Grumer before noon on Monday, April 29.

NOTE: Only Constant Air, not Building Services, should adjust the air-conditioning system between April 15 and April 26. Do not call Building Services with requests related to air-conditioning during this time. Constant Air's number is 555-3421.

Attachments can be mysterious to recipients. That's why you should identify on the attachment the original email it was attached to. Also, if recipients of a forwarded attachment receive the original email with it, the attachment itself doesn't require a lot of set or explanation.

With a Recommendation

822 Ocean View Drive,
Long Beach, California 90802
(714) 332-3978

Technical Memo 5270-33	Issued /Revised: Sept. 18, 2010
SUBJECT: Recommendation to Prime Windstream Fuselage With 680 (Not 780) to Please Customers and Cut Production Costs	
AUTHOR: Harry Roterman *H. Roterman* Department Production Supervisor	Reference(s): (a) Memo 5870-5-113 (b) Memo 3281-6-98
APPROVED: S. Kidd *S. Kidd* Department Production Manager	

Both the subject line and the opening paragraph summarize the key recommendation. See LETTERS and ORGANIZATION.

Rivlin 780 should not be used to prime the Windstream airplanes painted at the Foxx facility in Long Beach. We recommend continued priming with Rivlin 680 because it is:

- More appealing to customers
- More cost-effective
- Easier to apply

The bulleted list previews the three reasons for the recommendation. The memo addresses these three reasons in the same order as listed here. See ORGANIZATION.

Rivlin 680 More Appealing to Customers

Customers have rejected airplanes with unremovable Rivlin 780 stains. Rivlin 780 is impossible to mask; it seeps under even metal-foil tape and stains adjacent skin surfaces. Removing these stains requires hand or mechanical polishing, especially in critical areas. Often, such polishing is unsatisfactory, and customers notice it. Rivlin 680 leaves cleaner surfaces and appeals to customers.

Rivlin 680 More Cost-Effective

Unnecessary disposal costs for a single Windstream run around $650 when we use Rivlin 780 (Reference a). Rivlin 780 contains about six times as much chromium as Rivlin 680 and must therefore be disposed of via a tank truck instead of the sewer.

Rivlin 680 Easier to Apply

Labor and production delay would equal at least one 8-hour shift for each Windstream treated with 780. The whole painting process now takes only 3 days, so we would be adding another full day's costs.

Using Rivlin 780 in the areas to be painted does not eliminate the need for Rivlin 680 on the unpainted areas of the fuselage. Therefore, we would need to perform five extra steps to apply the 780: (1) Mask the skin area to remain unpainted, using metal-foil tape to prevent contact with Rivlin 680; (2) Apply Rivlin 780; (3) Rinse; (4) Allow to dry; (5) Remove metal-foil masking tape. After taking these steps, we would still need to apply Rivlin 680 to unpainted areas as usual (Reference b).

The paragraph list is clear primarily because of the numbered steps. A displayed list would have been an option, but the memo would have extended beyond a single page. See LISTS.

Summary: Continue to Prime With Rivlin 680

The performance benefits of Rivlin 780 (an insignificant 2 to 3 percent increase in durability and corrosion protection) do not warrant its use. We should consider staying with a uniform application of Rivlin 680 to appeal to customers and to avoid increased manufacturing costs.

Parentheses enclose backup or supporting information. See PARENTHESES.

Technical memos are valuable internal communications because they capture key decisions within the design and engineering processes. As in the above example, the memo includes, as attachments or merely as references, other memos or reports. Referencing allows this memo to come to the point.

The memo follows the four-box letter organization. Box 1 (paragraph 1) opens with a request or a statement of purpose. Box 2 (paragraph 1, second sentence) summarizes or lists the points to be covered. Box 3 (the remaining paragraphs, except for the final one) expands on the points listed in box 2. Box 4 (the final subheading and paragraph) closes with a review of the content and a final request for action. See ORGANIZATION.

Recommendation Memo

The subject line is very specific. Even though the reader might be averse to the request, the subject block should not be coy, misleading, or vague.

The opening paragraph sets up the request by providing sound support for it. The reader is being asked to do something he does not want to do; therefore, the writer must validate the request. See LETTERS and ORGANIZATION.

Repeating the 20 percent figure is helpful because it reinforces the writer's major point. See EMPHASIS and REPETITION.

To: Bob Conners
Cc:
Subject: Recommendation to Test the Toggles at a 20 Percent Level
Attachments: Alka-Seltzer test procedures
Arial 11 B I U

Please authorize a decrease in toggle testing from 100 percent to 20 percent of the nose landing gear transducers. Testing 100 percent of the transducers adds 12.5 hours to the production time of each lot.

Reasons for the Recommendation

1. Recent tests (100 percent of the transducers in each lot) have shown them to be of excellent quality. Tom Rogers has compiled the following test data. I believe he called you yesterday to summarize the data.

Lot Number	Transducers Tested	Transducers Passed
4	32	30
5	32	32
6	32	31

2. The three transducers that did not pass were only slightly outside the specification limits. The tests showed that none of the three failed to work, so they would not have caused a failure of the nose landing gear.
3. Quality fabrication and installation procedures are now our highest priority. We believe that doing things right during production makes final testing less and less necessary. The decrease to 20 percent now is a prudent decision; this level of testing will screen for any unexpected decrease in production quality.

By return memo, please authorize a decrease in toggle testing from 100 percent to 20 percent of the nose landing gear transducers. If I can provide further information, please let me know (ext. 6380).

Whenever possible, writers should open their memos by clearly stating their request.

In this case, Bob Conners has been adamant about 100 percent toggle testing, so the sender knows that he or she must follow the request immediately with solid evidence to support it. An optional approach would be for the sender to delay the request until later in the memo. We do not recommend this option. See MEMOS, LETTERS, and ORGANIZATION.

Status Report Memo

With an Outcome Orientation

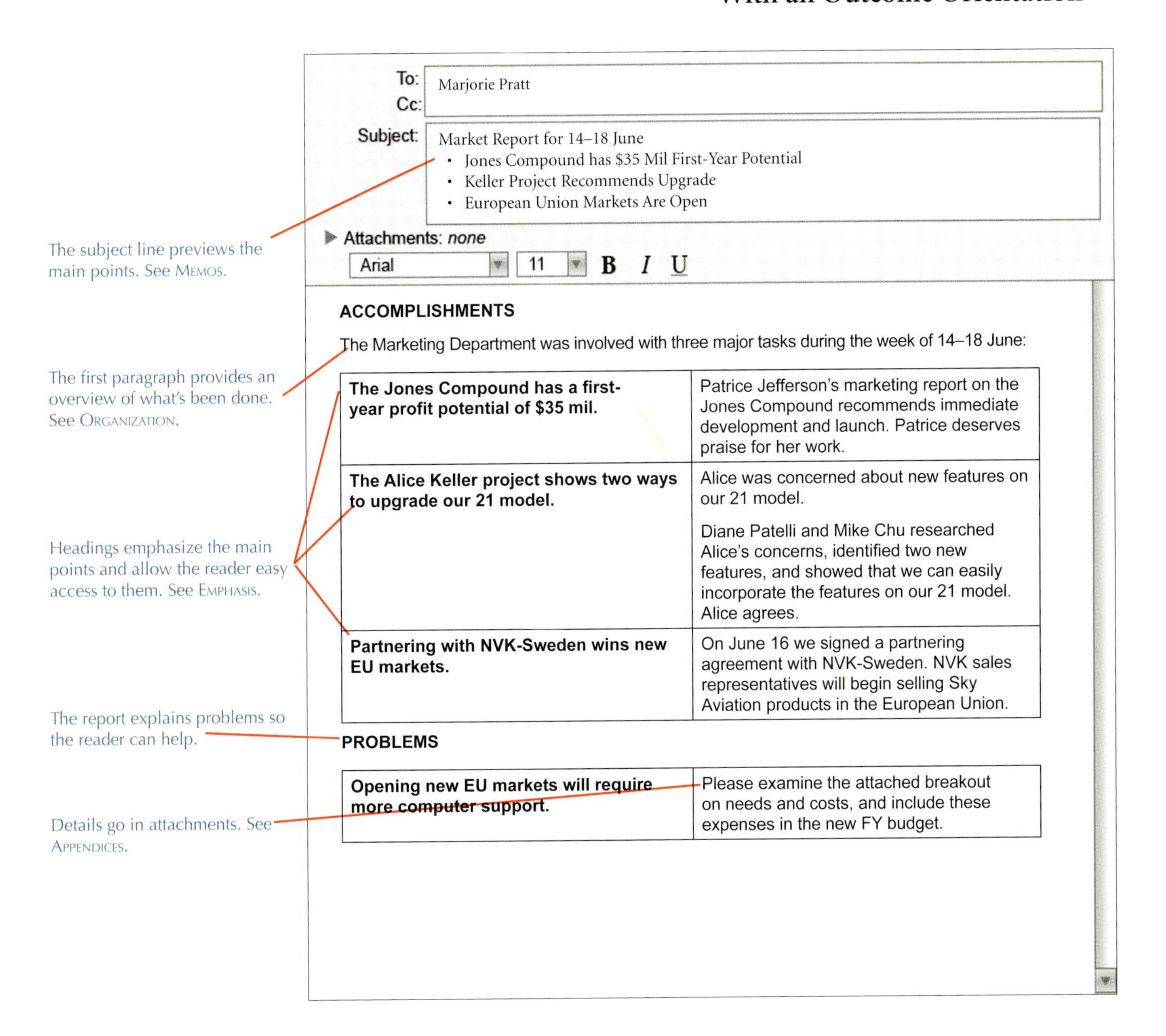

To: Marjorie Pratt
Cc:
Subject: Market Report for 14–18 June
- Jones Compound has $35 Mil First-Year Potential
- Keller Project Recommends Upgrade
- European Union Markets Are Open

▶ Attachments: *none*

ACCOMPLISHMENTS

The Marketing Department was involved with three major tasks during the week of 14–18 June:

The Jones Compound has a first-year profit potential of $35 mil.	Patrice Jefferson's marketing report on the Jones Compound recommends immediate development and launch. Patrice deserves praise for her work.
The Alice Keller project shows two ways to upgrade our 21 model.	Alice was concerned about new features on our 21 model. Diane Patelli and Mike Chu researched Alice's concerns, identified two new features, and showed that we can easily incorporate the features on our 21 model. Alice agrees.
Partnering with NVK-Sweden wins new EU markets.	On June 16 we signed a partnering agreement with NVK-Sweden. NVK sales representatives will begin selling Sky Aviation products in the European Union.

PROBLEMS

Opening new EU markets will require more computer support.	Please examine the attached breakout on needs and costs, and include these expenses in the new FY budget.

The subject line previews the main points. See Memos.
The first paragraph provides an overview of what's been done. See Organization.
Headings emphasize the main points and allow the reader easy access to them. See Emphasis.
The report explains problems so the reader can help.
Details go in attachments. See Appendices.

The report focuses on outcomes and action. It includes both accomplishments and problems so the reader has a complete picture and can help as needed. The format clearly identifies the main points and allows the reader to scan the document.

Safety Memo

With a Mild Reprimand

The specific subject line captures both issues raised in the memo.

This single-sentence opening paragraph concisely states the purpose of the message. It combines good and bad news, thus softening the impact of the reprimand. The contractions make the memo sound less formal, more personal. See PARAGRAPHS and TONE.

This closing paragraph contains a call to action. Normally, the call to action should come in the first paragraph, but as this memo is a reprimand, closing with the call to action might be more tactful.

To: Treion Muller
Cc: Cassidy Back, Jody Karr, Raju Venketapathy
Subject: Your accident on December 23, and the prevention of future accidents
Attachments: *none*

Arial 11 **B** *I* U

Treion, I'm pleased you weren't seriously injured in your accident on December 23, but I'm disappointed the accident occurred.

I understand that you were working as a substitute foreman on the Acid Line and that you stepped on a steel meter cover, slipped, and twisted your knee. It had been raining earlier in the afternoon, and the plate was wet. You were wearing safety shoes, but apparently your shoes are old and the tread is worn smooth.

As a manager, you should abide by company policies and procedures and be especially alert to hazardous conditions, such as wet flooring and worn treads on safety shoes. Those individuals working for you will follow your example. Please wear appropriate apparel and be more cautious in the future.

Personnel memos (or letters) are often hard to write, especially as in this case, when the memo contains even a mild reprimand. A proper tone is essential. As in this example, tone comes from several things: (1) positive information, as in the opening sentence; (2) a personal voice, using the addressee's name, pronouns, and contractions in the opening; and (3) a firm but courteous closing, with reasonable and honest requests for improvements and with the use of *please*. See PRONOUNS, MEMOS, and TONE.

Personnel Memo

With Suggested Prodedures

The opening paragraph states the point of the message. Beginning with a contraction creates an informal, conversational tone. See LETTERS, MEMOS, and TONE.

Some emphasis techniques are not available on email. Here the run-in headings are bolded and in all capital letters to highlight the suggestions list. Note that each listed item is parallel in structure. See LISTS and PARALLELISM.

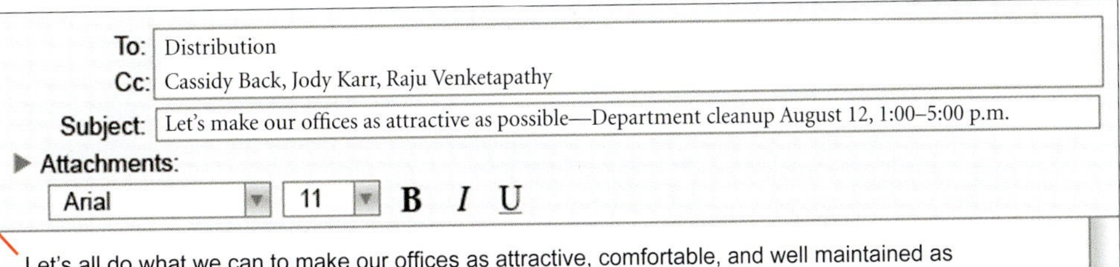

To: Distribution
Cc: Cassidy Back, Jody Karr, Raju Venketapathy
Subject: Let's make our offices as attractive as possible—Department cleanup August 12, 1:00–5:00 p.m.
Attachments:
Arial 11 B I U

Let's all do what we can to make our offices as attractive, comfortable, and well maintained as possible. Please arrange your schedules so you can devote August 12, 1:00–5:00 p.m., to cleaning and organizing your offices.

Offices should reflect our personalities, but they also should convey a professional impression to visitors. Please consider the following suggestions to ensure that your office reflects the Bedrock Mining & Milling standard of organization and cleanliness.

NAILS. Please use low-impact nails for hanging pictures. These nails will support most pictures and do less damage to the walls. My assistant has these if you need them.

CELLOPHANE TAPE. Please do not use it. Tape is a quick and easy way to get that poster, map, or note on the wall, but when it is removed, it removes paint and sometimes plaster.

POSTERS, MAPS, PICTURES. Please frame any that you intend to hang. Several inexpensive frame shops around town can make a $2.98 print look like a $200 lithograph. Family pictures likewise get special attention if attractively framed.

PLANT HANGERS. The secret to hanging things from ceilings is a little gadget called the "handi-hook," which is designed for suspended ceilings. It hooks over the metal separators and is easy to install or move. However, it can't support very heavy plants. I have a supply of handi-hooks.

GENERAL HOUSEKEEPING. Neatness takes a lot of effort, but it is worth it. We'll all be able to walk the halls without glancing into offices and wincing. Even more important, we'll be able to locate information more easily if the information is carefully labeled and filed.

Please, no boxes or files stacked on the floor. Let me know (ext. 431) if you have boxes or files you need sent to storage. Carefully label them, and I will arrange for pickup and delivery to the storeroom.

We have a beautiful facility, and with a little effort from us all, it will remain so. Remember—Department cleanup, August 12, 1:00–5:00 p.m.

Personnel notices often require a light touch. Employees rightly object to personnel notices that are too serious, impersonal, or even critical in tone (and content). Managers and supervisors sometimes have to send reminders and requests that are potentially negative (as this notice could be). The lighter tone helps writers convey negative information in a manner that readers will not find objectionable. See LETTERS, MEMOS, and TONE.

Response Memo

With Instructions

Using the reader's first name (*Diane*) is a good way to personalize your message. Optionally, you might insert the name in the initial sentence, enclosed by two commas. See Electronic Mail and Commas.

Paragraph 1 provides the background for paragraph 2, which summarizes the enrollment procedures. See Electonic Mail and Organization.

The list in paragraph 2 previews the information to follow. Screen 2 then follows the same sequence of topics. See Organization.

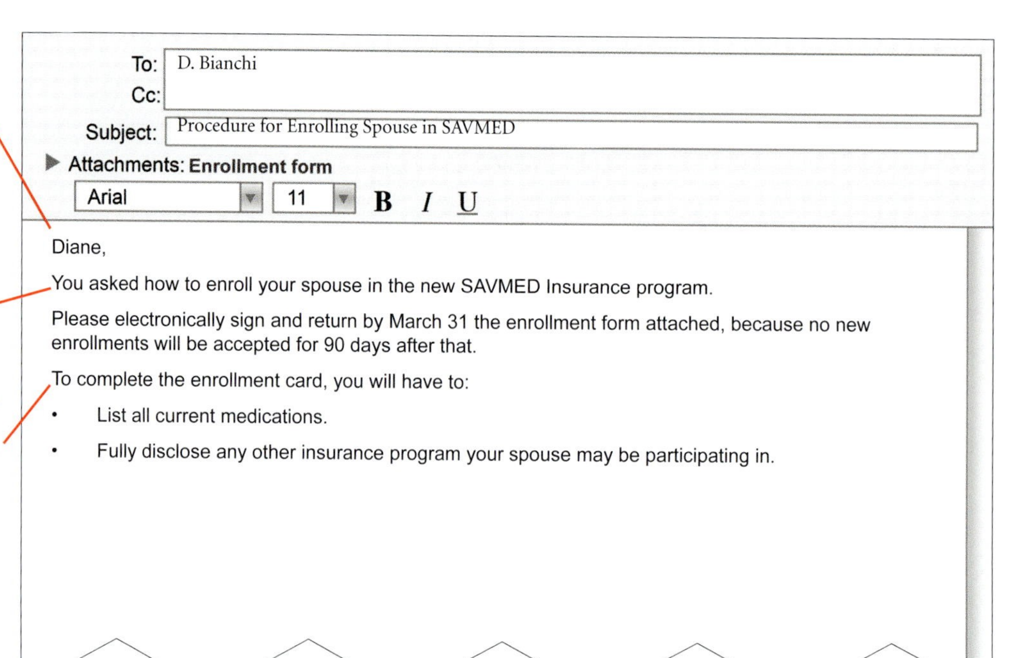

To: D. Bianchi
Cc:
Subject: Procedure for Enrolling Spouse in SAVMED
Attachments: **Enrollment form**

Diane,

You asked how to enroll your spouse in the new SAVMED Insurance program.

Please electronically sign and return by March 31 the enrollment form attached, because no new enrollments will be accepted for 90 days after that.

To complete the enrollment card, you will have to:

- List all current medications.
- Fully disclose any other insurance program your spouse may be participating in.

The format of the headings and the short paragraphs help make this page readable. Such readability features are just as important in email as in printed documents. See ELECTRONIC MAIL.

All Current Medications

Your doctors should list all types and dosages of medications currently prescribed for dependents.

Full Disclosure of Other Insurance Programs

You must name on the enrollment card any other insurance programs in which your spouse is currently enrolled at his employment (or in any other way).

To conclude, please sign and return the enrollment card by March 31 along with the records listed above. Please call me any time at ext. 0001 with questions.

In the last paragraph, "To conclude," signals that the message is at an end, so the reader should not have to look at additional screens. Optionally, you could use an informal closing like "Thank you" or "Have a good day." See ELECTRONIC MAIL.

Electronic mail (email) needs to be just as clear and as well written as printed documents. Often you might be tempted to dash off an email message without taking time to check it for accuracy or completeness. If you don't check your message, you may find yourself sending a follow-up email to cover missing facts or information. See ELECTRONIC MAIL.

Transmittal Memo

For Attachments

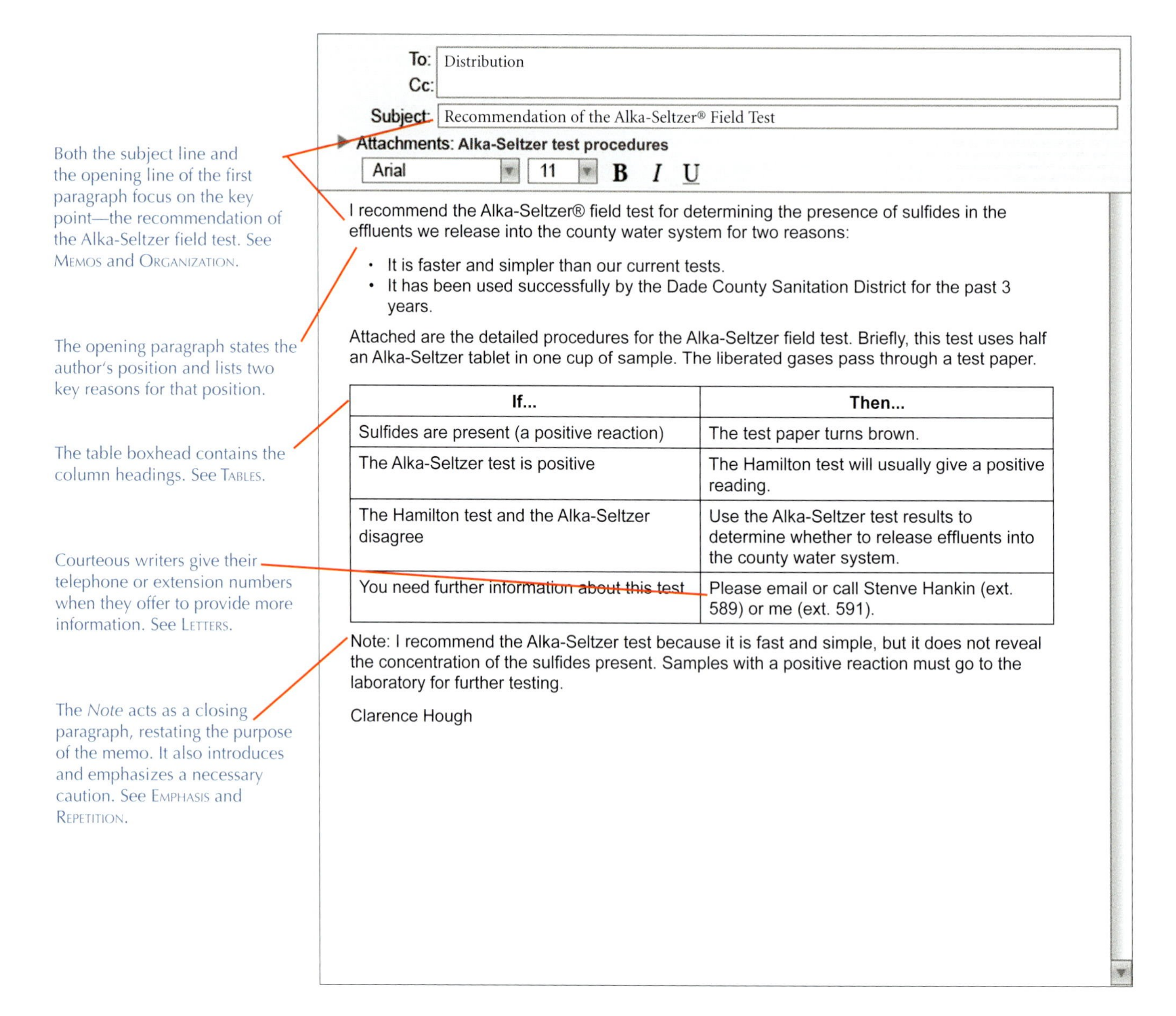

Both the subject line and the opening line of the first paragraph focus on the key point—the recommendation of the Alka-Seltzer field test. See MEMOS and ORGANIZATION.
The opening paragraph states the author's position and lists two key reasons for that position.
The table boxhead contains the column headings. See TABLES.
Courteous writers give their telephone or extension numbers when they offer to provide more information. See LETTERS.
The *Note* acts as a closing paragraph, restating the purpose of the memo. It also introduces and emphasizes a necessary caution. See EMPHASIS and REPETITION.
To: Distribution
Cc:
Subject: Recommendation of the Alka-Seltzer® Field Test
Attachments: Alka-Seltzer test procedures
Arial
11
B I U

I recommend the Alka-Seltzer® field test for determining the presence of sulfides in the effluents we release into the county water system for two reasons:

- It is faster and simpler than our current tests.
- It has been used successfully by the Dade County Sanitation District for the past 3 years.

Attached are the detailed procedures for the Alka-Seltzer field test. Briefly, this test uses half an Alka-Seltzer tablet in one cup of sample. The liberated gases pass through a test paper.

If...	Then...
Sulfides are present (a positive reaction)	The test paper turns brown.
The Alka-Seltzer test is positive	The Hamilton test will usually give a positive reading.
The Hamilton test and the Alka-Seltzer disagree	Use the Alka-Seltzer test results to determine whether to release effluents into the county water system.
You need further information about this test	Please email or call Stenve Hankin (ext. 589) or me (ext. 591).

Note: I recommend the Alka-Seltzer test because it is fast and simple, but it does not reveal the concentration of the sulfides present. Samples with a positive reaction must go to the laboratory for further testing.

Clarence Hough

A transmittal memo accompanies another document, such as a report or a procedure, and briefly explains the purpose and content of the attachment.

Start with a statement of the employer's business problem and a brief summary statement of your proposal to help solve the problem.

Use run-in informative headings to explain your experience and qualifications in solving such problems.

Quantify your accomplishments. Use actual or estimated numbers wherever possible.

Don't just list your degrees. Explain how your educational background contributes to solving the problem.

Marc Zeeman
15 Gruenwaldstr.
CH4004
Basel, Switzerland
Tel: +41 61 425 70 0
email: marcz@helvmail.ch

DIRECTOR OF REGULATORY AFFAIRS

At CodoPharm, it takes as long as 2 years for a new product application to be approved by the various medicine-control agencies such as the United States FDA. Every day out of the market costs about $2 million. With my background in improving regulatory processes, I would like to help CodoPharm reduce time to market by as much as 10 percent within 3 years.

How I have helped others:

Reduced average approval time at Milanoz from 20 months to 17 months. As director of regulatory documentation for 3 years, I instituted a "smart" Global Approval Process that met FDA requirements more precisely and cut review time by an average of 12 weeks. This translated into 60 days of market revenue, or roughly $24 million.

Gained $10 million in revenue for Grüssli Pharmaceuticals through simultaneous submissions. Grüssli was submitting drugs for approval in three different markets one after the other. In my 2 years as regulatory officer, I created a process for simultaneous submissions, thus cutting about 6 months off approval times.

My educational background in this field:

Earned LL.M. in World Commerce Law, University of Bern. My thesis research was a comparison of legal requirements for drug approvals in Europe, the United States, and Japan. I developed a strong understanding of the nuances that can hinder approval across these different agencies.

Most resumes simply list jobs and educational attainments in chronological order. This approach to resume writing is less effective than a "problem-solution" approach, in which you address real problems or opportunities your prospective employer faces. A problem-solution resume will distinguish you immediately from other applicants. Do careful research about the employer before writing a resume like this.

Resume layouts vary, but yours should be readable and uncluttered. Avoid irrelevant personal information like age, marital status, or outside interests.

Resume

Performance Format

Start with a brief summary of your unique value proposition and skills. What overall job have you done successfully in your career?

List accomplishments that will be meaningful to a specific employer. Rank them in order of importance to that employer. Use numbers wherever you can.

List jobs and degrees after listing your professional accomplishments.

CHRISTIANE CREER
cc@domail.ca

133 Yale Road
Northvale, BC V1B 2D3

Telephone/Messages:
(604) 555-3789

SENIOR SALES AND OPERATIONS EXECUTIVE

Sales leader with 10+ years of driving profitable growth through innovative sales and sales operations strategies. Performed as **Top National Salesperson** for two software-engineering companies.

SKILLS

Strategic Planning	Six Sigma & Analytics	Marketing Research
Strategic Alliances	Relationship Selling	Contract Management
CRM	Direct Marketing	Social Media

PERFORMANCE HIGHLIGHTS

- Boosted leads to 238 from <30 while cutting marketing costs 9%. (*AAA Consulting*)
- Created, in one year, a proactive marketing program that increased positive media coverage 650%. (*AAA Consulting*)
- Led cross-functional team that cut costs 50% by implementing preconfigured hosting solutions. (*BBB Software*)
- Increased the identification-stage pipeline 239% and the qualified-stage pipeline 249% through Portfolio Management program. (*SSS Consulting*)
- Developed a comprehensive sales/marketing program and built a culture of relationship selling—achieved 140% of the overall revenue plan. (*CCC Software*)
- Recruited 97 new accounts that delivered net revenue of $120K+ per month. (*CCC Software*)
- Restructured sales organization to build a high-caliber, market-driven team. Increased annual closing volume from $37.2M to $48.4M within 24 months. (*CCC Software*)

CAREER TRACK

- Director, Sales Operations, AAA Consulting, Vancouver, BC — 2006 – present
- Sales Manager, BBB Software, Toronto, ON — 2005 – 2006
- Client Partner, SSS Consulting, Chicago, IL — 2001 – 2005
- Business Development Manager, CCC Software, Montréal, QC — 1999 – 2001

EDUCATION

- MBA, Purdue University – Krannert School of Management, West Lafayette, IN 2005
- BS, University of Colorado, Boulder, CO – Computer Engineering 1999

A resume need not follow the traditional format of listing only jobs and educational attainments. Instead, resumes can be designed to emphasize qualifications, as the above resume does in its list of accomplishments. The best resume directly addresses the needs of the employer identified in research. For example, this resume appeals to a particular employer's need for someone who combines revenue-producing capabilities with creativity.

Frame your cover letter as a proposal to solve an employer's problem or capitalize on an opportunity. Your subject line should be an informative heading (see HEADINGS).

Make your opening direct, and refer to the position and source of information. Focus on the need you can fill.

Explain your understanding of the employer's problem or opportunity. Writing this paragraph requires research.

Give credible information about your ability to solve the problem. Use numbers even if you have to estimate.

Ask for an interview time.

Feb. 12, 2010

Haytham Arsad
Vice President, Marketing
Kali Palace Restaurants
7501 Bernard Ave.
Boston, MA 02110

Proposal to improve revenues and franchisee relationships as marketing liaison officer

You have an opening for a marketing liaison officer with your franchisees, according to massresto.com. I would like to meet with you and show how I can help you improve revenues and franchisee relationships at the same time.

Kali Palace has always had an excellent reputation among restaurant franchisees. Overall sales are flat, however, and some franchisees have expressed a desire for more marketing support from headquarters. I believe I can improve revenues by at least 10 percent in one year through a promotional approach that will go a long way toward resolving franchisee issues. This is my experience in conducting successful promotions:

- In my three years at Niko's Gyros, I led an event-based marketing strategy credited for an annual 15 percent increase in revenues. We held monthly chain-wide promotions based on seasonal themes. The most successful were Spring Greek Salad Days, Lamb for Easter, and the fall Greek Festival.
- Over five years at American Land restaurants, I successfully marketed and nearly doubled franchisee opportunities. The chain grew from 25 stores in 5 states to 47 stores in 14 states. One key to this success was a new 24/7 Web-based franchisee communication channel so that we could meet their needs on a dime. We also simplified both menu offerings and advertising.

My resume is attached. I appreciate your taking the time to review my proposal. I'd like to present it to you in further detail. Please let me know when we can talk.

Sincerely,
Nuri Patel
4145 Bazar St.
Suffolk, MA 02104
(617) 555-0372
Nuripatel@sluzen.net

The cover letter is more than a formality—it is your chance to show that you are a solution and not just another job-seeker. The purpose of the cover letter is to sell the reader on granting you an interview.

Many employers are more interested in your cover letter than in your resume because it tells them why you're applying and gives insight into your personality. Thus, you should use the cover letter as an opportunity to start a conversation with the employer about the value you can bring.

In your cover letter, you can expand or amplify important information that can only be reported in an abbreviated way in your resume. Be direct about your qualifications, but focus on employer needs and opportunities.

Always follow any interviews with a thank-you note.

Minutes

The subject line is quite specific. See HEADINGS and MEMOS.

An accurate record of participants is essential.

Action items emphasize the participants (with names boldfaced), their responsibilities, and the due dates. See BOLDFACE, EMPHASIS, and ORGANIZATION.

This overview supplies necessary background information (much as corrections to prior minutes would do).

The numbered list helps to highlight key information. See LISTS.

September 16, 2010

MINUTES OF THE SEPTEMBER 15 MEETING ON THE SHUTTLE SYSTEM FROM GREEN BRIAR TO LAKEVIEW CENTER

Attending: George Benson, Frank Houck, Martha Memmert, and Jeanne Skorut

Absent: Fred Householder

ACTION ITEMS

Martha Memmert Obtain bids from several other transportation companies. (Due October 1.)

Jeanne Skorut Develop an estimate of what it would cost for us to lease our own vans and hire drivers. (Due October 1.)

Frank Houck Study the feasibility of eliminating air-conditioning from the buses. (Due October 5.)

George Benson Discuss our work on the criteria with Fred Householder, who originally drafted the criteria. (Due October 1.)

AN OVERVIEW OF THE SHUTTLE SYSTEM CRITERIA

George Benson opened the meeting with a review of the criteria:

1. Buses must make one round trip every 2 hours (two buses to operate simultaneously), with trips starting at 7 a.m. and ending at 6 p.m. (Travel time is estimated to be about 50 minutes one way.)
2. Buses must be comfortable and attractive—to convey a good image for our company.
3. Buses must carry at least 12 passengers.
4. Buses must be economical to operate and maintain.

Minutes should highlight (1) actions during the meeting and (2) actions needed in the future (usually before the next meeting). Minutes should not attempt to capture everything that was discussed, and they deliberately do not record the meeting in strict chronological order. See ORGANIZATION and MEETINGS MANAGEMENT.

The bid information is necessary background to the following motions.

This motion highlights the name of the person making the motion, but the heading could identify the issue: Other Busing Systems. See HEADINGS.

These headings highlight the content of the motion, but the person's name is included in parentheses.

Minutes of the September 15 Meeting Page 2

MLT BUS SYSTEM'S PRELIMINARY BID

Frank Houck reviewed the preliminary bid from MLT Bus System. They proposed using two vehicles at a cost per hour of $37 (including the buses, their drivers, and maintenance).

If we maintained the proposed schedule of 11 hours a day, our weekly, monthly, and yearly rates would be as follows:

Weekly Fee	$2,035
Monthly Fee	$8,954
Yearly Fee	$107,448

George Benson's Motion
That we investigate other arrangements, given the costs of the MLT Bus System. Martha volunteered to investigate other companies, and Jeanne will develop an estimate of costs for leasing buses and hiring drivers.

POSSIBLE CHANGES TO THE CRITERIA

Number of Trips (motion by Martha Memmert)
That we use a single bus for most trips, with double buses on the trips at 7 a.m. and at noon. The motion passed, so Martha and Jeanne will include this new criterion in their reports.

Painting of the Buses (informal motion)
That we not paint the yellow MLT buses maroon (BCC's corporate color). Motion passed, saving a one-time cost per bus of $1,000.

Elimination of Air-Conditioning (motion by Frank Houck)
That we eliminate air-conditioning (also recommended in the MLT proposal). Initial installation costs would be $2,500 per bus, with weekly costs of $250 per bus (because of fuel, maintenance, etc.). Frank will investigate the problem and report no later than October 5.

Report to Fred Householder
George Benson agreed to report recommended criteria revisions to Fred, who originally developed the criteria. George will report on Fred's response by October 1.

Respectfully submitted,

Jeanne Skorut

Jeanne Skorut

139 Sequoia Park Way, Suite 303, Oakland, California 90022 (415) 555-7500 Fax (415) 555-7150

The headings, lists, short paragraphs, and brief sentences all contribute to the overall readability. See EMPHASIS.

Some repetition is inevitable at times, especially if the meeting is long and the issues complex. See REPETITION.

Job Description

Essential job functions describe the core of each position. These are the functions that must be done; the incumbent must be able to perform each function.

The nonessential job functions are the "nice-to-haves." These functions may be eliminated or modified for people with different skill-sets or for those employees requesting an accommodation due to a disability.

The functions or requirements in each section may be listed with bullets or numbers. Generally, numerical systems imply that the items are listed in order of importance. See Lists.

Job Title:	**Administrative Assistant**	**Job Code:**	**1380**
Department:	**Operations**	**Job Grade:**	**24**
Revision Date:	**March 3, 2011**	**FLSA:**	**Nonexempt**

Position Overview

Provide administrative and secretarial support for division managers. Create, edit, and produce written documentation. Establish division documentation standards.

Essential Job Functions

- Answer and screen incoming calls; take and deliver accurate messages.
- Write, edit, and proofread correspondence, reports, proposals, and other documents.
- Represent managers in their absence.
- Coordinate travel arrangements.
- Create presentation visuals using PowerPoint or other similar software.
- Create and maintain spreadsheets and generate weekly, monthly, and custom reports.
- Coordinate division meetings and functions.
- Coordinate internal and external speaking engagements.
- Complete all pre- and post-engagement work.

Nonessential Job Functions

- Assist with department budgeting.
- Prepare division purchase orders.
- Coordinate building maintenance and housekeeping.
- Schedule A/V equipment.

Requirements

- Thorough understanding of and ability to use up-to-date Microsoft Office programs, including Word, Excel, and PowerPoint.
- Minimum 5 years' secretarial experience.
- Excellent writing, editing, and proofreading skills.
- Type 70 wpm.
- Proven organization skills.
- Proven ability to work under pressure, meet aggressive deadlines, and make effective decisions.
- High school graduate.

Other Skills/Abilities

- Post-high school secretarial/business courses preferred.
- Background in accounting preferred.

NOTE: This job description is not intended to be all-inclusive. Employee may perform other related duties as required to meet the ongoing needs of the organization.

Clear and thorough job descriptions should become the basis for internal job postings and external job advertisements. Keep job descriptions up to date and use them as a tool in shared expectation reviews or performance reviews between employees and supervisors.

Companies have quite different policies about how and when to do performance reviews (often called shared expectation reviews). Because policies are so different, we do not provide any written examples of a performance review.

The bulleted list identifies the three major benefits covered in the proposal. The listed items reflect the organization both for the executive summary and for the main text of the proposal. See ORGANIZATION.

Each subheading ends with a section number in parentheses. These section numbers are cross-references to sections in the text of the proposal.

The marginal callouts highlight with sales language why Save-a-Lot Rental should be interested in Waterfall's assistance with management of their taxes. See PERSUASION.

Waterfall Financial Services

"We always know exactly what we're saving on taxes. There are no surprises when you work with WFS."

—Kim Lee
CFO, Thomas Corp.

You'll be in the best possible tax position consistent with proper business, personal, and financial planning with WFS. We believe taxpayers should pay only what the law requires.

Minimize all taxes with across-the-board strategic tax planning.

Get specialized tax counsel and quick solutions for your particular situation.

A PROPOSAL FOR SAVE-A-LOT RENTAL

Save Taxes With Tailored Solutions From Waterfall Financial Services (WFS)

With WFS, Save-a-Lot can:

- Save taxes through proper planning.
- Enhance executive income through tailored tax planning.
- Enjoy tailored national and international tax solutions.

Save taxes through proper planning (Section 2.0)

Save-a-Lot can minimize federal, state, and foreign taxes on business transactions with WFS continuous tax planning. You will get regular advice for maximizing credit (on, say, research activity and other areas), anticipating problems, and planning for opportunities, rather than just reacting to past events.

As Save-a-Lot expands (through acquisitions or mergers, for example), you'll need the timely, efficient tax planning we can give you as a result of our size and extensive practical experience with a large clientele.

Enhance executive income through tailored tax planning (Section 3.0)

Save-a-Lot executives can enhance their income and benefits with a tailored program to help them build capital through income tax savings, while you benefit from favorable funding arrangements and executive retention.

Enjoy tailored national and international tax solutions (Section 4.0)

Save-a-Lot will benefit from our frequent contact with national and district IRS offices as we obtain early answers to tax questions and rulings on tax aspects of proposed transactions. You'll avoid tax controversy and cut time and costs because of our expertise in resolving tax disputes efficiently and informally.

In a constantly changing tax environment, you need the specialized knowledge of our Washington WFS Office, specializing in more than 30 tax areas. Also, we have an international presence and experience in worldwide tax planning.

Our offices throughout the world would like to help Save-a-Lot with a full range of business, tax, and accounting services.

WATERFALL FINANCIAL SERVICES   214 BAYWAY BOULEVARD   ORLANDO, FLORIDA 32862-7201

Effective proposals usually include a well-written executive summary. Usually, as in this example, the executive summary highlights key benefits to the customer. Also, the summary introduces with cross-references key content in the main text of the proposal.

The many references to Save-a-Lot help tailor the proposal to Save-a-Lot readers, even though much of the text would be generic and usable for any client.

Executive Summary

For a Proposal (Training)

The preview of proposal sections repeats main themes and the wording of the executive summary. Readers of the executive summary can easily predict what the body of the proposal will contain. See REPETITION.

The marginal callouts highlight and emphasize key points. See PAGE LAYOUT and EMPHASIS.

The bar graph, along with its informative caption, visually captures a major point: the dramatic return on investment. See GRAPHS and CAPTIONS.

Proposal to Meet Tele-Corp Productivity Objectives by Partnering with FranklinCovey

This proposal contains these sections:

1.0 Executive Summary
2.0 Dramatic, Measurable Increases in Individual Productivity
3.0 A Focus on Tele-Corp Strategic Goals
4.0 FranklinCovey as the One Source for Productivity Training Solutions
5.0 Proposed Curriculum and Schedule
6.0 Investment Summary

1.0 Executive Summary

Tele-Corp is now taking steps to increase shareholder value and maximize return on resources. At a time when bottom-line performance is a top priority, Tele-Corp has identified a need to improve productivity in keeping with corporate goals.

Specifically, Tele-Corp desires to—

- Achieve dramatic, measurable increases in individual productivity.
- Focus everyone in the firm on achieving Tele-Corp strategic goals.
- Find one source for productivity training solutions.

FranklinCovey proposes to partner with Tele-Corp to accomplish these objectives.

1.1 Achieve dramatic, measurable increases in individual productivity (Section 2.0)

"Since FranklinCovey came here, Acme Company has had a 30% increase in quarterly sales."
—Don Atkinson, Vice President, Acme Company

Tele-Corp recognizes that its most important asset is its highly valued people and feels an urgent need to help them improve their performance. Tele-Corp realizes that effective time management is at the heart of increasing organizational efficiency and personal effectiveness. FranklinCovey *FOCUS* training will provide both the skills and tools needed to meet this need.

Tele-Corp will realize a measurable return on investment in FranklinCovey productivity, as figure 1 shows. FranklinCovey clients typically report substantial gains in employee productivity, which add up to major dollar savings.

Baker Financial invested $180 per employee in *FOCUS* training and gained $1,510 in productivity improvements per employee annually.

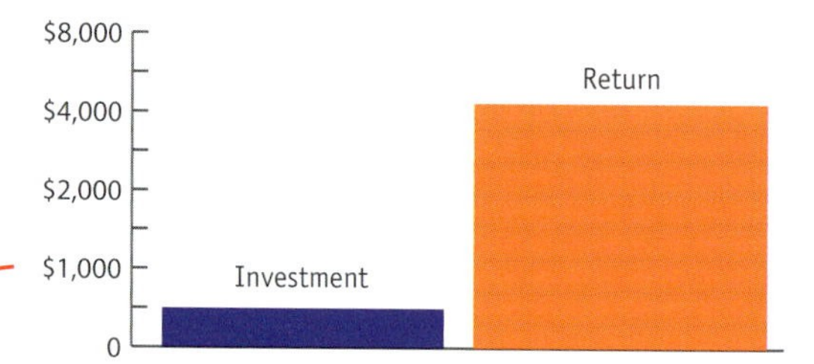

Figure 1. Return on Investment in *FOCUS* training. Baker Financial reported an annual productivity gain per employee of $1,510 on a $180 investment per employee. That's $8.39 return for every $1 invested in *FOCUS*.

A brief, highly readable executive summary has become the standard opening for most proposals. An executive summary combines persuasive sales information with a preview of the contents of the proposal. Often, as in the above example, each subhead in the executive summary parallels a section in the proposal.

Paragraphs are short and focused on a single point. See PARAGRAPHS.

"FranklinCovey is our sole source for everything we do. Each training session is consistently outstanding."
—Training Director, EPA

1.2 Focus everyone in the firm on achieving Tele-Corp strategic goals (Section 3.0)

Tele-Corp feels a need for increased focus on its highest priorities. FranklinCovey training provides the skills and tools to define and focus energy on achieving those highest priorities. Your chances of achieving the outcomes you want increase dramatically.

FranklinCovey training is always based on corporate values and the strategies to make corporate goals a reality. Participants receive tools that will empower them to focus every task they do every day on those corporate goals. Your people, and thus your organization, will be focused only on achieving your highest priorities!

Enjoy the benefits of a single training source:

- Deal with only one supplier instead of many.
- Be assured of world-class quality.
- Partner with a supplier who knows you and your needs.

1.3 Find one source for productivity training solutions (Section 4.0)

Tele-Corp wants to find solutions to productivity issues across all company systems—and to get training solutions from one source! Uniquely, FranklinCovey provides training to enhance productivity in crucial areas of your business.

Executive Summary

For an Audit

The major finding opens the summary, rather than coming at the end. See ORGANIZATION.

Paragraph 1 provides the essential background and references the full report, which is attached. This brief background is all that is necessary before a reader sees the recommendations in the table.

The table is an excellent tool for capturing the logic behind some crucial content. This table replaces three, four, or more paragraphs of text. See TABLES and EMPHASIS.

The conclusion totals the proposed financial savings and repeats the recommendation for a third time. Repetition is an essential feature of unambiguous text. See WRITING AND REVISING and REPETITION.

OSAGE GAS & ELECTRICAL COMPANY INC.

Major Account Servicing Audit MA41

1.0 Executive Summary

Major Finding: **Unsatisfactory Contracting and Collection Practices Expose Osage G&E to Significant Financial Losses**

We have concluded our audit of the internal controls over the major account contracting and collection functions for the year 2010 (final report attached). A random sampling of 150 deliveries was examined. Account management is unsatisfactory and exposes Osage G&E to a loss of as much as $2.5 million in fiscal 2011.

We recommend management's immediate attention to these principal concerns. Section references refer to sections in the final audit report.

Management should . . .	To improve controls over these unsatisfactory account servicing practices:
1. Motivate on-time payments with automatic discounts on future deliveries.	**Aging Accounts** A significant number (47 percent) of major account deliveries show more than 90 days past due for payment. Projected interest losses to Osage G&E over the next year: $1.1 million. (Section 3.2)
2. Instruct delivery directors to make no deliveries unless contracts are in hand first.	**Deliveries Without Contracts** Contracts were not issued in 8 percent of power deliveries to major accounts. Potential loss from payment defaults over one year: $1 million. (Section 3.3)
3. Require Financial Office to approve all discounts in advance.	**Noncompliance With Discounting Guidelines** Discounts of between 10 percent and 50 percent were given to selected major accounts where sales volume did not warrant them. Actual loss over the past year: $400,000. (Section 3.4)

Conclusion

Osage G&E might lose as much as $2.5 million in fiscal 2011 if control objectives are not met. We recommend immediate attention to the detailed findings and recommendations in the attached report.

4239 E. Jennings Avenue, Cedar Rapids, Iowa 54336

This executive summary opens with the major finding and the supporting recommendations. We recommend using this organization instead of the more traditional organization. Traditional audit reports often mirror the audit process, which begins with the collection of data, interviews with key employees, the collection of more data, and finally some interpretations. Following that sequence, the traditional report would present the audit findings on page 20, page 30, or later. See ORGANIZATION and SUMMARIES.

This initial page contains some cross-references to later sections of the audit report. These references would help readers to search and find information more efficiently. (We have not included these additional pages from the audit report.)

The title (heading) is a sales statement, not just the name of the company. See HEADINGS.

Paragraphs are short, both to emphasize main points and to fit into the narrow column for text. See PARAGRAPHS.

Marginal callouts emphasize (and repeat) key content from the main text in the righthand column. See PAGE LAYOUT and EMPHASIS.

The displayed list opens with capitals but does not use periods because the listed points are not complete sentences. See LISTS.

The shaded box is essentially an informal table (matrix) reflecting the logical links between the types of hazards and the various possible analyses. See GRAPHICS FOR DOCUMENTS and TABLES.

The final listed item is a summary sales statement.

Manage Risk

With Complete Hazards Evaluation Services

Facilities handling acutely hazardous materials need expert counsel when it comes to developing a risk-management and prevention program. Westerfals Waste Consulting's hazard evaluators expertly identify the potential for serious accidents and recommend the best means of prevention.

WWC's expertise in hazards evaluation is comprehensive, from system evaluation and hazard assessment through consequence analysis and dispersion modeling, with emphasis on risk management and risk communication.

Benefit from a Full-Scale Hazard and Operability (HAZOP) Study for Large Operating Facilities

Major hazards associated with large facilities can be managed with a full-scale HAZOP study that:

- **Identifies deviations from design intent using a line-by-line approach**
- **Identifies causes and consequences of deviations**
- **Recommends action to diminish frequency and consequences of potential accidents**
- **Examines facility operating, maintenance, and training procedures for ways to mitigate the risks and effects of accidents**

Obtain Quality Hazards Evaluation for Small-Scale Operations

WWC also specializes in these hazards-evaluation techniques tailored for design-phase or small-scale operations:

Hazards associated with . . .	Can be managed with this kind of analysis:
Conceptual or early design phase of a project	Preliminary Hazard Analysis
Proposed changes to facility	"What If" Analysis
Equipment failures	Failure Mode—Effects Analysis
Potential human error	Human Reliability Analysis

Count on Our Experience for Thorough Hazards Evaluation

WWC's team members merge experience in HAZOP study engineering and design engineering to provide our clients with complete hazards-evaluation services.

Westerfals Waste Consulting
2420 North Old Main
Boston, Massachusetts 02205
westerfalswaste.com

This marketing fact sheet deliberately uses text. Other versions might be more graphic, with only a few bulleted headline statements. This fact sheet could be sent to prospective clients as an attachment to a sales letter.

The format is designed to be open and inviting, and the text focuses on the two key benefits a client would gain from using Westerfals for waste-management problems.

Mission Statement

The mission sentence—actually an informative heading—is an effective, visual emphasis technique. See EMPHASIS and HEADINGS.

Each phrase in the mission sentence is broken out for separate analysis. This format breaks up what might have been a lengthy traditional paragraph. See PARAGRAPHS.

Boley Towers, Third Floor
2200 West Parkway Boulevard
New York, NY 10128
syndetoncorp.com

WE HELP PEOPLE THINK AND COMMUNICATE MORE PRODUCTIVELY

We help clients improve their intellectual productivity by helping them discover, define, and communicate valuable ideas and information.

We Help People . . .

Helping people is a simple, elegant approach, and in line with Syndeton's focus on helping people—individuals, teams, and organizations—gain control and be more productive. The focus on helping people, as opposed to customers, makes Syndeton a service-oriented company open to future service ideas.

. . . Think and Communicate . . .

This is our discriminator concept. The word *think* is a point of departure for raising issues with clients about leveraging intellectual resources, maximizing brain capital, and enhancing knowledge work. The link between thinking and communicating is backed by research on the nature of cognition and the fact that, in business, knowledge has no value unless it's shared. Furthermore, we benefit from the current interest in collaborative thinking and how to facilitate it.

. . . More Productively

This is our value. The issue for Syndeton, and for most clients, is increased productivity. The services we provide produce measurable results in terms of productivity. We help people think and communicate better, faster, and less expensively.

The Opportunity

Syndeton has the opportunity to provide significant value to people and organizations by helping them think and communicate better in accomplishing their critical goals. By helping people move beyond conventional ways of thinking and communicating, Syndeton can provide a breakthrough service to individuals, teams, and large organizations.

Mission statements are increasingly important because businesses and other organizations need a clear view of their core values and goals. Otherwise, employees will often waste time and money on peripheral tasks.

Successful mission statements must be brief (only one or two pages, at most). They should also be organized around a slogan or key sentence, as in the mission sentence that is the heading above.

Repetition of the mission sentence is proper, especially if page 2 is printed on the back of page 1. See REPETITION.

The numbered list, the boldface openings, and the short paragraphs produce a readable list of benefits. See EMPHASIS.

Mission Statement
Syndeton
Page 2

The Approach

We Help People Think and Communicate More Productively

Syndeton identifies market needs and trends to establish where it can provide significant value. Next, it designs consulting, training, projects, and product support to meet these needs. Finally, it organizes around these market needs.

The Benefits, as Seen by Our Clients

1. **Increase productivity.**

   Sky Aviation reduces costs 35 percent, improves sales quality. Syndeton partnered with this client in consulting and training to improve processes and internal capability to win new business more productively. The company increased their win rate on proposals from 50 percent to 73 percent, an increase of 23 percent. They also reduced their costs by 35 percent.

2. **Discover and leverage valuable ideas and information.**

   Bountiful Chemical Corporation, Inc., saves $900,000 by clarifying communication. Syndeton partnered with this customer to clarify critical ideas and procedures, facilitating a team process that allowed employees to design and leverage written policies and procedures. These procedures improve the thinking of employees, reduce risk and cost, and increase safety.

3. **Communicate the right message to the customers.**

   Midland Oil and Gas Operations, Inc., sharpens its message—customer wins up 300 percent. The Syndeton process of training and consulting was transferred into this client's environment. Syndeton trained 300 Midland Oil people to understand their customers' needs and to match their message directly in documents and presentations. The customer approval rate went from 20 percent to 80 percent over 3 years.

4. **Improve processes that facilitate collaboration and teamwork.**

   Osage Gas and Electrical Company's Syndeton-trained team outperforms others threefold. Osage contracted with Syndeton to instill processes and tools that brought a team together to productively solve critical issues. The Syndeton process brought simple but powerful standards for thinking, writing, and presenting.

Page 2 of this mission statement shifts to four success stories from actual clients. This shift to marketing or sales information is especially appropriate if the mission statement will be circulated to customers.

Procedure

For a Business Process

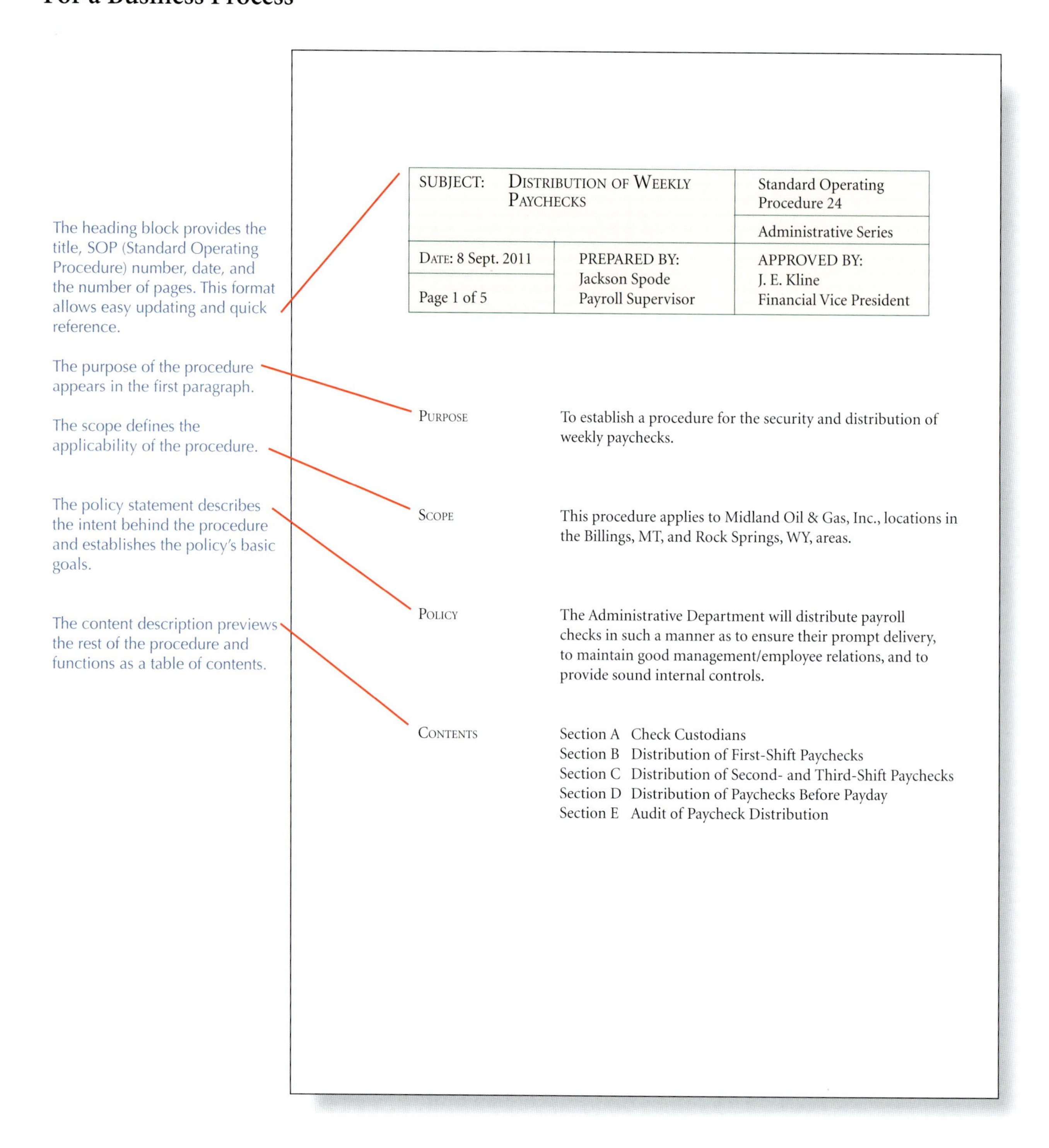

The heading block provides the title, SOP (Standard Operating Procedure) number, date, and the number of pages. This format allows easy updating and quick reference.

The purpose of the procedure appears in the first paragraph.

The scope defines the applicability of the procedure.

The policy statement describes the intent behind the procedure and establishes the policy's basic goals.

The content description previews the rest of the procedure and functions as a table of contents.

SUBJECT: Distribution of Weekly Paychecks		Standard Operating Procedure 24
		Administrative Series
Date: 8 Sept. 2011	PREPARED BY: Jackson Spode Payroll Supervisor	APPROVED BY: J. E. Kline Financial Vice President
Page 1 of 5		

Purpose — To establish a procedure for the security and distribution of weekly paychecks.

Scope — This procedure applies to Midland Oil & Gas, Inc., locations in the Billings, MT, and Rock Springs, WY, areas.

Policy — The Administrative Department will distribute payroll checks in such a manner as to ensure their prompt delivery, to maintain good management/employee relations, and to provide sound internal controls.

Contents

Section A Check Custodians
Section B Distribution of First-Shift Paychecks
Section C Distribution of Second- and Third-Shift Paychecks
Section D Distribution of Paychecks Before Payday
Section E Audit of Paycheck Distribution

Procedures should be as schematic as possible, with headings, lists, and imperative statements. These format techniques enhance readability and allow readers to find their particular responsibilities and actions.

Inevitably, procedures must be updated, so the numbering system, date of creation, and any revision date need to appear on each page.

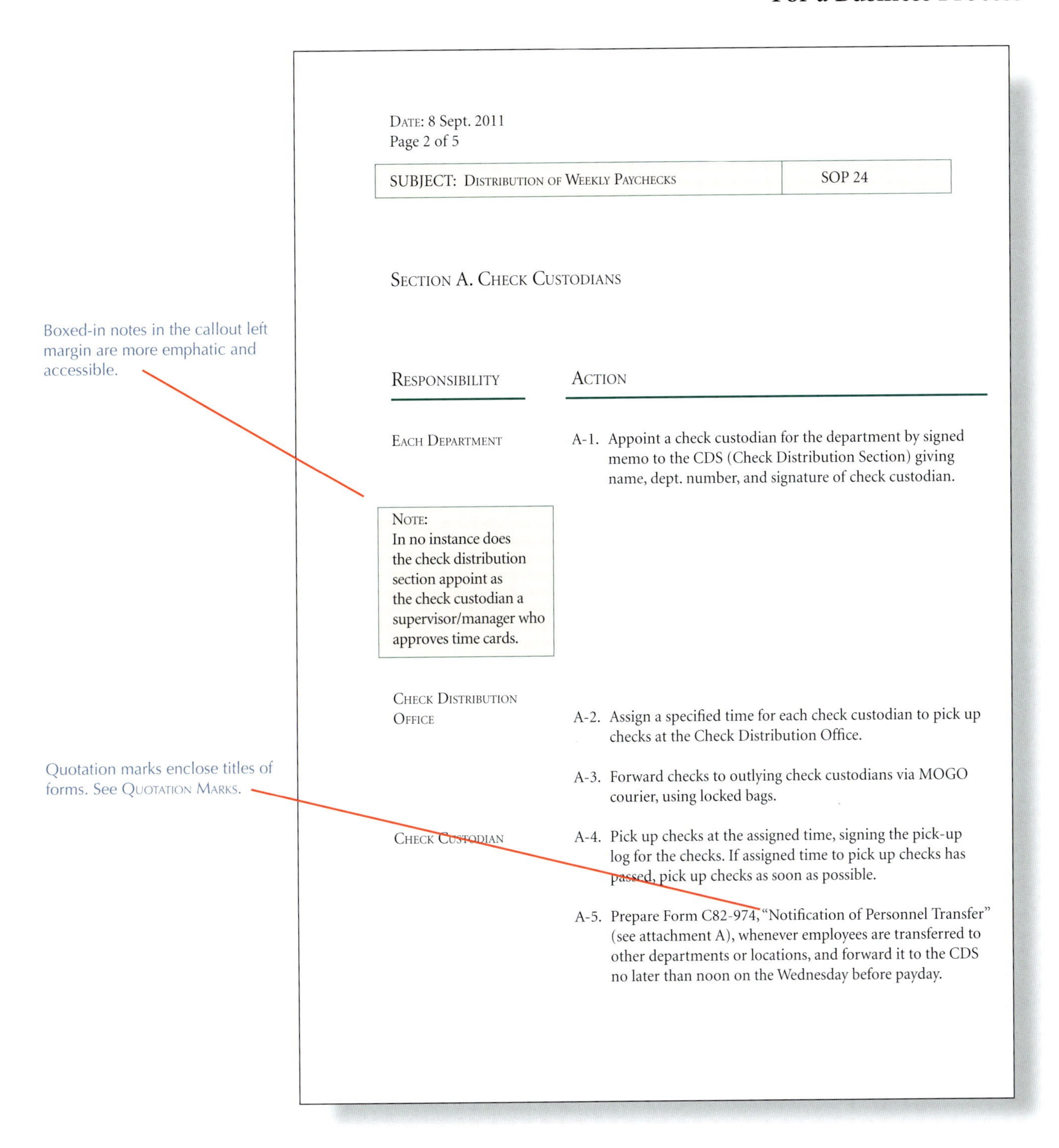

Boxed-in notes in the callout left margin are more emphatic and accessible.
Quotation marks enclose titles of forms. See QUOTATION MARKS.
DATE: 8 Sept. 2011
Page 2 of 5
SUBJECT: DISTRIBUTION OF WEEKLY PAYCHECKS
SOP 24
SECTION A. CHECK CUSTODIANS
RESPONSIBILITY
ACTION
EACH DEPARTMENT
A-1. Appoint a check custodian for the department by signed memo to the CDS (Check Distribution Section) giving name, dept. number, and signature of check custodian.
NOTE:
In no instance does the check distribution section appoint as the check custodian a supervisor/manager who approves time cards.
CHECK DISTRIBUTION OFFICE
A-2. Assign a specified time for each check custodian to pick up checks at the Check Distribution Office.
A-3. Forward checks to outlying check custodians via MOGO courier, using locked bags.
CHECK CUSTODIAN
A-4. Pick up checks at the assigned time, signing the pick-up log for the checks. If assigned time to pick up checks has passed, pick up checks as soon as possible.
A-5. Prepare Form C82-974, "Notification of Personnel Transfer" (see attachment A), whenever employees are transferred to other departments or locations, and forward it to the CDS no later than noon on the Wednesday before payday.

The most common problem with procedures is the passive voice. Writers list actions passively, and readers often don't know who is supposed to do what. A procedure that says "checks must be examined before delivery" does not indicate who is supposed to do the examining. Therefore, all steps in procedures must identify not only the action but also the person or the department responsible. See ACTIVE/PASSIVE.

Each step in the procedure opens with the person or department responsible for the step. Next comes the action associated with the step. Note that the action is stated in imperative sentences. See SENTENCES and ACTIVE/PASSIVE.

Procedure

For a Business Process

Using boldface, underlining, or capital letters emphasizes key warnings and cautions. See BOLDFACE, EMPHASIS, and UNDERLINING.

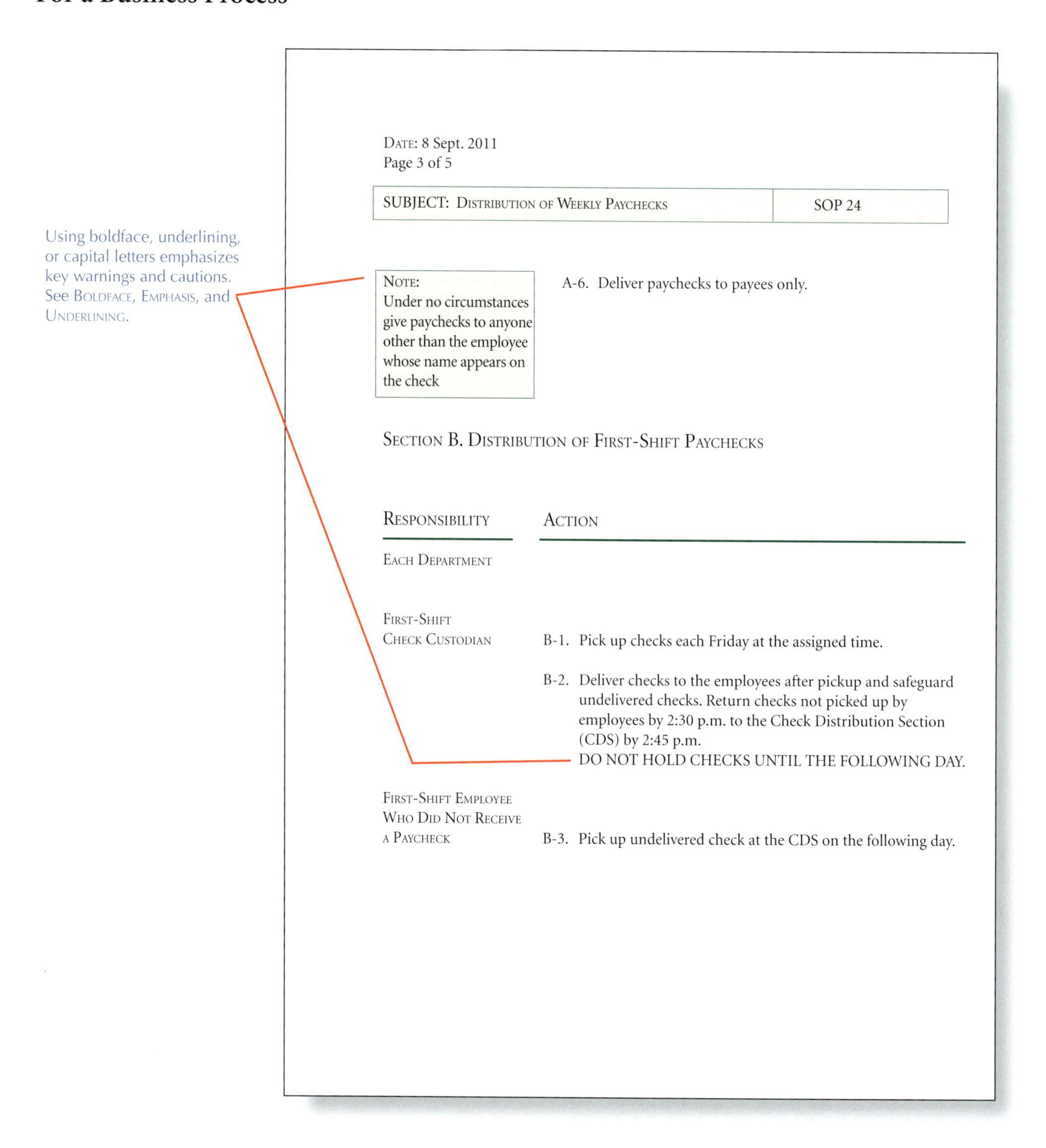
DATE: 8 Sept. 2011
Page 3 of 5

SUBJECT: DISTRIBUTION OF WEEKLY PAYCHECKS	SOP 24

NOTE:
Under no circumstances give paychecks to anyone other than the employee whose name appears on the check

A-6. Deliver paychecks to payees only.

SECTION B. DISTRIBUTION OF FIRST-SHIFT PAYCHECKS

RESPONSIBILITY	ACTION
EACH DEPARTMENT	
FIRST-SHIFT CHECK CUSTODIAN	B-1. Pick up checks each Friday at the assigned time.
	B-2. Deliver checks to the employees after pickup and safeguard undelivered checks. Return checks not picked up by employees by 2:30 p.m. to the Check Distribution Section (CDS) by 2:45 p.m. DO NOT HOLD CHECKS UNTIL THE FOLLOWING DAY.
FIRST-SHIFT EMPLOYEE WHO DID NOT RECEIVE A PAYCHECK	B-3. Pick up undelivered check at the CDS on the following day.

NOTE: This procedure would have several more pages. Because the other pages would add little to the model, we have omitted them.

Procedure

For a Technical Process

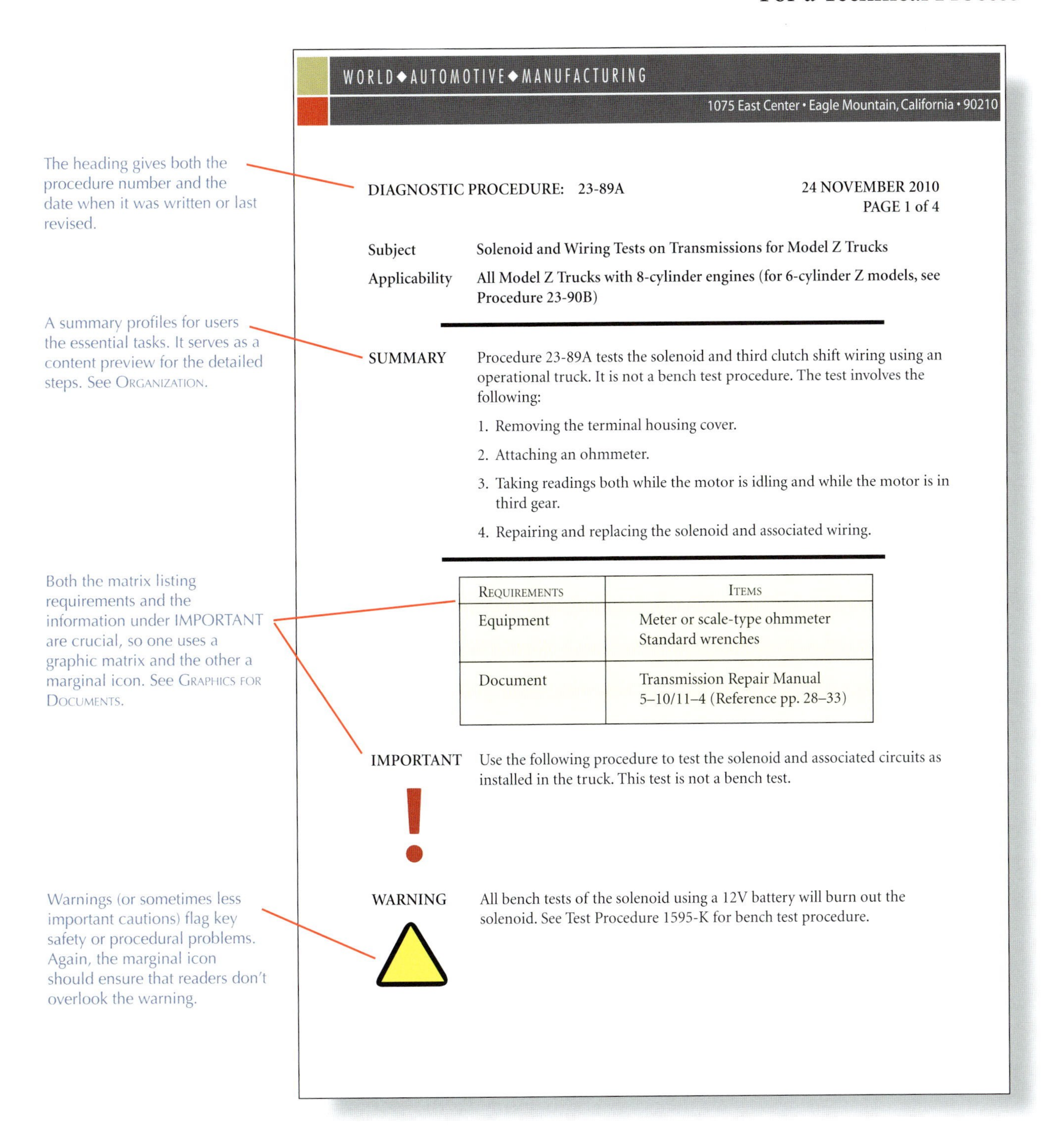

WORLD◆AUTOMOTIVE◆MANUFACTURING

1075 East Center • Eagle Mountain, California • 90210

DIAGNOSTIC PROCEDURE: 23-89A

24 NOVEMBER 2010
PAGE 1 of 4

Subject	Solenoid and Wiring Tests on Transmissions for Model Z Trucks
Applicability	All Model Z Trucks with 8-cylinder engines (for 6-cylinder Z models, see Procedure 23-90B)

SUMMARY Procedure 23-89A tests the solenoid and third clutch shift wiring using an operational truck. It is not a bench test procedure. The test involves the following:

1. Removing the terminal housing cover.
2. Attaching an ohmmeter.
3. Taking readings both while the motor is idling and while the motor is in third gear.
4. Repairing and replacing the solenoid and associated wiring.

Requirements	Items
Equipment	Meter or scale-type ohmmeter Standard wrenches
Document	Transmission Repair Manual 5–10/11–4 (Reference pp. 28–33)

IMPORTANT Use the following procedure to test the solenoid and associated circuits as installed in the truck. This test is not a bench test.

WARNING All bench tests of the solenoid using a 12V battery will burn out the solenoid. See Test Procedure 1595-K for bench test procedure.

The heading gives both the procedure number and the date when it was written or last revised.

A summary profiles for users the essential tasks. It serves as a content preview for the detailed steps. See Organization.

Both the matrix listing requirements and the information under IMPORTANT are crucial, so one uses a graphic matrix and the other a marginal icon. See Graphics for Documents.

Warnings (or sometimes less important cautions) flag key safety or procedural problems. Again, the marginal icon should ensure that readers don't overlook the warning.

A technical procedure must be both accurate and very readable. If possible, it should be 100 percent unambiguous—that is, its graphics and its text should present steps that no reader can misinterpret. See Writing and Revising.

For a Technical Process

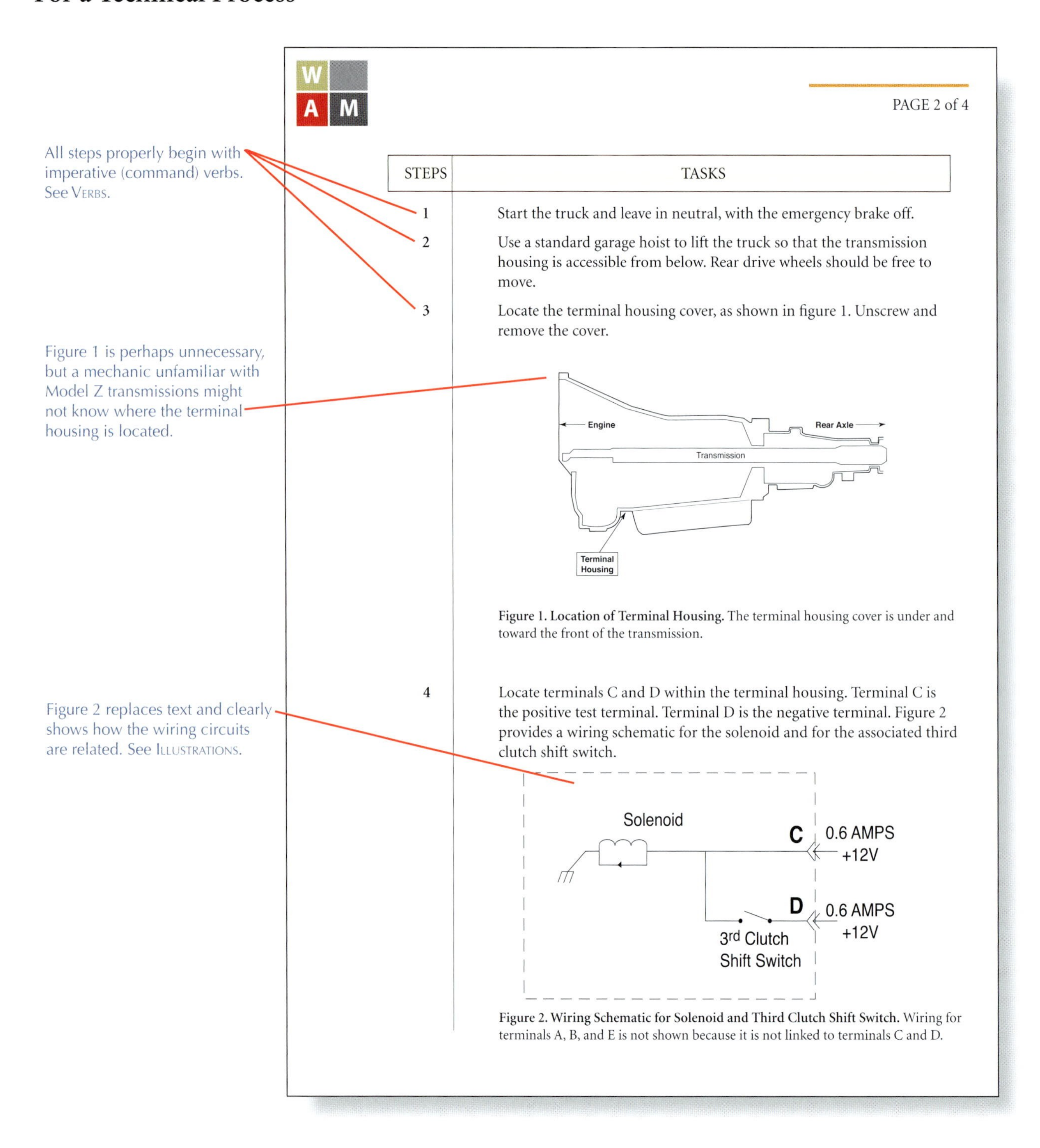

All steps properly begin with imperative (command) verbs. See VERBS.
Figure 1 is perhaps unnecessary, but a mechanic unfamiliar with Model Z transmissions might not know where the terminal housing is located.
Figure 2 replaces text and clearly shows how the wiring circuits are related. See ILLUSTRATIONS.
W
A M
PAGE 2 of 4
STEPS
TASKS
1
Start the truck and leave in neutral, with the emergency brake off.
2
Use a standard garage hoist to lift the truck so that the transmission housing is accessible from below. Rear drive wheels should be free to move.
3
Locate the terminal housing cover, as shown in figure 1. Unscrew and remove the cover.
Engine
Rear Axle
Transmission
Terminal Housing
Figure 1. Location of Terminal Housing. The terminal housing cover is under and toward the front of the transmission.
4
Locate terminals C and D within the terminal housing. Terminal C is the positive test terminal. Terminal D is the negative terminal. Figure 2 provides a wiring schematic for the solenoid and for the associated third clutch shift switch.
Solenoid
C
0.6 AMPS
+12V
D
0.6 AMPS
+12V
3rd Clutch
Shift Switch
Figure 2. Wiring Schematic for Solenoid and Third Clutch Shift Switch. Wiring for terminals A, B, and E is not shown because it is not linked to terminals C and D.

For clarity, each step in the procedure reflects a single task, and the tasks are carefully sequenced. The numbering of the steps allows for easy referencing and recall if one or more need to be repeated. See LISTS and NUMBERING SYSTEMS.

If/Then tables are valuable tools for capturing the common pattern of cause-and-effect actions. This table replaces a paragraph of text and is much more readable than a paragraph would be. See TABLES and PARAGRAPHS.

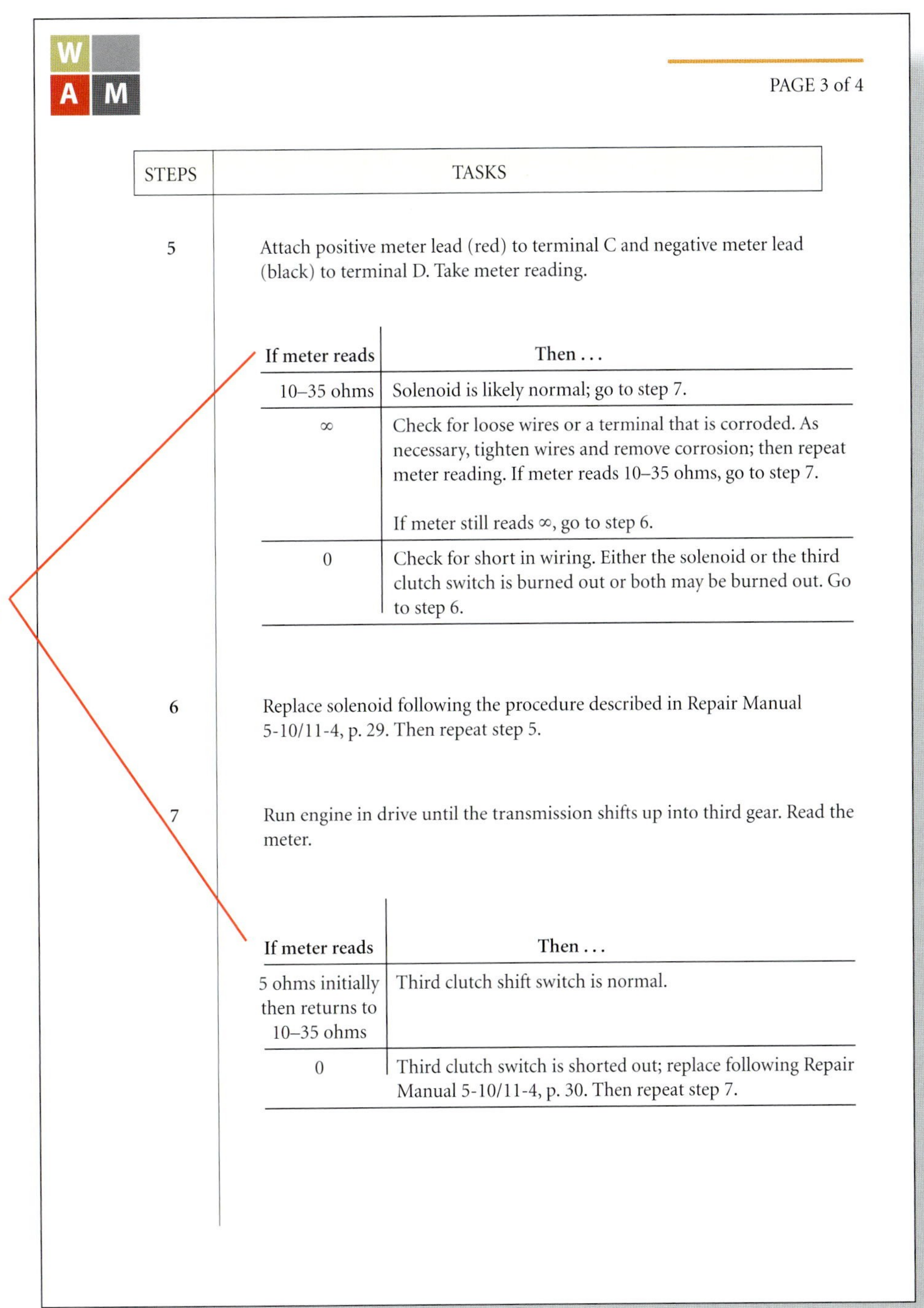
PAGE 3 of 4

STEPS	TASKS
5	Attach positive meter lead (red) to terminal C and negative meter lead (black) to terminal D. Take meter reading.

If meter reads	Then . . .
10–35 ohms	Solenoid is likely normal; go to step 7.
∞	Check for loose wires or a terminal that is corroded. As necessary, tighten wires and remove corrosion; then repeat meter reading. If meter reads 10–35 ohms, go to step 7. If meter still reads ∞, go to step 6.
0	Check for short in wiring. Either the solenoid or the third clutch switch is burned out or both may be burned out. Go to step 6.

STEPS	TASKS
6	Replace solenoid following the procedure described in Repair Manual 5-10/11-4, p. 29. Then repeat step 5.
7	Run engine in drive until the transmission shifts up into third gear. Read the meter.

If meter reads	Then . . .
5 ohms initially then returns to 10–35 ohms	Third clutch shift switch is normal.
0	Third clutch switch is shorted out; replace following Repair Manual 5-10/11-4, p. 30. Then repeat step 7.

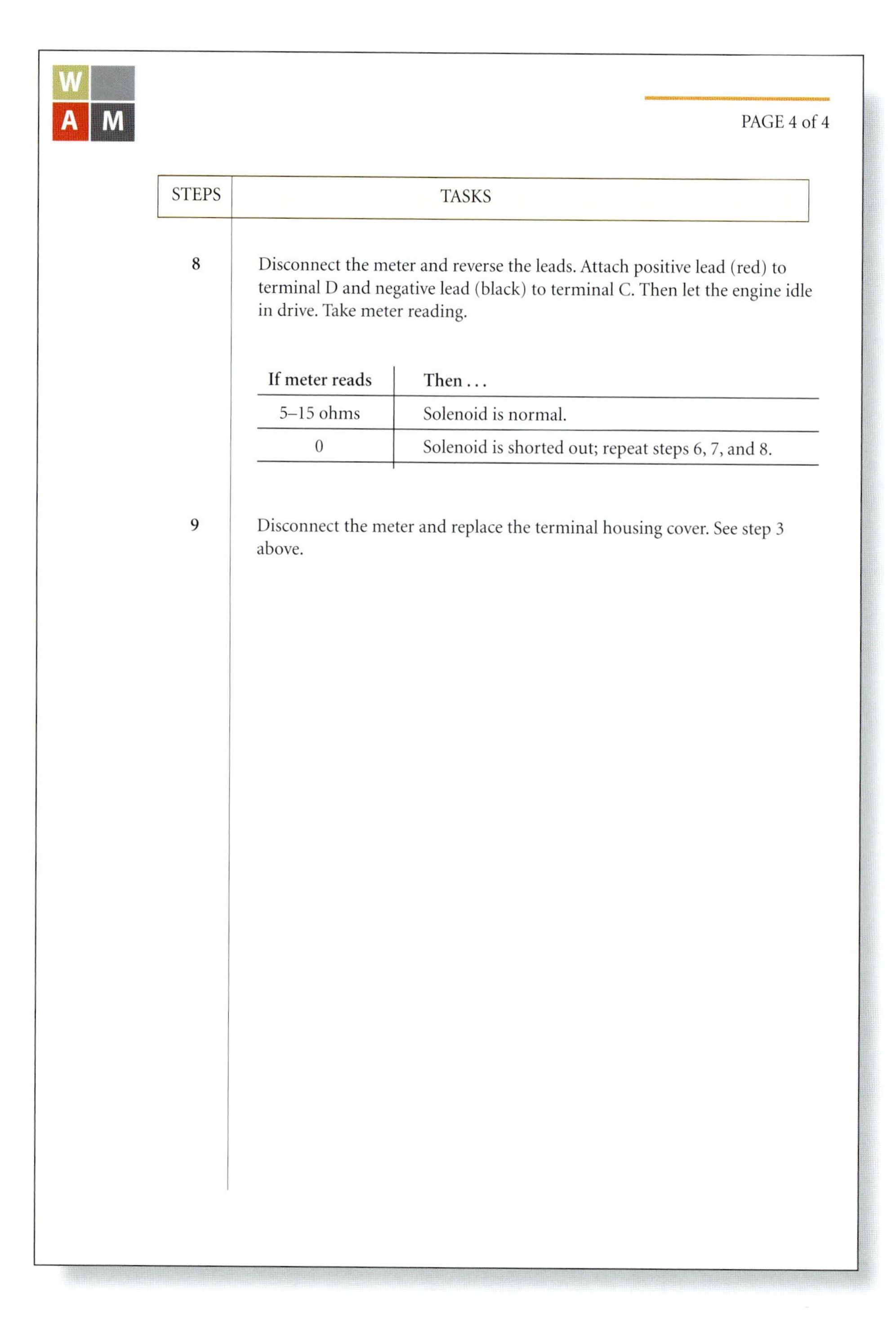

STEPS	TASKS
8	Disconnect the meter and reverse the leads. Attach positive lead (red) to terminal D and negative lead (black) to terminal C. Then let the engine idle in drive. Take meter reading.

If meter reads	Then . . .
5–15 ohms	Solenoid is normal.
0	Solenoid is shorted out; repeat steps 6, 7, and 8.

STEPS	TASKS
9	Disconnect the meter and replace the terminal housing cover. See step 3 above.

The title and associated data allow for careful cross-referencing and document storage and retrieval. The page notation (*1 of 3*) helps readers keep track of pages.

The one-page summary, although not asking for executive action, is almost an executive summary. See REPORTS and SUMMARIES.

The recommendations and their potential cost savings begin the summary.

This paragraph establishes the purpose as well as the original line of investigation. See MEMOS, LETTERS, and ORGANIZATION.

The listed findings are concise. The sentences are deliberately short. See SENTENCES and SCIENTIFIC/TECHNICAL STYLE.

822 Ocean View Drive,
Long Beach, California 90802
(714) 332-3978

RESEARCH AND DEVELOPMENT DEPARTMENT

Title	RECOMMENDED MODIFICATIONS TO PREVENT HOLE ELONGATION WHEN DRILLING THE G-175 STRUT FITTING		RDD Number 8795-3
Routing Status Routine	**Charge Number** 3-T3743-8042-286444	**Model Number(s)** G-175	**Page Number** 1 of 3

SUMMARY

We recommend a larger clamp foot (CF-8765-54-A) and a new Skylube application system (LA-5767-87) for use with all Acme 570 drills now being used. We estimate that these changes will save about $6,000 per year by eliminating the need to rework poorly drilled holes.

We have investigated the cause of the hole elongation in the drilling of the G-175 strut fitting. Three Acme Model 570 drills are currently used for this operation. We originally thought that improper sequencing of the feed and clamp-up system caused the elongation problem. In investigating this problem, we found the following:

1. Laboratory tests showed that one drill motor was unclamping with the drill still in the hole. We rebuilt the feed and clamp-up system to remedy this problem.
2. Further tests showed that the area of the clamp foot was too small to prevent the motor from rocking on its axis during drilling, so we designed a larger clamp foot.
3. Tests indicated that the drill motor feed rates were excessive. We corrected the feed rates on all three Acme drills.
4. The drill motors tended to stall right at the breakout of the drill, so we fabricated a Skylube application system. This modification prevents stalling.

Please see the attached discussion for a detailed report of our investigation.

Diane Metcalf 5/30/11
PREPARED BY DATE

Wallace Petersen 6/10/11
APPROVED BY DATE

Randall H. Volgel 6/12/11
APPROVED BY DATE

The summary (likely limited to one page) is a powerful technique for limiting documentation and for making technical reports more accessible. Many readers will not want or need to read more than the summary. If the summary contains the important details, these readers will be able to determine whether they need to read further. See REPORTS, SUMMARIES, and ORGANIZATION.

The brief introduction establishes the reason for and purpose of the investigation. As appropriate, relevant prior work and other background information might also appear in the introduction. See INTRODUCTIONS and REPORTS.

The displayed list highlights the proposed test and redesign program. This list actually repeats information covered under "Tests." See REPETITION and LISTS.

The headings are not specific, but they are probably standard. All research/investigation reports in this company have the same headings. The consistent format helps readers of many similar reports find information easily. See HEADINGS and ORGANIZATION.

822 Ocean View Drive,
Long Beach, California 90802
(714) 332-3978

INTRODUCTION

The Acme 570 drills were drilling many unsatisfactory holes in the strut fittings for the G-175 airplanes. The holes were often elongated or bell-mouthed, requiring reworking of the holes to an oversized diameter. Production Research initiated a program to determine the cause of these unsatisfactory holes.

PROGRAM APPROACH

To determine the cause of the problem, we developed a test and redesign program as follows:

1. Observe the Acme 570 drills in actual operation on a G-175.
2. Observe clamp-drill cycle with a high-speed TV camera.
3. Redesign the clamp foot to increase the clamp area and reduce flexing.
4. Prepare a recommendation for Production, including a revised drilling procedure and accompanying drawings.

TESTS

To determine the hole elongation, we brought one of the Acme 570 drills from the production line to the laboratory for testing. We used a high-speed TV camera to observe the clamp-drill cycle of the drill motor. The camera showed that this motor was unclamping with the drill still in the hole. We sent this drill to Small Tool Repair for the overhaul of the feed and clamp-up system.

Clamp Foot Test

The repaired Acme 570 was again tested and observed with the high-speed camera. The drill functioned well this time, but the clamp foot seemed to cover an insufficient area. The drill motor was able to move slightly during drilling. Apparently, vibration causes the drill to "migrate" during high-speed drilling, even though the bit tends to hold the drill in place. Pressure on the inside of the drill hole causes minute imperfections in the drilling circumference, which becomes exaggerated when the drill bites into one of these imperfections and causes it to elongate.

No. 8795-3
Page 2 of 4

The body of the report follows a scientific format rather than a managerial format. Thus, the conclusions and recommendations appear at the end rather than at the beginning. In most cases, the sequence and specificity of the headings in the body are almost irrelevant. Most readers will not read carefully beyond the summary, and readers familiar with the report format will already understand the content and organization of ideas appearing in the body. See ORGANIZATION.

822 Ocean View Drive,
Long Beach, California 90802
(714) 332-3978

The tests are explained in the chronological order in which they were performed. The chronological pattern helps readers follow the test sequence and therefore its logic. Note that each step ends with a conclusion or recommendation for further study.

To solve this problem, we designed a new clamp foot (CF-8765-54-A) with 57 percent greater surface area. Then we subjected the new clamp to 140 drill tests. During these tests, the drill did not migrate as before, nor did the hole elongate significantly, although some imperfections in the drilling holes were observable to the naked eye. (See attached figure 1.)

Drill Rate Test

We next checked the feed rate on the drill motor for drilling the 0.309 holes. The standard rate (4,300 rpm) produces a uniform drilling shaft so long as the drill bit is aligned precisely in the drill. But if the bit is not aligned precisely, the bit produces excessive vibration and hole elongation.

Most paragraphs are organized chronologically, from problem to result or finding. This pattern is common in technical and scientific reports. See PARAGRAPHS and SCIENTIFIC/TECHNICAL STYLE.

To determine an optimal drill speed, we tested the drive at five speed ranges: 2,000; 2,500; 3,000; 3,500; and 4,000 rpm. Rates below 3,000 rpm were unsatisfactory because the reduced speed created more friction and thus more heat. Rates above 3,500 rpm produced excessive vibration and hole elongation with drill bits not precisely aligned. So we repeated this test using another four speed ranges: 3,100; 3,200; 3,300; and 3,400 rpm.

Of these ranges, 3,400 rpm proved to be optimal. Further adjusting revealed that 3,460 rpm (±30 rpm) is the best compromise rate. Accordingly, we adjusted the driving speed to 3,460 rpm. (See attached figure 2.)

Lubrication Test

Finally, we noticed that the drill tended to stall just before the drill finished the hole. We designed a Skylube application system (LA-5767-87) to lubricate the bit. This system eliminated stalling problems. For further information on this application system, see RDD NO. 8799-6.

The conclusions and recommendations can be brief because they have already been covered first in the summary and later in the "Tests" section. See ORGANIZATION and REPORTS.

CONCLUSIONS AND RECOMMENDATIONS

The Acme 570 drill was returned to the Production line in mid-April. It had a new clamp foot, a correct feed rate, and the Skylube application system. We are continuing to monitor the performance of this drill, but preliminary results are promising.

We sent a recommendation memo (56984-76) to B. Worth recommending that all Acme 570 drills be modified with a larger clamp foot (CF-8765-54-A) and the new Skylube application system (LA-5767-87).

No. 8795-3
Page 3 of 4

822 Ocean View Drive,
Long Beach, California 90802
(714) 332-3978

Figure 1. Drill-Hole Elongation With Clamp Feet 53-B vs. 54-A

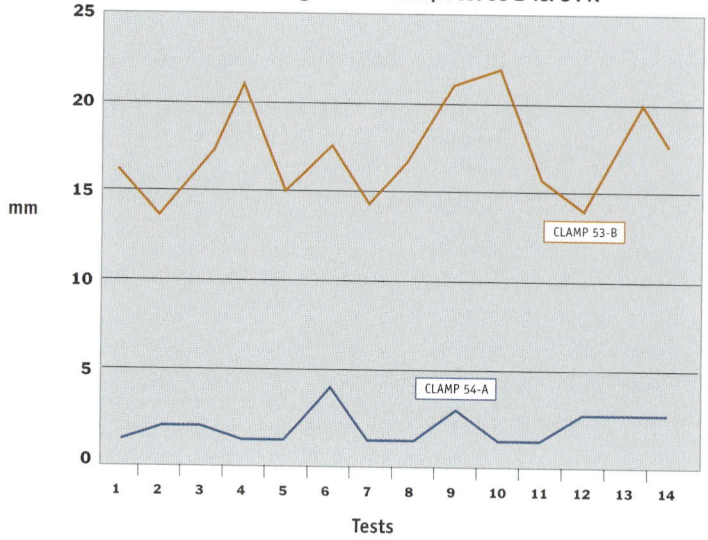

Drill holes with clamp foot CF-8765-53-B averaged 18 mm of elongation, while holes drilled with the new clamp foot CF-8765-54-A averaged 2.5 mm.

Figure 2. Drill-Hole Elongation at Various Drill Speeds

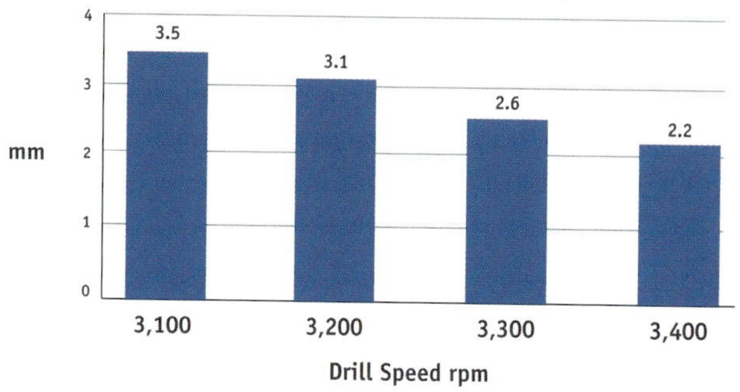

Drill holes became less elongated the higher the drill speed. In the end, setting the driving speed at 3,460 rpm proved optimal at 2.15 mm.

No. 8795-3
Page 4 of 4

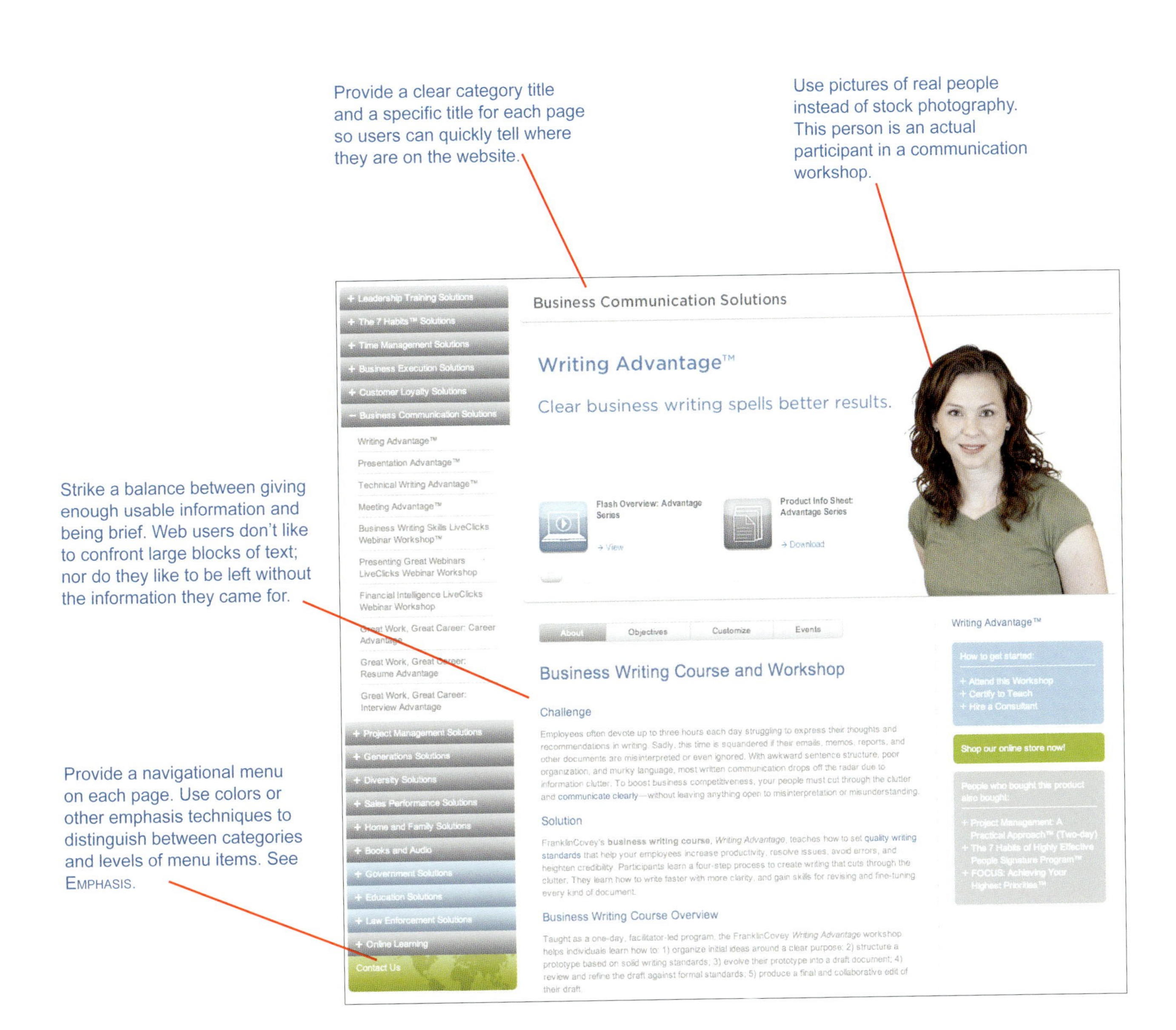

Provide a clear category title and a specific title for each page so users can quickly tell where they are on the website.
Use pictures of real people instead of stock photography. This person is an actual participant in a communication workshop.
Strike a balance between giving enough usable information and being brief. Web users don't like to confront large blocks of text; nor do they like to be left without the information they came for.
Provide a navigational menu on each page. Use colors or other emphasis techniques to distinguish between categories and levels of menu items. See EMPHASIS.
Business Communication Solutions
Writing Advantage™
Clear business writing spells better results.
Business Writing Course and Workshop
Challenge
Solution
Business Writing Course Overview

A business Web page serves the needs of customers, suppliers, and employees. Design the page the same way you would any customer-oriented document: make it easy to access the information the user wants. Most visitors to this page are interested in (1) the benefits and (2) the features of a service; so the "challenge-solution" structure works well for them.

Reading Internet pages is harder than reading print because screen resolution is only a fraction of that of a printed page. Therefore, design Web pages to be spacious as well as informative. Three columns of text are easier to read than one wide column.

Index

STYLE GUIDE
FIFTH EDITION

Using the Index

Indexes are often more annoying than useful. They omit key cross-references, or they refer readers to pages where the indexed word or phrase is missing. To address these problems, we have provided detailed cross-references. We have also tried to eliminate inaccurate or "ghost" references.

We used the following conventions as we compiled the index. Please take a moment to familiarize yourself with these conventions:

- Main entries from the Reference Glossary are printed in a large capital and small capitals. For example, ABBREVIATIONS, ACRONYMS, and ACTIVE/PASSIVE.
- Following the title of a main entry, the page number(s) in boldface refer to the main alphabetically arranged entries, as in this example: ADJECTIVES, **12–14.**
- Cross-references with only *See* and a page number indicate that what you are looking up appears under the *See* reference.
- Cross-references with *See also* and a page number indicate that you will find extra information on the pages listed, but that this information is not the main discussion. For example, under ADVERBS appears this reference: *See also* ADJECTIVES. Thus, you will find that the entry for ADJECTIVES discusses adverbs on p. 12, even though the main discussion of adverbs appears on p. 15.
- Model documents on pp. 358–420 are not indexed in detail. They are only listed once and then only by their title. These single references appear in boldface followed by a page number, as in this example: **Technical Report (model), 415–418.**

Index

A

Index

Index

Index

Index

Index

Index

Index

Index

Index

Index

Index

Index